GRAPH EDGE COLORING

GRAPH EDGE COLORING

Vizing's Theorem
and Goldberg's Conjecture

Michael Stiebitz
Diego Scheide
Bjarne Toft
Lene M. Favrholdt

A JOHN WILEY & SONS, INC., PUBLICATION

Library of Congress Cataloging-in-Publication Data:

Graph edge coloring : Vizing's theorem and Goldberg's conjecture / Michael Stiebitz ... [et al.].
    p. cm.
Includes bibliographical references and indexes.
ISBN 978-1-118-09137-1 (hardback)
1. Graph coloring. 2. Graph theory. I. Stiebitz, Michael, 1954–
QA166.247.G73 2012
511'.56—dc23   2011038045

10 9 8 7 6 5 4 3 2 1

To Vadim G. Vizing

CONTENTS

PREFACE

The **Edge Color Problem** (ECP) is to find the chromatic index $\chi'(G)$ of a given graph G, that is, the minimum number of colors needed to color the edges of G such that no two adjacent edges receive the same color. Edge coloring dates back to Peter Guthrie Tait's attempts around 1880 to prove the Four-Color Theorem. Tait [290] observed that coloring the countries of a map (i.e., of a plane, cubic and bridgeless graph G) with four colors is equivalent to coloring the boundaries (i.e., the edges of G) with three colors. Tait claimed that it is easily proved by induction that such a graph G can indeed be edge-colored using three colors, i.e., that $\chi'(G) = 3$. The statement became known as Tait's Theorem, but remained unproved until the Four-Color Theorem was finally resolved. Twenty years after Tait's claim a discussion in a French journal about Tait's Theorem prompted Petersen [242] to present his graph as an example showing that Tait's Theorem is false if the condition that G is planar is omitted.

Compared with vertex coloring, the theory of edge coloring has received less attention until relatively recently. However, much studied areas, such as map coloring, matching theory, factorization theory, Latin squares, and scheduling theory, all have strong connections to edge coloring.

The first monograph devoted to edge coloring, by Stanley Fiorini and Robin J. Wilson [100], appeared in 1977. It contains a stimulating and useful exposition. Like in many of the published papers on edge coloring, the book by Fiorini and Wilson deals mainly with edge colorings of simple graphs. When considering edge coloring theoretically, and also in connection with many scheduling problems, multigraphs, however, occur in a natural way, and multigraph edge coloring is a topic where further work needs to be done. One of the major unsolved problems on multigraphs is a conjecture (Conjecture 1.2 in Sect. 1.4) from around 1970 by Mark K. Goldberg, also posed independently by Paul D. Seymour. This monograph is an attempt to show the importance of this conjecture for both mathematical theory and algorithmics.

The cornerstones in the theory of edge coloring are (a) Shannon's and Vizing's bounds for the chromatic index in terms of the maximum degree and the maximum multiplicity, (b) the Adjacency Lemma for simple graphs as well as for multigraphs, and (c) several generalizations of the Adjacency Lemma. One aim of our book is to show that all these results may be derived from a common source: If a multigraph G has chromatic index $k + 1$ for $k \geq \Delta(G)$ and if $e = xy$ is a critical edge of G, i.e., $G - xy$ is k-edge-colorable, then there are $m \geq 2$ distinct neighbors $z_1, z_2, \ldots, z_m$ of x such that $z_1 = y$ and

$$\sum_{i=1}^{m} \big(d_G(z_i) + \mu_G(x, z_i) - k\big) \geq 2,$$

where $d_G(z)$ is the degree of z in G and $\mu_G(x, z)$ is the number of edges joining x and z in G. The proof of this fan inequality is based on the usual fan argument of Vadim G. Vizing from 1964; however, we define a fan in a slightly more general way than Vizing. This modification enables us to give a short proof in Chap. 2 (Theorem 2.1) of Vizing's Theorem for multigraphs. The above inequality is already implicitly contained in the work of Mark K. Goldberg [111, 114] and of Lars D. Andersen [5]. Moreover, the result in explicit form is the basis of the proof of Bruce Reed and Paul Seymour [253] of Vizing's conjecture, that Hadwiger's conjecture is true for line graphs of multigraphs (Theorem 4.56 in Sect. 4.8).

Vizing's bound was obtained independently by Ram Prakash Gupta during his Ph.D. studies, mostly at the Tata Institute of Fundamental Research in Bombay, 1965–1967, supervised by Sharadchandra Shankar Shrikhande, and stimulated by Claude Berge. Also, Gupta's proof was based on a variation of the fan idea (discovered independently by Gupta), and it was extended to locally bounded infinite graphs, i.e., infinite graphs with a finite maximum degree. In the thesis Gupta [120], like Vizing [297], discussed when the bound is best possible in terms of fixed values of the maximum degree and the maximum multiplicity, and he proposed a conjecture (♣ 4 in Chap. 9). In his diploma thesis from 2007, Diego Scheide [267, 272] rediscovered Gupta's conjecture and proved that its truth is a consequence of Goldberg's conjecture (Corollary 7.10 in Chap. 7).

We survey in Chap. 2 how the above fan inequality can be used to obtain both many well-known results and some new. We then go on and describe in Chap. 3 an alternative to Vizing fans, the so-called Kierstead paths, named after Henry A. Kierstead. Chapter 4 deals with the so-called Classification Problem for simple graphs, first treated by Vizing [297, 298, 299] in 1964. In 2000 Tashkinov [291] obtained a common generalization, Tashkinov trees, of Vizing fans and Kierstead paths. In

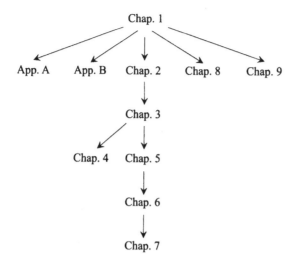

Figure 0.1 Chapter dependencies.

Chap. 5 we survey and prove Tashkinov's method and results, so far available only in Russian. We use this and an extension of Tashkinov's method to show in Chap. 5 that Goldberg's conjecture is asymptotically true (Corollary 5.20 in Sect. 5.3). Such an asymptotic result was first obtained in 1996 by Jeff Kahn [161] by probabilistic arguments; see also the fine book [229] by Michael Molloy and Bruce Reed on graph coloring and the probabilistic method. Our proof, in Chap. 5 and 6, due to Diego Scheide [268, 269, 271], is purely combinatorial and it gives an improved asymptotic bound. Also in Chap. 6, a proof for "the next" case of a parameterized version of Goldberg's conjecture (Theorem 6.5 in Sect. 6.1) is presented. Goldberg's conjecture supports the conjecture that there is a polynomial-time approximation algorithm for ECP which returns a solution within 1 of the optimum (Sect. 5.5 and Sect. 6.4). In Chap. 7 we discuss the question: For which values of the maximum degree and the maximum multiplicity does there exist a graph attaining Vizing's bound? In Chap. 8 we consider generalized edge colorings, where a color may appear more than once at a vertex. In Chap. 9 we list 20 pretty open problems related to edge colorings of graphs. The chapter is intended as an update of the similar chapter in the graph color problem book from 1995 by Tommy R. Jensen and Bjarne Toft [157]. The dependencies of the nine chapters may be illustrated as in Fig. 0.1.

Appendix A contains translations of Vizing's two fundamental papers, published in 1964 and 1965 in Russian in the journal Diskretnyi Analiz in Akademgorodok (close to Novosibirsk). The appendix also contains some biographical information about Vizing in its note section.

Appendix B is devoted to a thorough description of fractional edge coloring, related in particular to Chap. 6 on Goldberg's conjecture.

This book grew out of the 1998 M.Sc. thesis [90] of Lene Monrad Favrholdt at the University of Southern Denmark, supervised by Bjarne Toft. In the thesis, Vizing's fans and Kierstead's paths were treated, and in an attempt to unify the two into a common theory Favrholdt presented the fan inequality as a basic tool. The attempt to unify the two approaches failed, however, in the first round, but was carried through by Vladimir A. Tashkinov [291] in 2000. Our intention was at first to write a short survey paper, but the material grew and grew, one step being a technical report in 2006 from the University of Southern Denmark [91], and also with the help of the diploma thesis [267] and the doctoral thesis [269] of Diego Scheide at the Technical University of Ilmenau in 2008. The work of Scheide, supervised by Michael Stiebitz, resulted in several papers [268, 270, 271, 272, 273, 274], the contents of which are partially included in this book. Results of Scheide and Stiebitz [274] are included in Sect. 2.5 with permission from John Wiley & Sons Inc. Results of Scheide [271] and of Scheide and Stiebitz [272] are included in Sect. 6.3, respectively in Sect. 7.2, 7.3, and 7.4 with permission from Elsevier. We thank Wiley and Elsevier for these permissions.

We also wish to thank first of all Vladimir A. Tashkinov, who visited the Technical University of Ilmenau in November 2002 and shared his deep insight in edge coloring theory with the first author. We also wish to thank Vadim G. Vizing for interesting remarks and discussions at the University of Southern Denmark, in particular in September 2002. We have also benefitted from the continued support from our work places, the Mathematical Institute of the Technical University of Ilmenau, and the Department of Mathematics and Computer Science at the University of Southern Denmark. The mathematical library of the University of Southern Denmark and its efficient staff, in particular Tove Gundersen, provided unbounded help. Several other individuals have provided help and support, among them Oleg Borodin, Barbara Hamann, Tommy Jensen, Mark Goldberg, Gabi Käppler, Christian Klaue, Sasha Kostochka, Daniel Král', Jessica McDonald, Anna Schneider-Kamp, Jan-Sébastien Sereni, Robin Thomas, and Douglas Woodall. Also the staff of John Wiley and Sons gave efficient support; in particular we thank Susanne Steitz-Filler, Kellsee Chu, and Amy Hendrickson.

Finally, we are grateful for support from the Danish Council for Independent Research (FNU) for more than ten years, and from the Department of Mathematics at the London School of Economics and Political Science in the fall of 2011.

MICHAEL STIEBITZ
BJARNE TOFT

Odense (Denmark), Ilmenau (Germany), London (England)
August and November, 2011

CHAPTER 1

INTRODUCTION

We assume that the reader has a basic knowledge of graph theory. Concepts and notation not defined in this book will be used as in standard textbooks on graph theory.

1.1 GRAPHS

By a **graph** G we mean a finite undirected graph without loops, and possibly with multiple edges. The **vertex set** and the **edge set** of G are denoted by $V(G)$ and $E(G)$, respectively. Every edge of G is **incident** with two distinct vertices and the edge is then said to **join** these two vertices. For a vertex $x \in V(G)$, denote by $E_G(x)$ the set of all edges of G that are incident with x. Two distinct edges of G incident to the same vertex will be called **adjacent edges**. Furthermore, for $X, Y \subseteq V(G)$, let $E_G(X, Y)$ denote the set of all edges of G joining a vertex of X with a vertex of Y. When $Y = V(G) \setminus X$, then $E_G(X, Y)$ is called the **coboundary** of X in G and is denoted by $\partial_G(X)$. We write $E_G(x, y)$ instead of $E_G(\{x\}, \{y\})$. Two distinct vertices x, y of G with $E_G(x, y) \neq \emptyset$ are called **adjacent vertices** and **neighbors**. The set of all neighbors of x in G is denoted by $N_G(x)$, i.e., $N_G(x) = \{y \in V(G) \mid E_G(x, y) \neq \emptyset\}$.

Graph Edge Coloring: Vizing's Theorem and Goldberg's Conjecture,
First Edition. By M. Stiebitz, D. Scheide, B. Toft, and L. M. Favrholdt
Copyright © 2012 John Wiley & Sons, Inc.

The **degree** of the vertex $x \in V(G)$ is $d_G(x) = |E_G(x)|$, and the **multiplicity** of two distinct vertices $x, y \in V(G)$ is $\mu_G(x, y) = |E_G(x, y)|$. Let $\delta(G)$, $\Delta(G)$ and $\mu(G)$ denote the **minimum degree**, the **maximum degree**, and the **maximum multiplicity** of G, respectively. A graph G is called **simple** if $\mu(G) \leq 1$. A graph G is called **regular** and r-**regular** if $\delta(G) = \Delta(G) = r$, where $r \geq 0$ is an integer.

For a **subgraph** H of G, we write briefly $H \subseteq G$. For a graph G and a set $X \subseteq V(G)$, let $G[X]$ denote the subgraph of G **induced** by X, that is, $V(G[X]) = X$ and $E(G[X]) = E_G(X, X)$. Furthermore, let $G - X = G[V(G) \setminus X]$. We write $G - x$ instead of $G - \{x\}$. For $F \subseteq E(G)$, let $G - F$ denote the subgraph H of G with $V(H) = V(G)$ and $E(H) = E(G) \setminus F$. If $F = \{e\}$ is a singleton, we write $G - e$ rather than $G - \{e\}$.

If S is a sequence consisting of edges and vertices of a given graph G, then we denote by $V(S)$, respectively $E(S)$, the set of all elements of $V(G)$, respectively $E(G)$, that belong to the sequence S. Let G be a graph and let $S = (v_0, e_1, v_1, \ldots, v_{p-1}, e_p, v_p)$ be a sequence such that $v_0, \ldots, v_p$ are distinct vertices of G and $e_1, \ldots, e_p$ are edges of G, where we do not assume anything about incidences of the elements in S. For a vertex $v_i \in V(S)$, we define $Sv_i = (v_0, e_1, \ldots, e_i, v_i)$ and $v_i S = (v_i, e_{i+1}, \ldots, v_p)$.

By a path, a cycle, or a tree we usually mean a graph or subgraph rather than a sequence consisting of edges and vertices. There are only two exceptions to this: The Kierstead path (Sect. 3.1) and the Tashkinov tree (Sect. 5.1) are considered as sequences. If P is a **path** of length $p \geq 0$ with $V(P) = \{v_0, \ldots, v_p\}$ and $E(P) = \{e_1, \ldots e_p\}$ such that $e_i \in E_P(v_{i-1}, v_i)$ for $1 \leq i \leq p$, then we also write $P = \text{Path}(v_0, e_1, v_1, \ldots, e_p, v_p)$. Note that $\text{Path}(v_0, e_1, \ldots, e_p, v_p) = \text{Path}(v_p, e_p, \ldots, e_1, v_0)$; but the corresponding sequences are distinct, provided that $p \geq 1$. The vertices $v_0, \ldots, v_p$ of the path P are distinct and we say that v_0, v_p are the **endvertices** of the path P and that P is a path **joining** the vertices v_0 and v_p.

The **complete graph** on n vertices is denoted K_n, while the **cycle** on n vertices is denoted C_n. A K_3 (isomorphic to C_3) is often called a **triangle**. A cycle C_n is **odd** or **even**, depending on whether its order n is odd or even. As usual, the number of vertices of a graph is its **order**.

If G and H are two graphs with the same vertex set such that every pair (x, y) of distinct vertices satisfies $\mu_H(x, y) = 0$ if $\mu_G(x, y) = 0$ and $\mu_H(x, y) \geq \mu_G(x, y)$ otherwise, then H is called an **inflation graph** of G. If H is an inflation graph of G such that $\mu_H(x, y) = t \mu_G(x, y)$ for every $x, y \in V(G)$, where $t \geq 1$ is an integer, then we simply write $H = tG$ and call H a **multiple** of G. An inflation graph of a cycle C_n with $n \geq 3$ is also called a **ring graph**.

As usual, we shall write $\lfloor x \rfloor$ for the **lower integer part** of the real number x, and $\lceil x \rceil$ for the **upper integer part** of x.

1.2   COLORING PRELIMINARIES

A k-**edge-coloring** of a graph G is a map $\varphi : E(G) \to \{1, \ldots, k\}$ that assigns to every edge e of G a color $\varphi(e) \in \{1, \ldots, k\}$ such that no two adjacent edges of G

receive the same color. Denote by $C^k(G)$ the set of all k-edge-colorings of G. The **chromatic index** or **edge chromatic number** $\chi'(G)$ is the least integer $k \geq 0$ such that $C^k(G) \neq \emptyset$.

Let φ be a k-edge-coloring of G. For a color $\alpha \in \{1, \ldots, k\}$, the edge set $E_{\varphi,\alpha} = \{e \in E(G) \mid \varphi(e) = \alpha\}$ is called a **color class**. Then every vertex x of G is incident with at most one edge of $E_{\varphi,\alpha}$, i.e., $E_{\varphi,\alpha}$ is a **matching** of G (possibly empty). So, there is a one-to-one correspondence between k-edge-colorings of G and partitions $(E_1, \ldots, E_k)$ of $E(G)$ into k matchings (color classes); and the chromatic index of G is the minimum number of matchings into which the edge set of G can be partitioned.

A simple, but very useful recoloring technique for the edge color problem was developed by König [174], Shannon [284], and Vizing [297, 298]. Suppose that G is a graph and φ is a k-edge-coloring of G. To obtain a new coloring, choose two distinct colors α, β and consider the subgraph H with $V(H) = V(G)$ and $E(H) = E_{\varphi,\alpha} \cup E_{\varphi,\beta}$. Then every component of H is either a path or an even cycle and we refer to such a component as an (α, β)-**chain** of G with respect to φ. Now choose an arbitrary (α, β)-chain C of G with respect to φ. If we interchange the colors α and β on C, then we obtain a new k-edge-coloring φ' of G satisfying

$$\varphi'(e) = \begin{cases} \varphi(e) & \text{if } e \in E(G) \setminus E(C), \\ \beta & \text{if } e \in E(C) \text{ and } \varphi(e) = \alpha, \\ \alpha & \text{if } e \in E(C) \text{ and } \varphi(e) = \beta. \end{cases}$$

In what follows, we briefly say that the coloring φ' is obtained from φ by **recoloring** C, and we write $\varphi' = \varphi/C$. This operation is called a **Kempe change**. Furthermore, we say that an (α, β)-chain C has **endvertices** x, y if C is a path joining x and y.

Let G be a graph, let $F \subseteq E(G)$ be an edge set, and let $\varphi \in C^k(G - F)$ be a coloring for some integer $k \geq 0$. For a vertex $v \in V(G)$, define the two color sets

$$\varphi(v) = \{\varphi(e) \mid e \in E_G(v) \setminus F\} \text{ and } \overline{\varphi}(v) = \{1, \ldots, k\} \setminus \varphi(v).$$

We call $\varphi(v)$ the set of **colors present** at v and $\overline{\varphi}(v)$ the set of **colors missing** at v. Evidently, we have

$$|\overline{\varphi}(v)| = k - d_G(v) + |E_G(v) \cap F|. \tag{1.1}$$

For a color $\alpha \in \{1, \ldots, k\}$, let $m_{\varphi,\alpha}$ denote the number of vertices $v \in V(G)$ such that $\alpha \in \overline{\varphi}(v)$. Since the color class $E_{\varphi,\alpha}$ is a matching of G, we have $m_{\varphi,\alpha} = |V(G)| - 2|E_{\varphi,\alpha}|$. Consequently, we obtain

$$m_{\varphi,\alpha} \equiv |V(G)| \bmod 2 \tag{1.2}$$

for all colors $\alpha \in \{1, \ldots, k\}$ and, moreover, from (1.1)

$$\sum_{v \in V(G)} (k - d_G(v)) + 2|F| = \sum_{v \in V(G)} |\overline{\varphi}(v)| = \sum_{\alpha=1}^{k} m_{\varphi,\alpha}. \tag{1.3}$$

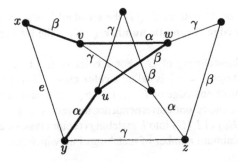

Figure 1.1   A graph G with a chain $P_x(\alpha, \beta, \varphi)$ (bold edges).

For a vertex set $X \subseteq V(G)$, we define

$$\overline{\varphi}(X) = \bigcup_{v \in X} \overline{\varphi}(v).$$

If $X = \{v_1, \ldots v_p\}$, then we also write $\overline{\varphi}(v_1, \ldots, v_p)$ instead of $\overline{\varphi}(X)$. The set X is called **elementary** with respect to φ if $\overline{\varphi}(u) \cap \overline{\varphi}(v) = \emptyset$ for every two distinct vertices $u, v \in X$. The set X is called **closed** with respect to φ if for every colored edge $f \in \partial_G(X)$ the color $\varphi(f)$ is present at every vertex of X, i.e., $\varphi(f) \in \varphi(v)$ for every $v \in X$. Finally, the set X is called **strongly closed** with respect to φ if X is closed with respect to φ and $\varphi(f) \neq \varphi(f')$ for every two distinct colored edges $f, f' \in \partial_G(X)$.

Let $\alpha, \beta \in \{1, \ldots, k\}$ be two distinct colors. For a vertex v of G, we denote by $P_v(\alpha, \beta, \varphi)$ the unique (α, β)-chain of $G - F$ with respect to φ that contains the vertex v. If it is clear that we refer to the coloring φ, then we just write $P_v(\alpha, \beta)$ rather than $P_v(\alpha, \beta, \varphi)$. If exactly one of the two colors α or β belongs to $\overline{\varphi}(v)$, then $P_v(\alpha, \beta)$ is a path, where one endvertex is v and the other endvertex is some vertex $u \neq v$ such that $\overline{\varphi}(u)$ contains either α or β. For two vertices $v, w \in V(G)$, the two chains $P_v(\alpha, \beta)$ and $P_w(\alpha, \beta)$ are either equal or vertex disjoint. For the coloring $\varphi' = \varphi/P_v(\alpha, \beta, \varphi)$, we have $\varphi' \in \mathcal{C}^k(G - F)$, since $\varphi \in \mathcal{C}^k(G - F)$. Furthermore, if x is not an endvertex of $P_v(\alpha, \beta, \varphi)$, then $\overline{\varphi}'(x) = \overline{\varphi}(x)$, else $\overline{\varphi}'(x)$ is obtained from $\overline{\varphi}(x)$ by interchanging α and β. We shall use these simple facts quite often without explicit mention.

Figure 1.1 shows the graph G obtained from the Petersen graph (see Fig. 1.2) by deleting one vertex as well as a 3-edge-coloring φ of $G - e$, where the three colors are α, β, γ. The graph G itself has chromatic index 4. Furthermore, $\overline{\varphi}(x) = \{\alpha, \gamma\}, \overline{\varphi}(y) = \{\beta\}, \overline{\varphi}(u) = \overline{\varphi}(v) = \overline{\varphi}(w) = \overline{\varphi}(z) = \emptyset$, and $P_x(\alpha, \beta)$ is a path of length 4 with vertex set $X = \{x, v, w, u, y\}$. The set X is elementary with respect to φ, but not closed.

If the condition $\varphi(e) \neq \varphi(e')$ for any two adjacent edges $e, e' \in E(G)$ is dropped from the definition of edge coloring, then φ is called an **improper edge coloring** of G. Accordingly, the term **proper edge coloring** is used in the graph theory literature

Figure 1.2   Three drawings of the Petersen graph.

in order to emphasize that the condition holds. In Chap. 8 we shall discuss some results about improper edge colorings.

A k-**vertex-coloring** of a graph G is an assignment of k colors to its vertices in such a way that adjacent vertices receive different colors. The minimum k for which a graph G has a k-vertex-coloring is called its **chromatic number**, denoted $\chi(G)$. A trivial lower bound for the chromatic number of a graph G is its **clique number** $\omega(G)$, that is, the maximum p for which G contains a complete graph on p vertices as a subgraph. On the other hand, every graph G satisfies $\chi(G) \leq \Delta(G) + 1$.

A graph with chromatic number at most k is also called a k-**partite graph**, where a 2-partite graph is also called a **bipartite graph**. The **complete bipartite graph** on two sets of n and m vertices is denoted $K_{n,m}$.

Clearly, a graph has chromatic number 0 if it has no vertices and chromatic number 1 if it has vertices, but no edges. A well-known result of König [174] states that a graph G is bipartite if and only if G contains no odd cycle.

For a graph G, the **line graph** of G, denoted $L(G)$, is the simple graph whose vertex set corresponds to the edge set of G and in which two vertices are adjacent if the corresponding edges of G have a common endvertex. Evidently, every edge coloring of G is a vertex coloring of $L(G)$, and vice versa; in particular, $\chi'(G) = \chi(L(G))$.

1.3 CRITICAL GRAPHS

By a **graph parameter** we mean a function ρ that assigns to each graph G a real number $\rho(G)$ such that $\rho(G) = \rho(H)$ whenever G and H are isomorphic graphs. A graph parameter ρ is called **monotone** if $\rho(H) \leq \rho(G)$ whenever H is a subgraph of G. Clearly, the set of all graph parameters form a real vector space with respect to the addition of functions and the multiplication of a function by a real number. Let ρ and ρ' be two graph parameters. If $\rho'(G) = c$ for every graph G, then instead of $\rho + \rho'$ we also write $\rho + c$. If $\rho(G) \leq \rho'(G)$ holds for every graph G, then we say that ρ' is an **upper bound** for ρ and ρ is a **lower bound** for ρ'.

Criticality is a general concept in graph theory and can be defined with respect to various graph parameters. The importance of the notion of criticality is that problems for graphs in general may often be reduced to problems for critical graphs whose structure is more restricted. Critical graphs (w. r. t. the chromatic number) were first defined and used by Dirac in 1951, in his Ph.D. thesis [71] and in Dirac [72].

Let ρ be a monotone graph parameter. A graph G is called ρ-**critical** if $\rho(H) < \rho(G)$ for every proper subgraph H of G. We say that $e \in E(G)$ is a ρ-**critical edge** if $\rho(G - e) < \rho(G)$. Evidently, in a ρ-critical graph every edge is ρ-critical.

Proposition 1.1 *Let ρ and ρ' be two monotone graph parameters. Then the following statements hold:*

(a) *Every graph G contains a ρ-critical subgraph H with $\rho(H) = \rho(G)$.*

(b) *If every ρ-critical graph H satisfies $\rho(H) \leq \rho'(H)$, then $\rho(G) \leq \rho'(G)$ for all graphs G.*

Proof: Since ρ is monotone, every graph G contains a minimal subgraph H with $\rho(H) = \rho(G)$. Obviously, H is ρ-critical. This proves (a). For the proof of (b), let G be an arbitrary graph. By (a), G contains a ρ-critical subgraph H with $\rho(H) = \rho(G)$. Then $\rho(H) \leq \rho'(H)$; and since ρ' is monotone, we obtain $\rho(G) = \rho(H) \leq \rho'(H) \leq \rho'(G)$. ∎

For convenience, we allow a graph G to be **empty**[1], i.e., $V(G) = E(G) = \emptyset$. In this case we also write $G = \emptyset$. For the empty graph G, define $\chi'(G) = \delta(G) = \Delta(G) = \mu(G) = 0$. So the empty graph is r-regular only for $r = 0$. If ρ is a monotone graph parameter, then the empty graph is ρ-critical; it is the only ρ-critical graph H with $\rho(H) = \rho(\emptyset)$.

By a **critical graph** we always mean a χ'-critical graph, and by a **critical edge** we always mean a χ'-critical edge. Clearly, an edge e of G is critical if and only if $\chi'(G - e) = \chi'(G) - 1$. For $k = 1, 2$, a graph G with $\chi'(G) = k$ is critical if and only if G is connected and has exactly k edges. It is also easy to show that a graph G with $\chi'(G) = 3$ is critical if and only if G is an odd cycle or G is a connected graph consisting of three edges that are all incident to the same vertex x of G. Furthermore, if a graph G satisfies $\chi'(G) \geq \Delta(G) + 1$, then $|V(G)| \geq 3$ and $\Delta(G) \geq 2$.

1.4   LOWER BOUNDS AND ELEMENTARY GRAPHS

Since, in an edge coloring, no two edges incident to the same vertex can have the same color, every graph G satisfies $\chi'(G) \geq \Delta(G)$. For a **fat triangle** of multiplicity μ, that is, a graph $T = \mu K_3$ consisting of three vertices pairwise joined by μ parallel edges, we obtain $\chi'(T) = 3\mu$ and $\Delta(T) = 2\mu$. This shows that the gap between the chromatic index and the maximum degree can be arbitrarily large.

Apart from the maximum degree there is another trivial lower bound for the chromatic index, sometimes called the **density** of G, written as $w(G)$. Consider a k-edge-coloring φ of G and a subgraph H of G with $|V(H)| \geq 2$. For every color α, the restricted color class $E_\alpha = E_{\varphi,\alpha} \cap E(H)$ is a matching of H. Consequently, $|E_\alpha| \leq \lfloor |V(H)|/2 \rfloor$ for every color α and, therefore, $|E(H)| \leq k \lfloor |V(H)|/2 \rfloor$.

---

[1]The empty graph is also called the null-graph.

This observation leads to the following definition of a parameter for graphs G with $|V(G)| \geq 2$, namely,

$$w(G) = \max_{H \subseteq G, |V(H)| \geq 2} \left\lceil \frac{|E(H)|}{\lfloor \frac{1}{2}|V(H)| \rfloor} \right\rceil. \tag{1.4}$$

For a graph G with $|V(G)| \leq 1$, define $w(G) = 0$. Then, clearly, w is monotone and every graph G satisfies

$$\chi'(G) \geq w(G). \tag{1.5}$$

As Scheinerman and Ullman [276] proved, the maximum in (1.4) can always be achieved for an induced subgraph H of G having odd order, provided that $|V(G)| \geq 3$. To see this, suppose that the maximum in (1.4) is achieved for a graph $H \subseteq G$ having even order. If $|V(H)| = 2$ this gives $w(G) = |E(H)|$, and hence $w(G) = \lceil 2|E(H')|/(|V(H')| - 1) \rceil$ for any subgraph H' of G with three vertices such that $H \subseteq H'$. Otherwise, $|V(H)| \geq 4$ and we argue as follows. Let v be a vertex of minimum degree in H and let $H' = H - v$. Then $|V(H')|$ is odd and $d_H(v) \leq 2|E(H)|/|V(H)| \leq \lceil 2|E(H)|/|V(H)| \rceil = w(G)$ and, therefore,

$$w(G) \geq \left\lceil \frac{2|E(H')|}{|V(H')| - 1} \right\rceil = \left\lceil \frac{2(|E(H)| - d_H(v))}{|V(H)| - 2} \right\rceil \geq \left\lceil \frac{2|E(H)|}{|V(H)|} \right\rceil = w(G),$$

which proves the claim. Hence, for a graph G with $|V(G)| \geq 3$, we have

$$w(G) = \max_{X \subseteq V(G), |X| \geq 3 \text{ odd}} \left\lceil \frac{2|E(G[X])|}{|X| - 1} \right\rceil. \tag{1.6}$$

Clearly, any graph G with $|V(G)| \leq 2$ satisfies $w(G) = \Delta(G) = \chi'(G)$. To see that the gap between $w(G)$ and $\chi'(G)$ can be arbitrarily large, consider the simple graph $G = K_{1,n}$ consisting of $n + 1$ vertices and n edges all incident to the same vertex. Then, for $n \geq 2$, we have $w(K_{1,n}) = 2$ and $\chi'(K_{1,n}) = \Delta(K_{1,n}) = n$. For the Petersen graph P, we have $\chi'(P) = 4$ and $\Delta(P) = w(P) = 3$. However, the situation seems to be different for graphs with $\chi'(G) > \Delta(G) + 1$.

A graph G is called an **elementary graph** if $\chi'(G) = w(G)$. The significance of this equation is that the chromatic index is characterized by a min–max equality. The following conjecture seems to have been thought of first by Goldberg [110, 114] around 1970 and, independently, by Seymour [280, 281].

Conjecture 1.2 (Goldberg [110] 1973, Seymour [281] 1979) *Every graph G such that $\chi'(G) \geq \Delta(G) + 2$ is elementary, i.e., $\chi'(G) = w(G)$.*

In this book, we will refer to Conjecture 1.2 briefly as Goldberg's conjecture. For a proof of this conjecture, it is sufficient to consider critical graphs. To see why, let G be an arbitrary graph with $\chi'(G) \geq \Delta(G) + 2$. Clearly, G contains a critical graph H with $\chi'(H) = \chi'(G)$. This implies that $\chi'(H) \geq \Delta(H) + 2$. If the graph H is known to be elementary, then also G is elementary, since in this case we have $w(G) \leq \chi'(G) = \chi'(H) = w(H) \leq w(G)$ and, therefore, $\chi'(G) = w(G)$.

Proposition 1.3 *Let G be a graph with $\chi'(G) = k + 1$ for an integer $k \geq \Delta(G)$. If G is critical and elementary, then the following statements hold:*

(a) $\chi'(G) = w(G) = \lceil |E(G)| / \lfloor \frac{1}{2}|V(G)| \rfloor \rceil$ *and $|V(G)|$ is odd.*

(b) *For every edge $e \in E(G)$ and every coloring $\varphi \in \mathcal{C}^k(G - e)$, we have $m_{\varphi,\alpha} = 1$ for all colors $\alpha \in \{1, \ldots, k\}$; i.e., the color α is missing at exactly one vertex of G.*

(c) $|E(G)| = k \lfloor \frac{1}{2}|V(G)| \rfloor + 1$.

Proof: Since $\chi'(G) \geq \Delta(G) + 1$, we have $|V(G)| \geq 3$ and $E(G) \neq \emptyset$. Since the graph G is both critical and elementary, every proper subgraph H of G satisfies $w(H) \leq \chi'(H) < \chi'(G) = w(G)$. Clearly, this implies that

$$\chi'(G) = w(G) = \left\lceil \frac{|E(G)|}{\lfloor \frac{1}{2}|V(G)| \rfloor} \right\rceil.$$

Consequently, $|V(G)|$ is odd, since otherwise

$$\chi'(G) = \lceil 2|E(G)|/|V(G)| \rceil \leq \Delta(G),$$

a contradiction. This proves (a). For the proof of (b), let $e \in E(G)$ and $\varphi \in \mathcal{C}^k(G-e)$. By (a) and (1.2), we have $m_{\varphi,\alpha} \equiv |V(G)| \equiv 1 \mod 2$. Hence, if $m_{\varphi,\alpha} \neq 1$ for some color α, then (1.3) implies that $\sum_{v \in V(G)}(k - d_G(v)) + 2 \geq k + 2$. Since $|V(G)|$ is odd, this yields $\lceil |E(G)|/\lfloor |V(G)|/2 \rfloor \rceil \leq k = \chi'(G) - 1$, a contradiction to (a). For the proof of (c), choose an edge $e \in E(G)$. Since G is critical, there is a coloring $\varphi \in \mathcal{C}^k(G-e)$. Then we deduce from (b) and (1.3) that $\sum_{v \in V(G)}(k-d_G(v))+2 = k$. Since $|V(G)|$ is odd, this implies that $|E(G)| = k\lfloor|V(G)|/2\rfloor + 1$. This completes the proof. ∎

The following result shows that elementary graphs and elementary sets are closely related to each other. This result is implicitly contained in the papers by Andersen [5] and Goldberg [114].

Theorem 1.4 *Let G be a graph with $\chi'(G) = k + 1$ for an integer $k \geq \Delta(G)$. If G is critical, then the following conditions are equivalent:*

(a) *G is elementary.*

(b) *For every edge $e \in E(G)$ and every coloring $\varphi \in \mathcal{C}^k(G - e)$, the set $V(G)$ is elementary with respect to φ.*

(c) *There is an edge $e \in E(G)$ and a coloring $\varphi \in \mathcal{C}^k(G - e)$ such that $V(G)$ is elementary with respect to φ.*

(d) *There is an edge $e \in E(G)$, a coloring $\varphi \in \mathcal{C}^k(G - e)$ and a set $X \subseteq V(G)$ such that X contains the two endvertices of e and X is elementary as well as strongly closed with respect to φ.*

Proof: That (a) implies (b) follows from Proposition 1.3(b). Evidently, (b) implies (c) and (c) implies (d) with $X = V(G)$. To prove that (d) implies (a), suppose that, for some edge $e \in E(G)$ and some coloring $\varphi \in C^k(G - e)$, there is a subset X of $V(G)$ such that both endvertices of e are contained in X and X is elementary as well as strongly closed with respect to φ. Let $H = G[X]$ and, for each color $\alpha \in \{1, \ldots, k\}$, let $E_\alpha = E_{\varphi,\alpha} \cap E(H)$. Since the edge e is uncolored and both endvertices of e belong to X, the set $\overline{\varphi}(X)$ is nonempty and $|X| \geq 2$.

First, consider an arbitrary color $\alpha \in \overline{\varphi}(X)$. Since X is elementary with respect to φ, color α is missing at exactly one vertex of H. Furthermore, since X is closed, no edge in $\partial_G(X)$ is colored with α. Since E_α is a matching of H, this implies that $|X| = |V(H)|$ is odd and $|E_\alpha| = \lfloor |V(H)|/2 \rfloor$. Now, consider an arbitrary color $\alpha \notin \overline{\varphi}(X)$. Then, clearly, color α is present at every vertex of $X = V(H)$. Since X is strongly closed, at most one edge of $\partial_G(X)$ is colored with α. Since E_α is a matching of H and $|V(H)|$ is odd, this implies that exactly one edge of $\partial_G(X)$ is colored with α and, moreover, $|E_\alpha| = \lfloor |V(H)|/2 \rfloor$, too. This proves that

$$|E(H)| = 1 + k \left\lfloor \frac{1}{2}|V(H)| \right\rfloor.$$

Since H is a subgraph of G with $|V(H)| \geq 2$, we then deduce that

$$w(G) \leq \chi'(G) = k + 1 = \left\lceil \frac{|E(H)|}{\lfloor \frac{1}{2}|V(H)| \rfloor} \right\rceil \leq w(G).$$

Therefore, G is an elementary graph. This shows that (d) implies (a). Hence the proof of Theorem 1.4 is complete.   ∎

Combining Proposition 1.3 and Theorem 1.4 together with the equations (1.2) and (1.3), we obtain the following result.

Corollary 1.5 *Let G be a critical graph with $\chi'(G) = k+1$ for an integer $k \geq \Delta(G)$. If $|V(G)|$ is odd, then $\sum_{v \in V(G)}(k - d_G(v)) + 2 \geq k$, where equality holds if and only if G is elementary. Furthermore, G is elementary if and only if $|V(G)|$ is odd and $\sum_{v \in V(G)}(k - d_G(v)) = k - 2$.*

Since it suffices to verify Goldberg's conjecture for the class of critical graphs, it follows from Corollary 1.5 that Goldberg's conjecture is equivalent to the following conjecture.

Conjecture 1.6 (Critical Multigraph Conjecture) *Every critical graph G with $\chi'(G) \geq \Delta(G) + 2$ is of odd order and satisfies*

$$2|E(G)| = (\chi'(G) - 1)(|V(G)| - 1) + 2.$$

We conclude this section with some basic facts about elementary sets that are useful for our further investigations.

Proposition 1.7 *Let G be a graph with $\Delta(G) = \Delta \geq 2$, let $e \in E_G(x,y)$ be an edge, and let $\varphi \in \mathcal{C}^k(G-e)$ be a coloring for an integer $k \geq \Delta$. If $X \subseteq V(G)$ is an elementary set with respect to φ such that both endvertices of e are contained in X, then the following statements hold:*

(a) $|X| \leq \frac{|\overline{\varphi}(X)|-2}{k-\Delta} \leq \frac{k-2}{k-\Delta}$, *provided that $k \geq \Delta + 1$.*

(b) $\sum_{v \in X} d_G(v) \geq k(|X|-1) + 2$.

(c) *Suppose that*

$$k + 1 > \frac{m}{m-1}\Delta + \frac{m-3}{m-1}$$

for an integer $m \geq 3$. Then $|X| \leq m - 1$ and, moreover, $|\overline{\varphi}(X)| \geq \Delta + 1$, provided that $|X| = m - 1$.

Proof: Since the set X is elementary with respect to $\varphi \in \mathcal{C}^k(G-e)$, we deduce that

$$\sum_{v \in X} |\overline{\varphi}(v)| = |\overline{\varphi}(X)| \leq k.$$

The edge $e \in E_G(x,y)$ being uncolored, for a vertex $v \in V(G)$, we have

$$|\overline{\varphi}(v)| = \begin{cases} k - d_G(v) + 1 & \text{if } v \in \{x,y\}, \\ k - d_G(v) & \text{otherwise.} \end{cases}$$

Then, since $x, y \in X$, we obtain

$$2 + |X|(k-\Delta) \leq 2 + \sum_{v \in X}(k - d_G(v)) = 2 + k|X| - \sum_{v \in X} d_G(v) = |\overline{\varphi}(X)| \leq k,$$

which implies (a) and (b). To prove (c), we first deduce from the hypothesis that $k - \Delta > (\Delta - 2)/(m-1)$. Since $\Delta \geq 2$ and $m \geq 3$, this implies that $k \geq \Delta + 1$. By (a), we then obtain

$$|X| \leq \frac{k-2}{k-\Delta} = 1 + \frac{\Delta - 2}{k - \Delta} < 1 + (m-1) = m$$

and, therefore, $|X| \leq m - 1$. Now, assume that $|X| = m - 1$. Then, by (a), we have $|\overline{\varphi}(X)| \geq (k-\Delta)|X| + 2 = (k-\Delta)(m-1) + 2 > \Delta$ and, therefore, $|\overline{\varphi}(X)| \geq \Delta + 1$. This completes the proof of (c). ∎

Let G be a critical graph with maximum degree Δ, and let $m \geq 3$ be an integer. Suppose that G satisfies

$$\chi'(G) > \frac{m}{m-1}\Delta + \frac{m-3}{m-1}.$$

Then $\chi'(G) = k + 1 \geq \Delta + 2 \geq 4$. Consequently, Goldberg's conjecture (that G is elementary) implies that $|V(G)| \leq m-1$ (Theorem 1.4, Proposition 1.7), respectively

$|V(G)| \leq m - 2$ if m is odd (Proposition 1.3). Thus the following conjecture, first posed by Jakobsen [156], may be seen as a weaker form of Goldberg's conjecture.

Conjecture 1.8 (Jakobsen [156] 1975) *Let G be a critical graph, and let*

$$\chi'(G) > \frac{m}{m-1}\Delta(G) + \frac{m-3}{m-1}$$

for an odd integer $m \geq 3$. Then $|V(G)| \leq m - 2$.

Thus for fixed $\Delta(G)$, or for fixed $\chi'(G)$, there are only finitely many critical graphs G with $\chi'(G) \geq \Delta(G) + 2$, assuming Goldberg's conjecture is true.

A **fat odd cycle**, i.e., a graph $G = \mu C_m$ for an odd integer $m \geq 3$, has for $\mu \equiv 1 \bmod (m-1)/2$

$$\chi'(G) = w(G) = \frac{m}{m-1}\Delta(G) + \frac{m-3}{m-1}$$

and it is critical with m vertices. Thus Conjecture 1.8 is in this sense best possible. To see why G is elementary and critical, note first that $|E(G)| = m\mu$, $\Delta(G) = 2\mu(G) = 2\mu$ and, by (1.5),

$$\chi'(G) \geq w(G) \geq \left\lceil \frac{2|E(G)|}{m-1} \right\rceil = \left\lceil \frac{2m\mu}{m-1} \right\rceil = \left\lceil \frac{m\Delta(G)}{m-1} \right\rceil.$$

By assumption, there are integers $\ell \geq 1$ and $p \geq 0$ such that $m = 2\ell + 1$ and $\mu = 1 + p\ell$. Then $|E(G)| = m\mu = \ell(2p\ell + p + 2) + 1$ and there is an integer $k \geq 2$ such that

$$k + 1 = \left\lceil \frac{m\Delta(G)}{m-1} \right\rceil = \frac{m}{m-1}\Delta(G) + \frac{m-3}{m-1} = 2p\ell + p + 3.$$

If e is an arbitrary edge of G, then it is easy to check that the remaining edge set of G can be partitioned into $k = 2p\ell + p + 2$ matchings, each having ℓ edges implying that $\chi'(G - e) \leq k$. Since G is connected and $\chi'(G) \geq w(G) \geq k + 1$, this implies that G is critical and $\chi'(G) = w(G) = k + 1$.

1.5 UPPER BOUNDS AND COLORING ALGORITHMS

The **Edge Color Problem** asks for an optimal edge coloring of a graph G, that is, an edge coloring with $\chi'(G)$ colors. Holyer [150] proved that the determination of the chromatic index is **NP**-hard, even for 3-regular simple graphs, where the chromatic index is either 3 or 4. Hence it is reasonable to search for upper bounds for the chromatic index, in particular for those bounds that are efficiently realized by a coloring algorithm. A graph parameter ρ is said to be an **efficiently realizable** upper bound for χ' if there exists an algorithm that computes, for every graph $G = (V, E)$, an edge coloring using at most $\rho(G)$ colors, where the algorithm has time complexity

t bounded from above by a polynomial in $|V|$ and $|E|$, that is, $t(G) \leq p(|E|, |V|)$ for some polynomial $p = p(x, y)$ over the real numbers in two variables.

Note that edge coloring algorithms may have an execution time polynomial in $|E|$, but being only pseudopolynomial in the number of bits needed to describe the graph, since edge multiplicities may be encoded as binary numbers, and the size of the input graph therefore may be of order less than the order of E.

A typical algorithm colors the edges of the input graph sequentially. Such an algorithm first fixes an edge order of the input graph, either an arbitrary order or one that satisfies a certain property. The core of the algorithm is a subroutine **Ext** that extends a given partial coloring of the input graph. The input of **Ext** is a tuple (G, e, x, y, k, φ), where G is the graph consisting of all edges that are already colored as well as the next uncolored edge $e \in E_G(x, y)$ with respect to the given edge order, and a coloring $\varphi \in C^k(G - e)$. The output of **Ext** is a pair (k', φ'), where $k' \in \{k, k + 1\}$ and $\varphi' \in C^{k'}(G)$.

Now, to explain how **Ext** works, a well-defined set $\mathcal{O}(G, e, \varphi)$ of so-called **test objects** will be introduced. A test object $T \in \mathcal{O}(G, e, \varphi)$ is usually a labeled subgraph of G that fulfills a certain property with respect to the uncolored edge e and the coloring $\varphi \in C^k(G - e)$. In most cases, we start with the test object that only consists of the uncolored edge e. When a test object $T \in \mathcal{O}(G, e, \varphi)$ is investigated, then, using an exhaustive case distinction, three basic outcomes are possible. The first possible outcome is that the vertex set $V(T)$ is not elementary with respect to φ; i.e., a color $\alpha \in \{1, \ldots, k\}$ is missing at two distinct vertices of T with respect to φ. In this case **Ext** returns (k, φ'), where the coloring $\varphi' \in C^k(G)$ is obtained from φ by Kempe changes, possibly involving more than one pair of colors in a small number of successive Kempe changes. The second possible outcome is that the vertex set $V(T)$ is elementary with respect to φ, but T cannot be enlarged. In that case e is colored with a new color resulting in a coloring $\varphi' \in C^{k+1}(G)$. Then **Ext** returns $(k + 1, \varphi')$. The third possible outcome is that the vertex set $V(T)$ is elementary with respect to φ, but T can be enlarged. Then an exhaustive search for a larger test object is needed. This process eventually terminates, because for sufficiently large test objects $T \in \mathcal{O}(G, e, \varphi)$, one of the first two cases has to be applicable.

To ensure that the subroutine **Ext**, and hence the algorithm, works correctly, we need a statement about the test objects of the following type.

(1) *Let G be a graph with $\chi'(G) = k + 1$ for some integer $k \geq \Delta(G)$, let $e \in E_G(x, y)$ be a critical edge of G, and let $\varphi \in C^k(G - e)$ be a coloring. Then the vertex set of each test object $T \in \mathcal{O}(G, e, \varphi)$ is elementary with respect to φ.*

This statement is equivalent to the statement that if $\varphi \in C^k(G - e)$ is a coloring and the vertex set of a test object $T \in \mathcal{O}(G, e, \varphi)$ is not elementary with respect to φ, then $\chi'(G) \leq k$, i.e., there is a coloring $\varphi' \in C^k(G)$. For the correctness of the algorithm it is, however, important that the proof of (1) is constructive and can be transformed into an efficient procedure for obtaining such a coloring $\varphi' \in C^k(G)$.

To control the number of colors used by a coloring algorithm of the above type, we need some further information about **maximal test objects**, which means test objects

$T \in \mathcal{O}(G, e, \varphi)$ that cannot be extended to some larger test object $T' \in \mathcal{O}(G, e, \varphi)$. For the proof of Goldberg's conjecture a statement of the following type would be sufficient.

(2) *Let G be a graph with $\chi'(G) = k + 1$ for some integer $k \geq \Delta(G) + 1$, let $e \in E_G(x, y)$ be a critical edge of G, and let $\varphi \in \mathcal{C}^k(G - e)$ be a coloring. Then the vertex set of each maximal test object $T \in \mathcal{O}(G, e, \varphi)$ is both elementary and strongly closed with respect to φ.*

Suppose our test objects satisfies (1) and (2) and we start our coloring algorithm with $k = \Delta(G) + 1$ colors. If the algorithm never uses a new color, then $\chi'(G) \leq \Delta(G) + 1$. Otherwise, let us consider the last call of **Ext** where we use a new color. The input is a tuple $(G', e, x, y, k, \varphi)$, where G' is a subgraph of G, $e \in E_{G'}(x, y)$, and $\varphi \in \mathcal{C}^k(G' - e)$. Since **Ext** returns a coloring $\varphi' \in \mathcal{C}^{k+1}(G')$, there exist a maximal test object $T \in \mathcal{O}(G', e, \varphi)$ such that $X = V(T)$ is elementary and strongly closed both with respect to φ. Clearly, the coloring algorithm terminates with a $(k + 1)$-edge-coloring of G implying $\chi'(G) \leq k + 1$. Now, let H be the subgraph of G with $V(H) = X$ and $E(H) = E(G[X]) \cap E(G')$. Then $E(H)$ consists of the uncolored edge e and all edges of G that are already colored and have both endvertices in X. Since X is elementary and strongly closed both with respect to $\varphi \in \mathcal{C}^k(G' - e)$, it then follows that $|X| = |V(H)| \geq 3$ is odd and $|E(H)| = 1 + k\lfloor|V(H)|/2\rfloor$ (see the proof of Theorem 1.4, the part where we show that (d) implies (a)). Consequently, we have $w(G) \geq w(H) \geq \lceil|E(H)|/(\lfloor|V(H)|/2\rfloor)\rceil \geq k + 1 \geq \chi'(G) \geq w(G)$ and, therefore, $\chi'(G) = w(G)$. Hence our algorithm colors the edges of G with at most $\max\{\Delta(G) + 1, w(G)\}$ colors.

Classical kinds of test objects are the fans, first used by Shannon [284] and by Vizing [297], the critical chains introduced independently by Andersen [5] and by Goldberg [111, 114], and the Kierstead paths introduced by Kierstead [166]. A more recent kind of test objects, namely Tashkinov trees, were invented by Tashkinov [291]. All these kinds of test objects satisfy (1), but up to now no test objects that fulfill both conditions (1) and (2) are known. A possible way out of this situation is to modify the subroutine **Ext** and to add further heuristics before using a new color. If the vertex set X of a maximal test object $T \in \mathcal{O}(G, e, \varphi)$ is both elementary and strongly closed with respect to φ, then we just color e with a new color. However, if X is elementary, but not strongly closed with respect to φ, it might be reasonable to use a small number of Kempe changes to obtain a better test object $T' \in \mathcal{O}(G, e', \varphi')$ and to continue with T' instead of T. We shall use this approach to get some partial results related to Goldberg's conjecture.

One obvious way to find an edge coloring of an arbitrary graph G with at least one edge is the following **greedy algorithm**: Starting from a fixed edge order $e_1, \ldots, e_m$ of G, we consider the edges in turn and color each edge e_i with the smallest positive integer not already used to color any adjacent edge of e_i among $e_1, \ldots, e_{i-1}$. Since no edge is adjacent to more than $2(\Delta(G) - 1)$ other edges, this simple greedy algorithm never uses more that $2\Delta(G) - 1$ colors. Hence, every graph G with $E(G) \neq \emptyset$ satisfies $\chi'(G) \leq 2\Delta(G) - 1$. Observe that this greedy strategy is the simplest version of a coloring algorithm that fits into our general approach; there is only one test object in $\mathcal{O}(G, e, \varphi)$, namely the graph consisting of the uncolored edge e and its two endvertices. As an immediate consequence, we obtain that 2Δ is an efficiently realizable upper bound for χ' (including the case $E(G) = \emptyset$). Since Δ is a lower bound for χ', this implies that $2\chi'$ is an efficiently realizable upper bound for χ'. Goldberg's conjecture supports the following suggestion by Hochbaum, Nishizeki, and Shmoys [146].

Conjecture 1.9 $\chi' + 1$ *is an efficiently realizable upper bound for* χ'.

The upper bound $2\Delta - 1$ on the number of colors used by the greedy algorithm is rather generous, and in most graphs there will be scope for an improvement of this bound by choosing a particularly suitable edge order to start with. Let us say that an **edge order** of a graph G is of **depth** p if each edge in this order is preceded by fewer than p of its adjacent edges. Clearly, if we start the greedy algorithm with an edge order of depth p, then the algorithm terminates with a p-edge-coloring. The least number $p \geq 1$ such that G has an edge order of depth p is called the **coloring index** $\mathrm{col}'(G)$ of G. Observe that the coloring index of a graph is nothing else than the so-called coloring number of its line graph. Obviously, every graph G with at least one edge satisfies $\mathrm{col}'(G) \leq 2\Delta(G) - 1$. For an edgeless graph G, we have $\mathrm{col}'(G) = 1$. It is also known (see, e.g., Jensen and Toft [158]), that an edge order $e_1, \ldots, e_m$ of depth $\mathrm{col}'(G)$ can be obtained by letting e_i be an edge having a minimum number of adjacent edges in the subgraph $G_i = G - \{e_{i+1}, \ldots, e_m\}$ for $i = m, m - 1, \ldots, 1$, where $G_m = G$. Hence, col' is an efficiently realizable upper bound for χ', obviously the best upper bound that can be realized by the greedy algorithm.

Finally, we discuss some implementation details. The time complexity t of our coloring algorithms has the form $t = t_1 + |E|t_2$, where t_1 is the time complexity for computing the required edge order of the input graph $G = (V, E)$ and t_2 is the (worst case) time complexity for one call of the subroutine **Ext**.

The running time t_2 depends on the manner in which the partial coloring is stored. As long as we are satisfied with an overall running time t that is polynomial in $|E|$ and $|V|$, one can use the approach by Hochbaum, Nishizeki, and Shmoys [146]. The idea is to combine the standard **incidence lists** for the vertices with the **same-color lists** for the colors. An edge $e \in E_G(u, v)$ receiving color α is stored in the two incidence lists for u and v, and in addition to that also in the same-color list for the color α. The elements in the corresponding three lists are linked to each other by pointers. Furthermore, a list of all uncolored edges is stored.

For the number of colors k, we may assume that $k = O(\Delta)$, where $\Delta = \Delta(G)$. Then, as explained in Hochbaum et al. [146], each set $\overline{\varphi}(x)$ can be found in time $O(\Delta)$ and, therefore, one can decide in time $O(\Delta)$ whether two vertices have a common missing color. Furthermore, it takes time $O(|V|)$ to find an (α, β)-chain $P = P_x(\alpha, \beta, \varphi)$. The colors of P can be interchanged in time $O(|V|)$, and updating the same-color list for the coloring $\varphi' = \varphi/P$ can be carried out in time $O(|V| + \Delta)$.

1.6 NOTES

Edge colorings of graphs were first considered in two short papers by Tait [290] published in the same proceedings between 1878 and 1880. Tait proved a theorem relating face colorings and edge colorings of plane graphs, i.e., graphs embedded in the plane or sphere. Tait's theorem deals with 3-regular graphs, which are also referred to as **cubic** graphs. A **cut-edge** or a **cut-vertex** of a graph is an edge or vertex whose deletion increases the number of components. A graph without cut-edges is also said to be a **bridgeless** graph. Tait's theorem says that if G is a bridgeless cubic plane (simple) graph, then G admits a 3-edge-coloring if and only if the faces of G can be colored with four colors such that adjacent faces receive different colors. Tait's result implies that the following three statements are equivalent.

(A) The faces of any bridgeless plane graph can be colored with four colors such that any two adjacent faces get different colors.

(B) Every bridgeless cubic planar simple graph G satisfies $\chi'(G) = 3$.

(C) Every bridgeless cubic planar graph G satisfies $\chi'(G) = 3$.

Tait did not prove any of the statements in his papers since he did not consider this necessary because of an already existing proof of (A) by Kempe [165].

The Four-Color Problem was first mentioned in writing in a letter from A. De Morgan to W. R. Hamilton, written in 1852 on the same day as De Morgan first heard about the problem from his student Frederic Guthrie, who had the problem from his brother Francis Guthrie. A proposed solution of the problem by Kempe [165] stood for more than a decade until it was refuted by Heawood [133]. Heawood proved, using Kempe's method, the Five-Color Theorem for planar graphs.

Statement (A) is equivalent to the same statement $(\tilde{A})$ with the words *faces* replaced by *vertices*, as already observed by A. B. Kempe:

$(\tilde{A})$ The vertices of any planar simple graph can be colored with four colors such that any two adjacent vertices get different colors.

It was however first with the famous paper of Brooks [33] that vertex coloring of general graphs became a topic of study. Brooks [33] proved that the complete graphs and the odd cycles are the only connected simple graphs whose chromatic number is larger than their maximum degree.

Even if Kempe's 1879 paper contained a serious flaw, it contained the idea of recoloring a connected component in the subgraph spanned by two colors, a so-called **Kempe chain**, by simply interchanging the two colors on the vertices of the component (we consider here vertex colorings rather than face colorings). This idea has since been a main tool in graph coloring theory, and, as explained in Sect. 1.2, also for edge colorings. Some recent results about Kempe changes and Kempe equivalence of edge colorings can be found in references [11, 219, 225].

The Four-Color Theorem was proved by K. Appel, W. Haken, and J. Koch [9, 10] in 1977, and later by Robertson, Sanders, Seymour, and Thomas [258] with an improved proof, essentially using the same approach as Appel, Haken, and Koch, but the proof is shorter and clearer, avoiding the problematic details of the original proof. In this way the Four - Color Problem has become the Four - Color Theorem.

In the 1890s there was some confusion about Tait's theorem. Some believed that Tait's theorem asserted that every bridgeless cubic simple graph is 3-edge-colorable. This motivated Petersen [242] to present, as a counterexample, the graph that has became famous as the Petersen graph (see Fig. 1.2).

Petersen [241], collaborating with James Joseph Sylvester, was the first mathematician who studied the problem of factorizing graphs in a general context. One of his fundamental results says that every bridgeless cubic graph G has a **perfect matching** M, i.e., $M \subseteq E(G)$ and every vertex x of G is incident with exactly one edge of M. That Petersen's result implies that every cubic graph has chromatic index 3 or 4 was pointed out by Sainte-Laguë [261] in 1926 (without a precise argument). In particular, the Petersen graph has chromatic index 4. We shall apply the elegant argument by Naserasr and Škrekovski [234] to prove the following slightly stronger statement.

(a) *Let P^* be the Petersen graph with one vertex deleted. Then $\chi'(P^*) = 4$.*

Proof of (a): Obviously, $\chi'(P^*) \leq 4$. Now, suppose there is a 3-edge-coloring of

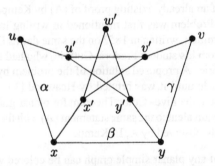

Figure 1.3 The graph P^*.

P^*. Let α, β, γ be the colors of the edges ux, xy, yv, respectively (see Fig. 1.3). Then α and γ may be equal, but $\beta \neq \alpha, \gamma$. Obviously, at each vertex of degree 3 each color must appear. Since β cannot appear on xx' or yy', color β appears on two distinct edges of the inner cycle $C' = (x', v', u', y', w', x')$, one of which must be $x'v'$ or $y'u'$. The same argument works for the colors α and γ. Since C' has only five edges, this implies that $\alpha = \gamma$. But then α has to appear on $u'v'$ and on two more edges of C', a contradiction. ∎

If we delete an arbitrary edge of P^*, then it is easy to show that the resulting subgraph has a 3-edge-coloring. Hence P^* is a critical graph.

The basic problem in the theory of graph factorization is decomposing a regular graph into other regular graphs on the same set of vertices. An r-**factor** of an arbitrary graph G is a **spanning subgraph** H of G, i.e., $H \subseteq G$ and $V(H) = V(G)$, such that H is r-regular. Evidently, a graph has a 1-factor if and only if it has a perfect matching. Another result of Petersen's fundamental paper [241] is his even factor theorem, that every $2r$-regular graph has a 2-factor. The statement that every bridgeless cubic graph has a 1-factor should perhaps also be formulated as a 2-factor theorem. As pointed out by Hanson, Loten, and Toft [129], every $(2r + 1)$-regular graph with at most $2r$ bridges has a 2-factor, thus the general Petersen theorem is about 2-factors rather than 1-factors.

As we know, every graph G satisfies $\Delta(G) \leq \chi'(G)$. In 1916 König [174] proved that equality holds for the class of bipartite graphs, and he deduced as a simple corollary that every regular bipartite graph has a perfect matching. König's proof uses induction on the number of edges and a simple recoloring argument. In particular, the proof yields an $O(mn)$ algorithm to find a $\Delta(G)$-edge-coloring for a bipartite graph G with n vertices and m edges. The algorithm is a simplified version of the coloring algorithm described in Sect. 1.5. The only test object in $\mathcal{O}(G, e, \varphi)$ is the graph consisting of the uncolored edge e and the two endvertices x, y of e. Since we have $k = \Delta(G)$ colors, we can chose two colors $\alpha \in \overline{\varphi}(x)$ and $\beta \in \overline{\varphi}(y)$. If $\alpha = \beta$, then we color e with α and continue with the next uncolored edge. Otherwise, we recolor $P = P_x(\alpha, \beta, \varphi)$. Since G contains no odd cycle, y does not belong to P and, therefore, for the coloring $\varphi' = \varphi/P$ we obtain $\alpha \in \overline{\varphi}'(x) \cap \overline{\varphi}'(y)$. Hence we can color e with α and continue with the next uncolored edge.

A simple, self-contained proof of König's theorem that does not use any alternating path argument was given by Rizzi [256].

In 1949 Shannon [284] proved that every graph G satisfies $\chi'(G) \leq \lfloor 3\Delta(G)/2 \rfloor$. The fat triangles are graphs for which Shannon's bound is attained. From Shannon's proof it also follows that $3\Delta/2$ is an efficiently realizable upper bound for χ'.

Shannon's proof uses induction and Kempe changes. He starts by remarking that if $\Delta(G) = 2r$ then G is a subgraph of a $2r$-regular graph. By Petersen's even factor theorem, this graph can be factorized into r 2-factors, each of which has a 3-edge-coloring. This immediately gives the desired result.

For $\Delta(G) = 2r + 1$ Shannon explains that there is a conjecture by Petersen that a $(2r + 1)$-regular bridgeless graph has a 1-factor. If true for graphs with at most one bridge this would imply the result like in the even case. But this conjecture is not true! Shannon did not know, but it is easy to construct a counterexample, namely a $(2r + 1)$-regular bridgeless graph without a 1-factor ($r \geq 2$), using Tutte's 1-factor criterion. For $\Delta(G) = 2r + 1$ one may, however, use a different factorization result. As explained above, it is now known that a $(2r + 1)$-regular graph with at most one bridge has a 2-factor. From this Shannon's theorem follows easily.

Shannon's own proof for the case $\Delta(G) = 2r + 1$ is by induction over the number of vertices. He removes a vertex x of degree $2r + 1$ from G, colors $G - x$ by induction using $3r + 1$ colors. Then he colors the edges incident at x one by one, using Kempe changes as the main tool.

It would seem more appropriate to use induction on the number of edges. Let $e \in E_G(x, y)$ be an edge of G and assume that $G - e$ is edge-colored with $3r + 1$ colors by induction. We want to extend this coloring $\varphi \in \mathcal{C}^{3r+1}(G - e)$ by including the remaining edge e.

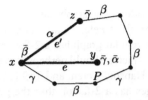

Figure 1.4   Shannon's Kempe change.

Since at most $2r$ edges incident with x are colored, $|\overline{\varphi}(x)| \geq r + 1$. Similarly, $|\overline{\varphi}(y)| \geq r + 1$. If $\overline{\varphi}(x)$ and $\overline{\varphi}(y)$ have a common color, then this color may be given to e and a $(3r + 1)$-edge-coloring of G is obtained. Hence we may assume that $\overline{\varphi}(x)$ and $\overline{\varphi}(y)$ are disjoint. Let α be a color in $\overline{\varphi}(y)$. Then α is present at x, i.e., $\alpha = \varphi(e')$ for an edge $e' \in E_G(x, z)$, where z is different from y. If a color $\beta \in \overline{\varphi}(x)$ is missing at z, then e' may be colored β and e may then be colored α. Hence we may assume that all colors from $\overline{\varphi}(x)$ are present at z. At z there are at most $2r + 1$ colors present, hence at most r colors from $\overline{\varphi}(y)$ belong to $\overline{\varphi}(z)$. This means that there is a color $\gamma \in \overline{\varphi}(y) \cap \overline{\varphi}(z)$. Let β be a color from $\overline{\varphi}(x)$, see Fig. 1.4; note that in Fig. 1.4, and from here on in this book, a bar above a color name means that the color is missing at a particular vertex. Consider the chain $P = P_x(\beta, \gamma)$. If the chain does not end at y, recolor P and color e by γ. If the chain does not end at z, recolor P, recolor e' by γ, and color e by α. One of the two cases must occur.

This proves Shannon's theorem. The proof is essentially Shannon's proof. He formulates it using a $(2n+1) \times (3n+1)$ 0-1-matrix with 1-entries showing the colors possible for the edges from x. He then rearranges columns and rows of the matrix (changes the order of colors and of edges) and makes Kempe changes, corresponding to the arguments above, to see that the matrix may be changed into one with a 1 in all places (i, i) of the matrix, thus showing that it is possible to extend a coloring of $G - x$ to include all edges from x also.

Following Tait, König, and Shannon, the next breakthrough was the theorem of Vizing [297, 298], obtained independently by Gupta [120]. This theorem, from 1964, says that $\chi'(G) \leq \Delta(G) + \mu(G)$ for every graph G.

By Vizing's result, the chromatic index of a simple graph G is either $\Delta(G)$ or $\Delta(G) + 1$. Vizing's proof yields a polynomial-time algorithm that colors the edges of any simple graph G with $\Delta(G) + 1$ colors. On the other hand, Holyer [150] proved that it is NP-complete to decide whether a cubic simple graph has chromatic index 3. These two results answer the edge coloring problem for the class of simple graphs – at least from an algorithmic point of view. Our knowledge about edge coloring of (multi)graphs, however, remains unsatisfactory. Goldberg's conjecture supports the conjecture that there is a polynomial-time algorithm that colors the edges of any graph G with $\chi'(G) + 1$ colors. Furthermore, Goldberg's conjecture implies that the only difficulty in determining the chromatic index of an arbitrary graph G in polynomial time is to distinguish between $\chi'(G) = \Delta(G)$ and $\chi'(G) = \Delta(G) + 1$.

CHAPTER 2

VIZING FANS

The fan argument was introduced by Vizing [297, 298] in order to prove that $\Delta + \mu$ is an upper bound for the chromatic index. However, the appropriate conclusion of the fan argument is not just Vizing's bound, but the so-called fan equation. The fan equation implies the classical bounds as well as the classical adjacency lemmas. We also discuss an extension of the fan equation, and we introduce the fan number as a new graph parameter.

2.1 THE FAN EQUATION AND THE CLASSICAL BOUNDS

Let G be a graph, let $e \in E_G(x, y)$ be an edge, and let $\varphi \in \mathcal{C}^k(G - e)$ be a coloring for some integer $k \geq 0$. A **multi-fan** at x with respect to e and φ is a sequence $F = (e_1, y_1, \ldots, e_p, y_p)$ with $p \geq 1$ consisting of edges $e_1, \ldots, e_p$ and vertices $y_1, \ldots, y_p$ satisfying the following two conditions:

(F1) The edges $e_1, \ldots, e_p$ are distinct, $e_1 = e$, and $e_i \in E_G(x, y_i)$ for $i = 1, \ldots, p$.

(F2) For every edge e_i with $2 \leq i \leq p$, there is a vertex y_j with $1 \leq j < i$ such that $\varphi(e_i) \in \overline{\varphi}(y_j)$.

Graph Edge Coloring: Vizing's Theorem and Goldberg's Conjecture,
First Edition. By M. Stiebitz, D. Scheide, B. Toft, and L. M. Favrholdt
Copyright © 2012 John Wiley & Sons, Inc.

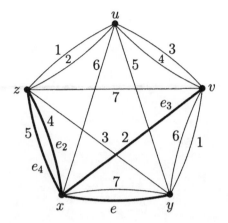

Figure 2.1 A multi-fan $F = (e, y, e_2, z, e_3, v, e_4, z)$ at x (bold edges).

Let $F = (e_1, y_1, \ldots, e_p, y_p)$ be a multi-fan at x. Since the vertices of F need not be distinct, the set $V(F) = \{y_1, \ldots, y_p\}$ may have cardinality smaller than p. For $z \in V(F)$, let $\mu_F(x, z) = |E_G(x, z) \cap \{e_1, \ldots, e_p\}|$.

Figure 2.1 shows a graph G together with a coloring $\varphi \in \mathcal{C}^7(G - e)$. For the sets of missing colors, we obtain $\overline{\varphi}(x) = \{1, 3\}, \overline{\varphi}(y) = \{2, 4\}, \overline{\varphi}(z) = \{6\}, \overline{\varphi}(u) = \{7\}$, and $\overline{\varphi}(v) = \{5\}$. Hence $V(G) = \{x, y, z, u, v\}$ is elementary and strongly closed with respect to φ. Furthermore, $F = (e, y, e_2, z, e_3, v, e_4, z)$ is a multi-fan at x with respect to e and φ. Clearly, $\chi'(G) \leq 8$ and it is easy to show that equality holds. This follows from the simple fact that $\chi'(G) \geq w(G) \geq \lceil |E(G)|/2 \rceil = 8$. Hence G is elementary. It is also easy to check that G is critical.

Theorem 2.1 *Let G be a graph with $\chi'(G) = k+1$ for an integer $k \geq \Delta(G)$, and let $e \in E_G(x, y)$ be a critical edge of G. Furthermore, let $F = (e_1, y_1, \ldots, e_p, y_p)$ be a multi-fan at x with respect to e and $\varphi \in \mathcal{C}^k(G - e)$. Then the following statements hold:*

(a) $\overline{\varphi}(x) \cap \overline{\varphi}(y_i) = \emptyset$ *for $i = 1, \ldots, p$.*

(b) *If $\alpha \in \overline{\varphi}(x)$ and $\beta \in \overline{\varphi}(y_i)$ for $1 \leq i \leq p$, then there is an (α, β)-chain with respect to φ having endvertices x and y_i.*

(c) *If $y_i \neq y_j$ for $1 \leq i, j \leq p$, then $\overline{\varphi}(y_i) \cap \overline{\varphi}(y_j) = \emptyset$.*

(d) *If F is a maximal multi-fan at x with respect to e and φ, then $|V(F)| \geq 2$ and $\sum_{z \in V(F)} (d_G(z) + \mu_F(x, z) - k) = 2$.*

Proof: In order to prove (a), assume that it is false and choose φ and F such that there is a color $\alpha \in \overline{\varphi}(x) \cap \overline{\varphi}(y_i)$ with i as small as possible. If $i = 1$, then we can color e with α, contradicting $\chi'(G) = k + 1$. Otherwise, $i \geq 2$ and for the color $\beta = \varphi(e_i)$ there is an index $j < i$ such that $\beta \in \overline{\varphi}(y_j)$ by (F2). Recolor e_i with

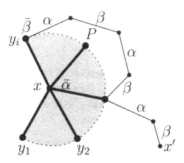

Figure 2.2   The (α, β)-chain P.

color α. This results in a new coloring $\varphi' \in \mathcal{C}^k(G - e)$. Then $(e_1, y_1, \ldots, e_j, y_j)$ is a multi-fan at x with respect to e and φ', where $\beta \in \overline{\varphi}'(x) \cap \overline{\varphi}'(y_j)$, contradicting the minimality of i. This proves (a).

In order to prove (b), assume that there is a counterexample. Choose i smallest. By (a), $\alpha \in \varphi(y_j)$ for $j = 1, \ldots, p$. Since $\beta \in \overline{\varphi}(y_i)$, the (α, β)-chain $P = P_{y_i}(\alpha, \beta)$ is a path, where one endvertex is y_i and the other endvertex is some vertex $x' \neq x$ (by assumption) and $x' \notin \{y_1, \ldots, y_i\}$, since otherwise, by (a), $\beta \in \overline{\varphi}(x')$, contradicting the minimality of i. Moreover, $E(P) \cap \{e_1, \ldots, e_i\} = \emptyset$, since x cannot be contained in P at all. Hence, the coloring $\varphi' = \varphi/P \in \mathcal{C}^k(G - e)$ satisfies $\varphi'(e_j) = \varphi(e_j)$ for $j = 2, \ldots, i$, $\overline{\varphi}'(y_j) = \overline{\varphi}(y_j)$ for $j = 1, \ldots, i - 1$ and $\overline{\varphi}'(y_i) = (\overline{\varphi}(y_i) \setminus \{\beta\}) \cup \{\alpha\}$. Consequently, $F' = (e_1, y_1, \ldots, e_i, y_i)$ is a multi-fan at x with respect to e and φ', where $\alpha \in \overline{\varphi}'(x) \cap \overline{\varphi}'(y_i)$, contradicting (a).

In order to prove (c), assume that there is a color $\beta \in \overline{\varphi}(y_i) \cap \overline{\varphi}(y_j)$. Since $k \geq \Delta(G)$, there is a color $\alpha \in \overline{\varphi}(x)$. Then, by (b), there is an (α, β)-chain with endvertices x and y_i as well as an (α, β)-chain with endvertices x and y_j both with respect to φ, contradicting $y_i \neq y_j$.

For the proof of (d), assume that F is maximal. First, we claim that the color sets $\Gamma = \{\varphi(e_2), \ldots, \varphi(e_p)\}$ and $\Gamma' = \bigcup_{z \in V(F)} \overline{\varphi}(z)$ are equal. By (F2), we have $\Gamma \subseteq \Gamma'$. Conversely, if $\beta \in \Gamma'$, then, by (a), $\beta \in \varphi(x)$ and, since we allow a multi-fan to have multiple edges, the maximality of F implies that $\beta \in \Gamma$. This proves the claim that $\Gamma = \Gamma'$. Now, by (c) and since $y \in V(F)$ is incident with the uncolored edge e, we conclude that

$$p - 1 = |\Gamma| = |\Gamma'| = \sum_{z \in V(F)} |\overline{\varphi}(z)| = 1 + \sum_{z \in V(F)} (k - d_G(z)).$$

Since $p = \sum_{z \in V(F)} \mu_F(x, z)$, this implies

$$\sum_{z \in V(F)} (d_G(z) + \mu_F(x, z) - k) = 2.$$

Since $k \geq \Delta(G)$, there is a color $\beta \in \overline{\varphi}(y_1)$. Then, by (a), there is an edge $e' \in E_G(x, y')$ with $\varphi(e') = \beta$, where $y' \neq y_1$. Since F is maximal, $y' \in V(F)$ and, therefore, $\{y', y_1\} \subseteq V(F)$ and $|V(F)| \geq 2$. ∎

We will refer to the equation in statement (d) of Theorem 2.1 as the **fan equation**. If G is a graph and $k \geq \Delta(G) + \mu(G)$, then the fan equation obviously fails for G. Hence, every critical graph G satisfies $\chi'(G) = k + 1 \leq \Delta(G) + \mu(G)$. By Proposition 1.1, this proves Vizing's theorem.

Theorem 2.2 (Vizing [297] 1964 and Gupta [120] 1967) *Every graph G satisfies* $\chi'(G) \leq \Delta(G) + \mu(G)$.

Theorem 2.3 *Let G be a graph with $\chi'(G) = k + 1$ for an integer $k \geq \Delta(G)$, and let $e \in E_G(x, y)$ be a critical edge of G. Then there is a vertex set Z such that $|Z| \geq 2$, $y \in Z \subseteq N_G(x)$,*

$$d_G(x) + d_G(y) - \mu_G(x, y) \geq k + 1, \tag{2.1}$$

and

$$\sum_{z \in Z} (d_G(z) + \mu_G(x, z) - k) \geq 2. \tag{2.2}$$

Furthermore, there are two distinct vertices $z_1, z_2 \in Z$ such that

(a) $d_G(z_1) + \mu_G(x, z_1) \geq \chi'(G)$,

(b) $d_G(z_1) + \mu_G(x, z_1) + d_G(z_2) + \mu_G(x, z_2) \geq 2\chi'(G)$, *and*

(c) $d_G(x) + d_G(z_1) + d_G(z_2) \geq 2\chi'(G)$.

Proof: Since $e \in E_G(x, y)$ is a critical edge of G and $\chi'(G) = k+1 \geq \Delta+1$, there exists a maximal multi-fan F at x with respect to e and a coloring $\varphi \in C^k(G - e)$. We then deduce from Theorem 2.1 that the set $Z = V(F)$ satisfies $|Z| \geq 2$ and $y \in Z \subseteq N_G(x)$. Furthermore, Theorem 2.1(a) implies that $\overline{\varphi}(x) \cap \overline{\varphi}(y) = \emptyset$. Hence all k colors are present at x or y and, therefore, $k = |\varphi(x) \cup \varphi(y)| = |\varphi(x)| + |\varphi(y)| - |\varphi(x) \cap \varphi(y)| \leq d_G(x) - 1 + d_G(y) - 1 - (\mu_G(x, y) - 1)$. This implies (2.1). Furthermore, since $\mu_F(x, z) \leq \mu_G(x, z)$, (2.2) is an immediate consequence of the fan equation. If $a_1, \ldots, a_m$ is a nonincreasing sequence of $m \geq 2$ integers with $\sum_{i=1}^{m} a_i \geq 2$, then, clearly, $a_1 \geq 1$ and $a_1 + a_2 \geq 2$. Hence (2.2) implies (a) as well as (b). Since $d_G(x) \geq \mu_G(x, z_1) + \mu_G(x, z_2)$, (b) implies (c). Thus the proof is complete. ∎

The inequality in statement (c) of Theorem 2.3 implies that $\chi'(G) \leq \lfloor \frac{3}{2}\Delta(G) \rfloor$ for every critical graph G. By Proposition 1.1, this proves Shannon's theorem.

Theorem 2.4 (Shannon [284] 1949) *Every graph G satisfies $\chi'(G) \leq \lfloor \frac{3}{2}\Delta(G) \rfloor$.*

For a graph G, define

$$\Delta^\mu(G) = \max\{d_G(z) + \mu_G(x, z) \mid E_G(x, z) \neq \emptyset\}$$

if G has at least one edge and $\Delta^\mu(G) = 0$ otherwise. Clearly, $\Delta(G) \leq \Delta^\mu(G) \leq \Delta(G) + \mu(G)$ and, moreover, $\Delta^\mu(G) = \Delta(G) + 1$, provided that G is a simple graph with at least one edge. Based on statement (a) in Theorem 2.3 we then conclude that $\chi'(G) \leq \Delta^\mu(G)$ for every critical graph G. By Proposition 1.1, this implies Ore's theorem.

Theorem 2.5 (Ore [238] 1967) *Every graph G satisfies $\chi'(G) \leq \Delta^\mu(G)$.*

We conclude this section with a new result that can also be easily deduced from Theorem 2.3 and that generalizes a result of Chetwynd and Hilton [50] about **almost simple** graphs, which are graphs G containing a vertex v such that the graph $G - v$ is simple. For an arbitrary graph G, define $\mu^-(G) = \min_{v \in V(G)} \mu(G - v)$ if $G \neq \emptyset$ and $\mu^-(G) = 0$ otherwise.

Theorem 2.6 *Every graph G satisfies $\chi'(G) \leq \Delta(G) + \mu^-(G)$.*

Proof: Since the graph parameters χ' and $\Delta + \mu^-$ are monotone, it suffices to prove the inequality for all critical graphs (Proposition 1.1). The proof is by contradiction. So assume that G is a critical graph satisfying $\chi'(G) \geq \Delta(G) + \mu^-(G) + 1$. Then G is connected and $|V(G)| \geq 3$. Furthermore, there is a vertex $v \in V(G)$ such that $\mu^-(G) = \mu(G - v)$. Let $X = V(G) \setminus \{v\}$ and $s = |X|$. Evidently, $s \geq 2$. Since G is connected and every edge of G is critical, we conclude from Theorem 2.3(b) that every vertex $x \in X$ has two distinct neighbors in G, denoted $z_1 = z_1(x)$ and $z_2 = z_2(x)$, such that

$$
\begin{aligned}
d_G(z_1) + \mu_G(x, z_1) + d_G(z_2) + \mu_G(x, z_2) &\geq 2\chi'(G) \\
&\geq 2\Delta(G) + 2\mu(G - v) + 2.
\end{aligned}
$$

This certainly implies that one of the two neighbors is equal to v, say $z_1 = v$. But then $z_2 \neq v$ and $\mu_G(x, z_2) \leq \mu(G - v)$, which gives

$$
\mu_G(x, v) + d_G(z_2) \geq \Delta(G) + \mu(G - v) + 2
$$

and so $\mu_G(x, v) = \mu_G(x, z_1) \geq \mu(G - v) + 2$. The last inequality holds for all $x \in X$. Since $s = |X| = |V(G)| - 1$, $s \geq 2$ and $z_2 \in X$, it follows that

$$
\begin{aligned}
d_G(z_2) &\geq \Delta(G) + \mu(G - v) + 2 - \mu_G(x, v) \\
&\geq d_G(v) + \mu(G - v) + 2 - \mu_G(x, v) \\
&\geq \mu_G(z_2, v) + \mu_G(x, v) + (s - 2)(\mu(G - v) + 2) \\
&\quad + \mu(G - v) + 2 - \mu_G(x, v) \\
&= \mu_G(z_2, v) + (s - 1)\mu(G - v) + 2(s - 2) + 2 \\
&\geq \mu_G(z_2, v) + (s - 1)\mu(G - v) + 2,
\end{aligned}
$$

which is, however, a contradiction. ∎

2.2 ADJACENCY LEMMAS

In the study of edge colorings of graphs, critical graphs are of particular interest. The reason for this is easy to see. On one hand, each graph G contains a critical graph H with $\chi'(H) = \chi'(G)$ as a subgraph. On the other hand, critical graphs have more structure than arbitrary graphs. Hence, by focusing on critical graphs, i.e., graphs G satisfying $\chi'(H) < \chi'(G)$ for every proper subgraph H of G, we keep the relevant information and often gain a better understanding.

The study of critical graphs was initiated by Vizing. In his third paper [299] about edge colorings he proved a result about the structure of critical simple graphs G with $\chi'(G) = \Delta(G) + 1$. This result became known as Vizing's Adjacency Lemma (Theorem 2.7). Goldberg [111] and Andersen [5] extended Vizing's result to critical graphs G with $\chi'(G) = \Delta^\mu(G)$ (Theorem 2.8). Both Theorems 2.7 and 2.8 are special cases of Theorem 2.9. This result is due to Choudum and Kayathri [61] and can be deduced easily from the fan equation, respectively from the **fan inequality** (2.2). The original proof by Choudum and Kayathri is based on a weaker version of the fan equation. This makes their proof longer and more complicated.

We finish this section with two further results, Theorems 2.10 and 2.11. Both are immediate consequences of the fan equation, too. Proofs of Theorem 2.10 with different approaches were presented by Goldberg [111] in 1974, by Andersen [5] in 1977, by Ehrenfeucht, Faber and Kierstead [78] in 1984, and by Chew [58] in 1997.

Theorem 2.7 (Vizing [299] 1965) *Let G be a simple graph with $\chi'(G) = \Delta(G) + 1$, and let $e \in E_G(x, y)$ be a critical edge of G. Then x is adjacent to at least $\Delta(G) - d_G(y) + 1$ vertices $z \neq y$ such that $d_G(z) = \Delta(G)$.*

Theorem 2.8 (Goldberg [111] 1974, Andersen [5] 1977) *Let G be a graph with $\chi'(G) = \Delta^\mu(G)$, and let $e \in E_G(x, y)$ be a critical edge of G. Then x is adjacent to at least $\Delta^\mu(G) - d_G(y) - \mu_G(x, y) + 1$ vertices $z \neq y$ such that $d_G(z) + \mu_G(x, z) = \Delta^\mu(G)$.*

Theorem 2.9 (Choudum and Kayathri [61] 1999) *Let G be a graph with $\chi'(G) = \Delta^\mu(G) - r \geq \Delta(G) + 1$, where $r \geq 0$ is an integer, and let $e \in E_G(x, y)$ be a critical edge of G. Let $D(x)$ be the number of vertices $z \neq y$ such that z is adjacent to x and $d_G(z) + \mu_G(x, z) \geq \Delta^\mu(G) - r = \chi'(G)$. Then*

$$D(x) \geq \left\lceil \frac{\chi'(G) - d_G(y) - \mu_G(x, y) + 1}{r + 1} \right\rceil.$$

Proof: Let $k = \chi'(G) - 1 = \Delta^\mu - r - 1$ with $\Delta^\mu = \Delta^\mu(G)$. Since $e \in E_G(x, y)$ is a critical edge of G, it follows from Theorem 2.3 that there is a set Z such that $|Z| \geq 2$, $y \in Z \subseteq N_G(x)$, and

$$\sum_{z \in Z} (d_G(z) + \mu_G(x, z) - k) \geq 2.$$

Let $m = k+2-d_G(y)-\mu_G(x,y) = \chi'(G)-d_G(y)-\mu_G(x,y)+1$ and $U = Z\setminus\{y\}$. Then

$$\sum_{z\in U}(d_G(z) + \mu_G(x,z) - k) \geq m.$$

If $m \leq 0$, then $D(x) \geq 0 \geq \lceil m/(r+1)\rceil$ and we are done. Otherwise, $m \geq 1$ and we argue as follows. For $z \in U$, let $a(z) = d_G(z) + \mu_G(x,z) - k$. Clearly, $a(z) \leq \Delta^\mu - k = r + 1$ for every $z \in U$. Consequently, if

$$W = \{z \in U \mid a(z) \geq 1\} = \{z \in U \mid d_G(z) + \mu_G(x,z) \geq k+1 = \Delta^\mu - r\},$$

then

$$\sum_{z\in W} a(z) \geq \sum_{z\in U} a(z) = \sum_{z\in U}(d_G(z) + \mu_G(x,z) - k) \geq m$$

and, therefore, $D(x) \geq |W| \geq \lceil\frac{m}{r+1}\rceil$. This completes the proof. ∎

Theorem 2.10 (Goldberg [111] 1974, Andersen [5] 1977) *Let G be a graph with $\chi'(G) = k + 1$ for an integer $k \geq \Delta(G)$, and let $e \in E_G(x,y)$ be a critical edge of G. Suppose that $d_G(z) + \mu_G(x,z) \leq k + 1$ for every vertex z adjacent to x in G. Then x is adjacent to at least $k - d_G(y) - \mu_G(x,y) + 2$ vertices $z \neq y$ such that $d_G(z) + \mu_G(x,z) = k + 1$.*

Proof: Since $e \in E_G(x,y)$ is a critical edge of G, it follows from Theorem 2.3 that there is a set Z such that $|Z| \geq 2$, $y \in Z \subseteq N_G(x)$, and

$$\sum_{z\in Z}(d_G(z) + \mu_G(x,z) - k) \geq 2.$$

Let $U = Z \setminus \{y\}$ and let W denote the set of all vertices $z \in N_G(x) \setminus \{y\}$ such that $d_G(z) + \mu_G(x,z) = k + 1$. By hypothesis $d_G(z) + \mu_G(x,z) \leq k + 1$ for all $z \in N_G(x)$. This implies that

$$|W| \geq \sum_{z\in U}(d_G(z) + \mu_G(x,z) - k) \geq k - d_G(y) - \mu_G(x,y) + 2,$$

which proves the theorem. ∎

Theorem 2.11 (Hilton and Jackson [137] 1987) *Let G be a graph with $\chi'(G) = k+1$ for an integer $k \geq \Delta(G)$, and let $e \in E_G(x,y)$ be a critical edge of G such that $d_G(y) + \mu_G(x,y) \leq k + 1$. Let $D(x)$ denote the number of edges of G joining x to vertices $z \neq y$ that satisfy $d_G(z) + \mu_G(x,z) \geq k + 1$. Then*

$$D(x) \geq 2k - d^*(x) - d_G(y) - \mu_G(x,y) + 2,$$

where $d^*(x) = \max\{d_G(z) \mid E_G(x,z) \neq \emptyset, z \neq y, d_G(z) + \mu_G(x,z) \geq k+1\}$.

Proof: Denote by W the set of all vertices $z \in V(G) \setminus \{y\}$ such that $E_G(z,x) \neq \emptyset$ and $d_G(z) + \mu_G(x,z) \geq k + 1$. Since $d_G(y) + \mu_G(x,y) - k \leq 1$, we deduce from Theorem 2.3 that $|W| \geq 1$ and so

$$D(x) = \sum_{z\in W} \mu_G(x,z) \geq \sum_{z\in W}(k - d_G(z)) + (k - d_G(y) - \mu_G(x,y) + 2).$$

Since $|W| \geq 1$ and $k - d_G(z) \geq k - d^*(x) \geq k - \Delta(G) \geq 0$ for every $z \in W$, this implies

$$D(x) \geq 2k - d^*(x) - d_G(y) - \mu_G(x,y) + 2.$$

Hence, the proof is complete. ∎

Clearly, $\chi'(G) \geq \Delta(G)$ for every graph G. The results in this section only deal with critical graphs G for which $\chi'(G) \geq \Delta(G) + 1$. The class of critical graphs G with $\chi'(G) = \Delta(G)$, however, can be easily characterized as follows. A graph G is called a **multi-star** if G is connected and if there is a vertex x, called the **center** of G, such that every edge of G is incident with x. Clearly, if G is a multi-star with center x, then $\chi'(G) = \Delta(G) = d_G(x)$. Furthermore, G is a critical graph with $\chi'(G) = \Delta(G)$ if and only if G is a multi-star.

2.3 THE SECOND FAN EQUATION

The fan equation provides information about the neighborhood of an endvertex of a critical edge in a graph whose chromatic index is at least $\Delta + 1$. In 2006, Kostochka and Stiebitz [177] established a new fan equation that also takes the second neighborhood of a vertex in a critical graph into account. This new fan equation is mainly based on a generalized definition of a multi-fan, where the color set $\overline{\varphi}(v)$ is replaced by a super set $C_{\varphi,x}(v)$. The proof of the second fan equation is quite similar to the proof of the first fan equation.

In the sequel, let G be a critical graph such that $\chi'(G) = k + 1$ for some integer $k \geq \Delta(G)$, let $e \in E_G(x)$ and $\varphi \in \mathcal{C}^k(G - e)$. Then, for a vertex v of G, we define the **restricted degree** $D_{G,x}(v)$ as the number of edges $e' \in E_G(v,w)$ such that $w = x$ or

$$d_G(w) > \frac{k - d_G(x)}{2} = \frac{\chi'(G) - 1 - d_G(x)}{2}.$$

Furthermore, let $C_{\varphi,x}(v)$ denote the set of all colors $\alpha \in \{1, \ldots, k\}$ such that $\alpha \in \overline{\varphi}(v)$ or there is an edge $e' \in E_G(v,w)$ with $w \neq x$, $\varphi(e') = \alpha$, and

$$d_G(w) \leq \frac{k - d_G(x)}{2}.$$

For a vertex v of G, we obviously have $|C_{\varphi,x}(v)| = |\overline{\varphi}(v)| + d_G(v) - D_{G,x}(v)$ and, therefore,

$$|C_{\varphi,x}(v)| = \begin{cases} k - D_{G,x}(v) + 1 & \text{if } e \in E_G(v), \\ k - D_{G,x}(v) & \text{otherwise.} \end{cases} \tag{2.3}$$

For the special vertex x, we have

$$D_{G,x}(x) = d_G(x) \quad \text{and} \quad C_{\varphi,x}(x) = \overline{\varphi}(x). \tag{2.4}$$

This follows from the fact that $d_G(x) + d_G(y) \geq k + 2$ whenever y is adjacent to x in G. Furthermore, for any two vertices u, v of G such that $d_G(z) \leq (k - d_G(x))/2$

for $z \in \{u, v\}$, we have

$$|\overline{\varphi}(u) \cap \overline{\varphi}(v)| \geq d_G(x), \quad |\overline{\varphi}(u) \cap \overline{\varphi}(v) \cap \overline{\varphi}(x)| \geq 1 \tag{2.5}$$

and, provided that $d_G(x) \leq k - 1$, we also have

$$|\overline{\varphi}(u) \cap \overline{\varphi}(x)| \geq 2. \tag{2.6}$$

This follows easily from the inclusion-exclusion principle and the facts that φ is a k-edge-coloring and $|\overline{\varphi}(w)| \geq k - d_G(w)$ for every vertex w of G.

A C-**fan** at x with respect to the edge e and the coloring φ is defined to be a sequence $F = (e_1, y_1, \ldots, e_p, y_p)$ with $p \geq 1$ consisting of edges $e_1, \ldots, e_p$ and vertices $y_1, \ldots, y_p$ satisfying the following two conditions:

(C1) The edges $e_1, \ldots, e_p$ are distinct, $e_1 = e$, and $e_i \in E_G(x, y_i)$ for $i = 1, \ldots, p$.

(C2) For every edge e_i with $2 \leq i \leq p$, there is a vertex y_j with $1 \leq j < i$ such that $\varphi(e_i) \in C_{\varphi, x}(y_j)$.

It should be noted that the definition of the C-fan resembles the definition of a multi-fan in the beginning of this chapter. The conditions (C1) and (F1) are the same, whereas the conditions (C2) and (F2) differ only in the underlying color sets for the vertices of the fan. The following result, due to Kostochka and Stiebitz [177], is similar to Theorem 2.1.

Theorem 2.12 (Kostochka and Stiebitz [177] 2006) *Let G be a critical graph with $\chi'(G) = k + 1$ for an integer $k \geq \Delta(G)$, and let $e \in E_G(x)$ be an edge of G. Furthermore, let $F = (e_1, y_1, \ldots, e_p, y_p)$ be a C-fan at x with respect to e and $\varphi \in \mathcal{C}^k(G - e)$. Then the following statements hold:*

(a) $C_{\varphi, x}(x) \cap C_{\varphi, x}(y_i) = \emptyset$ *for* $i = 1, \ldots, p$.

(b) *If $\alpha \in C_{\varphi, x}(x)$ and $\beta \in C_{\varphi, x}(y_i)$ for $1 \leq i \leq p$, then the (α, β)-chain with respect to φ that contains x also contains y_i and is a path, where one endvertex is x.*

(c) *If $y_i \neq y_j$ for $1 \leq i, j \leq p$, then $C_{\varphi, x}(y_i) \cap C_{\varphi, x}(y_j) = \emptyset$.*

(d) *If F is a maximal C-fan at x with respect to e and φ, then $|V(F)| \geq 2$ and $\sum_{z \in V(F)} (D_{G, x}(z) + \mu_F(x, z) - k) = 2$.*

Proof: Since G is a critical graph with $\chi'(G) = k + 1 \geq \Delta(G) + 1$, G is connected and has least 3 vertices. Consequently, $d_G(x) = k$ implies that each vertex v of G satisfies $D_{G, x}(v) = d_G(v)$ and hence $C_{\varphi, x}(v) = \overline{\varphi}(v)$. Then the result follows simply from Theorem 2.1. So, in what follows, suppose that $d_G(x) \leq k - 1$.

In order to prove (a), assume that it is false and choose G, e, φ and F such that there is a color $\alpha \in C_{\varphi, x}(x) \cap C_{\varphi, x}(y_i)$ with i as small as possible. By (2.4), we have $\alpha \in \overline{\varphi}(x)$.

First, consider the case that $i = 1$. If $\alpha \in \overline{\varphi}(y_1)$, then we can color $e \in E_G(x, y_1)$ with α, contradicting $\chi'(G) = k + 1$. Otherwise, $\alpha \in C_{\varphi, x}(y_1) \setminus \overline{\varphi}(y_1)$ and, therefore, there is an edge $e' \in E_G(y_1, u)$ such that $u \neq x$, $\varphi(e') = \alpha$, and $d_G(u) \leq (k - d_G(x))/2$. From (2.5) it follows then that there is a color $\beta \in \overline{\varphi}(x) \cap \overline{\varphi}(u)$. Since $k \geq \Delta(G)$ and the uncolored edge e is incident with y_1, there is also a color $\gamma \in \overline{\varphi}(y_1)$. Clearly, $\beta \neq \alpha$. By Theorem 2.1, it follows that $\beta \neq \gamma$ and the (β, γ)-chain $P_x(\beta, \gamma, \varphi)$ is a path whose endvertices are x and y_1. Then the (β, γ)-chain $P_u = P_u(\beta, \gamma, \varphi)$ is distinct from P_x and, moreover, P_u is a path where one endvertex is u. Consequently, for the coloring $\varphi' = \varphi/P_u$, we have $\gamma \in \overline{\varphi}'(y_1) \cap \overline{\varphi}'(u)$, $\alpha \in \overline{\varphi}'(x)$, and $\varphi'(e') = \alpha$. Now recolor e' with γ and color $e = e_1$ with α. This results in a coloring $\varphi'' \in \mathcal{C}^k(G)$, contradicting $\chi'(G) = k + 1$.

Now, consider the case that $i > 1$. For $2 \leq j \leq i$, let $\alpha_j = \varphi(e_j)$. Obviously, $\alpha_j \neq \alpha_{j'}$ for $2 \leq j < j' \leq i$ and, since $\alpha \in \overline{\varphi}(x)$, we have $\alpha \notin \{\alpha_2, \ldots, \alpha_i\}$. We claim that $\alpha_j \in C_{\varphi, x}(y_{j-1})$ for $2 \leq j \leq i$. Otherwise, choose the largest number $j_0 \in \{2, \ldots, i\}$ such that $\alpha_{j_0} \notin C_{\varphi, x}(y_{j_0-1})$ Then, by (C2), $\alpha_{j_0} \in C_{\varphi, x}(y_{j_1})$ with $j_1 \leq j_0 - 2$. But then $F' = (e_1, y_1, \ldots, e_{j_1}, y_{j_1}, e_{j_0}, y_{j_0}, \ldots, e_i, y_i)$ is a C-fan at x with respect to e and φ such that $\alpha \in C_{\varphi, x}(x) \cap C_{\varphi, x}(y_i)$, contradicting the minimality of i. This proves the claim.

If $\alpha_2 \in \overline{\varphi}(y_1)$, then color the edge $e_1 \in E_G(x, y_1)$ with α_2 and make the edge $e_2 \in E_G(x, y_2)$ uncolored. This results in a coloring $\varphi' \in \mathcal{C}^k(G - e_2)$. Then $F' = (e_2, y_2, \ldots, e_i, y_i)$ is a C-fan at x with respect to e_2 and φ' such that $\alpha \in C_{\varphi', x}(x) \cap C_{\varphi', x}(y_i)$, contradicting the minimality of i.

Otherwise, $\alpha_2 \in C_{\varphi, x}(y_1) \setminus \overline{\varphi}(y_1)$ and there is an edge $e' \in E_G(y_1, u)$ such that $u \neq x$, $\varphi(e') = \alpha_2$, and $d_G(u) \leq (k - d_G(x))/2$. Since $d_G(x) < k$, it follows then from (2.6) that there is a color $\beta \in \overline{\varphi}(x) \cap \overline{\varphi}(u)$ with $\beta \neq \alpha$. Furthermore, there is a color $\gamma \in \overline{\varphi}(y_1)$. We have $\gamma \notin \{a_2, \ldots, \alpha_i\}$, since otherwise $\gamma = \alpha_j$ for some j with $3 \leq j \leq i$ and $F' = (e_1, y_1, e_j, y_j, e_{j+1}, y_{j+1}, \ldots, e_i, y_i)$ would be C-fan at x with respect to e and φ such that $\alpha \in C_{\varphi, x}(x) \cap C_{\varphi, x}(y_i)$, contradicting the minimality of i. Since $\beta \in \overline{\varphi}(x)$, we have $\beta \notin \{\alpha_2, \ldots, \alpha_i\}$. By Theorem 2.1, it follows that $\gamma \notin \{\beta, \alpha\}$ and the chain $P_x(\beta, \gamma, \varphi)$ is a path whose endvertices are x and y_1. Then chain $P_u = P_u(\beta, \gamma, \varphi)$ is distinct from P_x and, moreover, P_u is a path where one endvertex is u. Consequently, for the coloring $\varphi' = \varphi/P_u$, we have $\gamma \in \overline{\varphi}'(y_1) \cap \overline{\varphi}'(u)$, $\alpha \in \overline{\varphi}'(x)$, and $\varphi'(e_2) = \varphi'(e') = \alpha_2$. Now recolor e' with γ, color e with α_2, and make the edge e_2 uncolored. This results in a coloring $\varphi'' \in \mathcal{C}^k(G - e_2)$. Then, since $\beta, \gamma \notin \{\alpha_2, \ldots, \alpha_i\}$, the sequence $F' = (e_2, x_2, \ldots, e_i, y_i)$ is a C-fan at x with respect to e_2 and φ'' such that $\alpha \in C_{\varphi'', x}(x) \cap C_{\varphi'', x}(y_i)$, contradicting the minimality of i. This completes the proof of (a).

In order to prove (b), assume that it is false and choose a counterexample G, e, φ and F with i as small as possible. Then $\alpha \in C_{\varphi, x}(x)$, $\beta \in C_{\varphi, x}(y_i)$ and the chain $P_x = P_x(\alpha, \beta, \varphi)$ does not contain y_i. By (2.4), we have $\alpha \in \overline{\varphi}(x)$. Hence P_x is a path and x is one of its endvertices. By (a), we have $\alpha \notin C_{\varphi, x}(y_j)$ for $1 \leq j \leq i$.

Since y_i is not contained in P_x, the chain $P' = P_{y_i}(\alpha, \beta, \varphi)$ is distinct from P_x. Then the coloring $\varphi' = \varphi/P'$ belongs to $\mathcal{C}^k(G - e)$ and the color α belongs to $\overline{\varphi}'(x) \cap C_{\varphi', x}(y_i)$.

We claim that $F' = (e_1, y_1, \ldots, e_i, y_i)$ is a C-fan at x with respect to e and φ'. Clearly, F' satisfies (C1). To verify that F' also satisfies (C2), consider the edge $e_j \in E_G(x, y_j)$ for some integer j with $2 \le j \le i$. Since F is a C-fan at x with respect to e and φ, for the color $\alpha' = \varphi(e_j)$, there is an integer $j' < j$ such that $\alpha' \in C_{\varphi,x}(y_{j'})$. Since $\alpha \in \overline{\varphi}(x)$, we have $\alpha' \ne \alpha$. If we also have $\alpha' \ne \beta$, then $\alpha' = \varphi'(e_j) \in C_{\varphi',x}(y_{j'})$ and we are done. Otherwise, $\alpha' = \beta$ and the edge e_j belongs to P_x. This implies that $\varphi'(e_j) = \varphi(e_j) = \beta$. Furthermore, $\beta \in C_{\varphi,x}(y_{j'})$ and, by the minimality of i, the vertex $y_{j'}$ belongs to P_x. Consequently, $\beta \in C_{\varphi',x}(y_{j'})$ and we are done, too. This proves the claim that F' is a C-fan at x with respect to e and φ'. Since $\alpha \in \overline{\varphi}'(x) \cap C_{\varphi',x}(y_i)$, this gives a contradiction to (a). Thus (b) is proved.

In order to prove (c), assume that, for two distinct vertices y_i and y_j, there is a color $\beta \in C_{\varphi,x}(y_i) \cap C_{\varphi,x}(y_j)$. Since $k \ge \Delta(G)$, the color set $C_{\varphi,x}(x) = \overline{\varphi}(x)$ is nonempty. From (b) it follows that, for every color $\alpha \in \overline{\varphi}(x)$, the chain $P_x(\alpha, \beta, \varphi)$ contains booth vertices y_i and y_j and is a path, where one endvertex is x. If $\beta \in \overline{\varphi}(y_i) \cap \overline{\varphi}(y_j)$, this is obviously impossible. Consequently, for one of the two vertices, say for y_i, we have $\beta \in C_{\varphi,x}(y_i) \setminus \overline{\varphi}(y_i)$. Then there is an edge $e' \in E_G(y_i, u)$ such that $u \ne x$, $\varphi(e') = \beta$, and $d_G(u) \le (k - d_G(x))/2$. Since G is critical and $\chi'(G) = k + 1$, Theorem 2.3 implies that the two adjacent vertices x and y_j satisfy $d_G(y_j) + d_G(x) \ge k + 2$ and, therefore, $u \ne y_j$. To reach a contradiction, we consider two cases.

Case 1: $\beta \in \overline{\varphi}(y_j)$. By (2.5), there is a color $\alpha \in \overline{\varphi}(x) \cap \overline{\varphi}(u)$. Then, since $u \ne y_j$, the (α, β)-chain P_x with respect to φ containing the vertex x can contain at most one of the two vertices y_i and y_j, a contradiction to (b).

Case 2: $\beta \in C_{\varphi,x}(y_j) \setminus \overline{\varphi}(y_j)$. Then there is an edge $e'' \in E_G(y_j, v)$ such that $v \ne x$, $\varphi(e'') = \beta$, and $d_G(v) \le (k - d_G(x))/2$. Since $u \ne y_j$, we have $v \ne u$. By (2.5), there is a color $\alpha \in \overline{\varphi}(x) \cap \overline{\varphi}(u) \cap \overline{\varphi}(v)$. Since $u \ne v$, the (α, β)-chain P_x with respect to φ containing the vertex x cannot contain both vertices y_i and y_j, a contradiction to (b).

Hence in both cases we arrive at a contradiction. This completes the proof of statement (c).

For the proof of (d), assume that F is maximal. First, we claim that the two color sets $\Gamma = \{\varphi(e_2), \ldots, \varphi(e_p)\}$ and $\Gamma' = \bigcup_{z \in V(F)} C_{\varphi,x}(z)$ are equal. By (C2), we have $\Gamma \subseteq \Gamma'$. Conversely, if $\beta \in \Gamma'$, then (a) and (2.4) imply that $\beta \notin C_{\varphi,x}(x) = \overline{\varphi}(x)$. Hence there is an edge $e' \in E_G(x, v)$ with $\varphi(e') = \beta$. Since F is maximal, it follows that $e' = e_j$ for some $j \in \{2, \ldots, n\}$ and, therefore, $\beta \in \Gamma'$. This proves the claim that $\Gamma = \Gamma'$. Since $e \in E_G(x, y_1)$, we then conclude from (c) and (2.3) that

$$p - 1 = |\Gamma| = |\Gamma'| = \sum_{z \in V(F)} |C_{\varphi,x}(z)| = 1 + \sum_{z \in V(F)} (k - D_{G,x}(z)).$$

Since $p = \sum_{z \in V(F)} \mu_F(x, z)$, this implies

$$\sum_{z \in V(F)} (D_{G,x}(z) + \mu_F(x, z) - k) = 2.$$

Since $k \geq \Delta(G)$ and the uncolored edge e is incident with y_1, there is a color $\beta \in \overline{\varphi}(y_1)$. Then, by (a) and (2.4), $\beta \notin \overline{\varphi}(x)$ and there is an edge $e' \in E_G(x, y')$ with $\varphi(e') = \beta$ and $y \neq y_1$. Since F is maximal, $\{y', y_1\} \subseteq V(F)$ and, therefore, $|V(F)| \geq 2$. This completes the proof of (d). ∎

We will refer to the equation in statement (d) of Theorem 2.12 as the **second fan equation**. The second fan equation implies immediately the next result; we can just repeat the proof of Theorem 2.3.

Theorem 2.13 (Kostochka and Stiebitz [177] 2006) *Let G be a critical graph with $\chi'(G) = k + 1$ for an integer $k \geq \Delta(G)$, and let $e \in E_G(x, y)$ be an edge of G. Then there is a vertex set Z such that $|Z| \geq 2$, $y \in Z \subseteq N_G(x)$,*

$$D_{G,x}(x) + D_{G,x}(y) - \mu_G(x, y) \geq k + 1,$$

and

$$\sum_{z \in Z}(D_{G,x}(z) + \mu_G(x, z) - k) \geq 2.$$

Furthermore, there are two vertices $z_1, z_2 \in Z$ such that

(a) $D_{G,x}(z_1) + \mu_G(x, z_1) \geq \chi'(G)$,

(b) $D_{G,x}(z_1) + \mu_G(x, z_1) + D_{G,x}(z_2) + \mu_G(x, z_2) \geq 2\chi'(G)$, *and*

(c) $D_{G,x}(x) + D_{G,x}(z_1) + D_{G,x}(z_2) \geq 2\chi'(G)$.

As shown in Sect. 2.2, many adjacency results can be easily deduced from the fan equation, respectively from the fan inequality (2.2). Based on the second fan inequality in Theorem 2.13, we can easily rewrite the proofs of the adjacency results in Sect. 2.2, to obtain restricted degree versions of these theorems.

Theorem 2.14 (Kostochka and Stiebitz [177] 2006) *Let G be a simple critical graph with $\chi'(G) = \Delta(G) + 1$, and let $e \in E_G(x, y)$ be an edge of G. Then x is adjacent to at least $\Delta(G) - D_{G,x}(y) + 1$ vertices $z \neq y$ such that $D_{G,x}(z) = \Delta(G)$.*

Theorem 2.15 (Kostochka and Stiebitz [177] 2006) *Let G be a critical graph with $\chi'(G) = \Delta^\mu(G) - r \geq \Delta(G) + 1$, where $r \geq 0$ is an integer, and let $e \in E_G(x, y)$ be an edge of G. Let $D(x)$ be the number of vertices $z \neq y$ such that z is adjacent to x and $D_{G,x}(z) + \mu_G(x, z) \geq \Delta^\mu(G) - r = \chi'(G)$. Then*

$$D(x) \geq \left\lceil \frac{\chi'(G) - D_{G,x}(y) - \mu_G(x, y) + 1}{r + 1} \right\rceil.$$

Theorem 2.16 (Kostochka and Stiebitz [177] 2006) *Let G be a critical graph with $\chi'(G) = k + 1$ for an integer $k \geq \Delta(G)$, and let $e \in E_G(x, y)$ be an edge of G. Suppose that $D_{G,x}(z) + \mu_G(x, z) \leq k + 1$ for every vertex z adjacent to x in G. Then x is adjacent to at least $k - D_{G,x}(y) - \mu_G(x, y) + 2$ vertices $z \neq y$ such that $D_{G,x}(z) + \mu_G(x, z) = k + 1$.*

2.4 THE DOUBLE FAN

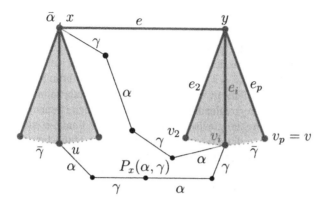

Figure 2.3 A double fan.

Theorem 2.17 *Let G be a graph with $\chi'(G) = k + 1$ for an integer $k \geq \Delta(G)$, let $e \in E_G(x, y)$ be a critical edge of G, and let $\varphi \in \mathcal{C}^k(G - e)$ be a coloring. Furthermore, let F^x be a multi-fan at x with respect to e and φ, and let F^y be a multi-fan at y with respect to e and φ. If $d_G(x) < k$ or $d_G(y) < k$, then $U = V(F^x) \cup V(F^y) \cup \{x, y\}$ is elementary with respect to φ.*

Proof: By symmetry, we may assume that $d_G(x) < k$. Suppose, on the contrary, that U is not elementary with respect to φ. By Theorem 2.1, $V(F^x) \cup \{x\}$ as well as $V(F^y) \cup \{y\}$ are elementary with respect to φ. Consequently, there is a vertex $u \in V(F^x) \setminus \{y\}$ and a vertex $v \in V(F^y) \setminus \{x\}$ such that $\overline{\varphi}(u) \cap \overline{\varphi}(v)$ contains a color γ. Since $v \in V(F^y) \setminus \{x\}$ and F^y is a multi-fan at y with respect to e and φ, there is a sequence $(v_1, \ldots, v_p)$ of distinct vertices and a sequence $(e_1, \ldots, e_p)$ of distinct edges such that, $v_1 = x$, $v_p = v$, $e_1 = e$, $e_i \in E_G(y, v_i)$ for $1 \leq i \leq p$, and $\varphi(e_i) \in \overline{\varphi}(v_{i-1})$ for $2 \leq i \leq p$. Hence $F' = (e, x, e_2, v_2, \ldots, e_p, v_p)$ is a multi-fan at y with respect to e and φ. Since $d_G(x) < k$, it follows that $\overline{\varphi}(x)$ contains a color α distinct from $\varphi(e_2)$. By Theorem 2.1, the set $\{x, y, v_2, \ldots, v_p\}$ is elementary with respect to φ. Hence the colors $\alpha, \varphi(e_2), \ldots, \varphi(e_p), \gamma$ are distinct. Since $\gamma \in \overline{\varphi}(u)$ and $u \in V(F^x)$, Theorem 2.1 then implies that $P_x(\alpha, \gamma)$ is a a a path whose endvertices are x and u (see Fig. 2.3). Consequently, for the coloring $\varphi' = \varphi/P_x(\alpha, \gamma) \in \mathcal{C}^k(G - e)$, we have $\gamma, \varphi(e_2) \in \overline{\varphi}'(x)$, $\varphi(e_i) = \varphi'(e_i) \in \overline{\varphi}'(v_{i-1})$ for $2 \leq i \leq p$, and $\gamma \in \overline{\varphi}'(v_p) = \overline{\varphi}'(v)$. Then F' is a fan at y with respect to e and φ' such that $V(F')$ is not elementary with respect to φ, a contradiction to Theorem 2.1. This completes the proof. ∎

Observe that the degree condition in Theorem 2.17 saying that $d_G(x)$ or $d_G(y)$ is smaller than k is obviously fulfilled if $k \geq \Delta(G) + 1$.

2.5 THE FAN NUMBER

Theorem 2.1 says, in particular, that if $\chi'(G) = k + 1$ for some integer $k \geq \Delta(G)$, then the vertex set of each multi-fan at x with respect to a critical edge $e \in E_G(x, y)$ and a coloring $\varphi \in \mathcal{C}^k(G - e)$ together with the vertex x form an elementary set with respect to φ. Since the proof of this statement is constructive, this yields a coloring algorithm of the type described in Sect. 1.5, where the set $\mathcal{O}(G, e, \varphi)$ of test objects consists of the sets $Z = V(F) \cup \{x\}$, where F runs through all multi-fans at x with respect to e and φ. This gives an $O(|E|(|V| + \Delta^2))$ algorithm and shows that Vizing's bound is efficiently realizable. However, similar to the bound $2\Delta - 1$ for the number of colors used by the greedy algorithm, Vizing's bound is rather generous, and in most graphs there will be scope for an improvement of this bound by choosing a particularly suitable edge order to start with. The best upper bound realizable by the greedy algorithm is the coloring index. To see what is the best upper bound realizable by a coloring algorithm using multi-fans as test objects, we introduce a new graph parameter, the so-called **fan number** of G, denoted fan(G).

Let G be a graph and let $k \geq 0$ be an integer. Denote by $\mathcal{F}_k(G)$ the set of all triples (x, y, Z) such that $x, y \in V(G)$, $y \in Z \subseteq N_G(x)$, $|Z| \geq 2$,

$$d_G(x) + d_G(y) - \mu_G(x, y) \geq k + 1,$$

and

$$\sum_{z \in Z} (d_G(z) + \mu_G(x, z) - k) \geq 2.$$

Obviously, $\ell \leq k$ implies that

$$\mathcal{F}_k(G) \subseteq \mathcal{F}_\ell(G). \tag{2.7}$$

Furthermore,

$$H \subseteq G \Rightarrow \mathcal{F}_k(H) \subseteq \mathcal{F}_k(G). \tag{2.8}$$

For a pair (x, y) of distinct vertices of G, let $\deg_G(x, y)$ be the smallest integer $k \geq 0$ such that there is no vertex set Z with $(x, y, Z) \in \mathcal{F}_k(G)$. Then $\deg_G(x, y)$ is called the **fan-degree** of the vertex pair (x, y) in G.

Note that if $E_G(x, y) = \emptyset$ or $|N_G(x)| \leq 1$, then the fan-degree of the vertex pair (x, y) satisfies $\deg_G(x, y) = 0$.

Now, suppose that $E_G(x, y) \neq \emptyset$ and $|N_G(x)| \geq 2$. For a vertex $z \in N_G(x)$, let $d(z) = d_G(z) + \mu_G(x, z)$. Then $d(z) \geq 2$ for every neighbor z of x and $N_G(x) \setminus \{y\}$ consists of $p \geq 1$ vertices $z_1, \ldots, z_p$, where the order is chosen such that $d(z_1) \geq d(z_2) \geq \cdots \geq d(z_p)$. For integers k, ℓ with $1 \leq \ell \leq p$, let

$$m_{k,\ell} = (d(y) - k) + \sum_{i=1}^{\ell} (d(z_i) - k).$$

Now, let S denote the set of all integers k such that $d_G(x) + d_G(y) - \mu_G(x, y) \geq k + 1 \geq 1$ and $m_{k,\ell} \geq 2$ for some integer $\ell \in \{1, \ldots, p\}$. Obviously, there is a set Z

such that $(x, y, Z) \in \mathcal{F}_k$ if and only if $k \in S$. If $q = \max S$, then $S = \{0, 1, \ldots, q\}$ and $\deg_G(x, y) = q + 1 \geq 2$.

This shows that the fan-degree $\deg_G(x, y)$ of a pair (x, y) of distinct vertices of G can be computed by a polynomial-time algorithm. Observe that $\deg_G(x, y)$ and $\deg_G(y, x)$ might be different.

It follows from (2.7) and (2.8) that if $H \subseteq G$ then every pair (x, y) of distinct vertices of H satisfies

$$\deg_H(x, y) \leq \deg_G(x, y). \tag{2.9}$$

For a graph G with at least one edge, let

$$\delta^{\mathrm{fa}}(G) = \min\{\deg_G(x, y) \mid x, y \in V(G),\ E_G(x, y) \neq \emptyset\} \tag{2.10}$$

be the **minimum fan-degree** of G. If G is an edgeless graph, we define $\delta^{\mathrm{fa}}(G) = 0$. Note that the parameter δ^{fa} is not monotone. However, the **fan number** defined by

$$\mathrm{fan}(G) = \max_{H \subseteq G} \delta^{\mathrm{fa}}(H) \tag{2.11}$$

is a monotone graph parameter. By the above remark, it follows that there is a polynomial-time algorithm that computes, for every graph G, the minimum fan-degree $\delta^{\mathrm{fa}}(G)$. In order to show that this is also true for the fan number, we describe an alternative way of computing the value $\mathrm{fan}(G)$.

Suppose that the graph G has at least one edge. Let $\ell = ((x_i, y_i) \mid i = 1, \ldots, m)$ be a sequence of pairs consisting of distinct vertices of G. For such a sequence ℓ, we define a sequence $(G_i \mid i = 1, \ldots, m + 1)$ of subgraphs of G by letting $G_1 = G$ and $G_{i+1} = G_i - E_{G_i}(x_i, y_i)$ for $i = 1, \ldots, m$. Then ℓ is said to be a **feasible sequence** for G if G_{m+1} is an edgeless graph and $E_{G_i}(x_i, y_i) \neq \emptyset$ for $i = 1, \ldots, m$. Now, we define $\mathrm{fan}'(G)$ to be the smallest integer p such that there exist a feasible sequence $\ell = ((x_i, y_i) \mid i = 1, \ldots, m)$ for G satisfying

$$\deg_{G_i}(x_i, y_i) \leq p \tag{2.12}$$

for $i = 1, \ldots, m$. Furthermore, a sequence $\ell = ((x_i, y_i) \mid i = 1, \ldots, m)$ is said to be an **optimal sequence** for G if ℓ is feasible and

$$\deg_{G_i}(x_i, y_i) = \delta^{\mathrm{fa}}(G_i) \tag{2.13}$$

for $i = 1, \ldots, m$. If G is an edgeless graph, define $\mathrm{fan}'(G) = 0$.

Theorem 2.18 *Every graph G satisfies*

$$\mathrm{fan}(G) = \mathrm{fan}'(G).$$

Furthermore, if G has at least one edge and if $\ell = ((x_i, y_i) \mid i = 1, \ldots, m)$ is an optimal sequence for G with $(G_i \mid i = 1, \ldots, m + 1)$ as the corresponding sequence of subgraphs, then

$$\mathrm{fan}(G) = \max_{1 \leq i \leq m} \deg_{G_i}(x_i, y_i).$$

Proof: First, consider the case that $\text{fan}(G) = 0$. Then $\delta^{\text{fa}}(H) = 0$ for every subgraph H of G. If G is edgeless, then $\text{fan}(G) = \text{fan}'(G) = 0$ and we are done. Otherwise, every subgraph H of G with at least one edge contains two adjacent vertices x, y such that $\deg_H(x, y) = 0$. This implies that there is a feasible sequence $\ell = ((x_i, y_i) \mid i = 1, \ldots, m)$ for G such that $\deg_{G_i}(x_i, y_i) = 0$ for $i + 1, \ldots, m$, where $(G_i \mid i = 1, \ldots, m + 1)$ is the corresponding sequence of subgraphs. Hence we are done, too.

Now, consider the case that $\text{fan}(G) \geq 1$. Then G has at least one edge and there is a feasible sequence $\ell = ((x_i, y_i) \mid i = 1, \ldots, m)$ for G such that $\deg_{G_i}(x_i, y_i) \leq \text{fan}'(G)$ holds for $i = 1, \ldots, m$. Let $(G_i \mid i = 1, \ldots, m + 1)$ be the sequence of subgraphs of G with $G_1 = G$ and $G_{i+1} = G_i - E_{G_i}(x_i, y_i)$ for $i = 1, \ldots, m$. By (2.11), there is a subgraph H of G such that $\text{fan}(G) = \delta^{\text{fa}}(H)$. Since $\delta^{\text{fa}}(H) = \text{fan}(G) \geq 1$, the graph H has at least one edge. Since ℓ is feasible, this implies that there is a smallest index i such that $E_H(x_i, y_i) \neq \emptyset$. Then H is a subgraph of G_i and, therefore, it follows from (2.9) that

$$\text{fan}(G) = \delta^{\text{fa}}(H) \leq \deg_H(x_i, y_i) \leq \deg_{G_i}(x_i, y_i) \leq \text{fan}'(G).$$

Now let $\ell = ((x_i, y_i) \mid i = 1, \ldots, m)$ be an optimal sequence for G, and let $(G_i \mid i = 1, \ldots, m + 1)$ be the corresponding sequence of subgraphs of G. Since ℓ is a feasible sequence for G, it follows that

$$\text{fan}'(G) \leq \max_{1 \leq i \leq m} \deg_{G_i}(x_i, y_i) = \max_{1 \leq i \leq m} \delta^{\text{fa}}(G_i) \leq \max_{H \subseteq G} \delta^{\text{fa}}(H) = \text{fan}(G),$$

which completes the proof of the theorem.   ∎

Theorem 2.18 implies that there is a polynomial-time algorithm that computes, for a given graph G, the fan number of G as well as an optimal sequence ℓ for G.

For a graph G, let

$$\text{Fan}(G) = \max\{\Delta(G), \text{fan}(G)\}.$$

Obviously, Fan is a monotone graph parameter and $\text{Fan}(G)$ can be computed by a polynomial-time algorithm. A complete proof of the following result can be found in Scheide [267], respectively in Scheide and Stiebitz [274].

Theorem 2.19 *The parameter* Fan *is an efficiently realizable upper bound for the chromatic index* χ'.

Sketch of Proof: Let us first show that Fan is an upper bound for χ'. Since both parameters are monotone, it suffices to show that every critical graph G satisfies $\chi'(G) \leq \text{Fan}(G)$ (Proposition 1.1). If $\chi'(G) = \Delta(G)$, this is evident. Now assume that $\chi'(G) = k + 1$ for some integer $k \geq \Delta(G)$. Then G is connected and has at least three vertices and, by Theorem 2.3, for every pair (x, y) of adjacent vertices, there is a vertex set Z, such that $(x, y, Z) \in \mathcal{F}_k(G)$ and hence $\deg_G(x, y) \geq k + 1$.

This implies that $\delta^{fa}(G) \geq k + 1$ and, therefore, $fan(G) \geq k + 1$. Consequently, $Fan(G) \geq k + 1 = \chi'(G)$.

Next, we prove that the upper bound Fan is efficiently realizable. First, it is not difficult to show that the proof of Theorem 2.1 can be transformed into an algorithm **VizExt**. The input of **VizExt** is a tuple (G, e, x, y, k, φ), where G is a graph, $e \in E_G(x, y)$ is an edge, $k \geq \Delta(G)$ is an integer, and $\varphi \in \mathcal{C}^k(G - e)$ is a coloring. The output of **VizExt** is either a coloring $\varphi' \in \mathcal{C}^k(G)$ or a maximal multi-fan F at x with respect to e and φ such that $V(F) \cup \{x\}$ is elementary with respect to φ and hence $(x, y, V(F)) \in \mathcal{F}_k(G)$ (see the proof of Theorem 2.3). Hence if $fan(G) \leq k$, the algorithm **VizExt** returns a k-edge-coloring of G.

The algorithm **VizExt** computes stepwise a finite sequence $(F^i)_{i=1,\ldots,p}$ of multi-fans at x with respect to e and φ with $F^i = (e_1, y_1, \ldots, e_i, y_i)$. In each step i the algorithm executes the following subroutine. First, the algorithm checks whether the following two conditions hold:

(1) $\overline{\varphi}(x) \cap \overline{\varphi}(y_i) = \emptyset$.

(2) $\overline{\varphi}(y_i) \cap \overline{\varphi}(y_j) = \emptyset$ whenever $j < i$ and $y_j \neq y_i$.

If (1) or (2) is violated, the algorithm computes a coloring $\varphi' \in \mathcal{C}^k(G)$ and returns φ'; see Case 1 and Case 2 for the details. Otherwise, the subroutine continues as follows. If there is an edge $e' \in E_G(x) \setminus \{e_1, \ldots, e_i\}$ such that $\varphi(e') \in \bigcup_{j=1}^{i} \overline{\varphi}(y_j)$, then F^i will be extended to the multi-fan $F^{i+1} = (e_1, y_1, \ldots, e_i, y_1, e_{i+1}, y_{i+1})$, where $e_{i+1} = e'$ and y_{i+1} is the endvertex of e' distinct from x. This can be done, since we allow a multi-fan to have multiple edges. Then we repeat the subroutine with i replaced by $i + 1$. If there is no such edge e', then F^i is a maximal multi-fan at x with respect to e and φ, and the algorithm returns the set $Z = V(F^i)$.

Clearly, the algorithm stops after a finite number of steps, say in step p with $F = F^p = (e_1, y_1, \ldots, e_p, y_p)$. To see that the algorithm works correctly, we distinguish three cases.

Case 1: Condition (1) is violated in step p. Then there is a color $\alpha \in \overline{\varphi}(x) \cap \overline{\varphi}(y_p)$. If $p = 1$, we color $e_1 = e$ with α and return the resulting coloring φ'. Otherwise, $p > 1$ and, since F is a multi-fan at x with respect to e and φ, there is a sequence $j_1, \ldots, j_q$ of integers such that $1 = j_1 < \cdots < j_q = p$ and $\varphi(e_{j_i}) \in \overline{\varphi}(y_{j_{i-1}})$ for $2 \leq i \leq q$. Let φ' be the coloring with $\varphi'(e_{j_q}) = \alpha$, $\varphi'(e_{j_{i-1}}) = \varphi(e_{j_i})$ for $2 \leq i \leq q$ and $\varphi'(f) = \varphi(f)$ for $f \in E(G) \setminus \{e_{j_1}, \ldots, e_{j_q}\}$. Clearly, $\varphi' \in \mathcal{C}^k(G)$ and the algorithm returns φ'.

Case 2: Condition (2) is violated in step p, but not condition (1). Then (1) holds for $1 \leq i \leq p$ and (2) holds for $1 \leq i \leq p - 1$. Furthermore, there is a color $\beta \in \overline{\varphi}(y_p) \cap \overline{\varphi}(y_j)$ with $j < p$ and $y_j \neq y_p$. Since $k \geq \Delta(G)$ and the uncolored edge e is incident with x, there is a color $\alpha \in \overline{\varphi}(x)$. Then $\alpha \notin \varphi(y_i)$ for $1 \leq i \leq p$ and if $\beta \in \overline{\varphi}(y_i)$ for some $i \in \{1, \ldots, p\}$, then $i = p$ or $y_i = y_j$. Note that $y_p \neq y_i$ for $1 \leq i < p$. We choose j such that $y_i \neq y_j$ for $1 \leq i < j$.

For $\ell \in \{p, j\}$, the chain $P_\ell = P_{y_\ell}(\alpha, \beta, \varphi)$ is a path, where one endvertex is y_ℓ and the other endvertex is some vertex $z_\ell \neq y_\ell$. Hence z_p or z_j is distinct from x and we choose ℓ smallest such that $z_\ell \neq x$. This implies that $z_\ell \notin \{y_1, \ldots, y_\ell\}$.

Moreover, $E(P_\ell) \cap \{e_1, \ldots, e_\ell\} = \emptyset$, since x cannot be contained in P_ℓ at all. Hence, the coloring $\varphi_1 = \varphi/P_\ell \in \mathcal{C}^k(G - e)$ satisfies $\varphi_1(e_i) = \varphi(e_i)$ for $i = 2, \ldots, \ell$, $\overline{\varphi}_1(y_i) = \overline{\varphi}(y_i)$ for $i = 1, \ldots, \ell - 1$ and $\overline{\varphi}_1(y_\ell) = (\overline{\varphi}(y_\ell) - \{\beta\}) \cup \{\alpha\}$. Consequently, $F' = (e_1, y_1, \ldots, e_\ell, y_\ell)$ is a multi-fan at x with respect to e and φ_1, where $\alpha \in \overline{\varphi}_1(x) \cap \overline{\varphi}_1(y_\ell)$. Now, as in Case 1, we can construct a coloring $\varphi' \in \mathcal{C}^k(G)$.

Case 3: F satisfies (1) and (2) in step p. Then (1) and (2) hold for $1 \le i \le p$ and F is a maximal multi-fan at x with respect to e and φ. To see that the algorithm works correctly, it now suffices to show that (x, y, Z) belongs to $\mathcal{F}_k(G)$, where $Z = V(F) = \{y_1, \ldots, y_p\}$.

As in the proof of Theorem 2.1(d), it follows that $|V(F)| \ge 2$ and

$$\sum_{z \in Z} (d_G(z) + \mu_G(x, z) - k) \ge \sum_{z \in Z} (d_G(z) + \mu_F(x, z) - k) = 2.$$

Finally, (1) implies that $\overline{\varphi}(x) \cap \overline{\varphi}(y) = \emptyset$. Hence all k colors are present at x or y and, therefore, $k = |\varphi(x) \cup \varphi(y)| = |\varphi(x)| + |\varphi(y)| - |\varphi(x) \cap \varphi(y)| \le d_G(x) - 1 + d_G(y) - 1 - (\mu_G(x, y) - 1)$ implying that $d_G(x) + d_G(y) - \mu_G(x, y) \ge k + 1$. Consequently, we obtain that $(x, y, Z) \in \mathcal{F}_k(G)$.

Now we can design an efficient algorithm **VizCol** that colors the edges of each graph G using at most $\mathrm{Fan}(G)$ colors.

VizCol(G):

1. **If** G is edgeless **then return** the empty coloring φ.
2. Compute $k = \mathrm{Fan}(G)$, an optimal sequence $\ell = ((x_i, y_i)|i = 1, \ldots, m)$ for G, and the corresponding sequence $(G_i|i = 1, \ldots, m + 1)$ of sub-graphs.
3. $G' \leftarrow G_{m+1}$ (the edgeless graph).
4. Let φ be the empty coloring of G'.
5. $i \leftarrow m$
6. **While** $i > 0$ **do:**
   (a) **For** every edge $e \in E_{G_i}(x_i, y_i)$ **do:**
       i. Add e to G'
       ii. $\varphi \leftarrow \mathrm{VizExt}(G', k, e, x_i, y_i, \varphi)$
   (b) $i \leftarrow i - 1$
7. **Return** φ.

End

The input of the algorithm **VizCol** is an arbitrary graph G and its output is a k-edge-coloring φ of G, where $k = \mathrm{Fan}(G)$. If the graph G is edgeless, this is evident. Now assume that G has at least one edge. For $i = m, m - 1, \ldots, 1$, we start the subroutine in 6 with a k-edge-coloring φ of $G' = G_{i+1} = G_i - E_{G_i}(x_i, y_i)$, where φ is the empty coloring in case of $i = m$. Then we add step by step the edges of $E_G(x_i, y_i)$ to G'. The resulting graph is G_i. If we add the edge $e \in E_G(x_i, y_i)$ to G', we have a k-edge-coloring $\varphi \in C^k(G' - e)$. Since ℓ is an optimal sequence for G, we obtain $\deg_{G'}(x_i, y_i) \leq \deg_{G_i}(x_i, y_i) = \delta^{\mathrm{fa}}(G_i) \leq \mathrm{fan}(G) \leq \mathrm{Fan}(G) = k$. Hence $\mathbf{VizExt}(G', k, e, x_i, y_i, \varphi)$ returns a k-edge-coloring of G'. This shows that the algorithm is correct.

If we take into account the implementation details discussed at the end of Sect. 1.5, it follows that **VizExt** has time complexity $O(|V| + \Delta + \Delta^2) = O(|V| + \Delta^2)$. Observe that Theorem 2.21 implies that $k = O(\Delta)$. Since an optimal sequence for a graph can be computed in polynomial time, we then conclude that the time complexity of the algorithm **VizCol** is bounded from above by a polynomial in $|V|$ and $|E|$, too. ∎

Theorem 2.19 implies in particular that $\chi'(G) \leq \mathrm{fan}(G)$ for every graph G with $\mathrm{fan}(G) \geq \Delta(G)$. The next result provides some information about the structure of fan-critical graphs. This result enables us to show that various upper bounds for the chromatic index are also upper bounds for the fan number.

Theorem 2.20 *Let G be a fan-critical graph with $\mathrm{fan}(G) = k + 1$ for an integer $k \geq 0$, and let $e \in E_G(x, y)$ be an edge of G. Then there is a vertex set Z such that $(x, y, Z) \in \mathcal{F}_k(G)$, i.e., $|Z| \geq 2$, $y \in Z \subseteq N_G(x)$, $d_G(x) + d_G(y) - \mu_G(x, y) \geq k + 1$, and*

$$\sum_{z \in Z}(d_G(z) + \mu_G(x, z) - k) \geq 2.$$

Furthermore, there are two vertices $z_1, z_2 \in Z$ such that

(a) $d_G(z_1) + \mu_G(x, z_1) \geq \mathrm{fan}(G)$,

(b) $d_G(z_1) + \mu_G(x, z_1) + d_G(z_2) + \mu_G(x, z_2) \geq 2\mathrm{fan}(G)$, *and*

(c) $d_G(x) + d_G(z_1) + d_G(z_2) \geq 2\mathrm{fan}(G)$.

Proof: Since fan is a monotone graph parameter and G is fan-critical, we conclude from (2.11) that $\mathrm{fan}(G) = \delta^{\mathrm{fa}}(G)$. Therefore, we have $\deg_G(x, y) \geq \delta^{\mathrm{fa}}(G) = k + 1 \geq 1$. This implies that there is a vertex set Z such that $(x, y, Z) \in \mathcal{F}_k(G)$, i.e., $|Z| \geq 2$, $y \in Z \subseteq N_G(x)$, $d_G(x) + d_G(y) - \mu_G(x, y) \geq k + 1$, and

$$\sum_{z \in Z}(d_G(z) + \mu_G(x, z) - k) \geq 2.$$

If $a_1, \ldots, a_m$ is a nonincreasing sequence of $m \geq 2$ integers with $\sum_{i=1}^m a_1 \geq 2$, then, clearly, $a_1 \geq 1$ and $a_1 + a_2 \geq 2$. This proves (a) as well as (b). Since $d_G(x) \geq \mu_G(x, z_1) + \mu_G(x, z_2)$, (b) implies (c). ∎

Theorem 2.21 *Every graph parameter $\rho \in \{\Delta + \mu, \frac{3}{2}\Delta, \Delta^\mu, \Delta + \mu^-\}$ is an upper bound for fan and hence an efficiently realizable upper bound for χ'.*

Proof: Let $\rho \in \{\Delta + \mu, \frac{3}{2}\Delta, \Delta^\mu, \Delta + \mu^-\}$. The aim is to show that $\mathrm{fan}(G) \leq \rho(G)$ for every graph G. Since ρ and fan are monotone, it follows from Proposition 1.1 that it is sufficient to prove this inequality for all fan-critical graphs G. Suppose, on the contrary, that there is a fan-critical graph G such that $\mathrm{fan}(G) \geq \rho(G) + 1 \geq 1$. If $\rho \in \{\Delta + \mu, \frac{3}{2}\Delta, \Delta^\mu\}$, this yields a contradiction to Theorem 2.20. If $\rho = \Delta + \mu^-$, then, based on Theorem 2.20(b), we can repeat the proof of Theorem 2.6 to obtain a contradiction, too. Hence every parameter $\rho \in \{\Delta + \mu, \frac{3}{2}\Delta, \Delta^\mu, \Delta + \mu^-\}$ is an upper bound for the fan number. Since Δ is a lower bound for ρ, this implies that ρ is also an upper bound for Fan. From Theorem 2.19 it follows then that ρ is an efficiently realizable upper bound for χ'. ∎

The following result is a counterpart to the adjacency result of Choudum and Kayathri for critical graphs (Theorem 2.9).

Theorem 2.22 *Let G be a* fan-*critical graph with* $\mathrm{fan}(G) = k + 1$ *for an integer $k \geq 0$, and let $e \in E_G(x, y)$ be an edge of G. Let $D(x)$ be the number of vertices $z \neq y$ such that z is adjacent to x and $d_G(z) + \mu_G(x, z) \geq \mathrm{fan}(G)$. Then*

$$D(x) \geq \left\lceil \frac{\mathrm{fan}(G) - d_G(y) - \mu_G(x, y) + 1}{\Delta(G) + \mu(G) - \mathrm{fan}(G) + 1} \right\rceil.$$

Proof: Since G is fan-critical and $\mathrm{fan}(G) = k + 1 \geq 1$, it follows from Theorem 2.20 that there is a vertex set Z such that $|Z| \geq 2$, $y \in Z \subseteq N_G(x)$, and

$$\sum_{z \in Z} (d_G(z) + \mu_G(x, z) - k) \geq 2.$$

Let $m = k + 2 - d_G(y) - \mu_G(x, y) = \mathrm{fan}(G) - d_G(y) - \mu_G(x, y) + 1$ and $U = Z \setminus \{y\}$. Then

$$\sum_{z \in U} (d_G(z) + \mu_G(x, z) - k) \geq m.$$

Observe that Theorem 2.21 implies that $\Delta(G) + \mu(G) - \mathrm{fan}(G) + 1$ is positive. Hence, if $m \leq 0$, then $D(x) \geq 0 \geq \lceil m/(\Delta(G) + \mu(G) - \mathrm{fan}(G) + 1) \rceil$ and we are done. Otherwise, $m \geq 1$ and we argue as follows. For $z \in U$, let

$$a(z) = d_G(z) + \mu_G(x, z) - k = d_G(z) + \mu_G(x, z) - \mathrm{fan}(G) + 1.$$

Clearly, $a(z) \leq \Delta(G) + \mu(G) - \mathrm{fan}(G) + 1$ for every $z \in U$. Consequently, if

$$W = \{z \in U \mid a(z) \geq 1\} = \{z \in U \mid d_G(z) + \mu_G(x, z) \geq \mathrm{fan}(G) = k + 1\},$$

then

$$\sum_{z \in W} a(z) \geq \sum_{z \in U} a(z) = \sum_{z \in U} (d_G(z) + \mu_G(x, z) - k) \geq m$$

and, therefore, $D(x) \geq |W| \geq \left\lceil \frac{m}{\Delta(G) + \mu(G) - \mathrm{fan}(G) + 1} \right\rceil$. Thus the proof is done. ∎

Corollary 2.23 *Let G be a simple* fan-*critical graph with* $\mathrm{fan}(G) \geq \Delta(G) + 1$, *and let $e \in E_G(x, y)$ be an edge of G. Then x is adjacent to at least $\Delta(G) - d_G(y) + 1$ vertices $z \neq y$ such that $d_G(z) = \Delta(G)$.*

2.6 NOTES

Our definition of a multi-fan at the beginning of this chapter differs slightly from the classical definition going back to Vizing [297, 298, 299]. We allow multiple edges and only require the color of an edge of the fan to be missing at some previous vertex of the fan (instead of missing exactly at the previous vertex). This change makes proofs easier and is essential for obtaining the fan equation in Theorem 2.1(d) and the fan inequality (2.2) in Theorem 2.3.

The fan equation and inequality are unifying results from which all the classical results on edge colorings seem easily derivable. The fan inequality appears in several earlier papers as part of proofs, rather than as a separate result of interest in its own right; versions of it can be found in the papers by Andersen [5], Goldberg [114], Hilton and Jackson [137], and Choudum and Kayathri [61].

There have been three independent papers that have explicitly mentioned the fan equation/inequality as an important result and tool, namely, in chronological order, the M.Sc. thesis of Favrholdt [90], the paper by Reed and Seymour [253], and the Ph.D. thesis of Cariolaro [37]. Cariolaro [38] goes even further in his analysis of fans in (multi)graphs by associating a directed walk in a certain directed graph to a fan, see also Cariolaro [39, 40].

The parameter $\mathrm{Fan} = \max\{\Delta, \mathrm{fan}\}$ seems to be the best upper bound for the chromatic index that can be obtained by the fan argument, respectively by the fan equation. All known upper bounds for the chromatic index that are derivatives of the fan equation are also upper bounds for Fan. In addition to the parameters mentioned in Theorem 2.21 there are two other such parameters in the literature. The first parameter was independently introduced by Andersen [5] and Goldberg [111, 114], namely

$$\mathrm{ag}(G) = \max\{\Delta(G), \max_{\mathbf{P}} \lfloor \tfrac{1}{2}(d_G(y) + \mu_G(x,y) + d_G(z) + \mu_G(x,z)) \rfloor\}$$

with $\mathbf{P} = \{(x,y,z) \,|\, x \in V(G), z \neq y \,\&\, z, y \in N_G(x)\}$. Based on Theorem 2.20(b) we conclude that every fan-critical graph G satisfies $\mathrm{fan}(G) \leq \mathrm{ag}(G)$. Since both parameters are monotone, Proposition 1.1 then implies that ag is an upper bound for fan and hence also for Fan. Theorem 2.19 then implies that ag is an efficiently realizable upper bound for χ'.

Another monotone parameter, the so-called **supermultiplicity** sm, was introduced by Kochol et al. [173]. For a graph G and two distinct vertices $x, y \in V(G)$, let

$$\mathrm{sm}_G(x,y) = \min\{d_G(y) + \mu_G(x,y), d_G(x) + d_G(y) - \mu_G(x,y)\}.$$

Let $k \geq \Delta(G)$ be an integer. We call x a k-**reducible** vertex of G if every neighbor y of x satisfies $\mathrm{sm}_G(x,y) \leq k$. Then the supermultiplicity $\mathrm{sm}(G)$ is the smallest integer $k \geq \Delta(G)$ for which there exists a labeling $x_1, \ldots, x_n$ of the vertices of G such that each x_i is a k-reducible vertex of the graph $G - \{x_1, \ldots, x_{i-1}\}$. If G is edgeless, then $\mathrm{sm}(G) = 0$. It is not difficult to show that sm is a monotone graph parameter. Now assume that G is a fan-critical graph. If $\mathrm{fan}(G) = 0$, then $\text{fan}(G) \le \text{sm}(G)$. Otherwise, $\text{fan}(G) \ge 1$ and we conclude from Theorem 2.20 that each vertex x has a neighbor z such that $d_G(z) + \mu_G(x,z) \ge \text{fan}(G)$ and $d_G(x) + d_G(z) - \mu_G(x,z) \ge \text{fan}(G)$ implying that $\text{sm}_G(x,z) \ge \text{fan}(G)$. Consequently, $\text{fan}(G) \le \text{sm}(G)$. By Proposition 1.1, this implies that sm is an upper bound for fan and hence an upper bound for Fan. Finally, Theorem 2.19 implies that sm is an efficiently realizable upper bound for χ'.

Hakimi and Schmeichel [128] introduced a graph parameter similar to sm and proved the following extension of Vizing's bound. Let G be a graph and let $S(G)$ denote the simple graph whose vertex set is VG and whose edge set consists of all pairs $\{x,y\}$ such that $\mu_G(x,y) \ge 2$. If $S(G)$ is a forest, then

$$\chi'(G) \le \Delta(G) + \lfloor (1 + \sqrt{4\mu(G) - 3})/2 \rfloor. \tag{2.14}$$

To prove (2.14), it suffices to consider the case that G is a critical graph with $\chi'(G) = k + 1$ for an integer $k \ge \Delta(G)$. Since $S(G)$ is a forest, there is an edge $e \in E_G(x,y)$ such that every vertex $z \in N_G(x) \backslash \{y\}$ satisfies $\mu_G(x,z) = 1$. Now, let F be a maximal multi-fan at x with respect to the edge e and a coloring $\varphi \in C^k(G-e)$. By (F1), every vertex, except possibly y, occurs at most once in the sequence F. Then $t = |\overline{\varphi}(y)|$ satisfies $t = k - d_G(y) + 1 \ge k - \Delta(G) + 1$ and, moreover, $|\overline{\varphi}(z)| \ge k - \Delta(G) \ge 1$ for $z \ne y$. By Theorem 2.1, $V(F) \cup \{x\}$ is elementary with respect to φ. Hence, for each color $\alpha \in \overline{\varphi}(y)$, there is an edge $e_\alpha \in E(F)$ such that $\varphi(e_\alpha) = \alpha$. Based on (F2), it is not difficult to prove that, for each color $\alpha \in \overline{\varphi}(y)$, there is a vertex $y_\alpha \in V(F)$ such that $\overline{\varphi}(y_\alpha) \subseteq \{\varphi(e') \mid e' \in E_G(x,y)\}$ and such that $y_\alpha \ne y_{\alpha'}$ whenever $\alpha \ne \alpha'$. Then the color set $\Gamma = \bigcup_{\alpha \in \overline{\varphi}(y)} \overline{\varphi}(y_\alpha)$ satisfies $\mu(G) - 1 \ge \mu_G(x,y) - 1 \ge |\Gamma| = \sum_{\alpha \in \overline{\varphi}(y)} |\overline{\varphi}(y_\alpha)| \ge t(k - \Delta(G)) \ge (k - \Delta(G) + 1)(k - \Delta(G))$. This implies (2.14).

The fan number is a direct descendent of the fan equation. It was first introduced by Scheide [267] in his diploma thesis. The results of Sect. 2.5 were first published in a paper by Scheide and Stiebitz [274]. The fan number seems to be the appropriate graph parameter associated with Vizing's fan argument. It requires more sophisticated coloring arguments to obtain a general upper bound for χ' that is better than Fan in some cases. Let us give an example of such a parameter, due to Hilton and Jackson [137].

Let G be a graph and let V_2 denote the set of all 2-element subsets of $V(G)$. We call $M \subseteq V_2$ an **independent set of multiple edges** if the sets in M are pairwise disjoint and if $E_G(x,y) \ne \emptyset$ whenever $\{x,y\} \in M$. Now, let

$$\Delta_M(G) = \max\{\Delta(G), \max_Q \lfloor \tfrac{1}{2}(d_G(y) + \mu_G(x,y) + d_G(z) + \mu_G(x,z)) \rfloor\}$$

$Q = \{(x,y,z) \mid x \in V(G), z \ne y \,\&\, z, y \in N_G(x), \{x,y\}, \{x,z\} \notin M\}$. To obtain a graph parameter, define $\Delta^o(G)$ to be the minimum of $\Delta_M(G)$ taken over all independent sets M of multiple edges of G. Then, as proved by Hilton and Jackson [137], every graph G satisfies $\chi'(G) \le \Delta^o(G)$. The proof is by contradiction and goes roughly as follows. Let G be a smallest counterexample, and let M be an independent set of multiple edges with $\Delta^o(G) = \Delta_M(G)$. Then G is critical. We now consider two cases. If there is a vertex x that is not covered by M, then we use the fan equation and conclude from Theorem 2.3(b) that $\Delta_M(G) \geq \chi'(G)$, a contradiction. Otherwise, for each 2-set $\{x,y\} \in M$, we choose an edge $e \in E_G(x,y)$. This results in a perfect matching F of G. Then we obviously have $\chi'(G - F) \geq \chi'(G) - 1$. It is also easy to check that $\Delta^o(G - F) \leq \Delta^o(G) - 1$. Since, by the choice of G, we have $\chi'(G - F) \leq \Delta^o(G - F)$, this yields that $\chi'(G) \leq \Delta^o(G)$, again a contradiction. This proves that Δ^o is an upper bound for χ', but we do not know whether this bound is efficiently realizable.

For an even cycle C, we have $\mathrm{Fan}(C) = 3$ and $\Delta^o(C) = 2$. This shows that Δ^o can be smaller than Fan. If a graph G has an independent set M of multiple edges, such that $\{x,y\} \in M$ whenever $\mu_G(x,y) \geq 2$, then $\Delta^o(G) \leq \Delta(G) + 1$, while for the seemingly similar bound of Andersen and Goldberg we only have $\mathrm{ag}(G) \leq \Delta(G) + (\mu(G) + 1)/2$.

Let G be a graph such that $\chi'(G) = \Delta(G) + \mu(G)$. Then G contains a critical subgraph H with $\chi'(H) = \chi'(G)$, which implies $\Delta(G) + \mu(G) = \chi'(G) = \chi'(H) \leq \Delta(H) + \mu(H)$ (Vizing's Theorem 2.2). Since the graph parameters Δ and μ are monotone, it follows that $\Delta(H) = \Delta(G)$, $\mu(H) = \mu(G)$ and $\chi'(H) = \Delta(H) + \mu(H)$. Now Theorem 2.10 tells us that there is an edge $e \in E_H(x,y)$ such that $d_H(x) = d_H(y) = \Delta(H)$ and $\mu_H(x,y) = \mu(H)$. Consequently, $e \in E_G(x,y)$, $d_G(x) = d_G(y) = \Delta(G)$, and $\mu_G(x,y) = \mu(G)$. This simple observation can be used to prove the following statement:

(a) *Let G be a graph and let $k \geq \Delta(G) + \mu(G)$ be an integer. Then there is a k-edge-coloring φ of G such that every edge $e \in E_G(x,y)$ with $\varphi(e) = k$ satisfies $d_G(x) = d_G(y) = \Delta(G)$ and $\mu_G(x,y) = \mu(G)$.*

Proof of (a): We apply induction on the number of edges of G. If G has no edge, then the statement is evident. If G has no edge $e \in E_G(x,y)$ such that $d_G(x) = d_G(y) = \Delta(G)$ and $\mu_G(x,y) = \mu(G)$, then by the above result G has a $(k-1)$-edge-coloring and we are done. Otherwise, choose an edge e satisfying these properties and let $G' = G - e$. Since $k \geq \Delta(G') + \mu(G')$, it follows from the induction hypothesis that G' has a desired k-edge-coloring φ'. Since $d_{G'}(x) = d_{G'}(y) = \Delta(G) - 1$, we have $\varphi'(e') \neq k$ for every edge $e' \in E_{G'}(x) \cup E_{G'}(y)$. Hence, if we color e with k we obtain a desired k-edge-coloring φ of G. ∎

A slightly weaker version of statement (a) was proved by Berge and Fournier [28]. Their proof, however, is from first principle, without using earlier edge coloring results. Hence it is an alternative proof of Vizing's bound, and it also yields an $O(|E|^2)$ algorithm to color the edges of each graph $G = (V, E)$ with at most $\Delta(G) + \mu(G)$ colors.

Two edge colorings $\varphi_1, \varphi_2 \in C^k(G)$ of a graph G are **Kempe equivalent** if φ_2 can be obtained from φ_1 by a sequence of Kempe changes, possibly involving more than one pair of colors in successive Kempe changes. Mohar [225] proved that every two k-edge-colorings of a simple graph G are Kempe equivalent, provided that $k \geq \chi'(G) + 2$. Whether this is also true for $k = \Delta(G) + 2$ seems unknown.

Mohar [225] and Asratian [11] proved that if G is a simple bipartite graph, then, for $k \geq \Delta(G) + 1$, all k-edge-colorings of G are Kempe equivalent. The complete bipartite graph $K_{p,p}$ (where p is a prime) has a p-edge-coloring in which every two color classes form a Hamiltonian cycle. This example shows that Mohar's result cannot be extended to Δ-edge-colorings of simple bipartite graphs. An example of a simple graph G with $\chi'(G) = \Delta(G) = 4$ and with two 4-edge-colorings that are not Kempe equivalent was first given by Vizing [298]. In the same paper he also asked whether every k-edge-coloring of a graph G such that $k \geq \chi'(G)$ is Kempe equivalent to an optimal edge coloring of G, that is an edge coloring with $\chi'(G)$ colors, see Problem ♣ 19 for more details.

For a k-edge-coloring φ of a graph G and a color $\alpha \in \{1, \ldots, k\}$, the color class $E_{\varphi,\alpha} = \{e \in E(G) \,|\, \varphi(e) = \alpha\}$ is a matching of G (possibly empty). The following result about the existence of a **balanced edge coloring** is well known.

(b) *Let G be a graph and let k be an integer such that $k \geq \chi'(G) \geq 1$. Then there is a k-edge-coloring φ of G such that for every color $\alpha \in \{1, \ldots, k\}$, we have*

$$\lfloor |E(G)|/k \rfloor \leq |E_{\varphi,\alpha}| \leq \lceil |E(G)|/k \rceil \tag{2.15}$$

Proof of (b): As $k \geq \chi'(G)$, there is a k-edge-coloring φ of G. Choose φ such that

$$\sum_{\alpha=1}^{k} |E_{\varphi,\alpha}|^2 \tag{2.16}$$

is minimum. Suppose that (2.15) is violated. Then there are two color classes $E_{\varphi,\alpha}$ and $E_{\varphi,\beta}$ with $|E_{\varphi,\alpha}| \geq |E_{\varphi,\beta}| + 2$. Consequently, there exists an (α, β)-chain P with respect to φ such that P is a path with endvertices x and y, where $\alpha \in \varphi(x) \cap \varphi(y)$ and $\beta \in \overline{\varphi}(x) \cap \overline{\varphi}(y)$. Then, for the coloring φ' obtained from φ by recoloring P, we have

$$
\begin{aligned}
|E_{\varphi',\alpha}|^2 + |E_{\varphi',\beta}|^2 &= (|E_{\varphi,\alpha}| - 1)^2 + (|E_{\varphi,\beta}| + 1)^2 \\
&= |E_{\varphi,\alpha}|^2 + |E_{\varphi,\beta}|^2 - 2|E_{\varphi,\alpha}| + 2|E_{\varphi,\beta}| + 2 \\
&< |E_{\varphi,\alpha}|^2 + |E_{\varphi,\beta}|^2.
\end{aligned}
$$

Since $E_{\varphi',\gamma} = E_{\varphi,\gamma}$ for all colors $\gamma \neq \alpha$, we have $\sum_{\alpha=1}^{k} |E_{\varphi',\alpha}|^2 < \sum_{\alpha=1}^{k} |E_{\varphi,\alpha}|^2$, contradicting the choice of φ. ∎

Another type of edge coloring was introduced by Burris and Schelp [35]. A k-edge-coloring φ of a graph G is called **vertex-distinguishing**, if $\varphi(u) \neq \varphi(v)$ for any two distinct vertices u, v of G. If G is a connected graph with $|V(G)| \geq 3$ and if $k \geq |E(G)|$, then G admits a vertex-distinguishing k-edge-coloring. The smallest integer k for which such a coloring exists is called the **vertex-distinguishing edge chromatic number** of G, denoted by $\chi'_{\text{vd}}(G)$. Bazgan et al. [16] proved that every simple graph G without isolated edges and with at most one isolated vertex satisfies $\chi'_{\text{vd}}(G) \geq |V(G)| + 1$. Further result about the parameter $\chi'_{\text{vd}}(G)$ for simple graphs G can be found in Balister et al. [14] and Bazgan et al. [17].

CHAPTER 3

KIERSTEAD PATHS

Kierstead [166] invented a new type of test objects for the edge coloring problem. He used this new type of test objects to give a strengthening of Vizing's result. His method can be used also to give an alternative coloring algorithm. Kierstead's algorithm is based on recoloring the edges of a path instead of recoloring the edges of a fan.

3.1 KIERSTEAD'S METHOD

Let G be a graph, let $e \in E_G(x, y)$ be an edge, and let $\varphi \in \mathcal{C}^k(G - e)$ be a coloring for some integer $k \geq 0$. A **Kierstead path** with respect to e and φ is defined to be a sequence $K = (y_0, e_1, y_1, \ldots, e_p, y_p)$ with $p \geq 1$ consisting of edges $e_1, \ldots, e_p$ and vertices $y_1, \ldots, y_p$ satisfying the following two conditions:

(K1) The vertices $y_0, \ldots, y_p$ are distinct, $e_1 = e$, and $e_i \in E_G(y_i, y_{i-1})$ for $1 \leq i \leq p$.

(K2) For every edge e_i with $2 \leq i \leq p$, there is a vertex y_j with $0 \leq j < i$ such that $\varphi(e_i) \in \overline{\varphi}(y_j)$.

Graph Edge Coloring: Vizing's Theorem and Goldberg's Conjecture,
First Edition. By M. Stiebitz, D. Scheide, B. Toft, and L. M. Favrholdt
Copyright © 2012 John Wiley & Sons, Inc.

If K is a Kierstead path, then it follows from (K1) that the corresponding graph with vertex set $V(K)$ and edge set $E(K)$ is indeed a path in G having endvertices y_0 and y_p.

In 1984, Kierstead [166] proved that, for every graph G with $\chi'(G) = k + 1$, the vertex set of any Kierstead path with respect to a critical edge $e \in E(G)$ and a coloring $\varphi \in \mathcal{C}^k(G - e)$ is elementary with respect to φ if $k \geq \Delta(G) + 1$. That Kierstead's argument also works in case of $k = \Delta(G)$, provided we add a degree condition, seems to have been noticed first by Zhang [317].

Theorem 3.1 (Kierstead [166] 1984) *Let G be a graph with $\chi'(G) = k + 1$ for an integer $k \geq \Delta(G)$, and let $e \in E(G)$ be a critical edge of G. If*

$$K = (y_0, e_1, y_1, \ldots, y_{p-1}, e_p, y_p)$$

is a Kierstead path with respect to e and a coloring $\varphi \in \mathcal{C}^k(G - e)$ such that $d_G(y_j) < k$ for $j = 2, \ldots, p$, then $V(K)$ is elementary with respect to φ.

Proof: For the proof we consider a minimal counterexample, that is, a pair (K, φ) satisfying the following conditions:

(a) $K = (y_0, e_1, y_1, \ldots, y_{p-1}, e_p, y_p)$ is a Kierstead path with respect to $e \in E(G)$ and $\varphi \in \mathcal{C}^k(G - e)$, where $d_G(y_j) < k$ for $j = 2, \ldots, p$.

(b) $V(K)$ is not elementary with respect to φ.

(c) $|V(K)|$ is minimum subject to (a) and (b).

Since $e_1 = e$ is uncolored and $e_1 \in E_G(y_0, y_1)$, the set $\{y_0, y_1\}$ is elementary with respect to φ. Otherwise, we could color e with some color $\alpha \in \overline{\varphi}(y_0) \cap \overline{\varphi}(y_1)$, contradicting $\chi'(G) = k + 1$. Hence (b) implies that $p \geq 2$.

It follows from (a) that $K' = (y_0, e_1, \ldots, e_{p-1}, y_{p-1})$ is a Kierstead path with respect to e and φ, where $d_G(y_j) < k$ for $j = 2, \ldots, p - 1$. Then (c) implies that $V(K')$ is elementary with respect to φ. Since $V(K)$ is not elementary with respect to φ (by (b)), we then conclude that there exists an element $i \in \{0, 1, \ldots, p - 1\}$ such that $\overline{\varphi}(y_i) \cap \overline{\varphi}(y_p) \neq \emptyset$. We refer to the maximal i as the *index* of the coloring φ and write $i = \text{index}(\varphi)$.

To arrive at a contradiction, we choose the minimal counterexample (K, φ) such that $i = \text{index}(\varphi)$ is maximum. We claim that $i = p - 1$. To prove the claim, assume that $i < p - 1$. Then there is a color $\alpha \in \overline{\varphi}(y_i) \cap \overline{\varphi}(y_p)$. Since the edge $e_1 = e$ is uncolored and $e_1 \in E_G(y_0, y_1)$, we have $|\overline{\varphi}(y_j)| = k - d_G(y_j) + 1$ for $j = 0, 1$ and $|\overline{\varphi}(y_j)| = k - d_G(y_j)$ otherwise. Since $k \geq \Delta(G)$ and $d_G(y_j) < k$ for $j = 2, \ldots, p$ (by (a)), we obtain that $\overline{\varphi}(y_j) \neq \emptyset$ for $j = 0, \ldots, p$. This implies, in particular, that there is a color $\beta \in \overline{\varphi}(y_{i+1})$. Since $y_{i+1} \in V(K')$ and $V(K')$ is elementary with respect to φ, the two colors α, β are distinct and, moreover, $\alpha \in \varphi(y_{i+1})$, $\beta \in \varphi(y_i)$ and $\alpha, \beta \in \varphi(y_h)$ for $h = 0, \ldots, i - 1$. Consequently $\varphi(e_h) \notin \{\alpha, \beta\}$ for $h = 2, \ldots, i + 1$ (by (K2)). Furthermore, we conclude that the (α, β)-chain $P = P_{y_{i+1}}(\alpha, \beta, \varphi)$ is a path, where one endvertex is y_{i+1} and the other endvertex is some vertex $z \in V(G) \setminus \{y_{i+1}\}$.

Hence $\varphi' = \varphi/P \in C^k(G - e)$. If $z = y_i$, then K is a Kierstead path with respect to e and φ', where $\alpha \in \overline{\varphi}'(y_{i+1}) \cap \overline{\varphi}'(y_p)$. Hence (K, φ') is a minimal counterexample with index$(\varphi') >$ index(φ), a contradiction to the choice of (K, φ). If $z \ne y_i$, then $K'' = (y_0, e_1, y_1, \ldots, e_{i+1}, y_{i+1})$ is a Kierstead path with respect to e and φ', where $\alpha \in \overline{\varphi}'(y_i) \cap \overline{\varphi}'(y_{i+1})$. Hence K'' is a Kierstead path with respect to e and φ', where $V(K'')$ is not elementary with respect to φ'. Since $|V(K'')| < |V(K)|$, this is a contradiction to (c). This proves the claim that $i = p - 1$.

Consequently, for the coloring φ, there is a color $\alpha \in \overline{\varphi}(y_{p-1}) \cap \overline{\varphi}(y_p)$. For the color $\beta = \varphi(e_p)$, there is a vertex y_j with $j < p$ such that $\beta \in \overline{\varphi}(y_j)$ (by (K2)). Since $e_p \in E_G(y_p, y_{p-1})$, we have $j < p - 1$. Recolor e_p with α. This results in a coloring $\varphi' \in C^k(G - e)$ such that $K' = (y_0, e_1, y_1, \ldots, e_{p-1}, y_{p-1})$ is a Kierstead path with respect to e and φ', where $\beta \in \overline{\varphi}'(y_{p-1}) \cap \overline{\varphi}'(y_j)$. Hence $V(K')$ is not elementary with respect to φ'. Since $|V(K')| < |V(K)|$, this is a contradiction to (c). The proof is now complete. ∎

Kierstead [166] used Theorem 3.1 to give an alternative proof of Vizing's bound in Theorem 2.2. Furthermore, he proved in [166] that if $\chi'(G) = \Delta(G) + \mu(G)$ for a graph G with $\mu(G) = \mu \ge 2$, then G contains a μ-**triangle**, that is, a subgraph T consisting of three vertices x, y, z such that $\mu_T(x, y) = \mu$, $\mu_T(y, z) = \mu - 1$ and $\mu_T(x, z) = 1$. A strengthening of this result is given in Chap. 5 (Theorem 5.6).

Theorem 3.1 and its proof tells us that Kierstead paths are suitable test objects for the type of coloring algorithms described in Sect. 1.5. If we use fans as test object, the subroutine **Ext** requires at most one Kempe change and at most one recoloring of a subsequence of edges of a fan. If we use Kierstead paths, however, the number of Kempe changes grow quadratic in the order of the paths.

Remark 3.2 *Let G be a graph, let $e \in E(G)$ be an edge, and let $\varphi \in C^k(G - e)$ be a coloring with $k \ge \Delta(G) + 1$. If K is Kierstead path with respect to e and φ such that $V = V(K)$ is not elementary with respect to φ, then a coloring $\varphi' \in C^k(G)$ can be derived from φ after $O(|V|^2)$ Kempe changes, in addition to coloring the edge e.*

That the number of colors used by an algorithm based on Kierstead paths as test objects is bounded from above by $\Delta + \mu$ follows from the fact that there is an inequality for Kierstead paths similar to the fan inequality. To see this, let G be a graph with $\chi'(G) = k + 1$ for an integer $k \ge \Delta(G) + 1$, and let $e \in E(G)$ be a critical edge of G. Furthermore, let $K = (y_0, e_1, y_1, \ldots, y_{p-1}, e_p, y_p)$ be a maximal Kierstead path with respect to e and a coloring $\varphi \in C^k(G - e)$. Then, by Theorem 3.1, $V(K)$ is elementary with respect to φ and, moreover, $p \ge 3$. Hence, for the vertex set $X = \{y_0, \ldots, y_{p-1}\}$, we have $\overline{\varphi}(X) \subseteq \varphi(y_p)$. Consequently, for every color $\alpha \in \overline{\varphi}(X)$, there is an edge $e_\alpha \in E_G(y_p)$ with $\varphi(e_\alpha) = \alpha$. Since $k \ge \Delta(G) + 1$ and K is a maximal Kierstead path with respect to e and φ, this implies that each edges e_α with $\alpha \in \overline{\varphi}(X)$ is incident with some vertex of X. Hence $|\overline{\varphi}(X)| \le |X|\mu(G) = p\mu(G)$. On the other hand, by Proposition 1.7(a), we have $|\overline{\varphi}(X)| \ge |X|(k - \Delta(G)) + 2 = p(k - \Delta(G)) + 2$. Clearly, this yields $p(\Delta(G) + \mu(G) - k) \ge 2$. This inequality fails obviously if $k \ge \Delta(G) + \mu(G)$. Hence $\chi'(G) \le \Delta(G) + \mu(G)$ and our coloring algorithm never uses more than

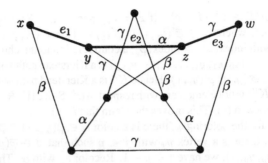

Figure 3.1 A Kierstead path $K = (x, e_1, y, e_2, z, e_3, w)$ (bold edges).

$\Delta + \mu$ colors. Under the assumption that G does not contain a $\mu(G)$-triangle, the above inequality can be strengthened to $p(\Delta(G) + \mu(G) - k - 1) \geq 1$; hence it already fails if $k \geq \Delta(G) + \mu(G) - 1$ and $\mu(G) \geq 2$.

3.2 SHORT KIERSTEAD'S PATHS

The degree condition in Theorem 3.1, saying that $d_G(y) < k$ for every vertex of the Kierstead path K except the first two vertices, is only needed if $k = \Delta(G)$. Figure 3.1 shows a critical graph G with $\chi'(G) = \Delta(G) + 1 = 4$, together with a Kierstead path K with respect to an uncolored edge e and a coloring $\varphi \in \mathcal{C}^3(G - e)$, such that $V(K)$ is not elementary with respect to φ. For such graphs, however, the following result has proved useful.

Theorem 3.3 (Kostochka and Stiebitz [177] 2006) *Let G be a graph with maximum degree Δ and $\chi'(G) = \Delta + 1$. Let $e \in E(G)$ be a critical edge and $\varphi \in \mathcal{C}^\Delta(G - e)$. If $K = (y_0, e_1, y_1, e_2, y_2, e_3, y_3)$ is a Kierstead path with respect to e and φ, then the following statements hold:*

   (a) $\overline{\varphi}(y_0) \cap \overline{\varphi}(y_1) = \emptyset$.

   (b) *If $d_G(y_2) < \Delta$, then $V(K)$ is elementary with respect to φ.*

   (c) *If $d_G(y_1) < \Delta$, then $V(K)$ is elementary with respect to φ.*

   (d) *If $\Gamma = \overline{\varphi}(y_0) \cup \overline{\varphi}(y_1)$, then $|\overline{\varphi}(y_3) \cap \Gamma| \leq 1$.*

Proof: Statement (a) follows from Theorem 3.1 and the fact that (y_0, e_1, y_1) is a Kierstead path with respect to e and φ. Note that each vertex v of G satisfies $|\overline{\varphi}(v)| = \Delta - d_G(v) + 1$ if $v \in \{y_0, y_1\}$ and $|\overline{\varphi}(v)| = \Delta - d_G(v)$ otherwise.

For the proof of (b), assume that $d_G(y_2) < \Delta$. If $\overline{\varphi}(y_3) \neq \emptyset$, then $d_G(y_3) < \Delta$ and $V(K)$ is elementary with respect to φ by Theorem 3.1. If $\overline{\varphi}(y_3) = \emptyset$, then Theorem 3.1 implies that $\{y_0, y_1, y_2\}$ is elementary with respect to φ and, therefore, $V(K)$ is elementary with respect to φ, too. This proves (b).

For the proof of (c), assume that $d_G(y_1) < \Delta$. Suppose, on the contrary, that $V(K)$ is not elementary with respect to φ. By (b), this implies that $d_G(y_2) = \Delta$, which gives $\overline{\varphi}(y_2) = \emptyset$. Since $\overline{\varphi}(y_0) \cap \overline{\varphi}(y_1) = \emptyset$ (by (a)), we then deduce that there is a color $\gamma \in \overline{\varphi}(y_3)$ such that $\gamma \in \overline{\varphi}(y_0) \cup \overline{\varphi}(y_1)$. Let $\alpha = \varphi(e_2)$ and $\beta = \varphi(e_3)$. Clearly, $\alpha \neq \beta \neq \gamma$. Since K is a Kierstead path with respect to e and $\varphi \in \mathcal{C}^{\Delta}(G - e)$, (K2) implies $\alpha \in \overline{\varphi}(y_0)$ and $\beta \in \overline{\varphi}(y_0) \cup \overline{\varphi}(y_1)$.

First consider the case that $\beta \in \overline{\varphi}(y_1)$. Color the edge $e = e_1$ with α, recolor the edge e_2 with β, and make the edge e_3 uncolored. This results in a coloring $\varphi' \in \mathcal{C}^{\Delta}(G - e_3)$ such that $\gamma, \beta \in \overline{\varphi}'(y_3)$, $\alpha \in \overline{\varphi}'(y_2)$, $\varphi'(e_2) = \beta$, $\varphi'(e_1) = \alpha$, $\gamma \in \overline{\varphi}'(y_0) \cup \overline{\varphi}'(y_1)$ and $\gamma \neq \beta$. Hence $K' = (y_3, e_3, y_2, e_2, y_1, e_1, y_0)$ is a Kierstead path with respect to e_3 and φ' such that $V(K') = V(K)$ is not elementary with respect to φ'. Since $d_G(y_1) < \Delta$, this is a contradiction to (b).

Now consider the case that $\beta \in \overline{\varphi}(y_0)$. If $\gamma \in \overline{\varphi}(y_1)$, then $P = P_{y_1}(\gamma, \beta)$ is a path whose endvertices are y_1 and y_0 (Theorem 2.1(b)). Then, clearly, neither y_3 nor e_3 belongs to P. Thus the coloring $\varphi' = \varphi/P$ belongs to $\mathcal{C}^{\Delta}(G - e)$ and we have $\alpha, \gamma \in \overline{\varphi}'(y_0)$, $\beta \in \overline{\varphi}'(y_1)$, $\varphi'(e_2) = \alpha$, $\varphi'(e_3) = \beta$, and $\gamma \in \overline{\varphi}'(y_3)$. Hence we are in the first case. If $\gamma \in \overline{\varphi}(y_0)$, then we argue as follows. Since $d_G(y_1) < \Delta$, there is a color $\delta \in \overline{\varphi}(y_1)$. By (a), $\delta \notin \{\alpha, \beta, \gamma\} \subseteq \overline{\varphi}(y_0)$. Consequently, Theorem 2.1(b) implies that $P' = P_{y_1}(\delta, \gamma)$ is a path with endvertices y_1 and y_0. Since $\gamma \in \overline{\varphi}(y_3)$, $y_3 \notin V(P')$. Let φ' be the coloring obtained from φ by recoloring P'. If $\gamma \neq \alpha$, then we have $\alpha, \beta, \delta \in \overline{\varphi}'(y_0)$, $\gamma \in \overline{\varphi}'(y_1) \cap \overline{\varphi}'(y_3)$, $\varphi'(e_2) = \alpha$, and $\varphi'(e_3) = \beta$. Hence we are in the previous subcase. If $\gamma = \alpha$, then e_2 belongs to P' and we have $\delta, \beta \in \overline{\varphi}'(y_0)$, $\alpha \in \overline{\varphi}'(y_1) \cap \overline{\varphi}'(y_3)$, $\varphi'(e_2) = \delta$, and $\varphi'(e_3) = \beta$. Hence we are in the previous subcase, too. This settles the second case. Hence the proof of (c) is complete.

In order to prove (d), assume that it is false. By (b) and (c), this implies that $d_G(y_1) = d_G(y_2) = \Delta$. Consequently, we have $|\overline{\varphi}(y_1)| = 1$ and $\overline{\varphi}(y_2) = \emptyset$. Hence there are two colors, say α and δ, such that $\overline{\varphi}(y_1) = \{\delta\}$ and $\varphi(e_2) = \alpha$. Clearly, $\alpha \neq \delta$. Since K is a Kierstead path with respect to e and φ, (K2) implies $\alpha \in \overline{\varphi}(y_0)$ and $\varphi(e_3) \in \Gamma$, where $\Gamma = \overline{\varphi}(y_0) \cup \overline{\varphi}(y_1)$. Now let $\mathcal{P}$ denote the set of all colorings $\varphi \in \mathcal{C}^{\Delta}(G - e)$ satisfying $\varphi(e_2) = \alpha \in \overline{\varphi}(y_0)$, $\overline{\varphi}(y_1) = \{\delta\}$, $\varphi(e_3) \in \Gamma$, and $|\overline{\varphi}(y_3) \cap \Gamma| \geq 2$. It follows from the assumption that $\mathcal{P}$ is non-empty. To arrive at a contradiction, we shall construct an appropriate coloring in $\mathcal{P}$.

First, we claim that there is a coloring $\varphi \in \mathcal{P}$ such that $\alpha \in \overline{\varphi}(y_3)$. To this end, choose an arbitrary coloring $\varphi \in \mathcal{P}$. There is nothing to prove if $\alpha \in \overline{\varphi}(y_3)$. Otherwise, we consider two cases. First, suppose that $\delta \in \overline{\varphi}(y_3)$. By Theorem 2.1(b), $P_{y_0}(\alpha, \delta)$ is a path with endvertices y_0 and y_1. Consequently, $P = P_{y_3}(\alpha, \delta)$ is a path, where one endvertex is y_3 and the other endvertex is some vertex $x \notin V(K)$. Furthermore, we have $\varphi(e_3) \neq \alpha, \delta$ and $e_2 \notin E(P)$. This implies that the coloring $\varphi' = \varphi/P$ belongs to $\mathcal{P}$ and $\alpha \in \overline{\varphi}'(y_3)$. Hence, φ' is a desired coloring. Now, suppose that $\delta \notin \overline{\varphi}(y_3)$. Then choose a color $\gamma \in \Gamma \cap \overline{\varphi}(y_3)$. Since $\alpha \notin \overline{\varphi}(y_3)$, we have $\gamma \neq \alpha, \delta$. Consequently, $\gamma \in \overline{\varphi}(y_0)$ and, by Theorem 2.1(b), $P_{y_0}(\gamma, \delta)$ is a path with endvertices y_0 and y_1. Therefore, $P' = P_{y_3}(\gamma, \delta)$ is a path, where one endvertex is y_3 and the other endvertex is some vertex $x \notin V(K)$. Then it is easy to check that the coloring $\varphi' = \varphi/P'$ belongs to $\mathcal{P}$ and satisfies $\alpha \notin \overline{\varphi}'(y_3)$ and

$\delta \in \overline{\varphi}'(y_3)$. Hence we can proceed as in the first case. This proves the claim that there is a coloring $\varphi \in \mathcal{P}$ such that $\alpha \in \overline{\varphi}(y_3)$.

Next, we claim that there is a coloring $\varphi \in \mathcal{P}$ such that $\alpha, \delta \in \overline{\varphi}(y_3)$. By the previous claim, we can choose a coloring $\varphi \in \mathcal{P}$ such that $\alpha \in \overline{\varphi}(y_3)$. If also $\delta \in \overline{\varphi}(y_3)$, we are done. Otherwise, since $|\Gamma \cap \overline{\varphi}(y_3)| \geq 2$, there is a color $\gamma \in \Gamma \cap \overline{\varphi}(y_3)$ such that $\gamma \neq \alpha, \delta$. Then $\gamma \in \overline{\varphi}(y_0)$ and Theorem 2.1(b) implies that $P_{y_0}(\gamma, \delta)$ is a path with endvertices y_0 and y_1. Therefore, $P = P_{y_3}(\gamma, \delta)$ is a path, where one endvertex is y_3 and the other endvertex is some vertex $x \notin V(K)$. Then it is easy to check that the coloring $\varphi' = \varphi/P$ belongs to $\mathcal{P}$ and $\alpha, \delta \in \overline{\varphi}'(y_3)$. This proves the claim.

To complete the proof, choose an arbitrary coloring $\varphi \in \mathcal{P}$ such that $\alpha, \delta \in \overline{\varphi}(y_3)$. Let $\beta = \varphi(e_3)$. Clearly, $\beta \notin \{\alpha, \delta\}$, $\beta \in \overline{\varphi}(y_0)$ (by (K2)), and $\beta \notin \overline{\varphi}(y_3)$. Since $\delta \in \overline{\varphi}(y_1)$, it follows from Theorem 2.1(b) that $P_{y_0}(\beta, \delta)$ is a path having endvertices y_0 and y_1. Hence $P = P_{y_3}(\beta, \delta)$ is a path, where one endvertex is y_3 and the other endvertex is some vertex $x \notin V(K)$. Furthermore, we have $e_3 \in E(P)$. Then $\varphi' = \varphi/P \in \mathcal{C}^\Delta(G - e)$ and we have $\alpha \in \overline{\varphi}'(y_0)$, $\delta \in \overline{\varphi}'(y_1)$, $\alpha \in \overline{\varphi}'(y_3)$, $\varphi'(e_2) = \alpha$, and $\varphi'(e_3) = \delta$. Consequently, the path $\mathrm{Path}(y_1, e_2, y_2, e_3, y_3)$ with endvertices y_1 and y_3 is the (α, δ)-chain with respect to φ' containing y_1, which gives a contradiction to Theorem 2.1(b). Hence the proof of (d) is complete. ∎

Let us conclude this section with a simple proposition that will be used in the next chapter.

Proposition 3.4 (Kostochka and Stiebitz [177] 2006) *Let G be a graph with maximum degree Δ and $\chi'(G) = \Delta + 1$. Furthermore, let $e \in E(G)$ be a critical edge and $\varphi \in \mathcal{C}^\Delta(G - e)$. Then the following statements hold:*

(a) *If $K = (y_0, e_1, y_1, \ldots, e_p, y_p)$ is a Kierstead path with respect to e and φ such that $d_G(y_i) < \Delta$ for $i \in \{2, \ldots, p\}$, then $\sum_{i=0}^{p} d_G(y_i) \geq p\Delta + 2$.*

(b) *If $K = (y_0, e_1, y_1, e_2,, y_2)$ is a Kierstead path with respect to e and φ, then $V(K)$ is elementary with respect to φ and $d_G(y_0) + d_G(y_1) + d_G(y_2) \geq 2\Delta + 2$.*

(c) *If $K = (y_0, e_1, y_1, e_2,, y_2, e_3, y_3)$ is a Kierstead path with respect to e and φ, then $d_G(y_1) + d_G(y_2) + d_G(y_3) \geq 2\Delta + 2$ and, moreover, $d_G(y_0) + d_G(y_1) + d_G(y_2) + d_G(y_3) \geq 3\Delta + 1$ with equality only if $d_G(y_1) = d_G(y_2) = \Delta$.*

Proof: Let $K = (y_0, e_1, y_1, \ldots, e_p, y_p)$ be a Kierstead path with respect to e and φ. Furthermore, let $D_p = \sum_{i=0}^{p} d_G(y_i)$. If $V(K)$ is elementary with respect to φ, then Proposition 1.7(b) implies that $D_p \geq p\Delta + 2$. Hence (a) follows from Theorem 3.1.

Now assume that $p = 2$. If $d_G(y_2) < \Delta$, then $V(K) = \{y_0, y_1, y_2\}$ is elementary with respect to φ (Theorem 3.1). Otherwise, $\overline{\varphi}(y_2) = \emptyset$ and $\overline{\varphi}(y_0) \cap \overline{\varphi}(y_1) = \emptyset$ (Theorem 2.1), which implies that $V(K)$ is elementary with respect to φ, too. Consequently, in booth cases, $D_2 \geq 2\Delta + 2$ (Proposition 1.7(b)), which proves statement (b).

Finally assume that $p = 3$. If $d_G(y_1) < \Delta$ or $d_G(y_2) < \Delta$, then Theorem 3.3 implies that $V(K)$ is elementary with respect to φ and, therefore, $D_3 \geq 3\Delta + 2$

(Proposition 1.7(b)) and so $d_G(y_1) + d_G(y_2) + d_G(y_3) \geq 2\Delta + 2$. Otherwise, $d_G(y_1) = d_G(y_2) = \Delta$ and, by Theorem 3.3, we have $\overline{\varphi}(y_0) \cap \overline{\varphi}(y_1) = \emptyset$ and $|\overline{\varphi}(y_3) \cap \Gamma| \leq 1$ with $\Gamma = \overline{\varphi}(y_0) \cup \overline{\varphi}(y_1)$. Hence $\Delta \geq |\overline{\varphi}(y_0, y_1, y_3)| \geq |\overline{\varphi}(y_0)| + |\overline{\varphi}(y_1)| + |\overline{\varphi}(y_3)| - 1$. Since $e \in E_G(y_0, y_1)$ is uncolored with respect to φ, this gives

$$\Delta \geq (\Delta - d_G(y_0) + 1) + (\Delta - d_G(y_1) + 1) + (\Delta - d_G(y_3)) - 1$$

and hence $d_G(y_0) + d_G(y_1) + d_G(y_3) \geq 2\Delta + 1$. Since $d_G(y_1) = d_G(y_2) = \Delta$, this gives $D_3 \geq 3\Delta + 1$ and $d_G(y_1) + d_G(y_2) + d_G(y_3) = 2\Delta + d_G(y_3) \geq 2\Delta + 2$. Hence (c) is proved.   ∎

3.3 NOTES

The proof of Theorem 3.1 is based on the standard argument due to Kierstead [166]. Kierstead's interest in edge coloring of multigraphs was mainly motivated by his research on vertex coloring of graphs, see references [166, 167, 168, 169]. Clearly, an edge coloring result can be always considered as a result about vertex coloring of line graphs.

For a simple graph G, let $\chi(G)$ and $\omega(G)$ denote the chromatic number and the clique number of G, respectively. For definitions of these two graph parameters see Sect. 1.2. Then Vizing's theorem for simple graphs says that any line graph G of a simple graph satisfies $\chi(G) \leq \omega(G) + 1$.

Beineke [20, 21] characterized line graphs of simple graphs in terms of forbidden induced subgraphs. For a set S of simple graphs, let $\mathcal{G}(S)$ denote the class of all simple graphs which do not contain any graph in S as an induced subgraph. Beineke proved that there is a particular set B_9 of nine simple graphs, including $K_{1,3}$ and K_5^- (the complete graph on 5 vertices with one edge deleted), such that $\mathcal{G}(B_9)$ is the class of line graphs of simple graphs. Hence Vizing's result about the chromatic index of simple graphs can be stated as follows: If G is a simple graph that does not have any of the nine graphs in B_9 as induced subgraphs, then $\chi(G) \leq \omega(G) + 1$. Translating Vizing's theorem to this form, it becomes very natural to ask whether the statement remains true if the set B_9 is replaced by a subset. That this is indeed the case was first shown by Choudum [60]. Javdekar [153] improved Choudum's result. A further improvement was obtained by Kierstead and Schmerl [168] and Kierstead [166]:

(a) *Every graph $G \in \mathcal{G}(\{K_{1,3}, K_5^-\})$ satisfies $\chi(G) \leq \omega(G) + 1$.*

As proved by Kierstead and Schmerl [168], statement (a) is equivalent to the statement that $\chi'(H) \leq \Delta(H) + 1$ for every graph H with $\mu(H) \leq 2$ and without a 2-triangle. A proof of the latter statement was given by Kierstead [166]. This shows that interesting results on vertex colorings of simple graphs may be derived from improved results on the chromatic index of graphs, see also references [167] and [169].

A set S of simple graphs is said to be a χ-**binding set** if every graph $G \in \mathcal{G}(S)$ satisfies $\chi(G) \leq f_S(\omega(G))$, where $f_S : \mathbb{N}_0 \to \mathbb{N}_0$ is a function depending on S. If this inequality holds for the function $f_S(x) = x + 1$, we say that S has the **Vizing property**. If S is the set consisting of all cycles with an odd number of vertices, then $\mathcal{G}(S)$ consists of all bipartite simple graphs and, therefore, S is a χ-binding set and it even has the Vizing property. However, every χ-binding finite set S of simple graphs contains at least one forest. This follows from the famous result of Erdős [80] saying that, for any two integers $k, g \geq 3$, there exists a simple graph G such that $\chi(G) = k$ and every cycle in G has at least g edges. On the other hand, Gyárfás [123] conjectured that $S = \{T\}$ is a χ-binding set for every tree T.

Randerath [251] proved in his dissertation that if a set $S = \{H\}$ consisting of a single simple graph H has the Vizing property then H is an induced subgraph of the path P_4 on four vertices. Furthermore, he proved that if the set $S = \{T, H\}$ consisting of two simple graphs T and H has the Vizing property, but neither the set $\{T\}$ nor the set $\{H\}$ has it, then one of the two, say T, has to be a tree not being an induced subgraph of P_4, and the other H has to belong to the set $\{K_3, K_4, K_4^-, K_5^-, K_3^+, K_4^*\}$, where K_3^+ is the graph obtained from K_3 by adding an additional vertex v and joining v to exactly one vertex of the K_3, and K_4^* is the graph obtained from K_4 by adding an additional vertex v and joining v to exactly two vertices of the K_4.

A simple graph G is called **perfect** if every induced subgraph H of G satisfies $\chi(H) = \omega(H)$. Chudnovsky et al. [63] proved in 2002 that a simple graph G is perfect if and only if neither G nor its complement $\overline{G}$ contains an odd cycle with at least 5 vertices as an induced subgraph. This result solved a longstanding conjecture made by Claude Berge around 1960, and it is one of the deepest results in graph theory.

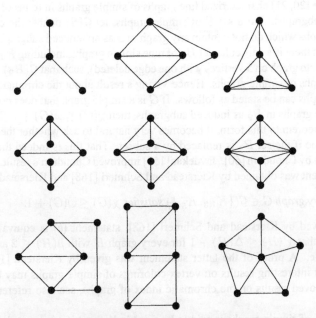

Figure 3.2   Beineke's class B_9 of forbidden induced subgraphs.

CHAPTER 4

SIMPLE GRAPHS AND LINE GRAPHS

Vizing considered (multi)graphs rather than just simple graphs, but he posed and was interested in the so-called **Classification Problem** mainly for simple graphs. For a simple graph G, Vizing's bound in Theorem 2.2 implies that $\Delta(G) \leq \chi'(G) \leq \Delta(G) + 1$. This leads to a natural classification of simple graphs into two classes. A graph G is said to be of **class one** if G is simple and $\chi'(G) = \Delta(G)$ and of **class two** if G is simple and $\chi'(G) = \Delta(G) + 1$.

4.1 CLASS ONE AND CLASS TWO GRAPHS

On one hand, it seems that graphs of class two are relatively scarce. Almost all simple graphs are of class one, as proved by Erdős and Wilson [85] in 1977. On the other hand, it is an NP-complete problem to decide whether a given 3-regular simple graph is of class one or two, as proved by Holyer [150] in 1981.

Consider an arbitrary graph G of class two, i.e., G is a simple graph with $\chi'(G) = \Delta(G)+1$. If H is a critical subgraph of G with $\chi'(H) = \chi'(G)$, then $\Delta(H) = \Delta(G)$, implying that H is a of class two. Otherwise, we would have $\Delta(H) \leq \Delta(G) - 1$ and, therefore, $\chi'(G) = \chi'(H) \leq \Delta(H) + 1 \leq \Delta(G) < \chi'(G)$, which is impossible. By Proposition 1.1, this implies the following result.

Graph Edge Coloring: Vizing's Theorem and Goldberg's Conjecture,
First Edition. By M. Stiebitz, D. Scheide, B. Toft, and L. M. Favrholdt

Proposition 4.1 (Vizing [299] 1965) *Every graph of class two contains a critical graph of class two with the same maximum degree as a subgraph.*

Proposition 4.2 (Vizing [299] 1965) *Every critical graph G of class two contains a critical graph G' of class two with maximum degree Δ' as a subgraph for each integer Δ' satisfying $2 \leq \Delta' \leq \Delta(G)$.*

Proof: By Proposition 4.1, there exists a critical graph $H \subseteq G$ of class two with maximum degree $\Delta = \Delta(G)$. Since $\Delta \geq \Delta' \geq 2$, there is an edge $e \in E_H(x, y)$. Then there is a coloring $\varphi \in C^\Delta(H - e)$ and we can choose a set Γ of Δ' colors such that $\Gamma \cap \overline{\varphi}(v) \neq \emptyset$ for $v = x, y$. Note that the color sets $\overline{\varphi}(x)$ and $\overline{\varphi}(y)$ are disjoint, but nonempty. We now consider the subgraph H' of H with $V(H') = V(G)$ and $E(H') = \{e' \in E(H) \mid e' = e \text{ or } \varphi(e') \in \Gamma\}$. Then $\Delta(H') \leq \Delta'$ and $\chi'(H') \geq \Delta' + 1$, since $\chi'(H) = \Delta + 1$. Consequently, H' is a graph of class two with maximum degree Δ'. By Proposition 4.1, there exists a critical graph $G' \subseteq H'$ of class two with maximum degree Δ'.   ∎

Most of the known structural properties of critical graphs of class two can be derived from Vizing's Adjacency Lemma. To give an example, we first introduce some further notation.

Let G be an arbitrary simple graph. The least positive integer k such that G has a vertex order in which each vertex is preceded by fewer than k of its neighbors is called the **coloring number** $\mathrm{col}(G)$ of G. Clearly, $\mathrm{col}(G) \leq \Delta(G) + 1$ and $\mathrm{col}(\emptyset) = 1$. The coloring number as an upper bound for the chromatic number was first defined and studied by Erdős and Hajnal [83] in 1966. Two years later Szekeres and Wilf [289] established another upper bound for the chromatic number of a simple graph G that turned out to be equal to the coloring number of G, namely the maximum minimum degree of the subgraphs of G plus 1.

That the coloring number of a simple graph G with $n \geq 1$ vertices can be computed by applying the following polynomial-time algorithm was first discovered by Matula [211] and, independently, by Finck and Sachs [96]. A **smallest last order** $v_n, v_{n-1}, \ldots, v_1$ of the vertices of G is obtained by letting v_i be a vertex of minimum degree in the subgraph $G_i = G - \{v_{i+1}, \ldots, v_n\}$ for $i = n, n-1, \ldots, 1$, where $G_n = G$. Then G satisfies

$$\mathrm{col}(G) = \max_{H \subseteq G} \delta(H) + 1 = \max_{1 \leq i \leq n} d_{G[v_1, \ldots, v_i]}(v_i) + 1.$$

The coloring number, equal to the Szekeres-Wilf number, was implicitly used by Vizing already in 1965.

Theorem 4.3 (Vizing [299] 1965) *Every simple graph G satisfies the following statements:*

   (a) *If G is a critical graph of class two, then $\Delta(G) \leq 2\,\mathrm{col}(G) - 3$.*

   (b) *If $\Delta(G) \geq 2\,\mathrm{col}(G) - 2$, then $\chi'(G) = \Delta(G)$.*

Proof: To prove (a), suppose on the contrary that there is a critical graph G of class two with $\Delta(G) \geq 2\mathrm{col}(G) - 2$. Let $d = \mathrm{col}(G)$. Since $\chi'(G) = \Delta(G) + 1$, $d \geq 2$ and so $\Delta(G) \geq 2d - 2 > d - 1$. Let $Y = \{y \in V(G) \mid d_G(y) \leq d - 1\}$. Since $\mathrm{col}(G) = d$ and $\Delta(G) > d - 1$, both sets Y and $V(G) \setminus Y$ are nonempty. Furthermore, the subgraph $H = G - Y$ contains a vertex x with $d_H(x) \leq d - 1$. Since $x \notin Y$, we have $d_G(x) \geq d$. Consequently, there is an edge $e \in E_G(x, y)$ for some vertex $y \in Y$. Now consider the set Z of all vertices $z \neq y$ such that z is adjacent to x in G and $d_G(z) = \Delta(G)$. Since $\Delta(G) > d - 1$, $Z \subseteq V(G) \setminus Y$, which gives $|Z| \leq d_H(x) \leq d - 1$. By Vizing's Adjacency Lemma (Theorem 2.7), however, $|Z| \geq \Delta(G) - d_G(y) + 1 \geq (2d - 2) - (d - 1) + 1 = d$. This contradiction proves statement (a).

In order to prove (b), let G be a simple graph with $\Delta(G) \geq 2\,\mathrm{col}(G) - 2$. If $\chi'(G) \neq \Delta(G)$, then, clearly, $\chi'(G) = \Delta(G) + 1$ and, by Proposition 4.1, there is a critical subgraph H of G with $\chi'(H) = \chi'(G)$ and $\Delta(H) = \Delta(G)$. Then $\Delta(H) = \Delta(G) \geq 2\,\mathrm{col}(G) - 2 \geq 2\,\mathrm{col}(H) - 2$, contradicting (a).   ∎

A vertex of an arbitrary graph G is called a **major vertex** of G if it has maximum degree in G; otherwise it is called a **minor vertex** of G.

It is easy to show that, for a critical graph G with $\chi'(G) \geq \Delta(G) + 1$, the vertex connectivity of G as well as the edge connectivity of G is at least two. That these two bounds are best possible even for critical graphs of class two having arbitrarily large chromatic index is also well known. Examples of such graphs can be found in [100], see also Sect. 4.7. However, Thomassen [293] proved that the maximum local edge connectivity of a graph of class two is equal to its maximum degree:

Theorem 4.4 (Thomassen [293] 2007) *Every graph G of class two contains two major vertices joined by $\Delta(G)$ edge-disjoint paths in G.*

Proof: Since every graph of class two contains a critical graph of class two with the same maximum degree, we may assume that G is a critical graph of class two. Then we have $\Delta(G) \geq 2$ and, by Vizing's Adjacency Lemma, every vertex of G is adjacent to at least two major vertices.

By an *edge-cut* of G we mean a pair (X, X') of two nonempty disjoint subsets of $V(G)$ such that $X \cup X' = V(G)$. An edge-cut (X, X') of G is called *good* if there are two major vertices $x \in X$ and $x' \in X'$ joined by $|E_G(X, X')|$ edge-disjoint paths in G. Since G has at least two major vertices, it follows from Menger's theorem that there is a good edge-cut in G. Among all good edge-cuts of G, let (X, X') be one with a minimum number m of major vertices in X. Since (X, X') is a good edge-cut, there are two major vertices $x \in X$ and $x' \in X'$ joined by $|E_G(X, X')|$ edge-disjoint paths in G.

Now we claim that $m = 1$. In order to prove the claim, assume it is false. Then, clearly, $m \geq 2$ and there are two distinct major vertices of G contained in X, say y and y'. Let H be the graph obtained from G by contracting the set X', that is, we replace in G the set X' by a new vertex z and join this new vertex to every vertex $u \in X$ by $|E_G(u, X')|$ edges. Let t denote the maximum number of edge-disjoint paths between y and y' in H. By Menger's theorem, there is a partition (Y, Y') of

$V(H)$ such that $Y \cap Y' = \emptyset$, $y \in Y$, $y' \in Y'$, and $|E_H(Y, Y')| = t$. By symmetry, we may assume that $z \in Y'$. Now, let $Y'' = (Y' \cup X') \setminus \{z\}$. Then $|E_G(Y, Y'')| = t$.

Since there are $|E_G(X, X')|$ edge-disjoint paths in G joining x and x', we conclude that, for every edge $e \in E_H(u, z)$ with $u \in X$, there is a path P_e joining u and x' in $G[\{u\} \cup X']$ and all these paths are edge-disjoint. Consequently, y and y' are joined by t edge-disjoint paths in G. This implies, in particular, that (Y, Y'') is a good edge-cut in G. Since $Y \subseteq X \setminus \{y'\}$, the number of major vertices of G contained in Y is less than m. This is a contradiction.

Hence, as claimed, $m = 1$ and, therefore, x is the only major vertex of G contained in X. Let d be the number of neighbors of x in G belonging to X. Since G is a simple graph, the major vertex x has $\Delta(G) - d$ neighbors in X'. By Vizing's Adjacency Lemma, every vertex of G is adjacent to at least two major vertices of G. Hence, each of the d neighbors of x belonging to X has at least one neighbor in X'. Consequently, $|E_G(X, X')| \geq \Delta(G)$, which implies that there are $\Delta(G)$ edge disjoint paths joining x and x' in the graph G. ∎

4.2 GRAPHS WHOSE CORE HAS MAXIMUM DEGREE TWO

For an arbitrary graph G and an integer $p \geq 0$, we denote by $G^{[p]}$ the subgraph of G induced by the set of all vertices having degree p in G. Instead of $G^{[\Delta(G)]}$ we briefly write $G^{[\Delta]}$. This graph is said to be the **major subgraph** of G or, as in several papers, the **core** of G.

The **Vizing-core** $VC(G)$ of a simple graph G is the unique maximal subgraph H of G such that H is empty or $2 \leq \delta(H) \leq \Delta(H) = \Delta(G)$ and every edge $e \in E_H(x, y)$ satisfies $|(N_H(x) \setminus \{y\}) \cap V(H^{[\Delta]})| \geq \Delta(H) - d_H(y) + 1$. As an immediate consequence of Vizing's Adjacency Lemma and Proposition 4.1 we obtain the following result.

Proposition 4.5 *Let G be a simple graph, and let H be the Vizing-core of G. Then G is of class one if and only if H is of class one.*

Proof: If G is of class one, then H is of class one, since H is empty or a subgraph of G with $\Delta(H) = \Delta(G)$. If G is of class two, then G contains a critical graph G' of class two with $\Delta(G') = \Delta(G)$. Then Vizing's Adjacency Lemma and the definition of H imply that $G' \subseteq H$ and, therefore, H is of class two. ∎

Remark 4.6 *The Vizing-core $VC(G)$ of a simple graph G can be constructed by repeated application of the following two rules:*

(1) *Suppose $x \in V(G)$ has at most one neighbor in $G^{[\Delta]}$. If $\Delta(G - x) < \Delta(G)$ then $VC(G) = \emptyset$ else $VC(G) = VC(G - x)$.*

(2) *Suppose $e \in E_G(x, y)$ satisfies $|N_G(x) \setminus \{y\} \cap V(G^{[\Delta]})| < \Delta(G) - d_G(y) + 1$. If $\Delta(G - e) < \Delta(G)$ then $VC(G) = \emptyset$ else $VC(G) = VC(G - e)$.*

If G is a simple graph whose major subgraph is a forest, then $VC(G) = \emptyset$ and, therefore, graph G is of class one. This simple consequence of Vizing's Adjacency

Lemma was rediscovered by Fournier [102, 103] between 1973 and 1977. There are various other sufficient conditions for a graph to be of class one, depending on its major subgraph (see, e.g., Hoffman and Rodger [148], Chetwynd and Hilton [49, 53, 54], or Niessen and Volkmann [236]). Many of these results are simple consequences of Vizing's Adjacency Lemma, respectively Proposition 4.5. Chetwynd, Hilton, and Hoffman [55] proved in 1989 that, for a simple graph H, there is a critical graph G of class two with $G^{[\Delta]} = H$ if and only if $\delta(H) \geq 2$.

Hilton and Zhao [142, 143] considered the problem of classifying graphs whose major subgraph has maximum degree at most 2. A graph G is said to be a **Hilton graph** if G is a connected graph of class two with $\Delta(G^{[\Delta]}) \leq 2$. All known Hilton graphs are elementary, except the graph P^* obtained from the Petersen graph by deleting one vertex. Hilton and Zhao [143] proposed the following conjecture.

Conjecture 4.7 (Hilton and Zhao [143] 1996) *Every Hilton graph $G \neq P^*$ is an elementary graph of class two.*

Figure 4.1   The only graph from $\mathcal{H}_2$ and the only graph from $\mathcal{H}_3$.

For an integer $p \geq 2$, let $\mathcal{H}_p$ denote the family of all graphs obtained from a complete bipartite graph $K_{p,q}$ with $3 \leq q \leq p+1$ and $p+q$ odd by inserting on the set of q independent vertices a 2-regular simple graph with q vertices and on the set of p independent vertices a $(p+1-q)$-regular simple graph with p vertices. It is straightforward to check that every graph in $\mathcal{H}_p$ is an elementary Hilton graph with maximum degree $p+2$. Before we prove that these are the only elementary Hilton graphs with maximum degree $p+2$, we need some basic properties of Hilton graphs.

Lemma 4.8 (Hilton and Zhao [142] 1992) *Every Hilton graph G satisfies the following statements:*

(a) *G is a critical graph of class two and $\delta(G^{[\Delta]}) = \Delta(G^{[\Delta]}) = 2$.*

(b) *$\delta(G) = \Delta(G) - 1$, or $\Delta(G) = 2$ and G is an odd cycle.*

(c) *Every vertex of G has at least two neighbors in $G^{[\Delta]}$.*

Proof: Let G be a Hilton graph with maximum degree Δ. Since $\chi'(G) = \Delta + 1$, we have $\Delta \geq 2$. Since G is simple and connected, $\Delta = 2$ implies that G is an odd cycle and we are done. Hence, in what follows, assume that $\Delta \geq 3$. Since $\Delta(G^{[\Delta]}) \leq 2$, this implies that the set of minor vertices $X = V(G) \setminus V(G^{[\Delta]})$ is nonempty.

Since G is of class two, there is a critical graph $H \subseteq G$ of class two with $\Delta(H) = \Delta$. Then $H^{[\Delta]}$ is a subgraph of $G^{[\Delta]}$ and, therefore, H is a Hilton graph with maximum degree Δ, too. Since H is critical, Vizing's Adjacency Lemma implies that every vertex of H has at least two neighbors in $H^{[\Delta]}$. Hence $\delta(H^{[\Delta]}) = \Delta(H^{[\Delta]}) = 2$. Since $\Delta \geq 3$, this implies that the set of minor vertices $Y = V(H) \setminus V(H^{[\Delta]})$ is nonempty. Let $y \in Y$ be a minor vertex of H. Then y is adjacent to some major vertex x of H. By Vizing's Adjacency Lemma, x is adjacent to at least $\Delta - d_H(y) + 1$ major vertices of H. Since x is in H adjacent to two major vertices, this implies that $d_H(y) = \Delta - 1$. Hence, to complete the proof, it suffices to show that $G = H$. Suppose this is false. Since G is connected, this implies that there is an edge $e \in E_G(u, v)$ such that $e \notin E(H)$ and $u \in V(H)$. Hence u is a minor vertex of H and $d_H(u) = \Delta - 1$. Consequently, u is a major vertex of G. Furthermore, u is adjacent to some major vertex x of H. Since x has two neighbors in $H^{[\Delta]}$ and H is a subgraph of G, this implies that x has at least three neighbors in $G^{[\Delta]}$. Since x is a major vertex of G and G is a Hilton graph, this is a contradiction. ∎

Lemma 4.9 *Let G be an elementary Hilton graph. Then $\Delta(G) = 2$ and G is an odd cycle, or $\Delta(G) \geq 4$ and $G \in \mathcal{H}_{\Delta(G)-2}$.*

Proof: Let G be an elementary Hilton graph and let $\Delta = \Delta(G)$. Then $\Delta \geq 2$ and hence $\Delta = p + 2$ for some integer $p \geq 0$. Clearly, $\Delta = 2$ implies that G is an odd cycle. Hence, in what follows, assume that $\Delta \geq 3$.

Let $A = V(G^{[\Delta]})$ and $B = V(G) \setminus A$. By Lemma 4.8, G is a critical graph of class two, $G[A] = G^{[\Delta]}$ is 2-regular, $B \neq \emptyset$ and each vertex in B is of degree $\Delta - 1$ in G and has at least two neighbors in A. Obviously, there is an edge $e \in E_G(x, y)$ such that $x, y \in A$. Since G is critical and $\chi'(G) = \Delta + 1$, there is a coloring $\varphi \in \mathcal{C}^{\Delta}(G - e)$. Since G is elementary, it follows from Theorem 1.4 that the vertex set $V = A \cup B$ of G is elementary with respect to φ. Since the uncolored edge joins two major vertices and each vertex in B has degree $\Delta - 1$, this implies that $|\overline{\varphi}(A)| = 2, |\overline{\varphi}(B)| = |B|$, and $|\overline{\varphi}(V)| = |\overline{\varphi}(A)| + |\overline{\varphi}(B)| \leq \Delta$. This implies that $|B| \leq \Delta - 2$. Since G is simple and $G[A] = G^{[\Delta]}$ is 2-regular, $q = |A| \geq 3$ and each vertex in A has $\Delta - 2$ neighbors in B. Hence $|B| = \Delta - 2 = p$ and each vertex in A is completely joined to B. Then $3 \leq q \leq \Delta - 1 = p + 1$ and $G[B]$ is a r-regular graph with $r = \Delta - q - 1 = p + 1 - q$. By Proposition 1.3, it follows that $|V| = p + q$ is odd. Consequently, $\Delta - 2 = p \geq 2$ and G belongs to $\mathcal{H}_p$. ∎

As an immediate consequence of Lemma 4.9 we deduce that Conjecture 4.7 is equivalent to the following conjecture.

Conjecture 4.10 *Let G be a Hilton graph. Then $\Delta(G) = 2$ and G is an odd cycle, or $\Delta(G) = 3$ and $G = P^*$, or $\Delta(G) \geq 4$ and $G \in \mathcal{H}_{\Delta(G)-2}$.*

Not much progress has been made since the conjecture was posed by Hilton and Zhao in 1996. A breakthrough was achieved in 2003, when D. Cariolaro and G. Cariolaro [41] settled the first nontrivial case $\Delta = 3$. Unfortunately, however, their method does not apply to any case $\Delta \geq 4$. Our proof for the case $\Delta = 3$ uses a different approach due to Král', Sereni, and Stiebitz [183].

First, we introduce some simplified notation for simple graphs. An edge of a simple graph G joining two vertices u, v is denoted by uv or vu. For each path P of a simple graph G we fix an orientation of P by choosing one of its endvertices as the first vertex of P, denoted by $x_1(P)$. If P consists of $p \geq 1$ vertices, then we denote by $x_1(P), \ldots, x_p(P)$ the vertices of P such that $P = \mathrm{Path}(x_1(P), x_1(P)x_2(P), x_2(P), \ldots, x_{p-1}(P), x_{p-1}(P)x_p(P), x_p(P))$. In this case we also write $P = (x_1(P), \ldots, x_p(P))$.

Theorem 4.11 (D. Cariolaro and G. Cariolaro [41] 2003) *The only Hilton graph with maximum degree $\Delta = 3$ is P^*, the Petersen graph with one vertex deleted.*

Proof: Let G be a Hilton graph with $\Delta(G) = 3$. Then Lemma 4.8 implies the following facts: G is a critical graph with $\chi'(G) = 4$, each major vertex of G is adjacent with two other major vertices and one minor vertex, and each minor vertex of G is of degree two and both of its neighbors are major vertices.

Since G is a critical graph with $\chi'(G) = 4$, every proper subgraph of G admits a 3-edge-coloring. In what follows, we denote the three colors of such a coloring by α, β and γ. Theorem 2.1 implies the following result.

Claim 4.11.1 *Let $uv \in E(G)$ and let φ be a 3-edge-coloring of $G - uv$. Then $\overline{\varphi}(u) \cap \overline{\varphi}(v) = \emptyset$. Furthermore, if $\delta \in \overline{\varphi}(u)$ and $\varepsilon \in \overline{\varphi}(v)$, then $P_u(\delta, \varepsilon, \varphi)$ is a path whose endvertices are u and v.*

In what follows, we will consider an arbitrary edge $xy \in E(G)$ such that x is a minor vertex of G. Then y is a major vertex of G. Let z denote the other neighbor of x in G. Clearly, z is a major vertex, too.

Let $\mathcal{P}$ be the set of all 5-tuples (φ, P, p, Q, q) such that the following conditions hold:

(1) φ is a 3-edge-coloring of $G - xy$ such that $\alpha, \beta \in \overline{\varphi}(x)$ and $\gamma \in \overline{\varphi}(y)$.

(2) $P = P_y(\alpha, \gamma, \varphi)$, $|V(P)| = p$, and $x_1(P) = y$.

(3) $Q = P_y(\beta, \gamma, \varphi)$, $|V(Q)| = q$, and $x_1(Q) = y$.

By Claim 4.11.1, $\mathcal{P} \neq \emptyset$ and, for every tuple $(\varphi, P, p, Q, q) \in \mathcal{P}$, we have $x_p(P) = x_q(Q) = x$ and, therefore, $x_{p-1}(P) = x_{q-1}(Q) = z$. Furthermore, $p, q \geq 3$ and $x_1(P), x_2(P), x_1(Q), x_2(Q)$ are all major vertices.

Claim 4.11.2 *There exists a tuple $(\varphi, P, p, Q, q) \in \mathcal{P}$ such that $x_3(P)$ or $x_3(Q)$ is a major vertex.*

Proof: Choose an arbitrary tuple $(\varphi, P, p, Q, q) \in \mathcal{P}$ and suppose that booth $x_3(P)$ and $x_3(Q)$ are minor vertices (see Fig. 4.2). Since G is a simple graph, either p or q is at least 4, say $q \geq 4$. Since $u = x_3(Q)$ is a minor vertex, $\overline{\varphi}(u) = \{\alpha\}$. Since $x_2(Q)$ is a major vertex, it follows that there is a neighbor v of $x_2(Q)$ such that $v \notin \{x_1(Q), x_3(Q)\}$ and v is a major vertex. Clearly, $P_u = P_u(\alpha, \gamma, \varphi)$ is

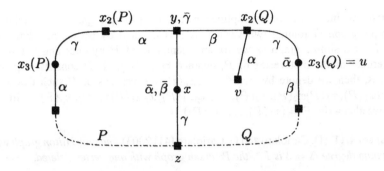

Figure 4.2   A tuple $(\varphi, P, p, Q, q) \in \mathcal{P}$.

disjoint from $P = P_y(\alpha, \gamma, \varphi)$ and $x_2(Q)v$ belongs to P_u. Then $\varphi' = \varphi/P_u$ is an 3-edge-coloring of $G - xy$ such that $\alpha, \beta \in \overline{\varphi}'(x)$, $\gamma \in \overline{\varphi}'(y)$, and $P = P_y(\alpha, \gamma, \varphi')$. Let $Q' = P_y(\beta, \gamma, \varphi')$ such that $x_1(Q') = y$. Then $x_3(Q') = v$ is a major vertex and the tuple $(\varphi', P, p, Q', |V(Q')|)$ belongs to $\mathcal{P}$ and has the desired property. This proves the claim. $\triangle$

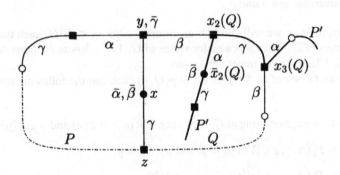

Figure 4.3   A tuple $(\varphi, P, p, Q, q) \in \mathcal{P}'$.

Let $\mathcal{P}^*$ be the set of all tuples $(\varphi, P, p, Q, q) \in \mathcal{P}$ such that $x_3(Q)$ is a major vertex. By Claim 4.11.2, $\mathcal{P}^* \neq \emptyset$. Consider a tuple $(\varphi, P, p, Q, q) \in \mathcal{P}^*$. Since $x_1(Q), x_2(Q)$ and $x_3(Q)$ are major vertices, there is a neighbor v of $x_2(Q)$ distinct from $x_1(Q)$ and $x_3(Q)$. We denote v by $\overline{x}_2(Q)$. Clearly, $\overline{x}_2(Q)$ is a minor vertex. Since $Q = P_y(\beta, \gamma, \varphi)$, we obtain that $\varphi(x_2(Q)\overline{x}_2(Q)) = \alpha$ and $\overline{\varphi}(\overline{x}_2(Q)) = \{\delta\}$, where $\delta \in \{\beta, \gamma\}$. If $\delta = \gamma$, then $Q' = P_{\overline{x}_2(Q)}(\beta, \gamma, \varphi)$ is disjoint from Q and the coloring $\varphi' = \varphi/Q'$ is a 3-edge-coloring of $G - xy$ such that $\alpha, \beta \in \overline{\varphi}'(x)$, $\gamma \in \overline{\varphi}'(y)$, and $P_y(\beta, \gamma, \varphi') = Q$. Furthermore, we have $\overline{\varphi}'(\overline{x}_2(Q)) = \{\beta\}$. This shows that there is a tuple $(\varphi, P', p', Q, q) \in \mathcal{P}^*$ such that $\overline{\varphi}(\overline{x}_2(Q)) = \{\beta\}$. Let $\mathcal{P}'$ denote the set of all such tuples in $\mathcal{P}^*$.

Claim 4.11.3 *Let* $T = (\varphi, P, p, Q, q) \in \mathcal{P}'$ *be an arbitrary tuple. Then* $x_2(Q) = x_k(P)$ *and* $x_3(Q) = x_{k-1}(P)$, *where* $3 \leq k \leq p - 3$. *Furthermore,* $\overline{x}_2(Q) = x_{k+1}(P)$.

Proof: Since $T \in \mathcal{P}'$, we have $\beta \in \overline{\varphi}(\overline{x}_2(Q))$ and $\varphi(x_2(Q)\overline{x}_2(Q)) = \alpha$. Now consider the chain $P' = P_{\overline{x}_2(Q)}(\alpha, \gamma, \varphi)$ (see Fig. 4.3). We claim that $P' = P$, where $P = P_y(\alpha, \gamma, \varphi)$. Otherwise, P and P' are disjoint and $\varphi' = \varphi/P'$ is a 3-edge-coloring of $G - xy$ such that $\alpha, \beta \in \overline{\varphi}'(x)$, $\gamma \in \overline{\varphi}'(y)$, and $P_y(\alpha, \gamma, \varphi') = P$. Then, $P_y(\beta, \gamma, \varphi') = (x_1(Q) = y, x_2(Q), \overline{x}_2(Q))$ where $\overline{x}_2(Q) \neq x$. This is a contradiction to Claim 4.11.1. Hence the claim that $P = P'$ is proved.

For the edge $e = x_2(Q)x_3(Q)$, we have $\varphi(e) = \gamma$ and $e \in E(P') = E(P)$. This implies that $x_2(Q) = x_k(P)$ for some integer k and $x_3(Q) \in \{x_{k-1}(P), x_{k+1}(P)\}$. If $x_3(Q) = x_{k+1}(P)$, then $\overline{x}_2(Q) = x_{k-1}(Q)$ and, therefore, $k \geq 4$. Now, uncolor the edge $x_k(P)x_{k-1}(P)$, color the edge $xx_1(P)$ by α and, for $i \in \{1, \ldots, k-2\}$, color the edge $x_i(P)x_{i+1}(P)$ by $\varphi(x_{i+1}(P)x_{i+2}(P))$. This results in a 3-edge-coloring φ' of $G - x_{k-1}(P)x_k(P)$ such that $\alpha \in \overline{\varphi}'(x_k(P))$, $\beta \in \overline{\varphi}'(x)$, and $\beta \in \overline{\varphi}'(x_{k-1}(P))$. Note that $x_{k-1}(P) = \overline{x}_2(Q)$ and $\beta \in \overline{\varphi}(\overline{x}_2(Q))$. Then $P^* = (x, x_1(P), x_k(P))$ equals $P_{x_k(P)}(\alpha, \beta, \varphi')$ and $x_{k-1}(P) \notin V(P^*)$, a contradiction to Claim 4.11.1. This contradiction shows that $x_3(Q) = x_{k-1}(P)$. Since $x_2(Q)\overline{x}_2(Q) \in E(P') = E(P)$, this implies that $\overline{x}_2(Q) = x_{k+1}(P)$ and $3 \leq k \leq p - 3$. In what follows, we call k the *index* of the tuple T.   △

Claim 4.11.4 *Let* $(\varphi, P, p, Q, q) \in \mathcal{P}'$ *be an arbitrary tuple with index* k. *Then* $x_3(P)$ *is a minor vertex and* $k \geq 5$.

Proof: Suppose that $x_3(P)$ is a major vertex. Since both $x_1(P)$ and $x_2(P)$ are major vertices, this implies that $x_2(P)$ is adjacent to some minor vertex $u \notin \{x_1(P), x_3(P)\}$. Then $\varphi(x_2(P)u) = \beta$ and $\overline{\varphi}(u) = \{\delta\}$, where $\delta \in \{\gamma, \alpha\}$.

First, suppose that $\delta = \alpha$. Since, by Claim 4.11.3, $v = \overline{x}_2(Q) = x_{k+1}(Q)$ and $\beta \in \overline{\varphi}(v)$, it follows then that the chain $P' = P_u(\alpha, \beta, \varphi)$ satisfies $P' = (u, x_2(P), x_1(P), x_k(P), v)$. Then, for the coloring $\varphi' = \varphi/P'$, we have $\gamma \in \overline{\varphi}(y) = \overline{\varphi}'(y)$, $\alpha, \beta \in \overline{\varphi}(x) = \overline{\varphi}'(x)$, and $\alpha \in \overline{\varphi}'(v)$. This implies that $P_x(\alpha, \gamma, \varphi')$ is a subpath of P with endvertices x and v, a contradiction to Claim 4.11.1.

Now, suppose that $\delta = \gamma$. Then $P' = P_u(\alpha, \gamma, \varphi)$ is disjoint from $P = P_y(\alpha, \gamma, \varphi)$. Hence, for the coloring $\varphi' = \varphi/P'$, we have $\alpha, \beta \in \overline{\varphi}'(x)$, $\gamma \in \overline{\varphi}'(y)$, and $P = P_y(\alpha, \gamma, \varphi')$. Let $Q' = P_y(\beta, \gamma, \varphi')$ with $x_1(Q') = y$ and $q' = |V(Q')|$. Then $x_j(Q') = x_j(Q)$ for $j = 1, 2, 3, 4$ and $\overline{x}_2(Q') = \overline{x}_2(Q)$. This implies that $(\varphi', P, p, Q', q') \in \mathcal{P}'$, where $\alpha \in \overline{\varphi}'(u)$. Hence, we obtain a contradiction to Claim 4.11.1 as in the previous case.

This proves that $x_3(P)$ is a minor vertex. Since $x_k(P) = x_2(Q)$ is a major vertex, we conclude that $k \geq 5$.   △

Claim 4.11.5 *Let* $(\varphi, P, p, Q, q) \in \mathcal{P}'$ *be an arbitrary tuple with index* k. *Furthermore, let* $u = x_3(P)$ *and* $v = x_{k+1}(P)$. *Then* $P_u(\beta, \gamma, \varphi)$ *is a path with endvertices* u *and* v.

Proof: It follows from Claim 4.11.3 and Claim 4.11.4 that $v = \bar{x}_2(Q)$ and u is a minor vertex. Hence $\overline{\varphi}(u) = \overline{\varphi}(v) = \{\beta\}$ and $P' = P_u(\beta, \gamma, \varphi)$ is a path, where one endvertex is u and the other endvertex is some vertex $u' \neq u$. Suppose that $u' \neq v$. Consider the coloring $\varphi' = \varphi/P'$. Since P' is disjoint from $Q = P_y(\beta, \gamma, \varphi)$, we have $\gamma \in \overline{\varphi}(y) = \overline{\varphi}'(y)$, $\alpha, \beta \in \overline{\varphi}(x) = \overline{\varphi}'(x)$, $\overline{\varphi}'(v) = \overline{\varphi}(v) = \{\beta\}$, and $P_y(\beta, \gamma, \varphi') = Q$. Let $P' = P_y(\alpha, \gamma, \varphi')$ with $x_1(P') = y$. Then Claim 4.11.1 implies that $(\varphi', P', |V(P')|, Q, q) \in \mathcal{P}'$. Furthermore, we have $x_2(P') = x_2(P)$ and $x_3(P')$ is the neighbor of $x_2(P)$ distinct from $x_3(P)$ and $x_1(P)$. Since $x_3(P)$ is a minor vertex, we obtain that $x_3(P')$ is a major vertex, a contradiction to Claim 4.11.4.   △

Claim 4.11.6 *Let* $(\varphi, P, p, Q, q) \in \mathcal{P}'$ *be an arbitrary tuple with index* k. *Then* $x_4(Q)$ *is a major vertex and* $k = 5$.

Proof: First, suppose that $w = x_4(Q)$ is a minor vertex. This implies that $w \neq x$ and $\overline{\varphi}(w) = \{\alpha\}$. Hence $P' = P_w(\alpha, \gamma, \varphi)$ is disjoint from P. Then, for the coloring $\varphi' = \varphi/P'$, we obtain $\gamma \in \overline{\varphi}(y) = \overline{\varphi}'(y)$, $\alpha, \beta \in \overline{\varphi}(x) = \overline{\varphi}'(x)$, and $P_w(\beta, \gamma, \varphi') = (w, x_3(Q), x_2(Q), x_1(Q))$. This is a contradiction to Claim 4.11.1. Hence $w = x_4(Q)$ is a major vertex.

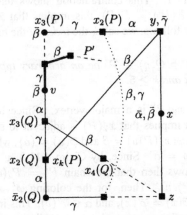

Figure 4.4   A tuple $(\varphi, P, p, Q, q) \in \mathcal{P}'$ with $k \neq 5$.

Now, suppose that $k \neq 5$. Then, by Claim 4.11.4, $k \geq 6$. This implies that $v = x_{k-2}(P) \neq x_3(P)$. Since the vertices $x_{k-1}(P)$, $x_k(P)$ and $x_4(Q)$ are major vertices, it follows that v is a minor vertex and hence $\overline{\varphi}(v) = \{\beta\}$ (see Fig. 4.4). Consider the chain $P' = P_v(\beta, \gamma, \varphi)$ and the coloring $\varphi' = \varphi/P'$. Then P' is distinct from $P_y(\beta, \gamma, \varphi) = Q$ and, by Claim 4.11.5, neither $x_3(P)$ nor $x_{k+1}(P)$ belongs to P'. Therefore, we have $\alpha, \beta \in \overline{\varphi}(x) = \overline{\varphi}'(x)$, $\gamma \in \overline{\varphi}(y) = \overline{\varphi}(y')$, $P_y(\beta, \gamma, \varphi') = Q$, and $\overline{\varphi}'(\bar{x}_2(Q)) = \overline{\varphi}(\bar{x}_2(Q)) = \{\beta\}$. Consequently, if $\tilde{P} = P_y(\alpha, \gamma, \varphi')$ with $x_1(\tilde{P}) = y$ and $\tilde{p} = |\tilde{P}|$, then Claim 4.11.1 implies that $(\tilde{P}, \tilde{p}, Q, q, \varphi') \in \mathcal{P}'$. Clearly, $\overline{\varphi}'(v) = \{\gamma\}$ and $\tilde{P}$ is distinct from $P_v(\alpha, \gamma, \varphi')$. This implies that neither $x_2(Q)$ nor $x_3(Q)$ belongs to $\tilde{P}$, a contradiction to Claim 4.11.3.   △

Claim 4.11.7 *Let* $(\varphi, P, p, Q, q) \in \mathcal{P}'$ *be an arbitrary tuple with index* k. *Then* $x_4(Q) = z$.

Proof: Assume that $x_4(Q) \neq z$. Then $v = x_5(Q)$ is distinct from $x = x_q(Q)$ and $\varphi(x_4(Q)v) = \gamma$. From Claim 4.11.3, Claim 4.11.4 and Claim 4.11.6 it follows that $x_3(P)$ is a minor vertex with $\overline{\varphi}(x_3(P)) = \{\beta\}$, $x_4(P) = x_3(Q)$, and $x_5(P) = x_2(Q)$.

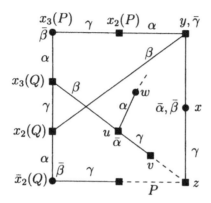

Figure 4.5   A tuple $(\varphi, P, p, Q, q) \in \mathcal{P}'$ with $x_4(Q) \neq z$.

First, consider the case that v is a major vertex. Since, by Claim 4.11.6, $u = x_4(Q)$ is a major vertex, u has a neighbor $w \notin \{x_3(Q), v\}$. Then $\varphi(uw) = \alpha$ and, since $x_3(Q)$ is a major vertex, w is a minor vertex and hence $\overline{\varphi}(w) = \{\delta\}$ with $\delta \in \{\beta, \gamma\}$ (see Fig. 4.5). If $\delta = \beta$, then $P_w = P_w(\beta, \alpha, \varphi)$ satisfies $P_w = (w, u, x_3(Q), x_3(P))$. Hence, for the coloring $\varphi' = \varphi/P_w$, we have $\gamma \in \overline{\varphi}(y) = \overline{\varphi}'(y)$, $\alpha, \beta \in \overline{\varphi}(x) = \overline{\varphi}'(x)$, $\overline{\varphi}'(x_3(P)) = \{\alpha\}$, and $P_y(\alpha, \gamma, \varphi') = (x_1(P), x_2(P), x_3(P))$. This is a contradiction to Claim 4.11.1. If $\delta = \gamma$, then we argue as follows. Since $\gamma \in \overline{\varphi}(w)$, the chain $P_w = P_w(\beta, \gamma, \varphi)$ is distinct from the chain $Q = P_y(\beta, \gamma, \varphi)$. By Claim 4.11.5, $x_3(P)$ does not belong to P_w. For the coloring $\varphi' = \varphi/P_w$, we then have $P'_w = P_w(\beta, \alpha, \varphi') = (w, u, x_3(Q), x_3(P))$ and, for the coloring $\varphi'' = \varphi'/P'_w$, we have $P_y(\alpha, \gamma, \varphi'') = (x_1(P), x_2(P), x_3(P))$. However, this is a contradiction to Claim 4.11.1.

It remains to consider the case that v is a minor vertex. Then $\overline{\varphi}(v) = \{\alpha\}$ and the chain $P' = P_v(\alpha, \gamma, \varphi)$ is distinct from $P = P_y(\alpha, \gamma, \varphi)$. Hence, for the coloring $\varphi' = \varphi/P'$, we have $\alpha, \beta \in \overline{\varphi}(x) = \overline{\varphi}'(x)$, $\gamma \in \overline{\varphi}(y) = \overline{\varphi}(y')$, and $P_y(\alpha, \gamma, \varphi') = P$. Let $Q' = P_y(\beta, \gamma, \varphi')$ be the chain with $x_1(Q') = y$ and let $q' = |Q'|$. Then $x_2(Q') = x_2(Q), x_3(Q') = x_3(Q), x_4(Q') = x_4(Q)$, $\overline{x}_2(Q') = \overline{x}_2(Q) = x_{k+1}(P)$, and $\overline{\varphi}'(x_{k+1}(P)) = \overline{\varphi}(x_{k+1}(P)) = \{\beta\}$. By Claim 4.11.1, x and y are the endvertices of the path Q'. Consequently, we have $(P, p, Q', q', \varphi') \in \mathcal{P}'$, where $x_4(Q') \neq z$ and $x_5(Q') = w$ is a major vertex. Then we can continue as in the former case to derive a contradiction.   △

Claim 4.11.8 *Let* $(\varphi, P, p, Q, q) \in \mathcal{P}'$ *be an arbitrary tuple with index* k. *Then* $k = 5$, $q = 5$, *and* $p = 9$.

Proof: By Claim 4.11.6 and Claim 4.11.7, $k = 5$ and $q = 5$. Let φ' be the coloring such that $\varphi'(e) = \varphi(e)$ for all edges $e \notin \{xy, xz\}$ and $\varphi'(xy) = \gamma$. Then the edge xz is uncolored and we can change the role of y and z. Let $P' = P_z(\alpha, \gamma, \varphi')$ with $x_1(P') = z$ and $Q' = P_z(\beta, \gamma, \varphi')$ with $x_1(Q') = z$. Then we have $P' = (z = x_{p-1}(P), x_{p-2}(P), \ldots, x_1(P), x_p(P) = x)$ and $Q' = (z, x_4(Q), x_3(Q), x_2(Q), x_1(Q), x_5(Q) = x)$ with $\overline{x}_2(Q') = x_3(P)$. Since $\overline{\varphi}'(x_3(P)) = \overline{\varphi}(x_3(P)) = \{\beta\}$, we can now apply Claim 4.11.6 to the tuple (φ', P', p, Q', q). Then we conclude that $p = 9$. $\triangle$

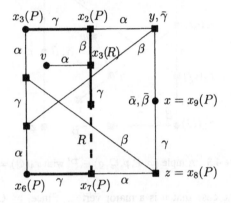

Figure 4.6 A tuple from $(\varphi, P, 9, Q, 5) \in \mathcal{P}'$.

Claim 4.11.9 *Let* $(\varphi, P, p, Q, q) \in \mathcal{P}'$ *be an arbitrary tuple with index* k. *Then* $x_2(P)x_7(P)$ *is an edge of* G *and, therefore,* $G = P^*$.

Proof: By Claim 4.11.8, $k = 5$, $q = 5$, and $p = 9$. By Claim 4.11.5, $x_3(P)$ and $x_6(P) = \overline{x}_2(Q)$ are the endvertices of a (β, γ)-chain R with respect to φ. Clearly, both edges $x_3(P)x_2(P)$ and $x_6(P)x_7(P)$ belong to the chain R.

Now, suppose that $x_2(P)x_7(P)$ is not an edge of G. Then $|V(R)| \geq 6$. Let $x_1(R) = x_3(P)$. Then $x_2(R) = x_2(P)$, and neither $x_3(R)$ nor $x_4(R)$ belongs to P. By Claim 4.11.4, $x_3(P)$ is a minor vertex. Consequently, $x_3(R)$ is a major vertex and $\varphi(x_2(R)x_3(R)) = \beta$. Hence $x_3(R)$ has a neighbor v not belonging to R (see Fig. 4.6). Clearly, $v \notin V(P)$ and $\varphi(vx_3(R)) = \alpha$. We may assume that v is a minor vertex. Otherwise, $v' = x_4(R)$ is a minor vertex, $v' \notin V(P)$, and $\varphi(v'x_3(R)) = \gamma$. Then the chain $P' = P_v(\alpha, \gamma, \varphi)$ is distinct from P and if we recolor this chain, we are in the same situation with v' instead of v.

Since v is a minor vertex, $\overline{\varphi}(v) = \{\delta\}$ with $\delta \in \{\beta, \gamma\}$. If $\delta = \gamma$, then the coloring $\varphi' = \varphi/R$ satisfies $\alpha, \beta \in \overline{\varphi}(x) = \overline{\varphi}'(x)$, $\gamma \in \overline{\varphi}(y) = \overline{\varphi}(y')$, and $P_y(\alpha, \gamma, \varphi') = (y, x_2(P), x_3(R), v)$. This is a contradiction to Claim 4.11.1. If $\delta = \beta$, then $P_v = P_v(\alpha, \beta, \varphi) = (v, x_3(R), x_2(P), x_1(P), x_5(P), x_6(P))$. Then the coloring $\varphi' = \varphi/P_v$ satisfies $\alpha, \beta \in \overline{\varphi}(x) = \overline{\varphi}'(x)$, $\gamma \in \overline{\varphi}(y) = \overline{\varphi}(y')$, and $P_y(\beta, \gamma, \varphi') = (y, x_2(P), x_3(P))$. This is a contradiction to Claim 4.11.1. Hence the claim is proved.   △

Since P^* is a Hilton-graph, this completes the proof of the theorem.   ■

4.3 SIMPLE OVERFULL GRAPHS

A graph G is said to be **overfull** if G satisfies the inequality

$$|E(G)| \geq \Delta(G)\lfloor \tfrac{1}{2}|V(G)|\rfloor + 1. \tag{4.1}$$

Proposition 4.12 *For an arbitrary graph G the following conditions are equivalent:*

(a) *G is overfull.*

(b) *G satisfies the following sequence of inequalities:*

$$\chi'(G) \geq w(G) \geq \left\lceil \frac{|E(G)|}{\lfloor \tfrac{1}{2}|V(G)|\rfloor} \right\rceil \geq \Delta(G) + 1. \tag{4.2}$$

(c) *G has odd order at least 3, and*

$$\sum_{v \in V(G)} (\Delta(G) - d_G(v)) \leq \Delta(G) - 2. \tag{4.3}$$

Proof: That (4.1) is equivalent to (4.2) follows from the definition of w and the fact that w is a lower bound for χ'. Evidently, every overfull graph G has odd order at least 3. Furthermore, it is easy to check that (4.1) is equivalent to (4.3) for each graph G of odd order at least 3.   ■

The next proposition shows that overfull simple graphs and elementary class two graphs are closely related to each other.

Proposition 4.13 *Let G be an arbitrary simple graph. Then the following statements hold:*

(a) *G is overfull if and only if*

$$w(G) = \left\lceil \frac{|E(G)|}{\lfloor \tfrac{1}{2}|V(G)|\rfloor} \right\rceil = \Delta(G) + 1. \tag{4.4}$$

(b) *G is an elementary graph of class two if and only if $w(G) = \Delta(G) + 1$.*

(c) *G is an elementary graph of class two if and only if G contains an overfull subgraph H with $\Delta(H) = \Delta(G)$.*

(d) *If G is a critical graph of class two, then G is an elementary graph if and only if G is overfull.*

Proof: Since G is simple, (1.5) and Theorem 2.2 imply that $w(G) \leq \chi'(G) \leq \Delta(G) + 1$. Hence, (a) follows from Proposition 4.12, and statement (b) is evident. Next we prove (c). Suppose first that G is an elementary graph of class two. Then (b) implies that $w(G) = \Delta(G) + 1$. Since w is a monotone graph parameter, the graph G contains a subgraph H such that

$$w(G) = \left\lceil \frac{|E(H)|}{\lfloor \frac{1}{2}|V(H)| \rfloor} \right\rceil = w(H).$$

Then $\Delta(H) + 1 \leq \Delta(G) + 1 = w(G) = w(H) \leq \Delta(H) + 1$, which gives $w(H) = \Delta(H) + 1$ and $\Delta(H) = \Delta(G)$. Then (a) implies that H is overfull. Suppose now that H is an overfull subgraph of G with $\Delta(H) = \Delta(G)$. Then (a) implies that $w(H) = \Delta(H) + 1$. Then we have $\Delta(G) + 1 = \Delta(H) + 1 = w(H) \leq w(G) \leq \Delta(G) + 1$ and, therefore, $w(G) = \Delta(G) + 1$. Then (b) implies that G is an elementary graph of class two. This proves (c).

For the proof of (d), assume that G is a critical graph of class two. If G is elementary, then Proposition 1.3(a) implies that

$$\Delta(G) + 1 = \chi'(G) = w(G) = \left\lceil \frac{|E(G)|}{\lfloor \frac{1}{2}|V(G)| \rfloor} \right\rceil.$$

Hence, it follows from (a) that G is overfull. Conversely, if G is overfull, then (a) and (b) imply that G is elementary. This proves (d).   ∎

The graph P^* (the Petersen graph with one vertex deleted) is an example of a critical graph of class two that is not elementary. For a proof of the fact that $\chi'(P^*) = \Delta(P^*) + 1 = 4$ see Sect. 1.6.

The following conjecture was presented by A. J. W. Hilton at the graph theory meeting at Sandbjerg, Denmark, June 1985. It was put in print in Chetwynd and Hilton [51, 52].

Conjecture 4.14 (Overfull Conjecture) *Every graph G of class two with $\Delta(G) > |V(G)|/3$ is elementary, i.e., G contains an overfull subgraph H with $\Delta(H) = \Delta(G)$.*

Recall that the chromatic index of the complete graph K_n $(n \geq 2)$ is either n or $n - 1$ and the chromatic index of the cycle C_n $(n \geq 2)$ is either 3 or 2 depending on whether n is odd or even.

Let G be a Δ-regular graph with n vertices. Then $\chi'(G) = \Delta$ if and only if G is **1-factorable**, i.e., G is the union of Δ edge disjoint 1-factors. Furthermore, Proposition 4.12 implies that G is overfull if and only if n is odd and ≥ 3. A simple Δ-regular graph with an even number of vertices, however, may contain an overfull subgraph with maximum degree Δ, as is shown by the example of a graph G consisting of two disjoint complete graphs K_n, when n is odd and ≥ 3.

Lemma 4.15 (Plantholt and Tipnis [247] 1991) *Let G be a Δ-regular graph of order $2n$ and with maximum multiplicity at most s, where $n, s \geq 1$ are integers. If $\Delta \geq sn$, then $w(G) \leq \Delta$.*

Proof: Suppose, on the contrary, that $w(G) \geq \Delta+1$. Then $\Delta = \Delta(G), |V(G)| \geq 3$ and, by (1.6), there is a vertex set $X \subseteq V(G)$ such that $|X|$ is odd, $|X| \geq 3$ and

$$w(G) = \left\lceil \frac{2|E(G[X])|}{|X| - 1} \right\rceil \geq \Delta + 1,$$

i.e.,

$$2|E(G[X])| > \Delta(|X| - 1),$$

which is equivalent to

$$|\partial_G(X)| = \sum_{x \in X}(\Delta - d_{G[X]}(x)) = \Delta|X| - 2|E(G[X])| < \Delta,$$

since G is Δ-regular. We may assume that $|X| \leq n$, since otherwise we can replace X by its complement $Y = V(G) \setminus X$. Note that $|Y|$ is odd and $\partial_G(Y) = \partial_G(X)$. Since G is a Δ-regular graph with $\mu(G) \leq s$ and $sn \leq \Delta$, it now follows that

$$
\begin{aligned}
\Delta|X| &= \sum_{x \in X} d_G(x) \\
&= 2|E(G[X])| + |\partial_G(X)| \\
&\leq s|X|(|X| - 1) + \Delta - 1 \\
&\leq sn(|X| - 1) + \Delta - 1 \\
&\leq \Delta(|X| - 1) + \Delta - 1 \\
&= \Delta|X| - 1,
\end{aligned}
$$

which is impossible.   ∎

In contrast to the lemma, any Δ-regular graph G of odd order $n \geq 3$ satisfies

$$w(G) \geq \left\lceil \frac{2|E(G)|}{n - 1} \right\rceil = \left\lceil \frac{\Delta n}{n - 1} \right\rceil = \Delta + \left\lceil \frac{\Delta}{n - 1} \right\rceil.$$

Lemma 4.15 shows that the overfull conjecture implies the following conjecture about 1-factorable regular simple graphs of even order and large degree. This conjecture appeared in Chetwynd and Hilton [49], but may go back to Dirac in the early 1950s.

Conjecture 4.16 (One-factorization Conjecture) *Let G be a Δ-regular simple graph of order $2n$. If $\Delta \geq n$, then G is 1-factorable.*

The conjecture is best possible, at least when n is odd. To see this, take the disjoint union of two complete graphs on n vertices. Another example is obtained by taking two disjoint copies of $K_n - e$, where e is any edge of K_n, and joining the corresponding endvertices of e in the two copies by edges.

A notable consequence of the one-factorization conjecture is that G or its complement is 1-factorable for any regular simple graph G of even order.

Let us introduce a useful notation. For an arbitrary graph G and a matching M of G, let $V(M)$ denote the set of vertices of G which are incident with some edge in M. If $v \in V(G) \setminus V(M)$, then we say that M **misses** the vertex v. Note that M is a perfect matching of G if and only if $V(M) = V(G)$, i.e., M misses no vertex of G. If M misses exactly one vertex of G, then M is said to be a **near-perfect matching** of the graph G.

The following result, which supports the one-factorization conjecture, was obtained by Chetwynd and Hilton [53] in 1989 as well as by Niessen and Volkmann [236] in 1990.

Theorem 4.17 (Chetwynd and Hilton [53] 1989, Niessen and Volkmann [236] 1990) *Let G be a Δ-regular simple graph of order $2n$. If $\Delta \geq (\sqrt{7} - 1)n$, then G is 1-factorable.*

Let G be a simple regular graph of order $2n$ and with maximum degree $\Delta \geq (\sqrt{7} - 1)n$. Let p denote the maximum number of paths of length two in $\overline{G}$, the complement of G, joining the same two vertices. As Chetwynd and Hilton [53] observed, $\overline{G}$ is regular of degree $D = 2n - 1 - \Delta$ and each vertex of $\overline{G}$ is the center of $\binom{D}{2}$ paths of length two. Hence the number of paths of length two in $\overline{G}$ is $2n\binom{D}{2}$, which certainly implies that

$$p \geq \frac{2n\binom{D}{2}}{\binom{2n}{2}} = \frac{(2n - \Delta - 1)(2n - \Delta - 2)}{2n - 1}.$$

Next, we claim that $\Delta \geq \frac{5}{3}n - \frac{1}{3}p - \frac{1}{6}$. If not, then the above inequality would give

$$\Delta < \frac{5}{6}(2n) - \frac{1}{3}\frac{(2n - \Delta - 1)(2n - \Delta - 2)}{2n - 1} - \frac{1}{6},$$

which is equivalent to

$$\Delta^2 + 2n\Delta - (6n^2 - 3/2) < 0.$$

This inequality obviously implies

$$\Delta < -n + \sqrt{7n^2 - 3/2} < (\sqrt{7} - 1)n,$$

which contradicts the hypothesis. This proves the claim. Hence, Theorem 4.17 is an immediate consequence of the following result.

Theorem 4.18 (Chetwynd and Hilton [53] 1989) *Let G be a Δ-regular simple graph of order $2n$, and let p be the maximum number of paths of length two in $\overline{G}$, the complement of G, joining the same two vertices. If $\Delta \geq \frac{5}{3}n - \frac{1}{3}p - \frac{1}{6}$, then G is 1-factorable.*

Proof: Let G be a simple regular graph of order $2n$ and with maximum degree $\Delta \geq \frac{5}{3}n - \frac{1}{3}p - \frac{1}{6}$. Our aim is to show that graph G is of a class one and hence 1-factorable. This is certainly true if $G = K_{2n}$. Hence, in what follows, we assume that $G \neq K_{2n}$, implying that $\Delta \leq 2n - 2$ and $n \geq 1$.

The definition of p implies that there are two vertices x and y in G such that $|V(G) \setminus (N_G(x) \cup N_G(y))| = p$. Since G is Δ-regular and $\Delta \leq 2n - 2$, the graph $G' = G - x$ satisfies $|V(G')| = 2n - 1$ and $\Delta(G') = \Delta$. If there is a coloring $\varphi \in \mathcal{C}^\Delta(G')$, then (1.2) and (1.3) imply that $m_{\varphi,\alpha} = 1$ for each color $\alpha \in \{1, \ldots, \Delta\}$. Therefore, φ can be extended to a Δ-edge-coloring of G. Hence, it suffices to show that G' is of class one. The essential idea, for proving that G' is of class one, is to delete an appropriate set of r matchings such that the resulting graph is of class one and has maximum degree $\Delta - r$. To show that the resulting graph is of class one its Vizing-core will be examined.

The vertex set of G' consists of the neighborhood $N_G(x)$ of x and the remaining vertex set $U = V(G) \setminus (N_G(x) \cup \{x\})$, where $U = V(G'^{[\Delta]})$ and $|U| = 2n - \Delta - 1$. Let $s = |U| - p - 1$. Since exactly p vertices of U are nonadjacent to y in G, there is a set $Y \subseteq U \cap N_G(y)$ such that $|Y| = s$. Furthermore, there is a set $X \subseteq N_G(x) \setminus (N_G(y) \cup \{y\})$ such that $|X| = s$. Observe that the vertex set U consists of the p vertices that are nonadjacent to y in G, the vertex set Y, and one additional vertex, denoted y^*, where $y^* = y$ or $y^* \in N_G(y)$ (depending on whether xy is an edge or not).

Let $H = G'[(U \cup X) \setminus \{y\}]$ and let $q = |V(H)|$. Then $q \leq |X| + |U| = s + |U| = 2|U| - p - 1 = 4n - 2\Delta - p - 3$. Let M_0 be a maximum matching of H, let $m = |M_0|$ and $t = q - 2m$. The q vertices of H are labeled $a_1, \ldots, a_{m+t}, b_1, \ldots, b_m$, the s vertices of X are labeled $x_1, \ldots, x_s$, and the s vertices of Y are labeled $y_1, \ldots, y_s$. The labels are chosen such that $M_0 = \{a_1 b_m, a_2 b_{m-1}, \ldots, a_m b_1\}$ and such that, for $1 \leq i \leq s - 1$, x_i comes before y_i in the linear order induced by the sequence $S = (a_1, \ldots, a_{m+t}, b_1, \ldots, b_m)$. Unless every edge in M_0 joins either two vertices of X or two vertices of Y, we could also suppose that x_s comes before y_s in S.

Let K denote the complete graph with vertex set $V(K) = V(H) \cup \{y\} = \{y, a_1, \ldots, a_{m+t}, b_1, \ldots, b_m\}$, let

$$E_1 = \{a_i a_j \mid 1 \leq i < j \leq m + t\} \cup \{a_i b_j \mid 1 \leq i \leq m, 1 \leq j \leq m - i + 1\}$$

and let $E_2 = E(K) \setminus E_1$. Note that $M_0 \subseteq E_1$ and so $M_0 \cap E_2 = \emptyset$.

Next, a partition of the edge set E_2 into q edge-disjoint matchings, denoted by $M_1^+, \ldots, M_q^+$, is defined as follows: For $1 \leq i \leq t$, let

$$M_i^+ = \{y a_i\} \cup \{b_m a_{i+1}, b_{m-1} a_{i+2}, \ldots, b_1 a_{m+i}\}$$

and, for $1 \leq i \leq m$, let

$$M_{t+i}^+ = \{y a_{t+i}\} \cup \{b_{m-h} a_{t+h+1+i} \mid 0 \leq h \leq m - i - 1\} \cup \{b_h b_{i+1-h} \mid 1 \leq h \leq \lfloor i/2 \rfloor\}$$

and

$$M_{t+m+i}^+ = \{y b_i\} \cup \{b_{m+1-h} b_{h+i} \mid 1 \leq h \leq \lfloor (m-i)/2 \rfloor\}.$$

It is not difficult to check that the edge set of the complete graph K satisfies

$$E(K) = E_1 \cup E_2 \text{ and } E_2 = M_1^+ \cup \cdots \cup M_q^+,$$

where $M_0 \subseteq E_1$, $E_1 \cap (M_1^+ \cup \cdots \cup M_q^+) = \emptyset$, and $M_i^+ \cap M_j^+ = \emptyset$ for $1 \leq i < j \leq q$. It is also easy to check that

$$|M_i^+| \leq \frac{1}{2}(q + 2 - 1)$$

for $1 \leq i \leq q$. It should be noted that, for every vertex $z \in V(H)$, there is exactly one matching M_i^+ containing yz and no vertex z' that comes before z in the sequence S is incident with an edge of the matching M_i^+.

From the matchings M_i^+ ($1 \leq i \leq q$) we now obtain new matchings M_i^* ($1 \leq i \leq q+1$). Clearly, $yy_s \in M_{i_0}^+$ for exactly one index i_0. If no edge in $M_{i_0}^+$ is incident with x_s, then put $M_i^* = M_i^+$ ($1 \leq i \leq q$) and $M_{q+1}^* = \emptyset$. If an edge in $M_{i_0}^+$ is incident with x_s, say e_0, then put $M_i^* = M_i^+$ if $i \in \{1, \ldots, q\} \setminus \{i_0\}$, $M_{i_0}^* = M_{i_0}^+ \setminus \{e_0\}$, and $M_{q+1}^* = \{e_0\}$. Observe that the latter case can only occur if x_s comes after w_s in S. Obviously, for $1 \leq i \leq q+1$,

$$|M_i^*| \leq \frac{1}{2}(q + 3 - i).$$

Clearly, each edge yy_i with $1 \leq i \leq s$ is contained in exactly one matching M_j^* and this matching then has no edge incident with x_i. Finally, let E^* be the set defined by $E^* = E(G[V(K)]) \setminus (E_1 \cup \{yy^*\})$ if $y \neq y^*$, and $E^* = E(G[V(K)]) \setminus E_1$ otherwise.

Claim 4.18.1 *There are edge-disjoint near-perfect matchings $F_1, \ldots, F_{q+1}$ of G' such that, for $1 \leq h \leq q + 1$, the following conditions hold:*

(a) $E^* \cap (M_1^* \cup \cdots \cup M_h^*) \subseteq F_1 \cup \cdots \cup F_h$.

(b) $M_0 \cap F_h = \emptyset$.

(c) *If $yy_i \in M_h^*$ for some $i \in \{1, \ldots s\}$, then $yy_i \in F_h$ and F_h misses x_i.*

(d) *If $yy_i \notin M_h^*$ for all $i \in \{1, \ldots s\}$, then F_h misses y.*

Proof: The proof is by induction on $h \geq 1$. So suppose that $F_1, \ldots, F_{h-1}$ have been chosen according to the rules. Let

$$M_h = (E^* \cap M_h^*) \setminus (F_1 \cup \cdots \cup F_{h-1}).$$

Let G_h be the graph with vertex set $V(G')$ and edge set $E(G') \setminus (F_1 \cup \cdots \cup F_{h-1} \cup M_0)$. To complete the proof, choose F_h to be a near perfect matching of G_h containing M_h such that F_h misses x_i if $yy_i \in M_h$ for some $i \in \{1, \ldots, s\}$ and missing y otherwise.

To see that F_h can be chosen in such a way, we apply Dirac's theorem and show that the graph $G'_h = G_h - V(M_h)$ contains a Hamilton cycle. Since $\Delta(G') = \Delta - 1$,

$$\delta(G'_h) \geq (\Delta - 1) - h - |V(M_h)|.$$

On the other hand, it follows that

$$
\begin{aligned}
\frac{1}{2}|V(G'_h)| &= \frac{1}{2}\{|V(G_h)| - |V(M_h)|\} \\
&= \frac{1}{2}\{2n - 1 - |V(M_h)|\}
\end{aligned}
$$

and

$$|V(M_h)| \leq |V(M_h^+)| \leq (q + 3 - h),$$

where $q \leq 4n - 2\Delta - p - 3$ (see above). Therefore, we obtain

$$
\begin{aligned}
\delta(G'_h) - \frac{1}{2}|V(G'_h)| &= \Delta - h - 1 - |V(M_h)| - n + \frac{1}{2} + \frac{1}{2}|V(M_h)| \\
&= \Delta - n - h - \frac{1}{2} - \frac{1}{2}|V(M_h)| \\
&\geq \Delta - n - h - \frac{1}{2} - \frac{1}{2}(q + 3 - h) \\
&= \Delta - n - \frac{1}{2}q - \frac{1}{2}h - 2 \\
&\geq \Delta - n - q - \frac{5}{2} \\
&\geq 3\Delta - 5n + p + \frac{1}{2} \\
&\geq 0,
\end{aligned}
$$

since $\Delta \geq \frac{5}{6}(2n) - \frac{1}{3}p - \frac{1}{6}$ by hypothesis. Hence $G'_h = G_h - V(M_h)$ contains a Hamilton cycle. If $yy_i \notin M_h^*$ for all $i \in \{1, \ldots, s\}$, then M_h misses y and, therefore, we can extend M_h to a near-perfect matching F_h such that F_h misses y. Otherwise, there is exactly one index $i \in \{1, \ldots, s\}$ such that yy_i belongs to M_h^*. Then x_i is missed by M_h^* as well as by M_h and we can extend M_h to a near-perfect matching F_h such that F_h misses x_i. This proves the claim $\triangle$

Claim 4.18.2 *The following statements hold:*

(a) $\tilde{G} = G' - (F_1 \cup \cdots \cup F_{q+1}) - y$ *is of a class one and* $\Delta(\tilde{G}) = \Delta - q - 1$.

(b) G' *is of class one.*

Proof: For each vertex $y_i \in Y$, the edge yy_i belongs to E^* and there is exactly one index $h \in \{1, \ldots, q+1\}$ such that $yy_i \in M_h^*$. In what follows we denote this index by $h(i)$. It follows from Claim 4.18.1 that $F_{h(i)}$ contains the edge yy_i and misses

x_i. Furthermore, if $I = \{1, \ldots, q+1\} \setminus \{h(1), \ldots, h(s)\}$, then $|I| = q + 1 - s$ and each matching F_h with $h \in I$ misses the vertex y.

For the proof of (a), we will show that the Vizing-core of $\tilde{G}$ is empty. By Proposition 4.5, this implies that $\tilde{G}$ is of class one. For a vertex $v \in V(\tilde{G})$, let $d(v) = d_{\tilde{G}}(v)$. For a vertex $u \in N_G(x)$, we have $d_{G'}(u) = \Delta - 1$ and, therefore, $d(u) = \Delta - 1 - q$ if $u \in X$ and $d(u) \leq \Delta - q - 2$ if $u \notin X$. For a vertex $w \in U \setminus \{y\}$, we have $d_{G'}(w) = \Delta$ and hence $d(w) = \Delta - q - 1$. Consequently, $\Delta(\tilde{G}) = \Delta - q - 1$ and $V(\tilde{G}^{[\Delta]}) = V(H) = \{a_1, \ldots, a_{m+t}, b_1, \ldots, b_m\}$. By Claim 4.18.1, all edges of M_0 are contained in $\tilde{G}^{[\Delta]}$ and all other edges belong to

$$E_1 = \{a_i a_j \mid 1 \leq i < j \leq m + t\} \cup \{a_i b_j \mid 1 \leq i \leq m, 1 \leq j \leq m - i + 1\}$$

Now consider the sequence $S = (a_1, \ldots, a_{m+t}, b_1, \ldots, b_m)$ consisting of all vertices of H. Since M_0 is a maximum matching of H, $A = \{a_{m+1}, \ldots, a_{m+t}\}$ is an independent set in G'. Suppose that the Vizing-core $\tilde{H}$ of $\tilde{G}$ is nonempty. Then $\Delta(\tilde{H}) = \Delta(\tilde{G})$ and, therefore, $\tilde{H}$ contains a vertex of $V(H)$. Hence we can choose a vertex $z \in V(\tilde{H}) \cap V(H)$ such that all other vertices of $V(H) \cap V(\tilde{H})$ comes before z in the sequence S. First, assume that $z = b_i$. Then $\{b_{i+1}, \ldots, b_m\} \cap V(\tilde{H}) = \emptyset$ and, therefore, $d_{\tilde{H}}(a_j) \leq \Delta(\tilde{H}) - 1$ for $j < m - i + 1$. Hence, all neighbors of z, except possibly a_{m-i+1}, are minor vertices of $\tilde{H}$, a contradiction. If $z = a_i$, then $\{b_1, \ldots, b_m\} \cap V(\tilde{H}) = \emptyset$ and, therefore, no vertex in $\{a_1, \ldots, a_m\}$ is a major vertex of $\tilde{H}$. Since $A = \{a_{m+1}, \ldots, a_{m+t}\}$ is an independent set in H, this implies that all neighbors of z in H are minor vertices, a contradiction. The proof of (a) is complete.

For the proof of (b), let us first consider the graph

$$G_1 = G' - (F_{h(1)} \cup \cdots \cup F_{h(s)}).$$

For $h \in I$, F_h is a perfect matching of $G_1 - y$. From (a) we then conclude that graph $G_1 - y$ is of class one and has maximum degree $D = \Delta - s$. Since $y y_i \in F_{h(i)}$ for $i = 1, \ldots, s$, we obtain that all neighbors of y, except possibly y^*, are minor vertices of G_1, where $\Delta(G_1) = D$. Hence, the Vizing-core of G_1 is contained in $G_1 - y$. Since $G_1 - y$ is of class one, it follows that the Vizing-core of G_1 is a of class one, too. By Proposition 4.5, this implies that G_1 is of a class one. Since $\Delta(G_1) = \Delta(G') - s$, we then conclude that G' is of class one. Hence the claim is proved. $\triangle$

Clearly, the proof of Theorem 4.18 is now complete.   ∎

Using the same approach as Chetwynd and Hilton for the proof of the above theorem, Niessen and Volkmann [236] obtained the following result. From this they also derived a proof of Theorem 4.17.

Theorem 4.19 Niessen and Volkmann [236] 1990) *A simple graph G is of class one, provided that*

(a) $|V(G)| = 2n$ *and* $\delta(G) \geq n + |V(G^{[\Delta]})| - 2$, *or*

(b) $|V(G)| = 2n + 1$ *and* $\delta(G) \geq n + |V(G^{[\Delta]})| + \min_{v \in V(G)} |N_G(v) \cap V(G^{[\Delta]})|$.

Perković and Reed proved that the one-factorization conjecture is asymptotically true. This result is a simple consequence of the following more general theorem. The proof combines the ideas developed by Chetwynd and Hilton with the probabilistic method.

Theorem 4.20 (Perković and Reed [240] 1997) *There exists a Δ_0 such that for all $\Delta \geq \Delta_0$ every Δ-regular simple graph G with $|V(G)| = 2n$ and $\Delta \geq n$ is 1-factorable, provided that G contains no subgraph H such that $\delta(H) \geq \Delta - \Delta^{39/40}$ and H is bipartite, or such that $|V(G)| - |V(H)| \geq \Delta - \Delta^{39/40}$.*

Theorem 4.21 (Perković and Reed [240] 1997) *For any $\varepsilon > 0$ there is an integer $N > 0$ such that every Δ-regular simple graph G satisfying $|V(G)| = 2n$, $n \geq N$, and $\Delta \geq (1 + \varepsilon)n$ is 1-factorable.*

Proof: Let $\varepsilon > 0$ be a real number and define $N = \lceil \max\{\Delta_0, 2^{39}\varepsilon^{-40}\}\rceil$, where Δ_0 is chosen according to Theorem 4.20. Now let G be a Δ-regular simple graph satisfying $|V(G)| = 2n$, $n \geq N$, and $\Delta \geq (1 + \varepsilon)n$. Then $\Delta \geq \Delta_0$ and we claim that G contains no subgraph H such that $\delta(H) \geq \Delta - \Delta^{39/40}$ and H is bipartite, or such that $|V(G)| - |V(H)| \geq \Delta - \Delta^{39/40}$.

Suppose on the contrary that G contains such a subgraph H. Then in booth cases we conclude that $|V(G)| \geq 2\Delta - 2\Delta^{39/40}$. Since $(1 + \varepsilon)n \leq \Delta < 2n$, this implies that

$$|V(G)| > 2(1 + \varepsilon)n - 2(2n)^{39/40} = 2n + 2n\varepsilon - 2(2n)^{39/40}.$$

Since $n \geq N \geq 2^{39}\varepsilon^{-40}$, it follows that $2n\varepsilon - 2(2n)^{39/40} \geq 0$ and, therefore, $|V(G)| > 2n = |V(G)|$, a contradiction. ∎

A graph G is said to be **just overfull** if G satisfies the equation

$$|E(G)| = \Delta(G)\lfloor \tfrac{1}{2}|V(G)|\rfloor + 1. \tag{4.5}$$

Obviously, each just overfull graph is overfull. Furthermore, it is easy to show that a simple graph G is just overfull if and only if G has odd order at least 3 and

$$\sum_{v \in V(G)} (\Delta(G) - d_G(v)) = \Delta(G) - 2. \tag{4.6}$$

Let G be a critical graph of class two. If G is elementary, then Proposition 1.3 implies that G is just overfull. Conversely, if G is just overfull, then Proposition 4.13(d) implies that G is elementary.

Let G be a Δ-regular graph of odd order. Then $\Delta \geq 2$ is even and Proposition 4.12 implies that G is overfull and G remains overfull as long as we delete no more than $(\Delta - 2)/2$ edges. If we delete exactly $(\Delta - 2)/2$ edges, then the resulting graph H becomes just overfull. If G is simple and $\Delta \geq |V(G)|/2$, then H contains no other overfull subgraph, by the following result. Hence the overfull conjecture implies that H is a critical graph of class two.

Lemma 4.22 (Niessen [235] 1994) *Let G be a simple overfull graph with maximum degree $\Delta \geq |V(G)|/2$. Then every overfull subgraph H of G with $\Delta(H) = \Delta$ satisfies $V(H) = V(G)$.*

Proof: Suppose that the overfull graph G contains a subgraph H such that H is overfull, $\Delta(H) = \Delta$, and $V(H) \neq V(G)$. By Proposition 4.12, $|V(G)| = 2n + 1$ and $|V(H)| = 2p + 1$. Then we have $n \geq p + 1$ and, therefore, $H' = G - V(H)$ is a nonempty subgraph of G. First, consider the number $m = |E_G(V(H), V(H'))|$ of edges joining a vertex of H with a vertex of H'. Since $\Delta(G) = \Delta$, it follows that

$$m = \sum_{v \in V(H)} (d_G(v) - d_H(v)) = \sum_{v \in H}(\Delta - d_H(v)) - \sum_{v \in H}(\Delta - d_G(v)).$$

Since H is overfull, Proposition 4.12(c) implies that the first sum is at most $\Delta - 2$. Since the second sum is at least zero, this yields $m \leq \Delta - 2$. Next, we count the number of edges in G. On one hand, since G is overfull, (4.1) yields $2|E(G)| \geq 2n\Delta + 2$. On the other hand, we have

$$
\begin{aligned}
2|E(G)| &= \sum_{v \in V(H)} d_G(v) + \sum_{v \in V(H')} d_G(v) \\
&= \sum_{v \in V(H)} d_G(v) + \sum_{v \in V(H')} (d_G(v) - d_{H'}(v)) + \sum_{v \in V(H')} d_{H'}(v) \\
&\leq (2p+1)\Delta + m + (2n - 2p)(2n - 2p - 1).
\end{aligned}
$$

Since $m \leq \Delta - 2$ and $2|E(G)| \geq 2n\Delta + 2$, this yields

$$2n\Delta + 2 \leq 2p\Delta + 2\Delta - 2 + 2(n - p)(2n - 2p - 1)$$

and hence

$$(n - p - 1)\Delta \leq (n - p)(2n - 2p - 1) - 2 < (n - p - 1)(2n - 2p + 1).$$

This inequality is not satisfied for $n = p + 1$, which implies $n > p + 1$ and hence $\Delta < 2n - 2p + 1$. Since $\Delta \geq |V(G)|/2$, we have $\Delta \geq n + 1$ and, therefore, $2p < n$. But now $2p = |V(H)| - 1 \geq \Delta \geq n + 1$ gives a contradiction. ∎

Conjecture 4.23 (Just Overfull Conjecture) *Let G be a simple graph with $\Delta(G) \geq |V(G)|/2$. Then G is a critical graph of class two if and only if G is just overfull.*

Theorem 4.24 (Hilton and Zhao [144] 1997) *The overfull conjecture implies the just overfull conjecture.*

Proof: First, suppose that G is a critical graph of class two. Then the overfull conjecture implies that G is elementary. By Proposition 1.3, this implies that G is just overfull. Conversely, suppose that G is just overfull. Then G is of class two and Lemma 4.22 implies that G contains no proper overfull subgraph H with $\Delta(H) = \Delta(G)$. Hence, by the overfull conjecture, $\chi'(H) < \chi'(G)$ for every proper subgraph H of G, that is, G is critical. ∎

4.4 ADJACENCY LEMMAS FOR CRITICAL CLASS TWO GRAPHS

In this and the next three sections we shall restrict our attention to critical graphs of class two. Critical graphs of class two have rather more structure than arbitrary graphs of class two, and every graph of class two contains a critical graph of class two with the same maximum degree as a subgraph. These facts can be used when proving results in relation to the Classification Problem.

Vizing's Adjacency Lemma (Theorem 2.7) describes an important property of critical graphs of class two, and Theorem 2.14 provides an extension of Vizing's result. In this section we shall present several other adjacency lemmas for critical graphs of class two.

First, we need to give some new definitions. Let G be an arbitrary graph. For a vertex set $X \subseteq V(G)$, let $N_G(X)$ denote the **neighborhood** of X in G, that is, the set of all vertices $v \in V(G)$ such that $E_G(v, x) \neq \emptyset$ for some vertex $x \in X$. Instead of $N_G(\{x_1, \ldots, x_n\})$ we simply write $N_G(x_1, \ldots, x_n)$. For a pair (x, y) of two distinct vertices of G, we define the **Kierstead set** of the pair (x, y) by

$$K_G(x, y) = \{z \in N_G(y) \setminus \{x\} \mid d_G(x) + d_G(y) + d_G(z) \geq 2\Delta(G) + 2\},$$

and the **Kierstead number** of the pair (x, y) by $\sigma_G(x, y) = |K_G(x, y)|$. The Kierstead number was first defined and studied by Woodall [309]. Observe that $\sigma_G(x, y)$ and $\sigma_G(y, x)$ might be different. If it is clear that we refer to the graph G, then we omit the index G. Recall, that for an edge $e \in E_G(x, y)$ of a simple graph G we also write $e = xy$ or $e = yx$.

Let G be a graph of class two with maximum degree Δ, and let e be a critical edge of G. Then there exists a coloring $\varphi \in \mathcal{C}^\Delta(G - e)$. Let $u \in V(G)$ be a vertex and let $\alpha \in \{1, \ldots, \Delta\}$ be a color. By an α-**neighbor** of u with respect to φ we mean a vertex v such that $uv \in E(G)$ and $\varphi(uv) = \alpha$. If such a vertex v exists, then it is unique and we write $v = u_{\alpha, \varphi}$ or $v = u_\alpha$ if it is clear that we refer to the coloring φ. Clearly, u_α exists if and only if $\alpha \in \varphi(u)$.

Many of the adjacency lemmas presented in this section can be derived from Theorem 3.3, respectively Proposition 3.4. An example is the following simple result.

Proposition 4.25 (Woodall [309] 2007) *Let G be a graph of class two with maximum degree Δ, and let xy be a critical edge of G. Furthermore, let $\varphi \in \mathcal{C}^\Delta(G - xy)$ be a coloring, and let $z \in N_G(x)$ be a vertex such that $\varphi(xz)$ belongs to $\overline{\varphi}(y)$. Then the following statements hold:*

(a) *$z \neq y$, $\{x, y, z\}$ is elementary with respect to φ, and $z \in K(y, x)$.*

(b) *For every color $\beta \in \overline{\varphi}(x)$, we have $y_\beta \in K(x, y)$ and $z_\beta \in K(x, z)$.*

Proof: First, we prove (a). Since $\overline{\varphi}(xz) \in \overline{\varphi}(y)$, $z \neq y$ and $K = (y, yx, x, xz, z)$ is a Kierstead path with respect to xy and φ. Hence Proposition 3.4(b) implies that $V(K) = \{x, y, z\}$ is elementary with respect to φ and $d_G(x) + d_G(y) + d_G(z) \geq 2\Delta + 2$. Thus $z \in K(y, x)$, which completes the proof of (a).

To prove (b), let $\beta \in \overline{\varphi}(x)$. By hypothesis, $\alpha = \varphi(xz) \in \overline{\varphi}(y)$. Since $\{x, y, z\}$ is elementary with respect to φ (by (a)), we have $\alpha \neq \beta$ and $\beta \in \varphi(y) \cap \varphi(z)$ and so the two vertices y_β and z_β do exist. Since $\varphi(yy_\beta) = \beta \in \overline{\varphi}(x)$, it follows from (a) that $y_\beta \in K(x, y)$. To see that $z_\beta \in K(x, z)$, color xy with α and make xz uncolored. This results in a coloring $\varphi' \in \mathcal{C}^\Delta(G - xz)$. The coloring φ' satisfies $\varphi'(zz_\beta) = \beta \in \overline{\varphi}(y) = \overline{\varphi}'(y)$ and, therefore, (a) implies that $z_\beta \in K(y, z)$. This proves (b). ∎

Proposition 4.26 *Let G be a graph of class two with maximum degree Δ, let xy be a critical edge of G, and let $\varphi \in \mathcal{C}^\Delta(G - xy)$. Then the following statements hold:*

(a) $\overline{\varphi}(x) \cap \overline{\varphi}(y) = \emptyset$ *and* $|\overline{\varphi}(x) \cup \overline{\varphi}(y)| = 2\Delta + 2 - (d_G(x) + d_G(y))$.

(b) $|\varphi(x) \cap \varphi(y)| = d_G(x) + d_G(y) - \Delta - 2$.

(c) *If $u \in N_G(x, y) \setminus \{x, y\}$ and $d_G(x) + d_G(y) + d_G(u) < 2\Delta + 2$, then we have* $|\overline{\varphi}(u) \cap (\overline{\varphi}(x) \cup \overline{\varphi}(y))| \geq |\varphi(x) \cap \varphi(y) \cap \varphi(u)| + 1 \geq 1$.

Proof: The edge xy being uncolored, we have $\overline{\varphi}(x) \cap \overline{\varphi}(y) = \emptyset$ and, therefore, $|\overline{\varphi}(x) \cup \overline{\varphi}(y)| = |\overline{\varphi}(x)| + |\overline{\varphi}(x)| = (\Delta - d_G(x) + 1) + (\Delta - d_G(y) + 1) = 2\Delta + 2 - (d_G(x) + d_G(y))$. This proves (a). Statement (b) is a simple consequence of (a) and the fact that $|\varphi(x) \cap \varphi(y)| + |\overline{\varphi}(x) \cup \overline{\varphi}(y)| = \Delta$.

For the proof of (c), assume that $d_G(x) + d_G(y) + d_G(u) < 2\Delta + 2$ for a vertex $u \in N_G(x, y) \setminus \{x, y\}$. Let $m = |\varphi(x) \cap \varphi(y) \cap \varphi(u)|$ and $p = |\overline{\varphi}(u) \cap (\overline{\varphi}(x) \cup \overline{\varphi}(y))|$. Using the hypothesis and statement (a), it follows that

$$
\begin{aligned}
\Delta - m &\geq |\overline{\varphi}(x) \cup \overline{\varphi}(y) \cup \overline{\varphi}(u)| = |\overline{\varphi}(x) \cup \overline{\varphi}(y)| + |\overline{\varphi}(u)| - p \\
&= (2\Delta + 2 - (d_G(x) + d_G(y))) + (\Delta - d_G(u)) - p \\
&= 3\Delta + 2 - (d_G(x) + d_G(y) + d_G(u)) - p \\
&> 3\Delta + 2 - (2\Delta + 2) - p = \Delta - p,
\end{aligned}
$$

which gives $p \geq m + 1$. This completes the proof. ∎

Proposition 4.27 (Woodall [309] 2007) *Let G be a graph of class two with maximum degree Δ, and let xy be a critical edge of G. Then the following statements hold:*

(a) $\Delta - d_G(x) + 1 \leq \sigma(x, y) \leq d_G(y) - 1$.

(b) *There are at least $\Delta - \sigma(x, y)$ vertices $z \in N_G(x) \setminus \{y\}$ for which there exists a coloring $\varphi \in \mathcal{C}^\Delta(G - xy)$ such that $\varphi(xz) \in \overline{\varphi}(y)$.*

Proof: Since the edge xy is critical, there is a coloring $\varphi \in \mathcal{C}^\Delta(G - xy)$. Since G is a simple graph, Proposition 4.25 implies that $\sigma(x, y) = |K(x, y)| \geq |\overline{\varphi}(x)| = \Delta - d_G(x) + 1$. On the other hand, we have $\sigma(x, y) = |K(x, y)| \leq |N_G(y) \setminus \{x\}| = d_G(y) - 1$. This proves (a).

For the proof of (b), consider the set Γ of all colors $\gamma \in \varphi(x)$ such that $\gamma \in \overline{\varphi}(y)$ or $y_\gamma \notin K(x, y)$. By Proposition 4.25, $K(x, y)$ contains all vertices y_β with $\beta \in \overline{\varphi}(x)$. Since G is a simple graph, this implies that $|\Gamma| = \Delta - \sigma(x, y)$.

Now, consider an arbitrary color $\gamma \in \Gamma$. Then $\gamma \in \varphi(x)$ and, therefore, the vertex $z = x_{\gamma, \varphi}$ exists. Since G is a simple graph, we have $z \in N_G(x) \setminus \{y\}$. We claim that there is a coloring $\varphi' \in \mathcal{C}^\Delta(G - xy)$ such that $\varphi'(xz) \in \overline{\varphi}'(y)$.

For the proof of the claim, we consider two cases. If $\gamma \in \overline{\varphi}(y)$, then the claim holds for $\varphi' = \varphi$. If $\gamma \notin \overline{\varphi}(y)$, then the vertex $u = y_{\gamma, \varphi}$ exists and u belongs to $N_G(y) \setminus \{x\}$, but not to $K(x, y)$. Hence $d_G(x) + d_G(y) + d_G(u) < 2\Delta + 2$. By Proposition 4.26(c), this yields $\overline{\varphi}(u) \cap (\overline{\varphi}(x) \cup \overline{\varphi}(y)) \neq \emptyset$. If there is a color $\alpha \in \overline{\varphi}(y) \cap \overline{\varphi}(u)$, then recolor the edge yu with α. This results in a coloring $\varphi' \in \mathcal{C}^\Delta(G - xy)$ such that $\varphi'(zx) = \varphi(zx) = \gamma \in \overline{\varphi}'(y)$ and we are done. If there is no such color, then there is a color $\beta \in \overline{\varphi}(x) \cap \overline{\varphi}(u)$. Clearly, $\beta \in \varphi(y)$ and, since the edge xy is uncolored, there is a color $\alpha \in \overline{\varphi}(y)$. Then Theorem 2.1 implies that $P = P_x(\alpha, \beta, \varphi)$ is a path joining x and y. Then, for the coloring $\varphi_1 = \varphi/P$, we have $\varphi_1 \in \mathcal{C}^\Delta(G - xy)$, $\varphi_1(zx) = \varphi_1(yu) = \gamma$, and $\beta \in \overline{\varphi}_1(y) \cap \overline{\varphi}_1(u)$. Then the coloring φ' obtained from φ_1 by recoloring yu with β belongs to $\mathcal{C}^\Delta(G - xy)$ and satisfies $\varphi'(zx) = \varphi_1(zx) = \gamma \in \overline{\varphi}'(y)$. This completes the proof of the claim. Thus (b) is proved. ∎

The following adjacency lemma is an enhancement of a result due to Sanders and Zhao [264]. The proof uses the fact that any critical edge can be extended to several Kierstead paths having four vertices.

Theorem 4.28 *Let G be a graph of class two with maximum degree Δ, and let xy be a critical edge of G. Furthermore, let Z denote the set of all vertices $z \in N_G(x) \setminus \{y\}$ for which there exists a coloring $\varphi \in \mathcal{C}^\Delta(G - xy)$ satisfying $\varphi(xz) \in \overline{\varphi}(y)$. Then the following statements hold:*

(a) $Z \subseteq K(y, x)$ *and* $|Z| \geq \Delta - \sigma(x, y) \geq \Delta - d_G(y) + 1$.

(b) *For each vertex $z \in Z$, there are at least $2\Delta + 1 - d_G(x) - d_G(y)$ vertices $u \in N_G(z) \setminus \{x\}$ such that $u = y$, or $d_G(x) = d_G(z) = \Delta$ and $d_G(u) \geq \Delta - d_G(y) + 1$, or $d_G(u) \geq 3\Delta - d_G(x) - d_G(y) - d_G(z) + 2$.*

Proof: Statement (a) follows from of Proposition 4.25(a) and Proposition 4.27(b).

For the proof of (b), consider an arbitrary vertex $z \in Z$. By the definition of Z, there is a coloring $\varphi \in \mathcal{C}^\Delta(G - xy)$ such that the color $\alpha = \varphi(xz)$ belongs to $\overline{\varphi}(y)$. By Proposition 4.26(a), the color set $\Gamma = \overline{\varphi}(x) \cup \overline{\varphi}(y)$ satisfies $|\Gamma| = 2\Delta - d_G(x) - d_G(y) + 2$.

First, we claim that $\Gamma \setminus \{\alpha\} \subseteq \varphi(z)$. Suppose this is not true. Then $\overline{\varphi}(z)$ contains a color $\beta \in \Gamma \setminus \{\alpha\}$. Clearly, $z \neq x, y$ and, therefore, $|\overline{\varphi}(z)| = \Delta - d_G(z)$. Consequently, $d_G(z) < \Delta$ and $K = (y, e, x, xz, z)$ is a Kierstead path with respect to e and φ, where $V(K)$ is not elementary with respect to φ, contradicting Theorem 3.1. This proves the claim. Therefore, for every color $\beta \in \Gamma \setminus \{\alpha\}$, the vertex z_β exists. Since G is a simple graph, for the set $U = \{z_\beta \mid \beta \in \Gamma \setminus \{\alpha\}\}$, we have $|U| = |\Gamma| - 1 = 2\Delta - d_G(x) - d_G(y) + 1$.

Now, consider an arbitrary color $\beta \in \Gamma \setminus \{\alpha\}$ such that $u = z_\beta$ is distinct from y. Since G is a simple graph, u is also distinct from x. Consequently, $K = (y, xy, x, xz, z, zu, u)$ is a Kierstead path with respect to xy and φ. Then Proposition 3.4(c) implies that $d_G(u) \geq 3\Delta - d_G(x) - d_G(y) - d_G(z) + 2$, or $d_G(x) = d_G(z) = \Delta$ and $d_G(u) \geq \Delta - d_G(y) + 1$. This completes the proof. ∎

Corollary 4.29 *Let G be a graph of class two with maximum degree Δ, and let xy be a critical edge of G. Then there are at least $\Delta - \sigma(x,y) \geq \Delta - d_G(y) + 1$ vertices $z \in N_G(x) \setminus \{y\}$ with $d_G(z) \geq 2\Delta - d_G(x) - d_G(y) + 2$ satisfying the following: There are at least $2\Delta + 1 - d_G(x) - d_G(y)$ vertices $u \in N_G(z) \setminus \{x\}$ such that $u = y$, or $d_G(x) = d_G(z) = \Delta$ and $d_G(u) \geq \Delta - d_G(y) + 1$, or $d_G(u) \geq 3\Delta - d_G(x) - d_G(y) - d_G(z) + 2$.*

The following strengthening of Vizing's Adjacency Lemma is due to Zhang [317] and can be easily deduced from Theorem 3.3, too.

Theorem 4.30 (Zhang [317] 2000) *Let G be a graph of class two with maximum degree Δ, and let xy be a critical edge of G. Then $d_G(x) + d_G(y) \geq \Delta + 2$ and equality implies that the following statements hold:*

(a) *If $z \in N_G(x,y) \setminus \{x,y\}$, then $d_G(z) = \Delta$.*

(b) *If $u \in N_G(N_G(x,y)) \setminus \{x,y\}$, then $d_G(u) \geq \Delta - 1$ and, moreover, $d_G(u) = \Delta$ provided that $d_G(x) < \Delta$ and $d_G(y) < \Delta$.*

Proof: By Vizing's Adjacency Lemma (Theorem 2.7), x is adjacent to at least $\Delta - d_G(y) + 1$ major vertices $z \neq y$. Since G is a simple graph, this implies that $d_G(x) + d_G(y) \geq \Delta + 2$. Now, assume that $d_G(x) + d_G(y) = \Delta + 2$.

To prove (a), let $z \in N_G(x,y) \setminus \{x,y\}$. By symmetry, we may assume that $z \in N_G(x) \setminus \{y\}$. Since x is adjacent to at least $\Delta - d_G(y) + 1$ major vertices distinct from y and G is simple, the assumption $d_G(x) = \Delta - d_G(y) + 2$ implies that z is a major vertex. This proves (a).

To prove (b), let us consider an arbitrary coloring $\varphi \in \mathcal{C}^\Delta(G - xy)$. Such a φ exists, since xy is a critical edge of G and $\chi'(G) = \Delta + 1$. Then, by Proposition 4.26(a), the color set $\Gamma = \overline{\varphi}(x) \cup \overline{\varphi}(y)$ satisfies $|\Gamma| = 2\Delta - d_G(x) - d_G(y) + 2 = \Delta$.

Now, let us consider an arbitrary vertex $u \in N_G(N_G(x,y)) \setminus \{x,y\}$. If $u \in N_G(x,y) \setminus \{x,y\}$, then (a) implies that $d_G(u) = \Delta$ and we are done. Otherwise, there is an edge $uz \in E(G)$ for some vertex $z \in N_G(x,y) \setminus \{x,y\}$. By symmetry, we may assume that $z \in N_G(x)$ and, therefore, there is an edge $xz \in E(G)$. Thus $\text{Path}(y, yx, x, xz, z, zu, u)$ is a path in G. Since $|\Gamma| = \Delta$, it follows that $\varphi(xz), \varphi(zu) \in \Gamma$. Consequently, $K = (y, yx, x, xz, z, zu, u)$ is a Kierstead path with respect to xy and φ. Then Theorem 3.3 implies that $|\overline{\varphi}(u) \cap \Gamma| \leq 1$ and, therefore, $d_G(u) = |\varphi(u)| \geq |\Gamma| - 1 = \Delta - 1$. If $d_G(x) < \Delta$, then Theorem 3.3 implies that $\overline{\varphi}(u) \cap \Gamma = \emptyset$ and, therefore, $d_G(u) = |\varphi(u)| \geq |\Gamma| = \Delta$. This completes the proof of (b). ∎

Theorem 4.31 (Woodall [309] 2007) *Let G be a graph of class two with maximum degree Δ, and let xy be a critical edge of G. Then there are at least $\Delta - \sigma(x, y) \geq \Delta - d_G(y) + 1$ vertices $z \in K(y, x)$ satisfying the following: $\sigma(x, y) + \sigma(x, z) \geq 2\Delta - d_G(x)$, and there are at least $2\Delta + 1 - d_G(x) - d_G(y)$ vertices $u \in N_G(z) \backslash \{x\}$ such that $u = y$, or $d_G(x) = d_G(z) = \Delta$ and $d_G(u) \geq \Delta - d_G(y) + 1$, or $d_G(u) \geq 3\Delta - d_G(x) - d_G(y) - d_G(z) + 2$.*

Proof: let Z denote the set of all vertices $z \in N_G(x) \backslash \{y\}$ for which there exists a coloring $\varphi \in \mathcal{C}^\Delta(G - xy)$ satisfying $\varphi(xz) \in \overline{\varphi}(y)$. It follows from Theorem 4.28(a) that $Z \subseteq K(y, x)$ and $|Z| \geq \Delta - \sigma(x, y) \geq \Delta - d_G(y) + 1$. By Theorem 4.28(b), for each vertex $z \in Z$, there are at least $2\Delta + 1 - d_G(x) - d_G(y)$ vertices $u \in N_G(z) \backslash \{x\}$ satisfying $u = y$, or $d_G(x) = d_G(z) = \Delta$ and $d_G(u) \geq \Delta - d_G(y) + 1$, or $d_G(u) \geq 3\Delta - d_G(x) - d_G(y) - d_G(z) + 2$. Hence, to complete the proof, it suffices to show that $\sigma(x, y) + \sigma(x, z) \geq 2\Delta - d_G(x)$ for all vertices $z \in Z$.

Let $\varphi \in \mathcal{C}^\Delta(G - xy)$ and let $z \in N_G(x) \backslash \{y\}$ such that $\alpha = \varphi(xz)$ belongs to $\overline{\varphi}(y)$. Now color the edge xy with α and make the edge xz uncolored. This results in a coloring $\varphi' \in \mathcal{C}^\Delta(G - xz)$ satisfying $\varphi'(v) = \varphi(v)$ for all vertices $v \in V(G) \backslash \{y, z\}$, $\varphi'(y) = \varphi(y) \cup \{\alpha\}$, and $\varphi'(z) = \varphi(z) \backslash \{\alpha\}$. In what follows, we write $\varphi' = \varphi(y \to z)$.

Claim 4.31.1 *Let $\varphi \in \mathcal{C}^\Delta(G - xy)$, let $z \in N_G(x) \backslash \{y\}$ such that $\alpha = \varphi(xz)$ belongs to $\overline{\varphi}(y)$, and let $\beta \in \varphi(x) \backslash \{\alpha\}$. Then the following statements hold:*

(a) *If $\beta \in \overline{\varphi}(y)$, then $z_\beta \in K(x, z)$.*

(b) *If $\beta \in \overline{\varphi}(z)$, then $y_\beta \in K(x, y)$.*

Proof: Assume first that $\beta \in \overline{\varphi}(y)$. By Proposition 4.25(a), $\beta \in \varphi(z)$. Since $\beta \neq \alpha$, $z_\beta \notin \{x, y\}$ and, therefore, $(y, yx, x, xz, z, zz_\beta, z_\beta)$ is a Kierstead path with respect to xy and φ. Then, by Proposition 3.4(c), $d_G(x) + d_G(z) + d_G(z_\beta) \geq 2\Delta + 2$ and so $z_\beta \in K(x, z)$. Thus (a) is proved.

Assume now that $\beta \in \overline{\varphi}(z)$. Let $\varphi' = \varphi(y \to z)$. Then $\varphi' \in \mathcal{C}^\Delta(G - xz)$, $y \in N_G(x) \backslash \{z\}$, and $\alpha = \varphi'(xy)$ belongs to $\overline{\varphi}'(z)$. Furthermore, $\beta \in \varphi'(x)$ and $\beta \in \overline{\varphi}'(z)$. Hence, we can apply statement (a) to the coloring φ' with z and y interchanged. This implies that that $y_{\beta, \varphi'} \in K(x, y)$. Since $y_{\beta, \varphi} = y_{\beta, \varphi'}$, this proves (b). $\triangle$

Claim 4.31.2 *Let $\varphi \in \mathcal{C}^\Delta(G - xy)$, let $z \in N_G(x) \backslash \{y\}$ such that $\alpha = \varphi(xz)$ belongs to $\overline{\varphi}(y)$, and let $\beta \in \varphi(x) \cap \varphi(y) \cap \varphi(z)$. Then the following statements hold:*

(a) *If $z_\beta \notin K(x, z)$, then $z_\beta \notin \{y, x\}$ and, moreover, $y_\beta \in K(x, y)$ or there is a color γ such that $\gamma \in \overline{\varphi}(x) \cap \overline{\varphi}(z_\beta) \cap \varphi(z) \cap \varphi(y_\beta)$*

(b) *If $y_\beta \notin K(x, y)$, then $y_\beta \notin \{z, x\}$ and, moreover, $z_\beta \in K(x, z)$ or there is a color γ' such that $\gamma' \in \overline{\varphi}(x) \cap \overline{\varphi}(y_\beta) \cap \varphi(y) \cap \varphi(z_\beta)$.*

Proof: For the proof of (a), assume that $z_\beta \notin K(x, z)$. Since $\beta \in \varphi(x) \cap \beta(y) \cap \varphi(z)$, the vertices $x_\beta, y_\beta, z_\beta$ do indeed exist. Since $\alpha \in \overline{\varphi}(y)$, $\alpha \neq \beta$. Since G is a simple graph and φ is an edge coloring of $G - xy$, the vertices $x_\beta, y_\beta, z_\beta$ are distinct and $x \notin \{y_\beta, z_\beta\}$. If $z_\beta = y$, then Proposition 4.25(a) implies that $d_G(x) + d_G(z) + d_G(z_\beta) \geq 2\Delta + 2$ and, therefore, $z_\beta \in K(x, z)$. This contradiction shows that $z_\beta \neq y$ and, therefore, $y_\beta \neq z$.

Since $z_\beta \notin K(x, z)$, $d_G(x) + d_G(z) + d_G(z_\beta) < 2\Delta + 2$. By Proposition 4.25(a), $\overline{\varphi}(x) \cap \overline{\varphi}(z) = \emptyset$. This implies that $\overline{\varphi}(z_\beta) \cap (\overline{\varphi}(x) \cup \overline{\varphi}(z)) \neq \emptyset$. Otherwise, since $\beta \notin \overline{\varphi}(x) \cup \overline{\varphi}(z) \cup \overline{\varphi}(z_\beta)$, we would obtain $\Delta - 1 \geq |\overline{\varphi}(x) \cup \overline{\varphi}(z) \cup \overline{\varphi}(z_\beta)| = |\overline{\varphi}(x)| + |\overline{\varphi}(z)| + |\overline{\varphi}(z_\beta)| = (\Delta - d_G(x) + 1) + (\Delta - d_G(z)) + (\Delta - d_G(z_\beta)) = 3\Delta + 1 - d_G(x) - d_G(z) - d_G(z_\beta) > \Delta - 1$, a contradiction.

First, assume that there is a color $\gamma \in \overline{\varphi}(z) \cap \overline{\varphi}(z_\beta)$. Then recolor the edge zz_β with γ, color the edge xy with α, and make the edge xz uncolored. This results in a coloring $\varphi' \in \mathcal{C}^\Delta(G - xz)$ such that $\alpha, \beta \in \overline{\varphi}'(z)$, $\varphi'(xy) = \alpha$, and $\varphi'(yy_\beta) = \beta$. Hence $(z, zx, x, xy, y, yy_\beta, y_\beta)$ is a Kierstead path with respect to xz and φ'. Then, by Proposition 3.4(c), we have $d_G(x) + d_G(y) + d_G(y_\beta) \geq 2\Delta + 2$ and, therefore $y_\beta \in K(x, y)$. Thus we are done.

Now, assume that $\overline{\varphi}(z) \cap \overline{\varphi}(z_\beta) = \emptyset$. Since $\overline{\varphi}(z_\beta) \cap (\overline{\varphi}(x) \cup \overline{\varphi}(z)) \neq \emptyset$, this implies that there is a color $\gamma \in \overline{\varphi}(x) \cap \overline{\varphi}(z_\beta)$. Clearly, we have $\gamma \in \varphi(z)$, $\gamma \neq \beta$, and, by Proposition 4.25(a), we also have $\gamma \neq \alpha$. If $\gamma \in \varphi(y_\beta)$ we are done. Otherwise, we have $\gamma \in \overline{\varphi}(y_\beta)$. Then we consider the chain $P = P_x(\beta, \gamma, \varphi)$ and the coloring $\varphi' = \varphi/P$. Clearly, P is a path, where one endvertex is x and the other endvertex is some vertex $u \neq x$. Furthermore, $\varphi' \in \mathcal{C}^\Delta(G - xy)$ and we have $\beta \in \overline{\varphi}'(x)$, $\alpha \in \overline{\varphi}'(y)$, and $\varphi'(xz) = \alpha$. If $\varphi'(yy_\beta) = \varphi(yy_\beta) = \beta$, then Proposition 4.25(b) implies that $y_\beta \in K(x, y)$ and we are done. Otherwise, $\varphi'(yy_\beta) \neq \varphi(yy_\beta)$ and, therefore, yy_β belongs to P. Since $\gamma \in \overline{\varphi}(y_\beta) \cap \overline{\varphi}(z_\beta)$, this implies that $u = y_\beta$ and so zz_β does not belong to P, which leads to $\varphi'(zz_\beta) = \varphi(zz_\beta) = \beta$. Since $\beta \in \overline{\varphi}'(x)$ and $\varphi'(xz) = \alpha \in \overline{\varphi}'(y)$, Proposition 4.25(b) implies that $z_\beta \in K(x, z)$, a contradiction. This completes the proof of (a).

For the proof of (b), assume that $y_\beta \notin K(x, y)$. Then, we can apply statement (a) to the coloring $\varphi' = \varphi(y \to z)$ with y and z interchanged. This proves (b). Δ

Claim 4.31.3 *Let $\varphi \in \mathcal{C}^\Delta(G - xy)$, let $z \in N_G(x) \setminus \{y\}$ such that $\alpha = \varphi(xz)$ belongs to $\overline{\varphi}(y)$, and let $\beta \in \varphi(x) \setminus \{\alpha\}$. Then $y_\beta \in K(x, y)$ or $z_\beta \in K(x, z)$.*

Proof: Suppose, on the contrary, that $y_\beta \notin K(x, y)$ and $z_\beta \notin K(x, z)$. It follows from Claim 4.31.1 that $\beta \in \varphi(x) \cap \varphi(y) \cap \varphi(z)$. To arrive at a contradiction, let us consider the set $\mathcal{P}$ of all colorings $\varphi' \in \mathcal{C}^\Delta(G - xy)$ satisfying $\varphi'(xz) = \alpha$, $\alpha \in \overline{\varphi}'(y)$, $\beta \in \varphi'(x)$, and $v_{\beta, \varphi'} = v_{\beta, \varphi}$ for $v \in \{x, y, z\}$. Clearly, the coloring φ belongs to $\mathcal{P}$. Note that every coloring $\varphi' \in \mathcal{P}$ satisfies $\beta \in \varphi'(x) \cap \varphi'(y) \cap \varphi'(z)$ because of Claim 4.31.1 and the assumption.

Next, we claim that there is a coloring $\varphi' \in \mathcal{P}$, such that $\alpha \in \overline{\varphi}'(y) \cap \overline{\varphi}'(y_\beta)$. To prove the claim, we start with the coloring φ. If $\alpha \in \overline{\varphi}(y) \cap \overline{\varphi}(y_\beta)$, we are done. Otherwise, we have $\alpha \in \overline{\varphi}(y) \cap \varphi(y_\beta)$. Since $\beta \in \varphi(x) \cap \varphi(y) \cap \varphi(z)$, it follows from Claim 4.31.2(b) that there is a color $\gamma' \in \overline{\varphi}(x) \cap \overline{\varphi}(y_\beta) \cap \varphi(y) \cap \varphi(z_\beta)$. Clearly,

$\gamma' \notin \{\alpha, \beta\}$. It now follows from Theorem 2.1 that $P = P_x(\alpha, \gamma', \varphi)$ is a path with endvertices x and y, which implies $xz \in E(P)$. Hence, $P' = P_{y_\beta}(\alpha, \gamma, \varphi)$ is a path disjoint from P, where y_β is an endvertex of P'. This implies that the coloring $\varphi' = \varphi/P'$ belongs to $\mathcal{P}$ and $\alpha \in \overline{\varphi}'(y) \cap \overline{\varphi}(y_\beta)$. Thus, the claim is proved.

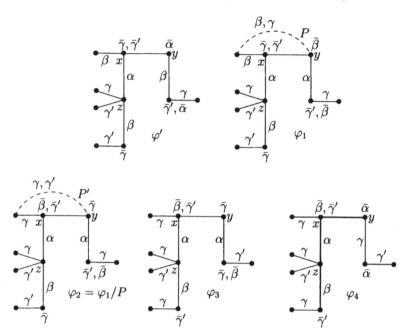

Figure 4.7   The colorings $\varphi', \varphi_1, \varphi_2, \varphi_3, \varphi_4 \in C^\Delta(G - xy)$.

Now, let $\varphi' \in \mathcal{P}$ be a coloring such that $\alpha \in \overline{\varphi}'(y) \cap \overline{\varphi}'(y_\beta)$. Then $\beta \in \varphi'(x) \cap \varphi'(y) \cap \varphi'(z)$. By Claim 4.31.2, the vertices $x, y, z, x_\beta, y_\beta, z_\beta$ are distinct and there is a color $\gamma \in \overline{\varphi}'(x) \cap \overline{\varphi}'(z_\beta) \cap \varphi'(z) \cap \varphi'(y_\beta)$ as well as a color $\gamma' \in \overline{\varphi}'(x) \cap \overline{\varphi}'(y_\beta) \cap \varphi'(y) \cap \varphi'(z_\beta)$. Clearly, the colors $\alpha, \beta, \gamma, \gamma'$ are distinct and we have $\gamma \in \varphi(y)$ and $\gamma' \in \varphi(z)$ (Proposition 4.25(a)) To get a contradiction, we shall construct several new colorings of $G - xy$ (Fig. 4.7). First, recolor the edge yy_β with color α to form a new coloring $\varphi_1 \in C^\Delta(G - xy)$. Then $\gamma \in \overline{\varphi}_1(x)$ and $\beta \in \overline{\varphi}_1(y) \cap \overline{\varphi}_1(y_\beta)$. Then $P = P_x(\beta, \gamma, \varphi_1)$ is a path joining x and y (Theorem 2.1). Thus the coloring $\varphi_2 = \varphi_1/P$ belongs to $C^\Delta(G - xy)$ and satisfies $\varphi_2(v) = \varphi_1(v)$ for $v \in V(G) \setminus \{x, y\}$, $\varphi_2(x) = (\varphi_1(x) \setminus \{\beta\}) \cup \{\gamma\}$, $\varphi_2(y) = (\varphi_2(y) \setminus \{\gamma\}) \cup \{\beta\}$, $\varphi_2(xz) = \varphi_2(yy_\beta) = \alpha$, and $\varphi_2(zz_\beta) = \beta$. Then $\gamma \in \overline{\varphi}_2(y) \cap \overline{\varphi}_2(z_\beta)$ and $\gamma' \in \overline{\varphi}_2(x) \cap \overline{\varphi}_2(y_\beta)$. By Theorem 2.1, this implies that $P' = P_x(\gamma, \gamma', \varphi_2)$ is a path joining x and y. Form a coloring φ_3 from φ_2 by interchanging colors γ and γ' on all edges not belonging to P'. Then φ_3 belongs to $C^\Delta(G - xy)$ and satisfies $\gamma \in \overline{\varphi}_3(y) \cap \overline{\varphi}_3(y_\beta)$, $\gamma' \in \overline{\varphi}_3(x) \cap \overline{\varphi}_3(z_\beta)$, $\varphi_3(xz) = \varphi_3(yy_\beta) = \alpha$, $\varphi_3(zz_\beta) = \beta$, and $\beta \in \overline{\varphi}_3(x) \cap \overline{\varphi}_3(y_\beta)$. Finally, form the coloring φ_4 from φ_3 by recoloring yy_β with γ. Then we have $\varphi_4 \in C^\Delta(G - xy)$, $\alpha \in \overline{\varphi}_4(y)$, $\beta \in \overline{\varphi}_4(x)$, $\varphi_4(xz) = \alpha$, and

$\varphi_4(zz_\beta) = \beta$. Hence $(y, yx, x, xz, z, zz_\beta, z_\beta)$ is a Kierstead path with respect to xy and φ_4. By Proposition 3.4(c), this gives $d_G(x) + d_G(z) + d_G(z_\beta) \geq 2\Delta + 2$ and so $z_\beta \in K(x, z)$. This contradiction completes the proof of Claim 4.31.3. $\triangle$

Now, consider an arbitrary vertex $z \in Z$. By the definition of Z, there is a coloring $\varphi \in \mathcal{C}^\Delta(G - xy)$ such that $\varphi(xz) \in \overline{\varphi}(y)$. Since G is a simple graph, it follows from Proposition 4.25 and Claim 4.31.3 that $\sigma(x, y) + \sigma(x, z) = |K(x, y)| + |K(x, z)| \geq 2|\overline{\varphi}(x)| + |\varphi(x)| - 1 = 2(\Delta - d_G(x) + 1) + (d_G(x) - 1) - 1 = 2\Delta - d_G(x)$. This completes the proof of the theorem.   ∎

Corollary 4.32 (Woodall [309] 2007) *Let G be a critical graph of class two with maximum degree Δ, let x be a vertex of G, and let*

$$p_{\min} = \min_{y \in N_G(x)} \sigma_G(x, y) - \Delta + d_G(x) - 1. \tag{4.7}$$

Then $0 \leq p_{\min} \leq d_G(x) - 2$, and x is adjacent to at least $d_G(x) - p_{\min} - 1$ vertices z such that $\sigma_G(x, z) \geq \Delta - p_{\min} - 1$.

Proof: That $0 \leq p_{\min} \leq d_G(x) - 2$ follows from Proposition 4.27(a) and the fact that $d_G(y) \leq \Delta$. If y achieves the minimum in (4.7), then Theorem 4.31 implies that there are at least $\Delta - \sigma_G(x, y) = d_G(x) - p_{\min} - 1$ vertices $z \in N_G(x) \setminus \{y\}$ for which $\sigma_G(x, z) \geq 2\Delta - d_G(x) - \sigma_G(x, y) = \Delta - p_{\min} - 1$.   ∎

Corollary 4.33 (Woodall [309] 2007) *Let G be a critical graph of class two with maximum degree Δ, let x be a vertex of G, and let*

$$p_{\min} = \min_{y \in N_G(x)} \sigma_G(x, y) - \Delta + d_G(x) - 1 \tag{4.8}$$

and

$$p = \min\{p_{\min}, \lfloor \tfrac{1}{2} d_G(x) \rfloor - 1\}. \tag{4.9}$$

Then x has at least $d_G(x) - p - 1$ neighbors y such that $\sigma_G(x, y) \geq \Delta - p - 1$.

Proof: If $p = p_{\min}$ the result reduces to Corollary 4.32. Otherwise, we have $p = \lfloor \tfrac{1}{2} d_G(x) \rfloor - 1 < p_{\min}$ and every vertex $y \in N_G(x)$ satisfies

$$\begin{aligned} \sigma_G(x, y) \;&>\; \Delta - d_G(x) + 1 + \lfloor \tfrac{1}{2} d_G(x) \rfloor - 1 \\ &\geq\; \Delta - \lfloor \tfrac{1}{2} d_G(x) \rfloor - 1 \\ &=\; \Delta - p - 2, \end{aligned}$$

by (4.8) and (4.9), which gives $\sigma_G(x, y) \geq \Delta - p - 1$.   ∎

Remark 4.34 *Let G be a graph with $\chi'(G) = k + 1$ for an integer $k \geq \Delta(G)$, let $e \in E_G(x, y)$ be a critical edge, and let $\varphi \in \mathcal{C}^k(G - e)$ be a coloring. Suppose that $\alpha, \beta \in \{1, \ldots, k\}$ are two colors such that $\alpha \in \overline{\varphi}(x)$ and $\beta \in \overline{\varphi}(y)$. Then,* *by Theorem 2.1, the chain $P = P_x(\alpha, \beta, \varphi)$ is a path joining x and y. Now let φ' denote the coloring obtained from φ by interchanging colors α and β for all edges not belonging to P. Then, for the coloring $\varphi' \in C^k(G - e)$, we have $\overline{\varphi}'(x) = \overline{\varphi}(x)$, $\overline{\varphi}'(y) = \overline{\varphi}(y)$, and $P = P_x(\alpha, \beta, \varphi')$. In what follows, we briefly write $\varphi' = \varphi(\alpha \leftrightarrow \beta)$ or $\varphi' = \varphi(\beta \leftrightarrow \alpha)$.*

The next three adjacency lemmas deal with vertices which happen to be in triangles. This is useful when dealing with planar graphs as in Sect. 4.9.

Theorem 4.35 (Sanders and Zhao [264] 2001) *Let G be a graph of class two with maximum degree Δ. Then there are no distinct vertices x, y, z such that xy is a critical edge, $xz \in E(G)$, $d_G(x) + d_G(y) + d_G(z) < 2\Delta + 2$, and xz is in at least $d_G(x) + d_G(y) - \Delta - 2$ triangles not containing y.*

Sketch of Proof: Suppose, on the contrary, that G is a graph of class two with maximum degree Δ containing such vertices x, y, z. Let T denote the set of all vertices u such that $u \neq y$ and xzu is a triangle. Then, by assumption, $|T| \geq d_G(x) + d_G(y) - \Delta - 2$.

Since xy is a critical edge of G, $C^{\Delta}(G - xy) \neq \emptyset$. In the sequel, let $\varphi \in C^{\Delta}(G - xy)$ be an arbitrary coloring and let $\alpha = \varphi(xz)$.

By hypothesis, $z \in N_G(x) \setminus \{y\}$ and $d_G(x) + d_G(y) + d_G(z) < 2\Delta + 2$. Hence, by Proposition 4.25(a), we have $\alpha \in \varphi(y)$ and, therefore, $\alpha \in \varphi(x) \cap \varphi(y) \cap \varphi(z)$. Then Proposition 4.26(c) implies $|\overline{\varphi}(z) \cap (\overline{\varphi}(x) \cup \overline{\varphi}(y))| \geq 2$.

Next, let us form an auxiliary graph H. The vertex set of H consists of all colors, except the color α. For each vertex $u \in T$, we introduce an edge in H joining $\varphi(ux)$ and $\varphi(uz)$. Since φ is an edge coloring of $G - xy$ and $\alpha = \varphi(xz)$, the graph H has $|T|$ edges and maximum degree at most two. By Proposition 4.26(b), the color set $\Gamma = (\varphi(x) \cap \varphi(y)) \setminus \{\alpha\}$ satisfies $|\Gamma| = d_G(x) + d_G(y) - \Delta - 3$ and, therefore, $|T| \geq |\Gamma| + 1$. Then we conclude that H contains a cycle having exactly one vertex not in Γ, or H contains a path, where the only vertices not in Γ are the two endvertices. Hence, there is a sequence $\alpha_1, \ldots, \alpha_r$ of colors and a sequence of distinct vertices $u_1, \ldots, u_{r-1} \in T$ such that $r \geq 2$, $\alpha_2, \ldots, a_{r-1} \in \Gamma$ are distinct colors, $\alpha_1, \alpha_r \in \overline{\varphi}(x) \cup \overline{\varphi}(y)$, $\varphi(u_i x) = \alpha_i$ and $\varphi(u_i z) = \alpha_{i+1}$ for $i = 1, \ldots, r-1$. Clearly, $\alpha_1 = \varphi(u_1 x)$ belongs to $\overline{\varphi}(y)$ and $\alpha_r = \varphi(u_{r-1} z)$ belongs to $\overline{\varphi}(x) \cup \overline{\varphi}(y)$. To obtain a contradiction, we consider three cases.

Case 1: $\alpha_1 \neq \alpha_r$ *and* $\alpha_r \in \overline{\varphi}(x)$. If $\alpha_1 \in \overline{\varphi}(z)$, then swapping the colors on $u_i x$ and $u_i y$ for $i = 1, \ldots, r-1$, and coloring xy with α_1 gives a Δ-edge coloring of G, contradicting $\chi'(G) = \Delta + 1$.

Now, assume that $\alpha_1 \notin \overline{\varphi}(z)$. If there exists a color $\beta \in \overline{\varphi}(z) \cap \overline{\varphi}(x)$, then $\varphi' = \varphi(\beta \leftrightarrow \alpha_1)$ is a coloring of $G - xy$ such that $\alpha_1 \in \overline{\varphi}'(z) \cap \overline{\varphi}'(y)$, $\alpha_r \in \overline{\varphi}'(x)$, $\varphi'(u_i x) = \varphi(u_i x)$ and $\varphi'(u_i z) = \varphi(u_i z)$ for $i = 1, \ldots, r-1$ (Remark 4.34). Thus, as before, swapping the colors on $u_i x$ and $u_i z$ for $i = 1, \ldots, r-1$, and coloring xy with α_1 gives a Δ-edge-coloring of G, contradicting $\chi'(G) = \Delta + 1$.

If $\overline{\varphi}(z) \cap \overline{\varphi}(x) = \emptyset$, then there are two colors $\gamma_1, \gamma_2 \in \overline{\varphi}(z) \cap \overline{\varphi}(y)$, because $|\overline{\varphi}(z) \cap (\overline{\varphi}(x) \cup \overline{\varphi}(y))| \geq 2$. Note that $\alpha_2, \ldots, \alpha_{r-1} \in \Gamma$ and $\gamma_1, \gamma_2 \notin \Gamma$. Then we construct the sequence $(\varphi_i)_{i=0}^{4}$ of colorings as follows: $\varphi_0 = \varphi, \varphi_1 = \varphi(\alpha_r \leftrightarrow \gamma_1)$,

$\varphi_2 = \varphi_1(\alpha_r \leftrightarrow \alpha_1)$, $\varphi_3 = \varphi_2(\alpha_r \leftrightarrow \gamma_2)$, and $\varphi_4 = \varphi_3(\alpha_r \leftrightarrow \gamma_1)$. Then $\varphi_i(u_1 x) = \alpha_1$ for $i = 0, 1, 2, 3, 4$ and $(\varphi_i(u_{r-1}z))_{i=0}^4 = (\alpha_r, \gamma_1, \gamma_1, \gamma_1, \alpha_r)$. For the coloring $\varphi_4 \in \mathcal{C}^\Delta(G - xy)$, we have $\alpha_1 \in \overline{\varphi}_4(z) \cap \overline{\varphi}_4(y)$, $\alpha_r \in \overline{\varphi}_4(x)$, $\varphi_4(u_i x) = \varphi(u_i x)$ and $\varphi_4(u_i z) = \varphi(u_i z)$ for $i = 1, \ldots, r - 1$. Thus, as before, swapping the colors on $u_i x$ and $u_i z$ for $i = 1, \ldots, r - 1$, and coloring xy with α_1 gives a Δ-edge-coloring of G, contradicting $\chi'(G) = \Delta + 1$. This settles Case 1.

Case 2: $\alpha_1 \neq \alpha_r$ and $\alpha_r \in \overline{\varphi}(y)$. If there is a color $\gamma \in \overline{\varphi}(x) \cap \overline{\varphi}(z)$, then let $\varphi' = \varphi/P_x(\gamma, \alpha_r, \varphi)$. If there is no such color, then, since $|\overline{\varphi}(z) \cap (\overline{\varphi}(x) \cup \overline{\varphi}(y))| \geq 2$, there is a color $\gamma' \in \overline{\varphi}(y) \cap \overline{\varphi}(z)$ such that $\gamma' \neq \alpha_1$ and, moreover, there is a color $\gamma \in \overline{\varphi}(x)$. Then let $\varphi_1 = \varphi(\gamma' \leftrightarrow \gamma)$ and $\varphi' = \varphi_1/P_x(\gamma, \alpha_r, \varphi_1)$.

In both cases, the coloring $\varphi' \in \mathcal{C}^\Delta(G - xy)$ satisfies $\alpha_r \in \overline{\varphi}'(x)$, $\alpha_1 \in \overline{\varphi}'(y)$, $\varphi'(xz) = \alpha$, $\varphi'(u_i x) = \varphi(u_i x)$ and $\varphi'(u_i z) = \varphi(u_i z)$ for $i = 1, \ldots, r - 1$. Hence, we obtain a contradiction as in Case 1.

Case 3: $\alpha_1 = \alpha_r$. Then, $r \geq 3$ and $\alpha_1 \in \overline{\varphi}(y)$. If there is a color $\gamma \in \overline{\varphi}(x) \cap \overline{\varphi}(z)$, then let $\varphi' = \varphi(\gamma \leftrightarrow \alpha_1)$. If there is no such color, then, since $|\overline{\varphi}(z) \cap (\overline{\varphi}(x) \cup \overline{\varphi}(y))| \geq 2$, there is a color $\delta \in \overline{\varphi}(y) \cap \overline{\varphi}(z)$ such that $\delta \neq \alpha_1$ and there is a color $\gamma \in \overline{\varphi}(x)$. Then let $\varphi_1 = \varphi(\gamma \leftrightarrow \delta)$ and $\varphi' = \varphi_1(\gamma \leftrightarrow \alpha_1)$.

In both cases, the coloring $\varphi' \in \mathcal{C}^\Delta(G - xy)$ satisfies $\overline{\varphi}(xz) = \alpha$, $\varphi'(u_i x) = \alpha_i$ for $i = 1, \ldots, r - 1$, $\varphi'(u_i z) = \alpha_{i+1}$ for $i = 1, \ldots, r - 2$, $\varphi'(u_{r-1}z) = \gamma \in \overline{\varphi}'(x)$, $\alpha_1 \in \overline{\varphi}'(y)$, and $\alpha, \alpha_2, \ldots, \alpha_{r-1} \in \varphi'(x) \cap \varphi'(y)$. Since $\gamma \neq \alpha_1$, we obtain a contradiction as in Case 1.

Hence, the proof of the theorem is complete. ∎

Theorem 4.36 (Sanders and Zhao [264] 2001) *Let G be a graph of class two with maximum degree Δ. Then there are no distinct vertices v, w, x, y, z such that xy is a critical edge, $d_G(w) \leq \Delta - 2$, $d_G(x) + d_G(y) \leq \Delta + 3$, $d_G(x) \geq 5$, $d_G(y) \geq 5$, and vwz and xyz are triangles.*

Sketch of Proof: Suppose, on the contrary, that G is a graph of class two with maximum degree Δ containing such vertices v, w, x, y, z. Since xy is a critical edge with $d_G(x) + d_G(y) \leq \Delta + 3$ and $w \in N_G(N_G(x, y)) \setminus \{x, y\}$ with $d_G(w) \leq \Delta - 2$, Theorem 4.30 implies that $d_G(x) + d_G(y) = \Delta + 3$. For $u \in \{x, y\}$, we have $d_G(u) \geq 5$ and, therefore, $d_G(u) \leq \Delta - 2$.

Since xy is a critical edge, there is a coloring $\varphi \in \mathcal{C}^\Delta(G - xy)$. Let $\Gamma = \overline{\varphi}(x) \cup \overline{\varphi}(y)$ and $\Gamma' = \varphi(x) \cap \varphi(y)$. By Proposition 4.26, we have $|\Gamma| = \Delta - 1$ and $|\Gamma'| = 1$. Let $\alpha = \varphi(xz)$, $\beta = \varphi(yz)$, $\gamma = \varphi(zv)$, and $\delta = \varphi(zw)$. Clearly, the colors $\alpha, \beta, \gamma, \delta$ are distinct and $\Gamma \cap \{\alpha, \beta\} \neq \emptyset$. By symmetry, we may assume that $\alpha \in \Gamma$ and hence $\alpha \in \overline{\varphi}(y)$. If $\delta \in \Gamma$, then (y, yx, x, xz, z, zw, w) is a Kierstead path with respect to xy and φ such that $d_G(x) + d_G(y) + d_G(z) + d_G(w) \leq (\Delta + 3) + \Delta + (\Delta - 2) = 3\Delta + 1$. Since $d_G(x) \leq \Delta - 2$, this is a contradiction to Proposition 3.4(c). Hence $\delta = \varphi(zw)$ is the only color in Γ' and, therefore, we have $\beta \in \overline{\varphi}(x)$ and $\gamma \in \Gamma$. By symmetry, we may assume that $\gamma \in \overline{\varphi}(x)$.

In what follows, let $\mathcal{P}$ denote the set of all colorings $\varphi' \in \mathcal{C}^\Delta(G - xy)$ such that $\varphi'(e) = \varphi(e)$ for all edges $e \in \{xz, yz, zv, zw\}$, $\beta, \gamma \in \overline{\varphi}'(x)$, $\alpha \in \overline{\varphi}'(y)$, and $\delta \in \varphi'(x) \cap \varphi'(y)$. Clearly, we have $\varphi \in \mathcal{P}$.

Let us consider an arbitrary coloring $\varphi' \in \mathcal{P}$. Then we have $\overline{\varphi}'(x) \cup \overline{\varphi}'(y) = \Gamma = \{1, \ldots, \Delta\} \setminus \{\delta\}$ and $\varphi'(x) \cap \varphi'(y) = \Gamma' = \{\delta\}$. Since $\delta = \varphi'(zw)$, we also have $\overline{\varphi}'(w) \subseteq \Gamma$, where $|\overline{\varphi}(w)| = \Delta - d_G(w) \geq 2$. For $u \in \{x, y\}$, we have $|\overline{\varphi}'(u)| = \Delta + 1 - d_G(u) \geq 3$.

First, we construct a coloring $\varphi_1 \in \mathcal{P}$ with $\beta \in \overline{\varphi}_1(w)$. If $\beta \in \overline{\varphi}(w)$, then let $\varphi_1 = \varphi$. So assume $\beta \notin \overline{\varphi}(w)$. If there is a color $\beta' \in \overline{\varphi}(w) \cap \overline{\varphi}(y)$, then let $\varphi_1 = \varphi(\beta \leftrightarrow \beta')$. Otherwise, $|\overline{\varphi}(w) \cap \overline{\varphi}(x)| \geq 2$ and there is a color $\beta' \in \overline{\varphi}(w) \cap \overline{\varphi}(x)$ with $\beta' \neq \gamma$. Then let $\varphi_1 = (\varphi(\beta' \leftrightarrow \alpha))(\alpha \leftrightarrow \beta)$. In all cases, we have $\varphi_1 \in \mathcal{P}$ and $\beta \in \overline{\varphi}_1(w)$. Now, we construct a coloring $\varphi_2 \in \mathcal{P}$ with $\varphi_2(vw) = \alpha$. If $\varphi_1(vw) = \alpha$, then let $\varphi_2 = \varphi_1$. Otherwise, $\varphi_1(vw) = \alpha' \neq \alpha$ and $\alpha' \in \overline{\varphi}_1(x) \cup \overline{\varphi}_1(y)$. If $\alpha' \in \overline{\varphi}_1(x)$, let $\varphi_2 = (\varphi_1(\beta \leftrightarrow \alpha))(\alpha' \leftrightarrow a)$. If $\alpha' \in \overline{\varphi}_1(y)$, let $\varphi_2 = (\varphi_1(\alpha' \leftrightarrow \beta))(\beta \leftrightarrow \alpha)$. In all cases, $\varphi_2 \in \mathcal{P}$ and $\varphi_2(vw) = \alpha$.

Let $\mathcal{P}'$ denote the set of all colorings $\varphi' \in \mathcal{P}$ such that $\varphi'(vw) = \alpha$. Clearly, $\varphi_2 \in \mathcal{P}'$. Let $\varphi' \in \mathcal{P}'$ be an arbitrary coloring. If $\gamma \in \overline{\varphi}'(w)$, then $P_x(\alpha, \gamma, \varphi') = \text{Path}(x, xz, z, zv, v, vw, w)$. Since $\alpha \in \overline{\varphi}'(y)$ and $\gamma \in \overline{\varphi}'(x)$, this is a contradiction to Theorem 2.1. Hence, we have $\gamma \in \varphi'(w)$. Now, we claim that $|\overline{\varphi}'(x) \cap \overline{\varphi}'(w)| \leq 1$. Suppose, on the contrary, that $|\overline{\varphi}'(x) \cap \overline{\varphi}'(w)| \geq 2$. Then color the edge xy by α and make xz uncolored. This results in a coloring $\varphi_1' \in \mathcal{C}^\Delta(G - xz)$ such that (x, xz, z, zv, v, vw, w) is a Kierstead path with respect to xz and φ_1', and, moreover, $|\overline{\varphi}_1'(x) \cap \overline{\varphi}_1'(w)| = |\overline{\varphi}'(x) \cap \overline{\varphi}'(w)| \geq 2$. This, however, is a contradiction to Theorem 3.3(d). Hence the claim is proved.

Next, we claim that there is a coloring $\varphi' \in \mathcal{P}'$ such that $\beta \in \overline{\varphi}'(w)$. To prove this, choose a coloring $\varphi' \in \mathcal{P}'$. If $\beta \in \overline{\varphi}'(w)$, then there is nothing to prove. So assume that $\beta \in \varphi'(w)$. Since we have $|\overline{\varphi}'(w) \cap \overline{\varphi}'(x)| \leq 1$, $\overline{\varphi}'(w) \subseteq \overline{\varphi}'(x) \cup \overline{\varphi}'(y)$, and $|\overline{\varphi}'(w)| \geq 2$, we conclude that there is a color $\alpha' \in \overline{\varphi}'(w) \cap \overline{\varphi}'(y)$. Since $\overline{\varphi}'(x) \cap \overline{\varphi}'(y) = \emptyset$ and $\delta \in \varphi'(x) \cap \varphi'(y)$, the colors $\alpha, \alpha', \beta, \gamma, \delta$ are distinct. Since $\beta \in \overline{\varphi}'(x)$ and $\alpha' \in \overline{\varphi}'(y) \cap \overline{\varphi}'(w)$, it follows then that the coloring $\varphi'' = \varphi'(\alpha' \leftrightarrow \beta)$ belongs to $\mathcal{P}'$ and $\beta \in \overline{\varphi}'(w)$. This proves the claim.

Finally, let us consider an arbitrary coloring $\varphi' \in \mathcal{P}'$ such that $\beta \in \overline{\varphi}'(w)$. As before, we conclude that there is a color $\alpha' \in \overline{\varphi}'(y) \cap \overline{\varphi}'(w)$. Furthermore, we have $|\overline{\varphi}'(x)| \geq 3$. Hence there is a color $\beta' \in \overline{\varphi}'(x) \setminus \{\gamma, \beta\}$. Obviously, the colors $\alpha, \alpha', \beta, \beta', \gamma, \delta$ are distinct. Since $\beta' \in \overline{\varphi}'(x)$ and $\alpha' \in \overline{\varphi}'(y) \cap \overline{\varphi}'(w)$, it follows then that the coloring $\varphi'' = \varphi'(\alpha' \leftrightarrow \beta')$ belongs to $\mathcal{P}'$ and $\beta, \beta' \in \overline{\varphi}'(w) \cap \overline{\varphi}''(x)$. This contradiction completes the proof. ∎

Theorem 4.37 (Sanders and Zhao [264] 2001) *Let G be a graph of class two with maximum degree Δ. Then there are no distinct vertices v, w, x, y, z such that xy is a critical edge, $d_G(v) \leq \Delta - 1$, $d_G(w) \leq \Delta - 1$, $d_G(x) + d_G(y) \leq \Delta + 3$, $d_G(x) \geq 4$, $d_G(y) \geq 4$, xyz is a triangle, and z is adjacent to v and w.*

Sketch of Proof: Suppose, on the contrary, that G is a graph of class two with maximum degree Δ containing such vertices v, w, x, y, z. Note that, for $u \in \{x, y\}$, we have $d_G(u) \geq 4$ and, therefore, $d_G(u) \leq \Delta - 1$. Since xy is a critical edge with $d_G(x) + d_G(y) \leq \Delta + 3$ and $v, w \in N_G(N_G(x, y)) \setminus \{x, y\}$ with $d_G(v), d_G(w) \leq \Delta - 1$, Theorem 4.30 implies that $d_G(x) + d_G(y) = \Delta + 3$.

Since xy is a critical edge, there is a coloring $\varphi \in \mathcal{C}^{\Delta}(G - xy)$. Let $\Gamma = \overline{\varphi}(x) \cup \overline{\varphi}(y)$ and $\Gamma' = \varphi(x) \cap \varphi(y)$. By Proposition 4.26, we have $|\Gamma| = \Delta - 1$ and $|\Gamma'| = 1$. Let $\alpha = \varphi(xz)$, $\beta = \varphi(yz)$, $\gamma = \varphi(zv)$, and $\delta = \varphi(zw)$. Clearly, the colors $\alpha, \beta, \gamma, \delta$ are distinct and $|\Gamma' \cap \{\alpha, \beta, \gamma, \delta\}| \le 1$.

Now, we claim that $\Gamma' \cap \{\gamma, \delta\} \ne \emptyset$. Suppose this is false. By symmetry, we may assume that $\alpha \in \Gamma$ and hence $\alpha \in \overline{\varphi}(y)$. Then (y, yx, x, xz, z, zv, v) is a Kierstead path with respect to xy and φ, where $d_G(x) < \Delta$. It follows now from Theorem 3.3 that $\{x, y, z, v\}$ is elementary with respect to φ. Since $|\Gamma| = \Delta - 1$, this implies that $\overline{\varphi}(v) = \{\delta_1\}$ and δ_1 is the only color in Γ'. The same argument can be applied to the Kierstead path (y, yx, x, xz, z, zw, w). Hence, we have $\overline{\varphi}(w) = \{\delta_1\}$. The colors $\alpha, \gamma, \delta, \delta_1$ are distinct and $\Delta \ge 5$. Hence there is a color $\beta' \in \Gamma \setminus \{\alpha, \gamma, \delta\}$ (possibly $\beta' = \beta$). Since $\beta' \ne \delta_1$, for $u \in \{v, w\}$, the chain $P_u = P_u(\beta', \delta_1, \varphi)$ is a path, where one endvertex is u and the other endvertex is some vertex $u' \ne u$. This implies that v' or w' does not belong to $\{x, y\}$, say $v' \notin \{x, y\}$. For the coloring $\varphi' = \varphi/P_v \in \mathcal{C}^{\Delta}(G - xy)$, we then obtain that (y, yx, x, xz, z, zv, v) is a Kierstead path with respect to xy and φ' such that $\beta' \in \overline{\varphi}'(v) \cap (\overline{\varphi}'(x) \cup \overline{\varphi}'(y))$, contradicting Theorem 3.3. This proves the claim that $\Gamma' \cap \{\gamma, \delta\} \ne \emptyset$.

By symmetry, we may assume that $\delta \in \Gamma'$, and hence $\gamma \in \Gamma$. Since δ is the only color not in Γ, we have $\alpha, \beta \in \Gamma$. Then $\alpha \in \overline{\varphi}(y)$, $\beta \in \overline{\varphi}(x)$, and, by symmetry, $\gamma \in \overline{\varphi}(x)$. Then the sequence (y, yx, x, xz, z, zv, v) is a Kierstead path with respect to xy and φ and, by Theorem 3.3, we obtain that $\overline{\varphi}(v) = \{\delta\}$. Since $|\overline{\varphi}(w)| = \Delta - d_G(w) \ge 1$, there is a color $\gamma' \in \overline{\varphi}(w)$.

First, consider the case when $\gamma' \in \overline{\varphi}(y)$. Then the colors $\beta, \gamma, \delta, \gamma'$ are distinct. Since $\overline{\varphi}(v) = \{\delta\}$, the chain $P_v = \mathcal{P}_v(\delta, \gamma', \varphi)$ is a path where one endvertex is v and the other endvertex is some vertex $v' \ne v$. If $v' \ne y$, then for the coloring $\varphi' = \varphi/P_v \in \mathcal{C}^{\Delta}(G - xy)$, we have that (x, xy, y, yz, z, zv, v) is a Kierstead path with respect to xy and φ' such that $\gamma' \in \overline{\varphi}'(v) \cap \overline{\varphi}'(y)$, contradicting Theorem 3.3. If $v = y$, then for the coloring $\varphi' = \varphi/P_w(\delta, \gamma', \varphi) \in \mathcal{C}^{\Delta}(G - xy)$, we obtain that δ is the only color in $\varphi'(x) \cap \varphi'(y)$, $\delta \ne \varphi'(vz) = \gamma$, and $\delta \ne \varphi'(zw)$, a contradiction to the above claim.

Finally, consider the case when $\gamma' \notin \overline{\varphi}(y)$. Then, $\gamma' \in \overline{\varphi}(x)$, and for the coloring $\varphi' = \varphi(\alpha \leftrightarrow \gamma') \in \mathcal{C}^{\Delta}(G - xy)$, we have $\delta \in \varphi'(x) \cap \varphi'(y)$, $\varphi'(zw) = \gamma$, $\overline{\varphi}(zw) = \delta$, and $\alpha \in \overline{\varphi}'(w) \cap \overline{\varphi}'(y)$. Hence, we obtain a contradiction as in the previous case. ∎

4.5 AVERAGE DEGREE OF CRITICAL CLASS TWO GRAPHS

On one hand, critical graphs of class two may have minimum degree 2. On the other hand, Vizing's Adjacency Lemma tells us that critical graphs of class two have many vertices of maximum degree, and that they are well distributed over the graph. This led Vizing to propose the following conjecture concerning the number of edges in critical class two graphs.

Conjecture 4.38 (Vizing [300] 1968) *If $G = (V, E)$ is a critical graph of class two with maximum degree Δ, then $|E| \ge \frac{1}{2}((\Delta - 1)|V| + 3)$.*

The above conjecture implies that the average degree of a critical graph of class two is greater than $\Delta - 1$. The first who established a lower bound for the average degree of a critical graph of class two, depending on its maximum degree, was S. Fiorini. He proved in reference [99], using Vizing's Adjacency Lemma, that every critical graph of class two $G = (V, E)$ with $\Delta(G) = \Delta$ satisfies $|E| \geq \frac{1}{4}(\Delta + 1)|V|$ if Δ is odd and $|E| \geq \frac{1}{4}(\Delta + 2)|V|$ if Δ is even. That Vizing's Adjacency Lemma leads to a better bound for the average degree of a critical class two graph was proved by Sanders and Zhao in 2002.

Theorem 4.39 (Sanders and Zhao [265] 2002) *If $G = (V, E)$ is a critical graph of class two with maximum degree Δ, then $|E| \geq \frac{1}{4}(\Delta + \sqrt{2\Delta - 1})|V|$.*

Woodall [310] proved that the above bound is the best possible that can be obtained by the use of Vizing's Adjacency Lemma alone. For large values of Δ, Woodall [309] improved this bound. Although his proof has the same structure as the proof of Theorem 4.39, it uses new adjacency lemmas in addition to Vizing's.

Theorem 4.40 (Woodall [309] 2007) *If $G = (V, E)$ is a critical graph of class two with maximum degree Δ, then $|E| \geq \frac{1}{3}(\Delta + 1)|V|$.*

Proof: Let $G = (V, E)$ be a critical graph of class two with maximum degree Δ and let $q = \frac{2}{3}(\Delta + 1)$. Then $\Delta \geq \delta(G) \geq 2$ and, therefore, $\Delta \geq q$. For a vertex $v \in V$, let $d(v) = d_G(v)$ and, moreover, let $d_q(v)$ be the number of neighbors z of v satisfying $d(z) < q$.

We now apply the discharging method. First, define a charge distribution M on V by $M(x) = d(x)$ for all $x \in V$. Starting from M, let each vertex $x \in V$ with $d(x) < q$ receive charge $rc(x, y)$ from each vertex $y \in N_G(x)$ with $d(y) > q$, where $rc(x, y) = (d(y) - q)/d_q(y)$. Denote the resulting charge distribution by M'. First, we show that if $x, y \in V$ are adjacent in G such that $d(x) < q$ and $d(y) > q$, then

$$rc(x, y) = \frac{d(y) - q}{d_q(y)} \geq \frac{d(y) - q}{d(x) + d(y) - \Delta - 1}. \tag{4.10}$$

The first equation follows from the definition of $rc(x, y)$ and the second inequality follows from the fact that $d_q(y) \leq d(y) - (\Delta - d(x) + 1) = d(x) + d(y) - \Delta - 1$, since y is adjacent to at least $\Delta - d(x) + 1$ major vertices by Vizing's Adjacency Lemma (Theorem 2.7).

Our discharging rule moves charge around, but the sum over all charges remains unchanged. Therefore,

$$2|E| = \sum_{x \in V} d(x) = \sum_{x \in V} M(x) = \sum_{x \in V} M'(x).$$

The goal is to prove that $M'(x) \geq q$ for every vertex $x \in V$, which gives the desired bound for $|E|$. That $M'(x) \geq q$ is certainly the case if $d(x) \geq q$. So assume that $x \in V$ and $d(x) < q$. It follows from (4.10) and the discharging rule that

$$M'(x) = M(x) + \sum_y rc(x, y) = d(x) + \sum_y \frac{d(y) - q}{d_q(y)}, \tag{4.11}$$

where the sum ranges over all neighbors y of x with $d(y) > q$.

Case 1: $d(x) = 2$. Then every vertex at distance 1 or 2 from x is a major vertex (Theorem 4.30), and so each neighbor of x sends $\Delta - q$ to x, which gives $M'(x) \geq 2 + 2(\Delta - q) = q$.

Case 2: $3 \leq d(x) \leq \Delta - q + 2$. If $y \in N_G(x)$, then Vizing's Adjacency Lemma implies that

$$d(y) \geq \Delta + 2 - d(x) \geq q.$$

By (4.11), this implies that

$$M'(x) = M(x) + \sum_{y \in N_G(x)} rc(x,y) = d(x) + \sum_{y \in N_G(x)} \frac{d(y) - q}{d_q(y)}. \qquad (4.12)$$

By the definition of the Kierstead set $K(x,y)$, each vertex $z \in K(x,y)$ satisfies

$$d(z) \geq 2\Delta - d(x) - d(y) + 2 \geq \Delta - d(x) + 2 \geq q.$$

Hence, since $d(x) < q$ and $K(x,y) \subseteq N_G(y)$, each vertex $y \in N_G(x)$ satisfies

$$1 \leq d_q(y) \leq d(y) - \sigma_G(x,y). \qquad (4.13)$$

Let p be defined as in Corollary 4.33, see (4.8) and (4.9). Then $p \leq \lfloor d(x)/2 \rfloor - 1$ and $\sigma_G(x,y) \geq \Delta - d(x) + p + 1$ for every $y \in N_G(x)$. By (4.13), this implies that

$$1 \leq d_q(y) \leq d(y) - \sigma_G(x,y) \leq d(y) - \Delta + d(x) - p - 1 \qquad (4.14)$$

for every $y \in N_G(x)$. On the other hand, Corollary 4.33 says that there is a set Y of $d(x) - p - 1$ vertices $y \in N_G(x)$ for which

$$\sigma_G(x,y) \geq \Delta - p - 1 \geq \Delta - d(x) + p + 1,$$

using $p \leq \lfloor d(x)/2 \rfloor - 1$ for the last inequality. By (4.13), this implies that

$$1 \leq d_q(y) \leq d(y) - \sigma_G(x,y) \leq d(y) - \Delta + p + 1 \qquad (4.15)$$

for each $y \in Y$. By Corollary 4.32 and Corollary 4.33, it follows that $p \geq 0$. Since $d(x) \geq 3$ and $p \leq \lfloor d(x)/2 \rfloor - 1$, we have $d(x) \geq p + 3$. Since $d(x) \leq \Delta - q + 2$, this implies that $q \leq \Delta - p - 1$. Consequently, the function

$$f(t) = (t - q)/(t - \Delta + p + 1)$$

is decreasing. Hence, by (4.10) and (4.15), we obtain that

$$rc(x,y) = \frac{d(y) - q}{d_q(y)} \geq \frac{d(y) - q}{d(y) - \Delta + p + 1} \geq \frac{\Delta - q}{p + 1} \qquad (4.16)$$

for each $y \in Y$. If $p = 0$, this yields $rc(x,y) \geq \Delta - q$ for every $y \in Y$ and $|Y| = d(x) - 1$. Consequently, $M'(x) \geq d(x) + (d(x) - 1)(\Delta - q) \geq 2 + 2(\Delta - q) = q$

and we are done. Otherwise, $p \geq 1$ and we argue as follows. Then $q \leq \Delta - d(x) + 2 \leq \Delta - d(x) + p + 1$ and, therefore, the function

$$\tilde{f}(t) = (t - q)/(t - \Delta + d(x) - p - 1)$$

is decreasing. Hence, by (4.10) and (4.14), we obtain that

$$rc(x, y) = \frac{d(y) - q}{d_q(y)} \geq \frac{d(y) - q}{d(y) - \Delta + d(x) - p - 1} \geq \frac{\Delta - q}{d(x) - p - 1} \qquad (4.17)$$

for each $y \in N_G(x)$. Combining (4.12) with (4.16) and (4.17), we obtain

$$
\begin{aligned}
M'(x) &= d(x) + \sum_{y \in Y} \frac{d(y) - q}{d_q(y)} + \sum_{y \in N_G(x) \backslash Y} \frac{d(y) - q}{d_q(y)} \\
&\geq d(x) + (d(x) - p - 1)\frac{\Delta - q}{p + 1} + (p + 1)\frac{\Delta - q}{d(x) - p - 1} \\
&\geq d(x) + (\xi + \xi^{-1})(\Delta - q) \geq 2 + 2(\Delta - q) = q,
\end{aligned}
$$

where $\xi = (d(x) - p - 1)/(p + 1)$. This settles Case 2.

Case 3: $\Delta - q + 2 < d(x) < q$. Our aim is to show that

$$M'(x) \geq d(x) + (\Delta - q + 1)\frac{\Delta - q}{d(x) - 1} \geq q. \qquad (4.18)$$

The latter inequality is obviously equivalent to $g(d(x)) \geq 0$, where

$$g(t) = (\Delta - q + 1)(\Delta - q) - (q - t)(t - 1)$$

is a quadratic function in t with minimum value when $t = \frac{1}{2}(q + 1) = \Delta - q + \frac{3}{2}$. Moreover, $g(t) = 0$ if $t = \Delta - q + 2$, and hence $g(t) \geq 0$ whenever $t = d(x) > \Delta - q + 2$, as is supposed. To show the first inequality, let m denote the number of major vertices adjacent to x. By (4.10), every major vertex y adjacent to x satisfies

$$rc(x, y) \geq \frac{\Delta - q}{d(x) - 1}. \qquad (4.19)$$

Hence, if $m \geq \Delta - q + 1$, then the first inequality in (4.18) follows from (4.11) and (4.19). So assume that $m < \Delta - q + 1 < d(x)$. Let $y \in V$ be a vertex adjacent to x. By Vizing's Adjacency Lemma, x is adjacent to at least $\Delta - d(y) + 1$ major vertices. Hence $\Delta - d(y) + 1 \leq m$ and, therefore, $d(y) \geq \Delta - m + 1 > q$. Consequently, we obtain that

$$rc(x, y) \geq \frac{d(y) - q}{d(x) + d(y) - \Delta - 1} \geq \frac{\Delta - m + 1 - q}{d(x) - m} \qquad (4.20)$$

for every vertex $y \in N_G(x)$. The first inequality follows from (4.10). The second inequality follows from the fact that the function $\tilde{g}(t) = (t - q)/(t + d(x) - \Delta - 1)$

is increasing, since $d(x) > \Delta - q + 1$ (by hypothesis). Using (4.11), (4.19), (4.20), and the fact that $1 \geq (\Delta - q)/(d(x) - 1)$ (by hypothesis), it follows that

$$
\begin{aligned}
M'(x) & \geq d(x) + m\frac{\Delta - q}{d(x) - 1} + (d(x) - m)\frac{\Delta - m + 1 - q}{d(x) - m} \\
& = d(x) + m\frac{\Delta - q}{d(x) - 1} + (\Delta - m + 1 - q) \\
& \geq d(x) + m\frac{\Delta - q}{d(x) - 1} + (\Delta - m + 1 - q)\frac{\Delta - q}{d(x) - 1} \\
& = d(x) + (\Delta - q + 1)\frac{\Delta - q}{d(x) - 1},
\end{aligned}
$$

as required. Clearly, this settles Case 3 and completes the proof of the theorem. ∎

For larger values of Δ, Woodall [309] improved the bound in Theorem 4.40 slightly, see the next theorem. The proof of this theorem is similar to the proof of Theorem 4.40, however, the proof uses not only Vizing's Adjacency Lemma and Corollary 4.33, but also Theorem 4.31, and it is also based on a modified discharging rule.

Theorem 4.41 (Woodall [309] 2007) *If $G = (V, E)$ is a critical graph of class two with maximum degree Δ, then*

$$
|E| \geq \begin{cases} \frac{1}{3}(\Delta + \frac{3}{2})|V| & \text{if } \Delta \geq 8, \\ \frac{1}{3}(\Delta + 2)|V| & \text{if } \Delta \geq 15. \end{cases}
$$

For small values of Δ, better bounds for the number of edges of critical graphs of class two with maximum degree Δ were established by several authors, see references [164, 190, 191, 192, 193, 197, 199, 200, 222, 223, 310, 315, 321]. In particular, Vizing's conjecture concerning the size of critical graphs of class two is known to be true if $\Delta \leq 5$ (see Jakobsen [155] for $\Delta = 3$, Fiorini and Wilson [100] for $\Delta = 4$, and Kayathri [164] for $\Delta = 5$). The next case $\Delta = 6$ was recently settled by Luo, Miao and Zhao [199]. On the other hand, Vizing's conjecture seems not to suggests the best possible bound and Woodall [310] proposed the following strengthening.

Conjecture 4.42 (Woodall [310] 2008) *If $G = (V, E)$ is a critical graph of class two with maximum degree Δ, then*

$$
|E| \geq \begin{cases} \frac{1}{2}(\Delta - 1 + \frac{3}{\Delta+1})|V| & \text{if } \Delta \text{ is even,} \\ \frac{1}{2}(\Delta - 1 + \frac{4}{\Delta+1})|V| & \text{if } \Delta \text{ is odd.} \end{cases}
$$

Woodall [310] emphasized, however, that it seems extremely unlikely that the minimum number of edges of a critical graph of class two is simply a multiple of its order.

4.6 INDEPENDENT VERTICES IN CRITICAL TWO GRAPHS

In the 1960s Vizing proposed several conjectures related to the Classification Problem. In his influential paper [300] Vizing wrote: *It appears perfectly natural that an n-vertex critical graph (of class two) cannot have more that $\frac{n}{2}$ pairwise nonadjacent vertices.*

A set of pairwise nonadjacent vertices of a given graph G is usually said to be an **independent set** of G. The maximum integer p such that G contains an independent set of cardinality p is called the **independence number** of G, denoted $\alpha(G)$. The following conjecture, which is known as Vizing's independence number conjecture, was published first in Vizing's paper [298].

Conjecture 4.43 (Independence Number Conjecture) *Every critical graph G of class two and of order n satisfies $\alpha(G) \leq \frac{1}{2}n$.*

That the bound for the independence number in Vizing's conjecture is best possible for every $\Delta \geq 2$ was explained in [32, 118]. Vizing [298] also conjectured that every critical graph of class two contains a 2-factor, which, if true, would clearly imply the independence number conjecture. Although Vizing's independence number conjecture is still unsolved, it is known to be true with $\frac{1}{2}$ replaced by $\frac{3}{5}$, as proved by Woodall [311]. The proof of the next result is reprinted from [311] (with small changes) with permission of John Wiley & Sons Inc.

Theorem 4.44 (Woodall [311] 2011) *If G is a critical graph of class two with maximum degree Δ and with order n, then $\alpha(G) < (3\Delta - 2)n/(5\Delta - 2)$. In particular, $\alpha(G) < \frac{3}{5}n$.*

Proof: Let G be a critical graph of class two with maximum degree Δ and with order n. Then $\chi'(G) = \Delta + 1$ and $\Delta \geq \delta(G) \geq 2$. Let X be a largest independent set in G, and let $Y = V(G) \setminus X$. Note that Y is not an independent set, since otherwise G would be bipartite and hence a graph of class one.

Our aim is to show that $|X| < (3\Delta - 2)n/(5\Delta - 2)$. To this end, we apply the discharging method. For a vertex x of G, let $d(x) = d_G(x)$ and $N(x) = N_G(x)$.

Let $X^- = \{x \in X \mid 2 \leq d(x) < \frac{1}{2}\Delta\}$, $X^+ = \{x \in X \mid \frac{1}{2}\Delta \leq d(x) < \Delta\}$, and $X^m = V(G^{[\Delta]})$, so that $V(G) = X^- \cup X^+ \cup X^m$. For each vertex $x \in X$, define $f_i(x) = g_i(d(x))$ ($i = 1, 2$), where

$$g_1(s) = \frac{2(\Delta - s)}{s} \text{ and } g_2(s) = \frac{\Delta - 2}{s - 1}.$$

Note that g_1 and g_2 are booth decreasing functions (in s). Furthermore, we have

$$g_2(s) - g_1(s) = \frac{(\Delta - 2)s - 2(\Delta - s)(s - 1)}{s(s - 1)} = \frac{(2s - \Delta)(s - 2)}{s(s - 1)}.$$

This is nonnegative, provided that $s \geq 2$ and $s \geq \frac{1}{2}\Delta$. Hence we obtain the following result.

Claim 4.44.1 *If $x \in X^+$, then $f_1(x) \leq f_2(x)$.*

Define three charge functions on $V(G)$ as follows:

$$M_0(x) = 0, \qquad M_1(x) = 2d(x), \qquad M_2(x) = 2\Delta \quad if \quad x \in X,$$
$$M_0(y) = 3\Delta - 2, \quad M_1(y) = \Delta - 2, \quad M_2(y) = 0 \qquad if \quad y \in Y.$$

Claim 4.44.2 $\sum_{v \in V(G)} M_1(v) < (3\Delta - 2)|Y|.$

Proof: Starting with the distribution M_0, let each vertex in X receive charge 2 from each of its neighbors in Y. Let the resulting charge distribution be called M_0^*. It is easy to see the $M_0^*(v) \geq M_1(v)$ for every vertex v of G, with strict inequality if v is a vertex of Y with fewer than Δ neighbors in X. There exist such vertices v, since Y is not an independent set in G. Thus, $\sum_{v \in V(G)} M_1(v) < \sum_{v \in V(G)} M_0^*(v) = \sum_{v \in V(G)} M_0(v) = (3\Delta - 2)|Y|.$ △

Starting with the charge distribution M_1, we will redistribute charges according to the following discharging rule (carried out in two steps):

1. Each vertex $y \in Y$ gives charge $f_1(x)$ to each vertex $x \in N(y) \cap X^+$.

2. Each vertex $y \in Y$ distributes its remaining charge equally among all vertices (if any) in $N(y) \cap X^-$.

Let the resulting charge distribution be M_1^*. We will prove that $M_1^*(v) \geq M_2(v)$ for each vertex v of G. By Claim 4.44.2, this implies that

$$2\Delta|X| = \sum_{v \in V(G)} M_2(v) \leq \sum_{v \in V(G)} M_1(v) < (3\Delta - 2)|Y| = (3\Delta - 2)(n - |X|),$$

from which it will follow immediately that $|X| < (3\Delta - 2)n/(5\Delta - 2)$, as required.

Claim 4.44.3 *If $v \in V(G) \setminus X^-$, then $M_1^*(v) \geq M_2(v)$.*

Proof: To prove this, it is helpful to compare the above discharging rule, the *actual discharging rule*, with the *equitable discharging rule* in which each vertex $y \in Y$ distributes its charge of $M_1(y) = \Delta - 2$ equally among all its neighbors (if any) in $X' = X^- \cup X^+$. If the minimum degree of a neighbor of y is d, then Vizing's Adjacency Lemma (Theorem 2.7) implies that y has at least $\Delta - d + 1$ neighbors of degree Δ, and hence at most $d - 1$ neighbors in X'. Thus, under the equitable discharging rule, each vertex $x \in N(y) \cap X'$ receives from y at least

$$\frac{\Delta - 2}{d - 1} \geq \frac{\Delta - 2}{d(x) - 1} = f_2(x).$$

By Claim 4.44.1, it follows that every vertex of $N(y) \cap X^+$ receives no more charge from y in step 1 of the actual discharging rule than it would receive under the equitable discharging rule, so that the remaining charge referred to in step 2 is nonnegative. It follows that $M_1^*(y) \geq 0 = M_2(y)$ for each $y \in Y$. It is easy to see that $M_1^*(x) = 2\Delta = M_2(x)$ for each vertex $x \in X^+ \cup X^m$. $\qquad \triangle$

It remains to consider the vertices in X^-. Fix a vertex $x \in X^-$ and let $s = d(x)$. Define

$$
\begin{aligned}
h(s,t) &= \frac{1}{s-t-1}(\Delta - 2 - tg_1(\Delta - s + 2)) \\
&= \frac{1}{s-t-1}\left(\Delta - 2 - t\frac{2(s-2)}{\Delta - s + 2}\right).
\end{aligned}
$$

Claim 4.44.4 *If t is a nonnegative integer and y is a neighbor of x such that $\sigma(x, y) \geq \Delta - s + t + 1$, then y gives x at least $h(s, t)$ in step 2.*

Proof: By the definition of $\sigma(x, y)$, y has $\sigma(x, y)$ neighbors with degree at least $2\Delta - d(x) - d(y) + 2 \geq \Delta - s + 2 > \frac{1}{2}\Delta$, since $x \in X^-$ and so $s = d(x) < \frac{1}{2}\Delta$. By Vizing's Adjacency Lemma, y has at least $\Delta - s + 1$ neighbors with degree Δ. So let L^m be the set of $\Delta - s + 1$ neighbors of y with degree Δ, and let L^+ be a set, disjoint from L^m, of t neighbors of y with degree at least $\Delta - s + 2$, which exists since $\sigma(x, y) \geq \Delta - s + t + 1$ by hypothesis. So $L^m \subseteq X^m \cup Y$ and $L^+ \subseteq X^+ \cup X^m \cup Y$. Then y gives nothing to any vertex in L^m, and in step 1 y gives at most $g_1(\Delta - s + 2)$ to each vertex in L^+. In the remainder of steps 1 and 2, y's remaining charge of at least $\Delta - 2 + tg_1(\Delta - s + 2)$ is divided among y's remaining $d(y) - \Delta + s - t - 1 \leq s - t - 1$ neighbors–not necessarily equally, but x, being in X^-, gets at least as much of it as any other neighbor of y. So the charge x receives from y is $\geq h(s, t)$, as required. $\qquad \triangle$

Let p be defined as in Corollary 4.33; see (4.8) and (4.9). Then there exists a set N^+ of $d(x) - p - 1 = s - p - 1$ neighbors y of x for which $\sigma(x, y) \geq \Delta - p - 1$. Let $N^- = N(x) \setminus N^+$. Then N^- contains $p + 1$ neighbors y of x, for each of which $\sigma(x, y) \geq \Delta - s + p + 1$, by the definition of p. Note that $N^+ \subseteq Y$, since X is an independent set. Now, applying Claim 4.44.4 to the vertices in N^- with $t = p$, and to the vertices in N^+ with $t = s - p - 2$, we see that x receives charge of at least $H(s, p)$ in step 2, where

$$
H(s, p) = (p + 1)h(s, p) + (s - p - 1)h(s, s - p - 2).
$$

To complete the prove, it suffices to show that $H(s, p) \geq 2(\Delta - s) = 2(\Delta - d(x))$, since this will imply that $M_1^*(x) \geq M_1(x) + H(s, p) \geq 2d(x) + 2(\Delta - d(x)) = 2\Delta = M_2(x)$, as required.

By the definition of h, $A = (p + 1)h(s, p)$ satisfies

$$
A = \frac{p + 1}{s - p - 1}\left(\Delta - 2 - p\frac{2(s - 2)}{\Delta - s + 2}\right)
$$

and $B = (s - p - 1)h(s, s - p - 2)$ satisfies

$$
B = \frac{s - p - 1}{p + 1}\left(\Delta - 2 - (s - p - 2)\frac{2(s - 2)}{\Delta - s + 2}\right).
$$

Let $r = p + 1$, so that $1 \leq r \leq \frac{1}{2}s$, since $0 \leq p \leq \frac{1}{2}d(x) - 1 = \frac{1}{2}s - 1$ by Corollary 4.33 and Proposition 4.27(a). Setting

$$b = \frac{2(s-2)}{\Delta - s + 2} \text{ and } a = \Delta - 2 + b,$$

we can write

$$H(s,p) = A + B = \frac{r(a - br)}{s - r} + \frac{(s - r)(a - b(s - r))}{r}.$$

The derivative of this with respect to r is

$$\frac{as - bs^2 + b(s - r)^2}{(s - r)^2} - \frac{as - bs^2 + br^2}{r^2} = \frac{as - bs^2}{(s - r)^2} - \frac{as - bs^2}{r^2}$$

This is zero if and only if $r = \frac{1}{2}s$ (unless $as - bs^2 = 0$, when $H(s,p)$ is independent of p). Thus $H(s,p)$, regarded as a function of p, has only one stationary point (for positive p), when $p + 1 = r = \frac{1}{2}s$. Substituting this value of p gives

$$H\left(s, \frac{1}{2}s - 1\right) = 2\left(\Delta - 2 - \frac{(s - 2)^2}{\Delta - s + 2}\right) \geq 2(\Delta - s),$$

where the inequality holds because $s < \frac{1}{2}\Delta$ and so

$$\frac{(s - 2)^2}{\Delta - s + 2} \leq s - 2.$$

To complete the proof (that $H(s,p) \geq 2(\Delta - s)$), we must also consider the other extreme value of p, namely $p = 0$, and show that $H(s,0) \geq 2(\Delta - s)$, i.e., we must show that

$$\frac{\Delta - 2}{s - 1} + (s - 1)\left(\Delta - 2 - \frac{2(s - 2)^2}{\Delta - s + 2}\right) \geq 2(\Delta - s). \tag{4.21}$$

This evidently holds with equality if $s = d(x) = 2$. So we may assume that $s \geq 3$. Since $s < \frac{1}{2}\Delta$, we can write $\Delta = 2s + D$, where $D \geq 1$. Ignoring the first term of (4.21), and dividing through by $s - 1$ and rearranging, it suffices to show that

$$2s + D - 2 - \frac{2(s + D)}{s - 1} - \frac{2(s - 2)^2}{s + D + 2} \geq 0. \tag{4.22}$$

Since the LHS of (4.22) is an increasing function of D, it suffices to verify (4.22) for $D = 1$, when the LHS becomes

$$2s - 1 - \frac{2(s + 1)}{s - 1} - \frac{2(s - 2)^2}{s + 3} = 11 - \frac{4}{s - 1} - \frac{50}{s + 3},$$

which is positive since $s \geq 3$. This completes the proof of the theorem. ∎

4.7 CONSTRUCTIONS OF CRITICAL CLASS TWO GRAPHS

With respect to edge coloring it is advantageous to study general graphs, even if one is primarily interested in simple graphs. In particular, for constructing critical graphs of class two, it has proved useful to involve critical graphs having multiple edges in such constructions. A graph satisfying $\chi' = \Delta$ is critical if and only if it is a multistar. Thus, in what follows we focus on critical graphs such that $\chi' \geq \Delta + 1$.

A critical graph of class two having maximum degree Δ and order n is called a (Δ, n)-**graph**. Clearly, there are no $(1, n)$-graphs, and the only $(2, n)$-graph is the cycle C_n when n is odd and $n \geq 3$.

Problem 4.45 *Let $\Delta \geq 3$ be an integer. Does there exists an integer N_Δ such that there is a (Δ, n)-graph for all integers $n \geq N_\Delta$?*

For constructing critical graphs, we first apply two simple graph operations, one of which is very common in graph theory and the other one was invented by Hajós [126] in 1961.

Splitting Operation. Let G be a simple graph, and let x be a vertex of G such that $d_G(x) = m \geq 2$ and $N_G(x) = \{x_1, \ldots, x_m\}$. We say that the simple graph H is obtained from G by **splitting** x into two vertices u and v ($u, v \notin V(G)$ and $u \neq v$) if

$$V(H) = V(G - x) \cup \{u, v\}$$

and

$$E(H) = E(G - x) \cup \{uv, ux_1, \ldots, ux_r, vx_{r+1}, \ldots, vx_m\}$$

for some r satisfying $1 \leq r < m$.

Hajós Construction. Suppose G_1, G_2 are disjoint graphs (i.e., $V(G_1) \cap V(G_2) = \emptyset$), $e_1 \in E_{G_1}(x_1, y_1)$ and $e_2 \in E_{G_2}(x_2, y_2)$ are edges. The **Hajós sum** of G_1 and G_2 with respect to (x_1, y_1) and (x_2, y_2) is the graph G obtained from the union of G_1 and G_2 by removing the edges e_1 and e_2, identifying x_1 and x_2 to one vertex x, and joining y_1 and y_2 by a new edge.

For $r, t \geq 1$, let K_r^t denote the **complete t-partite graph** having r vertices on each partition. Then $\Delta(K_r^t) = r(t-1)$ and $|V(K_r^t)| = rt$. Laskar and Hare [189] observed that K_r^t is a graph of class two if and only if $rt \geq 3$ is odd. Yap [314, 315, 316] proved that if $rt \geq 4$ is even then every graph H obtained from K_r^t by splitting a vertex into two vertices is a critical graph of class two. Let $S^2(K_r^t)$ denote the graph obtained from K_r^t by splitting a vertex into two vertices such that at least one of these two vertices has degree 2. For $m \geq 2$, $S^2(K_1^{2m})$ is a $(2m-1, 2m+1)$-graph with minimum degree 2 and $S^2(K_2^m)$ is a $(2m-2, 2m+1)$-graph with minimum degree 2. The graph $G = S^2(K_1^{2m})$ can be also obtained from K_{2m} by subdividing an edge, where to **subdivide** an edge e is to delete e, add a new vertex x, and join x to the endvertices of e. Then G is just overfull. That G is critical was first observed by Fiorini [98] in his thesis.

As we have seen in Sect. 4.3, if G is any graph obtained from K_{2r+1} by removing $r - 1$ edges, then G is just overfull and hence a graph of class two. The just overfull conjecture (Conjecture 4.14) implies that every such graph is critical, and Plantholt [243] has proved that this is indeed the case.

The Hajós sum is a very useful tool for analyzing χ-critical graphs, since it preserves not only the chromatic number but also criticality. As proved by Hajós [126], the Hajós sum of two disjoint χ-critical graphs with the same chromatic number k is a χ-critical graph with chromatic number k, too. That the Hajós sum can be also used to obtain new critical graphs from previously known ones, provided we add a degree condition, was proved by Jakobsen [154]. Observe that if G is the Hajós sum of G_1 and G_2 with respect to (x_1, y_1) and (x_2, y_2), then $d_G(x) = d_{G_1}(x_1) + d_{G_2}(x_2) - 2$ and $d_G(v) = d_{G_i}(v)$ for all $v \in V(G_i) \setminus \{x_i\}$ $(i = 1, 2)$.

Theorem 4.46 (Jakobsen [154] 1973) *Let G_1 and G_2 be two disjoint critical graphs such that $\chi'(G_1) = \chi'(G_2) = k + 1$, where $k \geq \Delta(G_i)$ for $i = 1, 2$. Furthermore, let G be the Hajós sum of G_1 and G_2 with respect to (x_1, y_1) and (x_2, y_2). If $d_{G_1}(x_1) + d_{G_2}(x_2) \leq k + 2$, then G is a critical graph with $\chi'(G) = k + 1$ and $k \geq \Delta(G)$.*

Proof: Observe that G is connected. Hence, in order to proof that G is a critical graph with chromatic index $k + 1$, it suffices to show that $\chi'(G) \geq k + 1$ and $\chi'(G - e) \leq k$ for all edges $e \in E(G)$. Let $e_1 \in E_{G_1}(x_1, y_1)$ and $e_2 \in E_{G_2}(x_2, y_2)$ be the two edges not belonging to G, and let $e' \in E_G(y_1, y_2)$ be the new edge.

Evidently, $\chi'(G) \geq k + 1$, since a k-edge-coloring of G would yield a k-edge-coloring of G_1 or G_2 by coloring e_1 or e_2 with the color of e', a contradiction.

Now, let e be an arbitrary edge of G. By symmetry, we may assume that $e \in E(G_1) \setminus \{e_1\}$ or $e = e'$. Suppose first that $e = e'$. Since both G_1 and G_2 are critical graphs and $d_{G_1 - e_1}(x_1) + d_{G_2 - e_2}(x_2) \leq k$, there are colorings $\varphi_1 \in C^{k-1}(G_1 - e_1)$ and $\varphi_2 \in C^{k-1}(G_2 - e_2)$ such that $\varphi_1(x_1) \cap \varphi_2(x_2) = \emptyset$. Hence by combining φ_1 and φ_2 a k-edge-coloring of $G - e'$ is obtained.

Suppose now that $e \in E(G_1) \setminus \{e_1\}$. Since G_1 is critical, there is a coloring $\varphi_1 \in C^k(G - e)$, say with $\varphi_1(e_1) = \alpha$. Since G_2 is critical, there is a coloring $\varphi_2 \in C^k(G - e_2)$. Then we have $|\varphi_1(x_1)| = d_{G_1}(x_1)$, $|\varphi_2(x_2)| = d_{G_2}(x_2) - 1$, and $\overline{\varphi}_2(x_2) \cap \overline{\varphi}_2(y_2) = \emptyset$. Since $d_{G_1}(x_1) + d_{G_2}(x_2) - 1 \leq k + 1$, we may permute the colors in φ_2 such that $\varphi_1(x_2) \cap \varphi_2(x_2) = \{\alpha\}$ and $\alpha \in \overline{\varphi}_2(y_2)$. Hence by combining φ_1 and φ_2 a k-edge-coloring of $G - e'$ is obtained. ∎

Corollary 4.47 (Jakobsen [154] 1973) *Let G_1 and G_2 be two disjoint critical graphs of class two with the same maximum degree Δ, and let G be the Hajós sum of G_1 and G_2 with respect to (x_1, y_1) and (x_2, y_2). If $d_{G_1}(x_1) + d_{G_2}(x_2) \leq \Delta + 2$, then G is a critical class two graph with maximum degree Δ.*

The degree condition restricts the application of the Hajós construction. However, if there exists a critical graph H of class two with maximum degree Δ and minimum degree 2, we obtain an infinite sequence $(G_p)_{p \in \mathbb{N}}$ of critical graphs of class two each with maximum degree Δ as follows. Let z be a vertex of degree 2 in H, and let x, y be the two neighbors of z in H. Furthermore, let $H^1, \ldots, H^p$ be p disjoint copies of $H - z$. The graph G_p is obtained from the union of $H^1, \ldots, H^p$ and one additional vertex z^* by adding one edge between the copy of y in H^i and the copy of x in H^{i+1} ($i = 1, \ldots, p-1$), one edge between the copy of x in H^1 and z^*, and one edge between the copy of y in H^p and z^* (see Fig. 4.8). Evidently, for each $p \geq 1$, the graph G_{p+1} is the Hajós sum of G_p and H. In what follows, we write $G_p = HS^p(H, x, y, z)$ or briefly $G_p = HS^p(H)$.

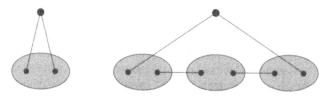

Figure 4.8   A graph H and the graph $HS^3(H)$.

For a graph G, let us define the **deficiency** $\mathrm{def}(G)$ of G by

$$\mathrm{def}(G) = \sum_{v \in V(G)} (\Delta(G) - d_G(v)).$$

Clearly, $\mathrm{def}(G) \geq 0$ and equality holds if and only if G is regular. By Proposition 4.12, it follows that G is overfull if and only if G has an odd number of vertices and $\mathrm{def}(G) \leq \Delta(G) - 2$.

Parity Argument. Let G be a graph, let $F \subseteq E(G)$ be an edge set, and let $\varphi \in \mathcal{C}^k(G - F)$ be a coloring. For a color $\alpha \in \{1, \ldots, k\}$, let $m_{\varphi,\alpha}$ be the number of vertices $v \in V(G)$ satisfying $\alpha \in \overline{\varphi}(v)$. Then, see (1.2) and (1.3), we have $m_{\varphi,\alpha} \equiv |V(G)| \bmod 2$ for each color α, and

$$\sum_{v \in V(G)} (k - d_G(v)) + 2|F| = \sum_{\alpha=1}^{k} m_{\varphi,\alpha}.$$

Proposition 4.48 (Fiorini and Wilson [100] 1977) *Let G be a critical graph with* $\chi'(G) = \Delta(G) + 1$. *If $|V(G)|$ is odd then* $\mathrm{def}(G) \geq \Delta(G) - 2$, *else* $\mathrm{def}(G) \geq 2(\Delta(G) - \delta(G)) + 2$.

Proof: Let $\Delta = \Delta(G)$. Clearly, G contains an edge, say $e \in E_G(x, y)$. Since G is critical, there is a coloring $\varphi \in \mathcal{C}^\Delta(G - e)$. By the parity argument, we have

$$\mathrm{def}(G) + 2 = \sum_{\alpha=1}^{\Delta} m_{\varphi,\alpha}.$$

Hence, if $|V(G)|$ is odd, this yields $\mathrm{def}(G) \geq \Delta - 2$. Now, assume that $|V(G)|$ is even. The edge e being uncolored, we have $\overline{\varphi}(x) \cap \overline{\varphi}(y) = \emptyset$. Furthermore, for each color $\alpha \in \overline{\varphi}(x) \cup \overline{\varphi}(y)$, we have $m_{\varphi,\alpha} \geq 2$. Since G is connected, we may choose e such that $d_G(x) = \delta(G)$. Then $|\overline{\varphi}(x) \cup \overline{\varphi}(y)| = |\overline{\varphi}(x)| + |\overline{\varphi}(y)| = 2\Delta + 2 - (d_G(x) + d_G(y)) \geq \Delta + 2 - \delta(G)$. Thus the parity argument implies that $\text{def}(G) \geq 2(\Delta + 2 - \delta(G)) - 2 = 2(\Delta(G) - \delta(G)) + 2$. ∎

Clearly, every critical graph with $\chi' \geq \Delta + 1$ has at least three vertices. The graph $T = T(r,s,t)$ consisting of three vertices x, y, z such that $\mu_K(x,y) = r$, $\mu_K(y,z) = s$, and $\mu_K(x,z) = t$ is a critical graph with $\chi'(K) = r + s + t$ and $\Delta(K) = \max\{r + s, s + t, r + t\}$, provided that $r, s, t \geq 1$.

Let H be a $(3,n)$-graph, and let xy be an edge of H such that $d_H(x) = 2$. Now, let G be the graph obtained from $H - xy$ by adding two new vertices x', y' and by adding the edges $xx', x'y', xy'$, and yy'. Then G is the Hajós sum of H and $T = T(1,1,2)$ with respect to the pair (x, y) and the pair (z, y') of T with $\mu_T(z, y') = 2$. Hence G is a $(n + 2, 3)$-graph and we can repeat the process with G and the edge $x'y'$. As H we can take the the $(3,5)$-graph $S^2(K_1^4)$ obtained from the complete graph K_4 by subdividing an edge (see Fig. 4.9). This yields an infinite sequence of critical graphs of class two each with $\Delta = 3$. This shows that there exists a $(3,n)$-graph G for every odd integer $n \geq 5$ such that $\text{def}(G) = \Delta - 2 = 1$. As H we can also take the $(3,22)$-graph obtained by Goldberg [113], shown in Fig. 4.10. This yields another infinite sequence of critical graphs of class two each with $\Delta = 3$, showing that there exists a $(3,n)$-graph G for every even integer $n \geq 22$ such that $\text{def}(G) = 2(\Delta - \delta) + 2 = 4$. This provides a positive answer to Problem 4.45 at least for $\Delta = 3$ and with $N_3 = 22$.

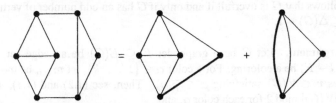

Figure 4.9   The Hajós sum of $S^2(K_2^2)$ and $T(1,1,2)$.

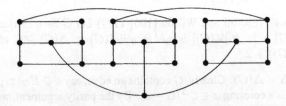

Figure 4.10   Goldberg's $(3,22)$-graph.

Let G be a graph, and let $X \subseteq V(G)$ be a vertex set. Let H be the graph obtained from G by **contracting** the set X, that is, we replace in G the set X by a new vertex z and join this new vertex to every vertex $u \in V(G) \setminus X$ by $|E_G(u, X)|$ edges. In this case we write $H = G/X$. Instead of $G/(V(G) \setminus X)$ we also write G/X^c. Clearly, we have $\chi'(G) \le \max\{\chi'(G/X), \chi'(G/X^c)\}$. The vertex set X is called k-**contractible** if $|X|$ is odd, $X \ne V(G)$, and $\sum_{v \in X}(k - d_{G[X]}(v)) \le k$. The following statement generalizes a result due to Grünewald and Steffen [117].

Proposition 4.49 *Let G be a critical graph with $\chi'(G) = k + 1$ for an integer $k \ge \Delta(G)$, and let $X \subseteq V(G)$ be a k-contractible vertex set of G. Then $G' = G/X$ is a critical graph with $\Delta(G') \le k$ and $\chi'(G') = \chi'(G)$.*

Proof: Since $\Delta(G) \le k$ and X is k-contractible, it follows that $|\partial_G(X)| \le \sum_{v \in X}(k - d_{G[X]}(v)) \le k$ and the graph $G' = G/X$ satisfies $\Delta(G') \le k$. Since G is critical, G is connected and so G' is also connected. Hence, it suffices to show that $\chi'(G') \ge k + 1$ and $\chi'(G' - e) \le k$ for all edges $e \in E(G')$.

Since X is a proper subset of $V(G)$, we have $\chi'(G[X]) \le k$. Now, consider an arbitrary coloring $\varphi \in \mathcal{C}^k(G[X])$. By the parity argument, $m_{\varphi,\alpha} \equiv |X| \equiv 1 \bmod 2$ for each color α and $\sum_\alpha m_{\varphi,\alpha} = \sum_{v \in X}(k - d_{G[X]}(v)) \le k$, since X is k-contractible. This implies that $m_{\varphi,\alpha} = 1$ for all colors $\alpha \in \{1, \ldots, k\}$ and, therefore, $\sum_{v \in X}(k - d_{G[X]}(v)) = k$. Since $|\partial_G(X)| \le k$, we conclude that φ can be extended to a k-edge-coloring of. Consequently, we have $\chi'(G/X^c) \le k$ and, therefore, $\chi'(G') = \chi'(G/X) \ge k + 1$.

Now, consider an arbitrary edge e of G'. Such an edge corresponds to an edge e' of $G - E(G[X])$. Since G is critical, there is a coloring $\varphi' \in \mathcal{C}^k(G - e)$. Hence, the restriction of φ' to X yields a k-edge-coloring φ of $G[X]$. Since $m_{\varphi,\alpha} = 1$ for each color α, no two edges in $\partial_G(X)$ receive the same color. Hence, φ' yields a k-edge-coloring of $G' - e$. ∎

The next operation was invented by Grünewald and Steffen [117]. This operation allows us to construct simple critical graphs from general critical graphs.

The Grünewald–Steffen Construction. Let G be a graph, let $v \in V(G)$ be a vertex, and let $k \ge \Delta(G)$ be an integer. Let $K = K_{k,k-1}$ be the complete bipartite graph disjoint from G, and let A be the set of k independent vertices of K, each of degree $k-1$. For each vertex $u \in N_G(v)$ choose a subset N_u of A such that all these sets are pairwise disjoint and $|N_u| = |E_G(u,v)|$. Let G' be the graph obtained from $G - v$ and K by adding one edge between u and x whenever $u \in N_G(v)$ and $x \in N_u$. The graph G' is called a k-**extension** of G and we write $G' = ET_k(G,v)$.

Theorem 4.50 (Grünewald and Steffen [117] 1999). *Let G be a critical graph with $\chi'(G) = k + 1$ for an integer $k \ge \Delta(G)$, and let $v \in V(G)$ be a vertex. Then $G' = ET_k(G,v)$ is a critical graph with $\Delta(G') = k$ and $\chi'(G') = \chi'(G)$.*

Proof: Clearly, $\Delta(G') = k$ and G' is connected. Hence, it suffices to show that $\chi'(G') \ge k + 1$ and $\chi'(G' - e) \le k$ for all edges $e \in E(G')$. We use the same notation as in the description of the Grünewald–Steffen construction. Note that $V(K)$ is a k-contractible vertex set of G'.

To show that $\chi'(G') \ge k + 1$, suppose this is false. Then there is a coloring $\varphi \in \mathcal{C}^k(G')$. Based on the parity argument, we conclude that the edges in $\partial_{G'}(V(K))$ are colored pairwise differently. Hence, φ yields a k-edge-coloring of $G'/V(K) = G$. This contradiction shows that $\chi'(G') \geq k + 1$.

Now, let us consider an arbitrary edge $e \in E(G')$. To show that $\chi'(G' - e) \leq k$, we consider two cases.

Case 1: $e \notin E(K)$. Then e corresponds to an edge of G and, therefore, there is a coloring $\varphi_1 \in \mathcal{C}^k(G' - e - E(K))$ such that all edges in $F' = \partial_{G'}(V(K)) \setminus \{e\}$ are colored differently. By König's theorem, we have $\chi'(K_{k,k}) = k$. Hence, there is a coloring $\varphi_2 \in \mathcal{C}^k(K)$ such that each of the k colors is missing at exactly one vertex of A. Since $|F' \cap E_G(x)| \leq 1$ for all vertices $x \in A$, by permuting the colors of φ_2 if necessary, $\varphi = \varphi_1 \cup \varphi_2$ yields a k-edge-coloring of $G' - e$.

Case 2: $e \in E(K)$. First, we choose an edge $e_1 \in F = \partial_{G'}(V(K))$. Then e_1 joins a vertex $x_1 \in A$ with a vertex of $N_G(v)$ and, moreover, e_1 corresponds to an edge $f \in E_G(v)$. Since $\chi'(G - f) = k < \chi'(G)$, we then conclude that there is a coloring $\varphi_1 \in \mathcal{C}^k(G' - E(K))$ such that the edges of $F' = F \setminus \{e_1\}$ are colored differently and $\varphi_1(e_1) = \varphi_1(e_2) = \alpha$ for some edge $e_2 \in F'$. Then e_2 joins a vertex $x_2 \in A$ with a vertex of $N_G(v)$, where $x_2 \neq x_1$. Since $k \geq \Delta(G)$, there is a color $\beta \in \{1, \ldots, k\}$ such that $\varphi_1(e') \neq \beta$ for all edges $e' \in F$.

Now, let K' be the graph obtained from K by adding a vertex y and joining y to all vertices of A by one edge. Since $K' = K_{k,k}$, there is a Hamilton cycle C in K such that all three edges e, yx_1, yx_2 belong to C. Hence we find an edge-coloring φ' of C using the two colors α, β such that $\varphi'(yx_1) = \alpha$ and $\varphi'(yx_2) = \beta$. By König's theorem, we also find an edge-coloring φ'' of $K' - E(C)$ using the remaining $k - 2$ colors. Let $\varphi^* = \varphi' \cup \varphi''$ and let φ_2 be the restriction of φ^* to $V(K) = V(K') \setminus \{y\}$. Then $\varphi_2 \in \mathcal{C}^k(K)$ and each of the k colors is missing at exactly one vertex of A. Furthermore, $\alpha \in \overline{\varphi}(x_1)$, $\beta \in \overline{\varphi}(x_2)$ and, moreover, $P_{x_1}(\alpha, \beta, \varphi_2) = P_{x_2}(\alpha, \beta, \varphi_2) = C - y$. Now, make the edge e be uncolored and let $\varphi_3 \in \mathcal{C}^k(K - e)$ be the resulting coloring. Since $e \in C - y$, for the coloring $\varphi_4 = \varphi_3/P_{x_2}(\alpha, \beta, \varphi_3)$, we then have $\alpha \in \overline{\varphi}_4(x_1) \cap \overline{\varphi}_4(x_2)$ and $\overline{\varphi}_2(x) \subseteq \overline{\varphi}_4(x)$ for all vertices $x \in A \setminus \{x_1, x_2\}$. Since $|F \cap E_G(x)| \leq 1$ for all vertices $x \in A$, by permuting the colors $\{1, \ldots, k\} \setminus \{\alpha, \beta\}$ in φ_4 if necessary, $\varphi = \varphi_1 \cup \varphi_4$ yields a k-edge-coloring of $G' - e$.   ∎

If we identify two nonadjacent vertices u, v of a graph G, then the resulting graph G' satisfies $\chi'(G') \geq \max\{\chi'(G), d_G(u) + d_G(v)\}$. Yap [316] was the first who noticed that this operation sometimes preserves not only the chromatic index, but also criticality.

Yap's Construction. Let H be a graph with maximum degree $\Delta \geq 4$ and with two vertices x, u of degree 2, and let $y \in N_G(x) \setminus \{u\}$. Let H' be a disjoint copy of H with the corresponding vertices x', y', u'. Let G' be the Hajós sum of H and H' with respect to (x, y) and (x', y'). Then the graph G obtained from G' by identifying u and u' is denoted by $YA(H, x, y, u)$, see Fig. 4.11.

Theorem 4.51 (Yap [316] 1986) *Let H be a critical with maximum degree $\Delta \geq 4$ and chromatic index $\Delta + 1$. Furthermore, let $x, u \in V(H)$ be two vertices of degree*

2, and let $y \in N_H(x) \setminus \{u\}$. Then $G = YA(H, x, y, u)$ is a critical graph with $\Delta(G) = \Delta$ and $\chi'(G) = \Delta + 1$.

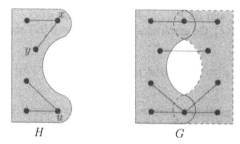

$$H \qquad\qquad\qquad G$$

Figure 4.11   A graph H and the graph $G = YA(H, x, y, u)$.

Evidently, Theorem 4.51 is an immediate consequence of the following slightly more general result.

Theorem 4.52 *Let G_1 and G_2 be disjoint critical graphs both with maximum degree $\Delta \geq 4$ and chromatic index $\Delta + 1$. For $i = 1, 2$, let x_i be a vertex of degree 2 in G_i, let $y_i \in N_{G_i}(x_i)$, and let $u_i \in V(G_i) \setminus \{x_i, y_i\}$. Furthermore, let G' be the Hajós sum of G_1 and G_2 with respect to (x_1, y_1) and (x_2, y_2). If $d_{G_1}(u_1) + d_{G_2}(u_2) \leq \Delta$, then the graph G obtained from G' by identifying u_1 and u_2 is a critical graph with $\Delta(G) = \Delta$ and $\chi'(G) = \Delta + 1$.*

Proof: We use x_2 in order to label the vertex of G' obtained by identifying x_1 and x_2. Furthermore, we denote the edge in $E_{G_2}(x_2, y_2) \setminus E(G')$ by e_2.

By Theorem 4.46, G' is a critical graph with $\Delta(G') = \Delta$ and $\chi'(G') = \Delta + 1$. Hence, we have $\chi'(G) \geq \chi'(G') \geq \Delta + 1$. Now, let $e \in E(G)$ be an arbitrary edge and let f be the corresponding edge in $E(G')$. In order to show that $\chi'(G-e) \leq \Delta$, it suffices to show that there is a coloring $\varphi \in C^\Delta(G'-f)$ such that $\varphi(u_1) \cap \varphi(u_2) = \emptyset$. By symmetry, we may assume that $f \in E(G_1)$ or f is the only edge in $E_{G'}(y_1, y_2)$.

Since G' is critical, we can choose a coloring $\varphi \in C^\Delta(G'-f)$ such that $m = |\varphi(u_1) \cap \varphi(u_2)|$ is minimum. If $m = 0$ we are through. So suppose that $m \geq 1$. Since $d_{G'}(u_1) + d_{G'}(u_2) \leq \Delta$, this implies that there is a color $\alpha \in \overline{\varphi}(u_1) \cap \overline{\varphi}(u_2)$ as well as a color $\gamma \in \varphi(u_1) \cap \varphi(u_2)$. Then the chain $P = P_{u_2}(\alpha, \gamma, \varphi)$ is a path joining u_2 and u_1, since otherwise $\varphi' = \varphi/P$ would be a coloring in $C^\Delta(G' - f)$ such that $|\varphi'(u) \cap \varphi'(v)| \leq m - 1$, contradicting the choice of φ. Hence, P contains the vertex x_2 or the only edge in $E_{G'}(y_1, y_2)$.

Clearly, the restriction φ_2 of φ to the edge set of $G_2 - e_2$ belongs to $C^\Delta(G_2 - e_2)$, where $|\overline{\varphi}_2(x_2)| = \Delta - 1$ and $|\overline{\varphi}_2(y_2)| \geq 1$. Since $\chi'(G_2) = \Delta + 1$, we have $\overline{\varphi}_2(x_2) \cap \overline{\varphi}_2(y_2) = \emptyset$. Consequently, exactly one color, say β, is missing at y_2 with respect to φ_2 and $\overline{\varphi}_2(x_2) = \{1, \ldots, \Delta\} \setminus \{\beta\}$. Obviously, the restriction P_2 of the chain P to the graph $G_2 - e_2$ satisfies $P_2 = P_{u_2}(\alpha, \gamma, \varphi_2)$. Since P contains x_2 or the edge in $E_{G'}(y_1, y_2)$, it follows that P_2 is a path, where one endvertex is u_2 and the other endvertex is either x_2 or y_2. Clearly, this implies that $\beta \in \{\alpha, \gamma\}$. Since the vertices x_2, y_2, u_2 are distinct, we then conclude that $P_{x_2}(\alpha, \gamma, \varphi_2)$ is a path joining x_2 and a vertex different from y_2. Since $\beta \in \overline{\varphi}_2(y_2)$ and $\{\alpha, \gamma\} \setminus \{\beta\} \subseteq \overline{\varphi}_2(x_2)$, this is a contradiction to Theorem 2.1. ∎

Let G be a graph, and let $F_1, \ldots, F_p$ be matchings of G. Then we denote by $H = G + F_1 + \cdots + F_p$ the graph satisfying $V(H) = V(G)$ and

$$\mu_H(x, y) = \mu_G(x, y) + \sum_{i=1}^{p} |E_G(x, y) \cap F_i|$$

for every pair (x, y) of distinct vertices of G. We say that H is obtained from G by **adding matchings** $F_1, \ldots, F_p$. If $F_i = F$ for $i = 1, \ldots, p$, then we also write $G + p \cdot F$ instead of $G + F_1 + \cdots + F_p$.

Proposition 4.53 (Grünewald and Steffen [117] 1999). *Let G be a graph obtained from the Petersen graph P by adding $k \geq 0$ perfect matchings, and let $v \in V(P)$ be any vertex. Then $G' = G - v$ is a critical graph with $\Delta(G') = k + 3$ and $\chi'(G') = k + 4$.*

Proof: The graph G is a Δ-regular graph with $\Delta = k + 3$ and $\chi'(G) \leq \chi'(P) + k \leq 4 + k = \Delta + 1$. Let C be any induced 5-cycle of G. It is known that each perfect matching of P contains exactly two edges of C. Hence, G is not 1-factorable, which implies $\chi'(G) = \Delta + 1$. Now, let e be any edge of G. Clearly, e corresponds to an edge of P and, since $P - v$ is a Hilton graph, Lemma 4.8 implies that $P - v - e$ is 3-edge-colorable. Therefore, we have $\chi'(G - v - e) \leq k + 3 = \Delta$. ∎

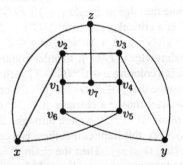

Figure 4.12   The Petersen graph P.

The Grünewald–Steffen Graph Family. For any integer $\Delta \geq 4$, we will construct a critical graph G_Δ having maximum degree Δ, chromatic index $\Delta + 1$, even order, and minimum degree 2. This graph family appeared first in a paper by Grünwald and Steffen [117].

We label the Petersen graph P as in Figure 4.12 and choose two perfect matchings, namely $F_1 = \{v_2v_3, v_1v_6, v_4v_7, v_5x, zy\}$ and $F_2 = \{v_2v_3, v_4v_5, v_1v_7, v_6y, zx\}$. For $1 \leq \ell \leq \Delta - 2$, let $P_\Delta^\ell = P + (\ell - 1) \cdot F_1 + (\Delta - \ell - 2) \cdot F_2$. By Proposition

4.53, $H_\Delta^\ell = P_\Delta^\ell - z$ is a critical graph with maximum degree Δ and chromatic index $\Delta + 1$. We shall mark the vertices of H_Δ^ℓ with an upper index ℓ, where all these graphs are assumed to be disjoint. Note that the degree function of $H = H_\Delta^\ell$ satisfies $d_H(v_7^\ell) = \Delta - 1$, $d_H(x^\ell) = \ell + 1$, $d_H(y^\ell) = \Delta - \ell$, and $d_H(v) = \Delta$ for the remaining vertices v of H.

For $1 \le \ell \le \Delta - 2$, a graph G_Δ^ℓ is constructed recursively as follows. Let $G_\Delta^1 = H_\Delta^1$. For $\ell \ge 2$, let G_Δ^ℓ be the Hajós sum of $G_\Delta^{\ell-1}$ and H_Δ^ℓ with respect to $(y^{\ell-1}, v_3^{\ell-1})$ and (x^ℓ, v_2^ℓ), where the vertex obtained by identifying $y^{\ell-1}$ and x^ℓ is labeled by $y^{\ell-1,\ell}$. The degree function then satisfies

- $d_{G_\Delta^{\ell-1}}(x^1) = 2$,

- $d_{G_\Delta^{\ell-1}}(y^{\ell-1}) = d_{H_\Delta^{\ell-1}}(y^{\ell-1}) = \Delta - \ell + 1$, and $d_{H_\Delta^\ell}(x^\ell) = \ell + 1$.

It follows from Theorem 4.46 that G_Δ^ℓ is a critical graph with maximum degree Δ and chromatic index $\Delta + 1$. For $\ell = \Delta - 2$, G_Δ^ℓ is a critical with two vertices of degree 2, namely x^1 and $y^{\Delta-2}$.

Finally, let $G_\Delta = YA(G_\Delta^{\Delta-2}, x^1, v_2^1, y^{\Delta-2})$. It follows from Theorem 4.51 that G_Δ is a critical graph having maximum degree Δ, chromatic index $\Delta + 1$, even order, and minimum degree 2.

Theorem 4.54 (Grünewald and Steffen [117] 1999) *For each $\Delta \ge 4$, there are infinitely many critical graphs G of class two such that $\Delta(G) = \Delta$, $|V(G)|$ is even, and $\delta(G) = 2$.*

Proof: Let $G' = G_\Delta^{\Delta-2}$ be the critical graph constructed above. By reapplication of the Δ-extension, a simple graph G'' with two vertices of degree 2 is obtained. By Theorem 4.50, G'' is a critical class two graph with maximum degree Δ. It follows then from Theorem 4.51 that $H = YA(G'')$ is a critical class two graph with maximum degree Δ and minimum degree 2. Then $(HS^{2p+1}(H))_{p \in \mathbb{N}}$ is an infinite family of critical class two graphs, each with maximum degree Δ, even order, and minimum degree 2. ∎

Theorem 4.55 (Grünewald [115] 2000) *For each $\Delta \ge 5$, there is a critical graph G such that $\Delta(G) = \Delta$, $\chi'(G) = \Delta + 1$, $|V(G)| = 20$, and $\delta(G) = 2$.*

Sketch of Proof: Let $G' = G_\Delta^3$ be the critical graph constructed above. Then G' has $n' = 25$ vertices. Let $X^1 = \{v_6^1, y^{1,2}, v_5^2\}$ and $X^2 = \{v_6^2, y^{2,3}, v_5^3\}$. For $\ell = 1, 2$, we conclude that X^ℓ is a Δ-contractible vertex set in G'. Hence, Proposition 4.49 implies that $H = (G/X^1)/X^2$ is a critical graph with $n = 21$ vertices. Then the graph G obtained from H by identifying x^1 and y^3 is a critical graph of order 20. ∎

4.8 HADWIGER'S CONJECTURE FOR LINE GRAPHS

Hadwiger's conjecture [124] from 1943 is the major unsolved problem in graph coloring theory. The conjecture deals with vertex colorings of simple graphs and suggests a far-reaching generalization of the Four-Color Theorem. Toft [294] gave a comprehensive survey of the history of the conjecture and the many partial results that have been obtained.

Let G be a simple graph. Recall that the **chromatic number** $\chi(G)$ of G is the minimum number of colors needed to color the vertices of G in such a way that no two adjacent vertices are assigned the same color. The **Hadwiger number** $\eta(G)$ of G is the maximum number k such that the complete graph K_k is a **minor** of G, i.e., there are k pairwise disjoint subsets $X_1, \ldots, X_k$ of $V(G)$ such that $G[X_i]$ is connected for $1 \le i \le k$ and $E_G(X_i, X_j) \ne \emptyset$ for $1 \le i < j \le k$.

Hadwiger's conjecture asserts that $\chi(G) \le \eta(G)$ for every simple graph G. This conjecture seems difficult to attack and might be false for the class $\mathcal{G}$ of all simple graphs. Therefore, it is reasonable to search for interesting subclasses of $\mathcal{G}$ for which Hadwiger's conjecture can be proved. An appropriate candidate for such a subclass was already proposed by Vizing [300], namely the class of line graphs. That Hadwiger's conjecture is true for all line graphs of simple graphs is easily deduced from Vizing's bound in Theorem 2.2. That the conjecture is also true for line graphs of graphs having multiple edges was proved by Reed and Seymour [253]. Surprisingly, the proof is short and uses mainly the fan equation/inequality.

Recall that the **line graph** of an arbitrary graph G, written $L(G)$, is the simple graph whose vertices are the edges of G, and where two vertices are joined by an edge if and only if the corresponding edges of G are adjacent. Hence, if $H = L(G)$ and $x \in V(G)$, then $E_G(x) \subseteq V(H)$ and $H[E_G(x)]$ is a complete graph. It is clear, that if G is a graph with at least one edge, then there is a one-to-one correspondence between the edge colorings of G and the vertex colorings of $L(G)$ implying that $\chi'(G) = \chi(L(G))$. Furthermore, it is not difficult to show, that, for an arbitrary graph G, we have $\eta(L(G)) \ge k$, provided that G contains k subtrees $T_1, \ldots, T_k$, each with at least one edge, satisfying $E(T_i) \cap E(T_j) = \emptyset$ and $V(T_i) \cap V(T_j) \ne \emptyset$ for $1 \le i < j \le k$. Such a set of subtrees of G will be briefly called a **cluster**. The Hadwiger number of $L(G)$ is, in fact, the maximum number k such that G contains a cluster of k subtrees. That Hadwiger's conjecture holds for the class of line graphs follows now from the following result.

Theorem 4.56 (Reed and Seymour [253] 2004) *Every graph G with $\chi'(G) = \ell$ for some integer $\ell \ge 0$ contains a cluster of ℓ subtrees, i.e., Hadwiger's conjecture is true for all line graphs.*

Proof: Let G be a possible counterexample such that $|V(G)| + |E(G)|$ is minimum. Then $\chi'(G) = k + 1$, where $k \ge \Delta(G) \ge 1$, and G does not contain a cluster of $k + 1$ subtrees. By the choice of G, it follows that G is critical.

First, let x, y be two distinct vertices of G and let $d = \min\{d_G(x), d_G(y)\}$. Furthermore, let s be the maximum number of edge-disjoint paths in G between x and y. We claim that $s = d$. Suppose that the claim is false. Then $s < d$ and, by Menger's theorem, there is a partition (X_1, X_2) of $V(G)$ such that $X_1 \cap X_2 = \emptyset$, $x \in X_1, y \in X_2$, and $|E_G(X_1, X_2)| = s$. Since $s < d$, both sets X_1 and X_2 have at least two vertices. For $i = 1, 2$, let G_i denote the graph obtained from G by contracting the set X_i, that is, we replace in G the set X_i by a new vertex x_i and join this new vertex to every vertex $z \in V(G) \setminus X_i$ by $|E_G(z, X_i)|$ edges. Then $d_{G_i}(x_i) = s$ and $|V(G_i)| + |E(G_i)| < |V(G)| + |E(G)|$ for $i = 1, 2$. If both graphs G_1 and G_2 are ℓ-edge-colorable, then G is ℓ-edge-colorable, too. Consequently, for one of these two graphs, say for G_1, we have $\chi'(G_1) \geq k + 1$. By the choice of G, it follows then that G_1 has a cluster of $k + 1$ subtrees. Since in G we have s edge-disjoint paths joining x and y, we conclude that, for every edge $e \in E_{G_1}(u, x_1)$ with $u \in X_2$, there is a path P_e joining u and x in $G[\{u\} \cup X_1]$ and all these paths are edge-disjoint. Consequently, the cluster of $k + 1$ subtrees in G_1 can be modified in order to obtain a cluster of $k + 1$ subtrees for the graph G, a contradiction. Hence the claim is proved.

Now, let x be a vertex of maximum degree in G. Since every edge of G is critical, it follows from Theorem 2.3(b) that there are two neighbors z_1, z_2 of x such that $d_G(z_1) + \mu_G(x, z_2) \geq k + 1$. Then, for $d = \min\{d_G(z_1), d_G(x)\}$, we have $d + \mu_G(x, z_2) \geq k + 1$. By the above claim, G contains d edge-disjoint paths between z_1 and x. Then there are d edge-disjoint paths between z_1 and $\{x, z_2\}$, where no edge between x and z_2 belongs to any of them. Consequently, these d paths together with the edges between x and z_2 form a cluster of at least $k + 1$ subtrees in G, a contradiction. This contradiction completes the proof. ∎

For a simple graph G, the **Hajós number** $\eta^o(G)$ is the maximum number k such that G contains a subdivision of the complete graph K_k, i.e., there are k vertices in G, joined pairwise by internally disjoint paths. Evidently, every simple graph G satisfies $\eta^o(G) \leq \eta(G)$.

In the 1960s the conjecture that $\chi(G) \leq \eta^o(G)$ for every simple graph G became very popular among graph theorists. Usually, this conjecture is attributed to G. Hajós, but the origin of this conjecture is unclear. Dirac knew the conjecture already in the 1950s. Dirac [73] verified the conjecture for all graphs G with $\chi(G) \leq 4$. For over a quarter of a century, it was believed by many graph theorists that the conjectures of Hadwiger and Hajós were in principle of the same type. For graphs with chromatic number at most 4, these two conjectures are indeed equivalent. For graphs with chromatic number 5 Hajós' conjecture implies the Four-Color Theorem. However, in 1979 Catlin [43] proved that Hajós' conjecture fails for graphs G with $\chi(G) \geq 7$ even in the class of line graphs. One of the classical counterexamples presented by Catlin is $G = L(3 \cdot C_5)$. By Theorem 6.3, $\chi(G) = \chi'(3 \cdot C_5) = 8$, but $\eta^o(G) \leq 7$. Whether Hajós' conjecture holds for graphs G with $5 \leq \chi(G) \leq 6$ is still open.

In 1981 Erdős and Fajtlowicz [82] proved that Hajós conjecture fails for almost all graphs. In 2005, Thomassen [292] obtained new interesting classes of counterexamples. On the other hand, Thomassen [293] proved that Hajós' conjecture is true for line graphs of simple graphs. This result is an immediate consequence of Theorem 4.4.

Theorem 4.57 (Thomassen [293] 2007) *Let $G = L(H)$ be the line graph of the simple graph H with $\Delta(H) = p \geq 1$. If H is of class one, then $\chi(G) = p$ and $K_p \subseteq G$. If H is of class two, then $\chi(G) = p + 1$ and G contains a complete graph K_p and a vertex x such that x is joined to K_p by p paths having only the vertex x in common. In particular, $\chi(G) \leq \eta^o(G)$.*

Recall that the **clique number** $\omega(G)$ of a graph G is the maximum number k such that K_k is a subgraph of G. It is easy to see that $\omega(G) \leq \chi(G) \leq \Delta(G) + 1$ for every simple graph G. On the other hand, there is no upper bound for the chromatic number in terms of the clique number, that is, there are triangle free graphs with arbitrarily large chromatic number. The following result due to King, Reed and Vetta [172], however, provides an upper bound for the chromatic number of a line graph in terms of the clique number and the maximum degree.

Theorem 4.58 (King, Reed, and Vetta [172] 2007) *If $G = L(H)$ is the line graph of an arbitrary graph H, then*

$$\chi(G) \leq \left\lceil \frac{\Delta(G) + \omega(G) + 1}{2} \right\rceil.$$

That Theorem 4.58 holds not only for line graphs, but for simple graphs in general, was conjectured by Reed [252] in 1998.

The proof of Theorem 4.58 has two cases. Let $G = L(H)$ be the line graph of the graph $H = (V, E)$ with $\Delta(H) \geq 1$. If $|V| \leq 2$, then $\omega(G) = \Delta(H)$. Otherwise,

$$\omega(G) = \max\{\Delta(H), \max_{X \subseteq V, |X|=3} |E(H[X])|\}.$$

First, the authors consider the case when $\Delta(G) \geq \frac{3}{2}\Delta(H) - 1$. By a result of Nishizeki and Kashiwagi [237] (see also Corollary 6.7), we have

$$\chi(G) = \chi'(H) \leq \max\{\lfloor 1.1\Delta(H) + 0.8\rfloor, \lceil \chi'^*(H)\rceil\},$$

where $\chi'^*(H)$ is the fractional edge chromatic number of H. Molloy and Reed [229] proved a fractional analogue of Theorem 4.58 for arbitrary graphs implying that

$$\lceil \chi'^*(H)\rceil \leq \left\lceil \frac{\Delta(G) + \omega(G) + 1}{2} \right\rceil.$$

Consequently, we have

$$\chi(G) \leq \max\left\{ \lfloor 1.1\Delta(H) + 0.7\rfloor, \left\lceil \frac{\Delta(G) + \omega(G) + 1}{2} \right\rceil\right\}.$$

Since $\Delta(G) \geq \frac{3}{2}\Delta(H) - 1$ and $\omega(G) \geq \Delta(H) \geq 1$, this implies

$$\lceil (\Delta(G) + \omega(G) + 1)/2\rceil \geq \lceil (5/4)\Delta(H)\rceil \geq \lfloor 1.1\Delta(H) + 0.8\rfloor,$$

and so we are done.

It remains to consider the case when $\Delta(G) < \frac{3}{2}\Delta(H) - 1$. Then, as proved by King et al. [172], there is a maximal independent set S in G containing a vertex of every maximal complete subgraph of G. This implies that $G' = G - S$ satisfies $\omega(G') = \omega(G) - 1$ and $\Delta(G') \leq \Delta(G) - 1$. Thus the desired result follows by induction.

4.9 SIMPLE GRAPHS ON SURFACES

For a closed **surface S**, i.e., a compact, connected 2-dimensional manifold without boundary, let

$$\Delta(\mathbf{S}) = \max \left\{ \Delta(G) \mid G \text{ is a graph of class two that is embeddable in } \mathbf{S} \right\}.$$

That the maximum always exists was first proved by Vizing [299]; the existence can be easily deduced by combining Theorem 4.3 and a well known result due to Heawood [133]. He determined an upper bound for the coloring number

$$\text{col}(\mathbf{S}) = \max\{\text{col}(G) \mid G \text{ is a simple graph that is embeddable in } \mathbf{S}\}$$

of an arbitrary surface **S**. Theorem 4.3 implies that

$$\Delta(\mathbf{S}) \leq 2\text{col}(\mathbf{S}) - 3 \tag{4.23}$$

for every surface **S**. For the sphere **S**, we have $\text{col}(\mathbf{S}) \leq 6$ and hence $\Delta(\mathbf{S}) \leq 9$. An example of a planar graph of class two with maximum degree 5 is the graph obtained from the icosahedron by subdividing any edge. Obviously, this yields a planar simple graph with maximum degree 5. That this also yields a graph of class two follows from the fact that this graph is just overfull. On the other hand, Vizing [299] proved that every planar simple graph with maximum degree 8 is of class one. Hence, $5 \leq \Delta(\mathbf{S}) \leq 7$ for the sphere **S** and Vizing [300] conjectured that the right answer for the sphere is five or six. That planar simple graphs of maximum degree 7 are of class one was proved, independently, by Grünewald [116], by Sanders and Zhao [264], and by Zhang [317].

Theorem 4.59 (Grünewald [116], Sanders and Zhao [264], Zhang [317]) *Every planar simple graph G with $\Delta(G) \geq 7$ satisfies $\chi'(G) = \Delta(G)$.*

Proof: By Proposition 4.2, it suffices to show that there exists no planar critical graph of class two with maximum degree 7. Suppose this is false and let G be such a planar graph. We may fix an embedding of G on the sphere having vertex set V, edge set E, and face set F. Since G is a critical graph of class two with maximum degree $\Delta = 7$, we conclude that $|V| \geq 8$, G has vertex connectivity at least 2, and the boundary of each face $f \in F$ forms a cycle in G. For $v \in V$, let $d(v)$ denote the degree of v in G and, for $f \in F$, let $d(f)$ denote the degree of f with respect to the embedding of G. Note that $d(f)$ is the length of the cycle that bounds f. Since G is a critical graph of class two with maximum degree $\Delta = 7$, we have $2 \leq d(v) \leq 7$ for each vertex $v \in V$ and $d(f) \geq 3$ for each face $f \in F$. By a *big face* we mean a face $f \in F$ with $d(f) \geq 4$. If $d(t) = k$ for a vertex or face $t \in V \cup F$, then t is said to be a *k-vertex* or a *k-face*. A *k*-vertex adjacent to a vertex $x \in V$ is also said to be a *k-neighbor* of x. By an (i, j, k)-*face* we mean a 3-face incident with distinct vertices x, y, z such that $d(x) = i, d(y) = j$, and $d(z) = k$.

To arrive at a contradiction, we apply the discharging method. We define an initial charge distribution ch on $V \cup F$ by

$$ch(x) = \begin{cases} 6 - d(x) & \text{if } x \in V, \\ 6 - 2d(f) & \text{if } x \in F. \end{cases}$$

Then it follows by Euler's formula that

$$\sum_{t \in V \cup F} ch(t) = 6|V| - 2|E| + 6|F| - 4|E| = 6(|V| - |E| + |F|) = 12. \quad (4.24)$$

Starting with the initial charge distribution ch, we redistribute the charge by application of the following 11 discharging rules.

1. Each 2-vertex x sends 1 to each 7-neighbor of x.

2. Each 2-vertex x sends 2 to each big face incident with x.

3. Each 3-vertex x sends s to each 7-neighbor y of x, where $s = 1$ if xy is incident with two 3-faces and $s = \frac{1}{2}$ otherwise.

4. Each k-vertex x with $3 \leq k \leq 6$ sends $s = \frac{1}{2} + m\frac{1}{4}$ to each big face f incident with x, where m is the number of 7-vertices y such that xy is incident with f.

5. Each 3-vertex x sends 1 to each 7-vertex z such that z is not adjacent to x, but there is a 6-vertex adjacent to both x and z.

6. Each 4-vertex x adjacent to a 5-vertex sends $\frac{2}{3}$ to each 7-neighbor of x.

7. Each 4-vertex x not adjacent to a 5-vertex sends s to each 7-neighbor y of x, where $s = \frac{3}{5}$ if xy is incident with two 3-faces and $s = \frac{1}{5}$ otherwise.

8. Each 4-vertex x sends $\frac{2}{5}$ to each 6-neighbor of x.

9. Each 5-vertex x adjacent to a 4-vertex sends $\frac{1}{3}$ to each 7-neighbor of x.

10. Each 5-vertex x not adjacent to a 4-vertex sends s to each vertex y such that xy is an edge incident with two 3-faces and $d(y) \geq 6$, where $s = \frac{2}{5}$ if xy is incident with exactly one $(5,5,7)$-face and $s = \frac{1}{5}$ otherwise.

11. Each 6-vertex x not adjacent to a 3-vertex sends s to each 7-neighbor y of x such that xy is incident with two 3-faces, but not incident with two $(6,7,7)$-faces, where $s = \frac{2}{5}$ if xy is incident with two $(4,6,7)$-faces and $s = \frac{1}{5}$ otherwise.

In what follows, we refer to the eleven rules as R1, ... , R11. Let ch' denote the resulting new charge distribution. Each rule moves charge around, but the sum over all charges remains unchanged, i.e.,

$$\sum_{t \in V \cup F} ch'(t) = \sum_{t \in V \cup F} ch(t) = 12.$$

We claim that $ch'(t) \leq 0$ for each $t \in V \cup F$, which gives the required contradiction. To this end, we mainly use Vizing's Adjacency Lemma (VAL), see Theorem 2.7. By VAL, for every edge xy of G, the number of 7-neighbors of x distinct from y is at least $8 - d(y)$ and, moreover, $d(x) + d(y) \geq 9$. Let $t \in V \cup F$ be an arbitrary element. We will use the following abbreviations: $s^+(t)$ denotes the total charge that t receives according to the discharging rules, $s^-(t)$ denotes the total charge that t sends out according to the discharging rules, and $s(t) = s^+(t) - s^-(t)$. Clearly, $ch'(t) = ch(t) + s(t)$ and for the proof it suffices to show that $s(t) \leq -ch(t)$.

Claim 4.59.1 *Every face $f \in F$ satisfies $s(f) \leq -ch(f) = 2(d(f) - 3)$.*

Proof: Let $f \in F$ be an arbitrary face. Note that $d(f) \geq 3$ and $s^-(f) = 0$, i.e., $s(f) = s^+(f)$. If $d(f) = 3$, then $s^+(f) = 0$ and we are done. Now suppose that $d(f) = 4$. If one of the four vertices incident with f is a 2-vertex, then VAL implies that the remaining three vertices incident with f are 7-vertices. Then $s^+(f) \leq 2$ by R2 and we are done. Otherwise all four vertices incident with f have degree at least 3 and $s^+(f) \leq 2$ by R4 as required. Finally, suppose that $d(f) = k$ for $k \geq 5$. If f is incident with a 2-vertex x and if xyz is the sequence of vertices in the facial walk of f, then VAL implies that both y and z are 7-vertices. By R2 and R4, it follows that in average each vertex incident with f sends at most $\frac{2}{3}$ to f, which gives $s^+(f) \leq 2k/3 \leq 2(k-3) = 2(d(f) - 3)$ since $k \geq 5$. This completes the proof of the claim.   △

Claim 4.59.2 *Every vertex $v \in V$ satisfies $s(v) \leq -ch(v) = d(v) - 6$.*

Proof: Let $v \in V$ be an arbitrary vertex. Then $2 \leq d(v) \leq 7$ and we distinguish four cases.

Case 1: $2 \leq d(v) \leq 4$. Then $s^+(v) = 0$, i.e., $s(v) = -s^-(v)$, and it suffices to show that $s^-(v) \geq 6 - d(v)$.

First, assume $d(v) = 2$. By VAL, the two neighbors of v are both 7-vertices. Since G is simple and certainly not K_3, v is incident with at least one big face. It now follows from R1 and R2 that $s^-(v) \geq 2 + 2 = 4$, as required.

Next, assume $d(v) = 3$. By VAL, each neighbor of v has degree at least 6 and at least two neighbors have degree 7. If v is adjacent to a 6-vertex u, then VAL implies that u has at least five 7-neighbors and so $s^-(v) \geq 3$ by R5. Otherwise, each of the three neighbors of v is a 7-vertex and $s^-(v) \geq 3$ by R3 and R4. This settles the subcase $d(v) = 3$

Finally, assume $d(v) = 4$. By VAL, each neighbor of v has degree at least 5 and at least two neighbors have degree 7. If v has a 5-neighbor, then VAL implies that v has three 7-neighbors and $s^-(v) \geq 2$ by R6, as required. Otherwise, v has a ≥ 2 7-neighbors and $b = 3 - a$ 6-neighbors, which gives $s^-(v) \geq a\frac{3}{5} + b\frac{2}{5} \geq 2$ by R4, R7 and R8, as required.

Case 2: $d(v) = 5$. Since $s^+(v) = 0$, it suffices to show that $s^-(v) \geq 1$. By VAL, each neighbor of v has degree at least 4 and at least two neighbors have degree 7. If v has a 4-neighbor, then VAL implies that v has four 7-neighbors and $s^-(v) \geq 4\frac{1}{3} \geq 1$ (R9), so we are through. Otherwise, each neighbor of v has degree at least 5. If v is incident with at least one big face, then $s^-(v) \geq 1$ (R4 and R10). Otherwise, v is incident with five 3-faces. If all five neighbors of v have degree at least 6, then $s^-(v) \geq 5\frac{1}{5} = 1$ (R4 and R10), and we are through. If v has a 5-neighbor, then VAL implies that v has at least three 7-neighbors and hence at most two 5-neighbors. If v has two 5-neighbors, then $s^-(v) \geq 2\frac{2}{5} + \frac{1}{5} = 1$ by R10, and we are done. If v

has only one 5-neighbor z, then R10 implies that $s^-(v) \geq \frac{2}{5} + 3\frac{1}{5} = 1$, and we are through. This settles Case 2.

Case 3: $d(v) = 6$. It suffices to show that $s^-(v) \geq s^+(v)$. By VAL, each neighbor of v has degree at least 3. Let a and b denote the number of 4-neighbors and 5-neighbors of v, respectively. By R8 and R10, vertex v receives at most $a\frac{2}{5} + b\frac{1}{5}$ and no other rules send charge into v, i.e., $s^+(v) \leq a\frac{2}{5} + b\frac{1}{5}$.

If $a = b = 0$ this gives $s^+(v) = 0 \leq s^-(v)$ and we are done. If $a \geq 1$ or $b \geq 1$, then VAL implies that each neighbor of v has degree at least 4. If $a \geq 1$, then v has at least four 7-neighbors (VAL) and, therefore, $a + b \leq 2$. By R4 and R11, it follows that $s^-(v) \geq s^+(v)$. Otherwise, we have $a = 0$ and $b \geq 1$. Then, by VAL, v has at least three 7-neighbors and, therefore, $b \leq 3$. Furthermore, VAL implies that no edge incident with v is incident with two $(5, 5, 6)$-faces. Then, based on R4 and R11, we conclude that $s^-(v) \geq s^+(v)$.

Case 4: $d(v) = 7$. Since $s^-(v) = 0$, it suffices to show $s^+(v) \leq 1$. Note that no 7-vertex sends charge to v.

If v has a 2-neighbor u, then the remaining six neighbors are 7-vertices (VAL) and $s^+(v) = 1$ by R1, which gives the desired result. If v has a 6-neighbor x which has a 3-neighbor y that is not adjacent to v, then Theorem 4.30(b) implies that each vertex in $N_G(v) \setminus \{x\}$ is a 7-vertex, and $s^+(v) = 1$ (R5 and R11), as required. If v has a neighbor x with $d(x) \in \{4, 5\}$ that is adjacent to a vertex y with $d(x) + d(y) = 9$, then Theorem 4.30(b) implies that each vertex $v \in N_G(v) \setminus \{x, y\}$ is a 7-vertex, and $s^+(v) \leq 2/3 + 1/3 = 1$ (R6 and R7).

If v has a 3-neighbor x, then VAL implies that v has a neighbor $y \neq x$ with $3 \leq d(y) \leq 7$ and each vertex $u \in N_G(v) \setminus \{x, y\}$ is a 7-vertex. If $d(y) = 3$, then Theorem 4.35 implies that neither vx nor vy is incident with two 3-faces, and it follows then from R3 that $s^+(v) \leq 1/2 + 1/2 = 1$, as required. If $d(y) = 7$, then only x sends charge 1 to v by R3 and we are done. It remains to consider the case $4 \leq d(y) \leq 6$. If $d(y) \in \{4, 5\}$, then we may assume that y has no neighbor z with $d(z) + d(y) = 9$, since the opposite was already considered. Thus v receives charge only from x according to R3 and from y according to R7, R10, or R11. If $4 \leq d(y) \leq 5$, then VAL implies that $xy \notin E(G)$ and Theorem 4.35 implies that vy is not incident with a 3-face. Hence, if $d(y) = 5$ the only charge v receives is at most 1 from x. If $d(y) = 4$, then Theorem 4.35 implies that vx is not incident with two 3-faces, and the only charge v receives is $\frac{1}{2}$ from x and $\frac{1}{5}$ from y. If $d(y) = 6$, then y is adjacent to x, vy is not incident with two 3-faces, or vy is incident with two $(6, 7, 7)$-faces. Hence, the only charge v receives is at most 1 from x. This settles the case that v has a 3-neighbor.

Next, suppose that v has a 4-neighbor, but no 3-neighbor. We may assume that no 4-neighbor of v is adjacent to a 5-vertex, since the opposite was already considered. Let N denote the set of all neighbors z of v such that $d(z) < 7$. By VAL, v has at least four 7-neighbors, and $|N| \leq 3$. First, assume that one 4-neighbor x of v sends $\frac{3}{5}$ to v according to R7. Then vx is incident with two 3-faces. It follows from Theorem 4.35 that x is the only 4-neighbor of v, and so $d(y) \geq 5$ for $y \in N \setminus \{x\}$. Furthermore, Theorem 4.36 implies that v is not incident with a $(7, 5, 5)$-face. This implies $s^+(v) \leq 3/5 + 2(1/5) = 1$ (R7, R9, R10 and R11). Now, assume that no 4-neighbor of v sends $\frac{3}{5}$ to v. Then $s^+(v) \leq 1/5 + 2(2/5) = 1$ (R7, R9, R10, and R11).

Finally, suppose that each neighbor of v has degree at least five. We may assume that no 5-neighbor of v is adjacent to a 4-vertex, since the opposite was already considered. If there is 5-neighbor x of v which sends $\frac{2}{5}$ to v according to R10, then vx is incident with two 3-faces and incident with exactly one $(5,5,7)$-face. Then Theorem 4.37 implies that v has at most three neighbors with degree at most 6, which gives $s^+(v) \leq \frac{1}{5} + 2\frac{2}{5} = 1$ (R9, R10, R11). If there is no 5-neighbor x of v which sends $\frac{2}{5}$ to v, then VAL implies that at most five neighbors of v have degree at most 6, and so $s^+(v) \leq 5\frac{1}{5} = 1$ (R9, R10, R11). This settles Case 4 and completes the proof of the claim. $\triangle$

Thus, the proof of the theorem is complete. ∎

Surfaces can be classified according to their Euler characteristic and orientability. The **orientable surfaces** are the surfaces $\mathbf{S}_g$, for $g \geq 0$, obtained from the sphere by attaching g handles. The **nonorientable surfaces** are the surfaces $\mathbf{N}_h$, for $h \geq 1$, obtained by taking the sphere with h holes and attaching h Möbius bands along their boundary to the boundaries of the holes. Then $\mathbf{N}_1$ is the projective plane, $\mathbf{N}_2$ is the Klein bottle, etc. The **Euler characteristic** $\varepsilon(\mathbf{S})$ of the surface $\mathbf{S}$ is $2 - 2g$ if $\mathbf{S} = \mathbf{S}_g$ and $2 - h$ if $\mathbf{S} = \mathbf{N}_h$. The classification theorem for surfaces states that every closed surface is homeomorphic to either $\mathbf{S}_h$ or $\mathbf{N}_h$, for a suitable value of h.

Consider a simple graph $G = (V, E)$ that is embedded on a surface $\mathbf{S}$ of Euler characteristic $\varepsilon = \varepsilon(\mathbf{S})$. Euler's Formula tells us that $|V| - |E| + |F| \geq \varepsilon$, where F is the set of faces and with equality holding if and only if every face is a 2-cell. Therefore, $|E| \leq 3|V| - 3\varepsilon$ if $|V| \geq 3$. For $\varepsilon \leq 1$, this implies that $\mathrm{col}(G) \leq H(\varepsilon)$, where

$$H(\varepsilon) = \left\lfloor \frac{7 + \sqrt{49 - 24\varepsilon}}{2} \right\rfloor .$$

Consequently, $\mathrm{col}(\mathbf{S}) \leq H(\varepsilon)$ if $\varepsilon = \varepsilon(\mathbf{S}) \leq 1$. This result was proved by Heawood [133] in 1890. For every surface $\mathbf{S}$ distinct from the sphere or the Klein bottle, the Heawood number $H(\varepsilon)$ is, in fact, the maximum coloring number of graphs that can be embedded on $\mathbf{S}$, where the maximum is attained by the complete graph on $H(\varepsilon)$ vertices. This landmark result, which was conjectured by Heawood [133], is due to Ringel [254] and Ringel and Youngs [255]. For the sphere we have $\mathrm{col}(\mathbf{S}_0) = 6$ and $H(2) = 4$ and for the Klein bottle we have $\mathrm{col}(\mathbf{N}_2) \leq H(0) = 7$.

By (4.23) and Heawood's result, every surface $\mathbf{S}$ of Euler characteristic $\varepsilon \leq 1$ satisfies $\Delta(\mathbf{S}) \leq 2H(\varepsilon) - 3$. In 1970 Mel'nikov [220] improved this bound to

$$\Delta(\mathbf{S}) \leq \max \left\{ \left\lfloor \frac{11 + \sqrt{25 - 24\varepsilon}}{2} \right\rfloor, \left\lfloor \frac{8 + 2\sqrt{52 - 18\varepsilon}}{3} \right\rfloor \right\},$$

provided that $\varepsilon = \varepsilon(\mathbf{S}) \leq 0$.

For an integer $\varepsilon \leq 2$, let $J(\varepsilon) = \lfloor 3 + \sqrt{13 - 6\varepsilon} \rfloor$. One can easily check that either $J(\varepsilon) = H(\varepsilon)$ or $J(\varepsilon) = H(\varepsilon) - 1$. The next result shows that Vizing average degree conjecture (Conjecture 4.38) implies that $J(\varepsilon(S)) - 1 \leq \Delta(S) \leq J(\varepsilon(S))$ for every surface S distinct from the Klein bottle and with $\varepsilon(S) \leq 0$.

Proposition 4.60 *Let* S *be a surface distinct from the Klein bottle and of Euler characteristic* $\varepsilon \leq 1$. *Then the following statements hold:*

(a) $\Delta(S) \geq H(\varepsilon) - 1$.

(b) $\Delta(S) \leq J(\varepsilon)$, *provided that* $\varepsilon \leq 0$ *and that Conjecture 4.38 holds.*

Proof: Statement (a) follows from the fact that the complete graph K on $H(\varepsilon)$ vertices can be embedded on S. If $H(\varepsilon)$ is odd, then K is of class two. If $H(\varepsilon)$ is even, then the graph obtained from K by inserting a new vertex in any edge of K is of class two. Hence, in both cases, we obtain a graph of class two with maximum degree $H(\varepsilon) - 1$. This proves (a).

In order to prove (b), assume that it is false. Then there is a critical graph $G = (V, E)$ of class two such that G can be embedded on S and $\Delta(G) \geq J(\varepsilon) + 1$. Since $\varepsilon \leq 0$, $\Delta(G) \geq 7$. If Conjecture 4.38 holds, then $2|E| \geq (\Delta(G) - 1)|V| + 3$. On the other hand, $2|E| \leq 6|V| - 6\varepsilon$, which implies that $(\Delta(G) - 7)|V| + 3 + 6\varepsilon \leq 0$. Since $\Delta(G) \geq 7$ and $|V| \geq \Delta(G) + 1$, this gives $(\Delta(G) - 7)(\Delta(G) + 1) + 3 + 6\varepsilon \leq 0$ and, therefore, $\Delta(G) \leq 3 + \sqrt{13 - 6\varepsilon}$. Since $\Delta(G)$ is an integer, this implies that $\Delta(G) \leq J(\varepsilon)$, a contradiction. ∎

Up to know, exact values for $\Delta(S)$ are only known if $-3 \leq \varepsilon(S) \leq 0$. Sanders and Zhao [266] proved that $\Delta(S) = 6$ if $\varepsilon(S) = 0$. Luo and Zhao [202] proved that $\Delta(S) = 7$ if $\varepsilon(S) = -1$ and $\Delta(S) = 8$ if $\varepsilon(S) = -2, -3$. Consequently, each surface $S \in \{N_2, N_3, N_4, N_5, S_1, S_2\}$ satisfies $\Delta(S) = J(\varepsilon(S))$.

4.10 NOTES

Let $\mathcal{G}$ be the class of all simple graphs. From the proof of Vizing's bound it follows that there is a polynomial time algorithm that colors the edges of any graph $G \in \mathcal{G}$ with at most $\Delta(G) + 1 \leq \chi'(G) + 1$ colors. By Holyer's result [150], either $P = NP$ or finding an optimal edge coloring for graphs in $\mathcal{G}$ cannot always be done in polynomial time. From an algorithmic point of view, these two results solves the Classification Problem.

Proposition 4.5, saying that a simple graph is of class one if and only if its Vizing-core is of class one, seems to have been folklore, at least from the second half from the 1980s. The Vizing-core is the maximal subgraph satisfying Vizing's Adjacency Lemma and it might be useful to involve other adjacency lemmas for critical graphs of class two.

Since the Classification Problem is difficult for the class $\mathcal{G}$ of all simple graphs, it is reasonable to search for interesting subclasses of $\mathcal{G}$ for which the Classification Problem admits a good solution. One such class was proposed by Hilton and Zhao [143], namely the class $\mathcal{G}'$ of all simple graphs G with $\Delta(G^{[\Delta]}) \leq 2$ (Conjecture 4.7, respectively Conjecture 4.10). For graphs with maximum degree 2, this conjecture is evidently true. That the conjecture also holds for graphs with maximum degree 3 was proved by D. Cariolaro and G. Cariolaro [41] (Theorem 4.11). Their proof uses induction on the number of vertices. We gave an alternative proof to show that the only connected graph $G \in \mathcal{G}'$ of class two with $\Delta(G) = 3$ is the graph P^* obtained from the Petersen graph by deleting one vertex. This proof by Kr ál' et al. [183] is based on an analysis of the critical chains, that means the Kempe chains arising in a 3-edge-coloring of $G - e$ and joining the two endvertices of the uncolored edge e. The next case would be to prove that the only connected graph $G \in \mathcal{G}'$ of class two with $\Delta(G) = 4$ is K_5^-, i.e., the complete graph K_5 with one edge deleted. Using the critical chain method, it can be proved [183] that $V(G) \setminus V(G^{[\Delta]})$ is an independent set for any such graph.

The overfull conjecture implies that, for the subclass of $\mathcal{G}$ consisting of all simple graphs G with $\Delta(G) > |V(G)|/3$, the Classification Problem can be solved in polynomial time. This follows from the results in Sect. 6.1 about the fractional chromatic number. However, very little is known about the overfull conjecture. That the overfull conjecture holds for simple graphs G with $\Delta(G) \geq |V(G)| - 3$ was proved in a series of papers [243, 244, 47, 48, 51, 52]. The graph P^* shows that the conjecture is not always true if $\Delta(G) = |V(G)|/3$.

That the overfull conjecture implies the one-factorization conjecture relies on Lemma 4.15. For simple graphs, this result was obtained by Hilton [136] as well as by Niessen and Volkmann [236]. The one-factorization conjecture states that every regular graph of even order $2n$ and degree $\lambda(2n)$, where $\lambda \geq 1/2$, is of class one and hence 1-factorable. That the conjecture holds for $\lambda \geq (\sqrt{7} - 1)/2 \approx 5/6$ was proved by Chetwynd and Hilton [53] as well as by Niessen and Volkmann [236]. The above proof of Theorem 4.18 is due to Chetwynd and Hilton [53]. As far as we know, the coefficient $(\sqrt{7} - 1)/2$ in this result has not been improved so far. Some partial improvements have been made recently by Cariolaro and Hilton [42]. At the 22nd British Combinatorial Conference 2009, D. Cariolaro, A. J. W. Hilton, and E. R. Vaughan announced a small improvement of the coefficient.

Much attention has been paid to the theory of edge coloring of critical graphs of class two, in particular, to the function $e_\Delta = e_\Delta(n)$, that is, the minimum number of edges possible in a (Δ, n)-graph, where the value is infinite if no such graph exists, see [22, 65, 98, 99, 100, 125, 155, 164, 190, 191, 192, 193, 197, 199, 200, 222, 223, 224, 309, 310, 315, 318, 321]. Clearly, there are no $(1, n)$-graphs, and the only $(2, n)$-graph is C_n for odd $n \geq 3$. Hence $e_2(n) = n$ for odd $n \geq 3$. The best lower bounds for the function e_Δ with $3 \leq \Delta \leq 12$ are as follows: $e_3(n) \geq \frac{4}{3}n$ (Jakobsen [155]), $e_4(n) \geq \frac{7}{4}n$ (Miao and Pang [222]), $e_5(n) \geq \frac{15}{7}n$ (Woodall [310]), $e_6(n) \geq \frac{5}{2}n$ (Zhao [321]) respectively $e_6(n) \geq \frac{5}{2}n + \frac{3}{2}$ (Luo, Miao and Zhao [199]), $e_7(n) \geq \frac{17}{6}n$ (Zhao [321]), $e_8(n) \geq \frac{19}{6}n$ (Li [191]), $e_9(n) \geq \frac{7}{2}n$ (Li [191]), $e_{10}(n) \geq 4n$ (Li and Li [190]), $e_{11}(n) \geq \frac{17}{4}n$ (Li and Li [190]), and $e_{12}(n) \geq \frac{46}{10}n$ (S. Li and X. Li [190]). Recently Li [193] proved that $e_\Delta(n) \geq \frac{5\Delta+4}{14}n$ provided that $8 \leq \Delta \leq 17$. For larger values of Δ, the best lower bounds are due to Woodall [310, 309], see Theorem 4.40 and Theorem 4.41. The above proof of Theorem 4.40 is due to Woodall [309].

All known lower bounds for the function $e_\Delta(n)$ are obtained by applying appropriate adjacency lemmas for critical graphs of class two. This book contains several adjacency lemmas for critical graphs of class two (Sect. 2.2, Sect. 2.3, Sect. 2.4, Sect. 3.2, and Sect. 4.4). Our list, however, is far from being complete. The investigation of the function $\varepsilon_\Delta(n)$ for small values of Δ is based on very special adjacency lemmas. For example, in the paper by Luo et al. [199] it is proved that if G is a critical graph of class two with maximum degree $\Delta \geq 5$, then every vertex $x \in V(G)$ having degree 3 in G is adjacent to two major vertices which are not adjacent to any vertex with degree at most $\Delta - 2$ except to the vertex x.

Let us consider the special case $e_3 = e_3(n)$. Let G be a $(3, n)$-graph. Let n_2 and n_3 denote the number of vertices of G of degree 2 and 3, respectively. It follows from Theorem 4.30 that $n = n_2 + n_3$ and $n_3 \geq 2n_2$. From this we get $n_3 \geq 2n_2$, and therefore, $n_2 \leq \frac{1}{2}n$ and $n_3 \geq \frac{2}{3}n$. Eventually, we get $|E(G)| = \frac{1}{2}(2n_2 + 3n_3) = \frac{1}{2}(2n + n_3) \geq \frac{4}{3}n$. Consequently, we obtain that $e_3(n) \geq \frac{4}{3}n$. This argument was presented by Jakobsen [155]. In this paper he also proved Theorem 4.30 for the class of $(3, n)$-graphs. He also noticed that the graph P^* is a $(3, 9)$ graph with 12 edges. Then $HS^p(P^*)$ is a $(3, n)$-graph with $n = 8p + 1$ and with $m = 11p + 1$ edges, where $p \in \mathbb{N}$. If H is the $(3, 22)$-graph with 31 edges found by Goldberg (Fig. 4.10), then $HS^p(H)$ is a $(3, n)$-graph with $n = 21p + 1$ and with $m = 30p + 1$ edges, where $p \in \mathbb{N}$. In Sect. 4.7 we presented, for every integer $n \geq 22$, a $(3, n)$-graph with m edges, where $m = (3n - 1)/2$ if n is odd and $m = (3n - 4)/2$ if n is even. Hence, for every integer $n \geq 22$, we have

$$e_3(n) \leq \begin{cases} 11(n-1)/8 + 1 & \text{if } n \equiv 1 \bmod 8, \\ 30(n-1)/21 + 1 & \text{if } n \equiv 1 \bmod 21, \\ \lfloor 3(n-1)/2 \rfloor + 1 & \text{otherwise.} \end{cases}$$

Proposition 4.48 implies that every (Δ, n) graph G with $\Delta \geq 3$ satisfies $|E(G)| \leq \frac{1}{2}(n-1)\Delta + 1$ if n is odd, and $|E(G)| \leq \frac{1}{2}(n-1)\Delta + \delta(G) - 1$ if n is even. This implies, in particular, that G is not regular.

For several years it was a common belief of many graph theorists that every critical graph with $\chi' \geq \Delta + 1$ must have an odd number of vertices. This conjecture became known as the "Critical Graph Conjecture" (see references [57, 141]). The conjecture was formulated by Jakobsen [156] and, independently, by Beineke and Wilson [23]. To support this conjecture, Jakobsen [155] showed that there are no $(3, n)$-graphs for $n \in \{4, 6, 8, 10\}$. Further evidence to support the conjecture was obtained by Beineke and Fiorini [22], and Fiorini [98]. They showed that any (Δ, n)-graph with even $n \leq 10$ must have a 1-factor. This result enabled them to show that there is no (Δ, n)-graph with even $n \leq 10$. The conjecture was eventually disproved by Goldberg [113], who showed that the graph in Fig. 4.10 is a $(3, 22)$-graph. Somewhat later Chetwynd [45] and Fiol [97] constructed a $(4, 18)$-graph. Fiol's graph was obtained by removing a triangle from the line graph of Isaacs' flower snark J_3. For a general discussion of these and other snarks see Chetwynd and Wilson [56]. A **snark** is a bridgeless cubic graph of class two; to avoid trivial cases, it is usual to require that snarks cannot be reduced by certain constructions. The Petersen graph is the smallest snark. The name "snark", used in a mathematical context, is due to

Gardner [106] and refers to Lewis Carroll's poem *The Hunting of the Snark* from 1876. Snark hunting is still popular, at least among graph theorists. One reason for this is that certain conjectures are known to be true in general, provided that they can be established for snarks. Moreover, finding snarks is a difficult matter, and the Four-Color Theorem tells us that there are no planar snarks. A snark is never critical, but it contains a critical graph of class two with maximum degree $\Delta = 3$. It is known that line graphs of snarks are of class two, but the line graph of the line graph of the Petersen graph is of class one, a 6-edge-coloring of this graph is presented in reference [141]. After the presentation of a $(4, 18)$-graph by Chetwynd and Fiol it took more that 20 years before Grünewald and Steffen [117] constructed, for each $\Delta \geq 4$, a critical graph of class two having maximum degree Δ and even order. Grünewald [115] also disproved the so-called weak critical graph conjecture (see Jensen and Toft [157, Problem 12.4]), but the critical multigraph conjecture (Conjecture 1.6) is still unsolved.

Vizing's independence number conjecture was proved by Grünewald and Steffen [118] for critical graphs of class two with many edges, including all overfull graphs, and it was proved by Luo and Zhao [201] whenever $n \leq 2\Delta$; similar results may also be found in references [203, 221]. For an arbitrary critical graph G of class two having order n, Brinkman et al. [32] proved that $\alpha(G) \leq \frac{2}{3}n$. This was improved by Luo and Zhao [205] to $\alpha(G) \leq \frac{5}{8}n$ and by Woodall [311] to the so far best $\alpha(G) \leq \frac{3}{5}n$. The above proof of Theorem 4.44 is Woodall's original proof from reference [311]. Since $\alpha(G) \leq n - m/\Delta$ for any graph G with n vertices, m edges and maximum degree Δ, the average degree conjecture (Conjecture 4.38) supports the independence number conjecture, at least asymptotically.

The book of Jensen and Toft [157] contains a list of around 200 problems related to vertex colorings of simple graphs as well as to edge coloring of graphs. If we intend to study a certain conjecture about vertex coloring of simple graphs one can start to investigate this conjecture for the class of line graphs of either simple graphs or arbitrary graphs. Then this conjecture becomes a statement about edge colorings of graphs and very often such statements can be proved or disproved. In Sect. 4.8 we considered Hadwiger's conjecture, Hajós' conjecture, and Reed's conjecture for the class of line graphs. Another candidate is the Erdős-Lovás-Tihany conjecture (see Jensen and Toft [157, Problem 5.12]). This conjecture can be easily established for line graphs of arbitrary graphs (see Kostochka and Stiebitz [178]). If one can establish a vertex coloring conjecture for the class of line graphs, the next step would be to investigate this conjecture for a larger class of graphs, e.g., the class of quasi-line graphs, the class of claw-free graphs, or the class of graphs with independence number 2. That Hadwiger's conjecture holds for quasi-line graphs was proved by Chudnovsky and Fradkin [62]. For claw-free graphs or graphs with independence number 2, however, Hadwiger's conjecture remains unsolved.

That Hadwiger's conjecture is true for all line graphs (Theorem 4.56) was obtained independently in 2002 by M. Stiebitz, B. Toft, and V. G. Vizing. Their proof was based on the fan equation and was similar to the proof by Reed and Seymour [253]. A third independent proof was obtained by C. Thomassen (oral communication).

Ding [70] established an edge version of Hadwiger's conjecture. He proved that every graph G with $\chi'(G) = m \geq 1$ contains a minor which is a multi-star with m edges or a graph consisting on three vertices and m edges.

Many recent papers [34, 87, 145, 188, 191, 195, 196, 200, 202, 204, 223, 263, 264, 266, 312, 313, 317, 322] are related to the Classification Problem for graphs embedded on a given surface, in particular to Vizing's planar graph problem. Vizing [300] asked the following question: Does there exists a simple planar graph G with $\Delta(G) = 6$ or $\Delta(G) = 7$ for which $\chi'(G) = \Delta(G) + 1$. Theorem 4.59 tells us that there is no such graph with $\Delta = 7$ and hence $5 \leq \Delta(S_0) \leq 6$. However, the case $\Delta = 6$ remains unsolved. It is a common belief that there is no planar graph of class two with maximum degree $\Delta = 6$, and several authors (see, e.g., references [147, 202, 264]) claim that Vizing made such a conjecture. Even if Vizing never directly formulated such a conjecture, there is a much stronger conjecture due to Seymour [282]. He posed the conjecture that every planar graph of class two is elementary or, equivalently, contains an overfull subgraph with the same maximum degree. By Euler's formula, this conjecture implies that there exists no planar graph of class two with maximum degree 6, and hence $\Delta(S_0) = 5$. Partial results supporting this implication can be found in references [34, 188, 196, 322]. Hoffman, Mitchem, and Schmeichel [147] listed a number of other implications of Seymour's conjecture, among them the Four-Color Theorem and the existence of a polynomial time algorithm determining the chromatic index of any simple planar graph.

The above proof of Theorem 4.59 that every simple planar graph with maximum degree seven is of class one is due to Sanders and Zhao [264]. The two other proofs, given by Grünewald [116] and by Zhang [317], are also based on the discharging method, but both proofs are longer and more complicated. All three proofs use Vizing's Adjacency Lemma and several other adjacency lemmas. As Grünewald [116] pointed out there are simple planar graphs with maximum degree $\Delta = 7$ which fulfill the conclusion of Vizing's Adjacency Lemma (Theorem 2.7). One can take any simple planar graph G with $\Delta(G) = 7$ such that each minor vertex of G has degree 3, each major vertex is adjacent to exactly two minor vertices, and each minor vertex is adjacent to three major vertices.

As Sanders and Zhao noticed their proof of Theorem 4.59 can be also applied to simple projective graphs, that means simple graphs embedded on N_1. Since equation (4.24) has an analog for simple projective graphs with 12 replaced by 6, Sanders and Zhao proved that $\Delta(N_1) \leq 6$, improving the earlier bound $\Delta(N_1) \leq 7$ due to Mel'nikov [220]. Similar results on some other surfaces were obtained in [145, 191, 313]. The proof of Proposition 4.60 shows how a good lower bound for the average degree of a critical graph of class two can be used to obtain an upper bound for the parameter $\Delta(S)$.

A result concerning the chromatic index whose proof mainly relies on Vizing's fan argument can quite often be transformed into a result about the fan number. For instance, Scheide [267] proved that $\mathrm{fan}(G) = \Delta(G)$, provided that G is a simple graph with $\Delta(G) \geq 2\mathrm{col}(G) - 2$ or G is a simple planar graph with $\Delta(G) \geq 8$.

CHAPTER 5

TASHKINOV TREES

Tashkinov [291] obtained a common generalization, Tashkinov trees, of Vizing fans and Kierstead paths. Tashkinov trees form a very useful type of test objects for the edge coloring problem. Tashkinov [291] invented his method to give an alternative proof of a result by Nishizeki and Kashiwagi [237], supporting Goldberg's conjecture (see Chap. 6). Based on Tashkinov's method, Scheide [271] established Goldberg's conjecture asymptotically, see Theorem 5.19 respectively Corollary 5.20 in Sect. 5.3, and he proved the "next step" of Goldberg's conjecture (Theorem 6.5 in Chap. 6).

5.1 TASHKINOV'S METHOD

Let G be a graph, let $e \in E_G(x, y)$ be an edge, and let $\varphi \in \mathcal{C}^k(G - e)$ be a coloring for some integer $k \geq 0$. A **Tashkinov tree** with respect to e and φ is a sequence $T = (y_0, e_1, y_1, \ldots, e_p, y_p)$ with $p \geq 1$ consisting of edges $e_1, \ldots, e_p$ and vertices $y_0, \ldots, y_p$ satisfying the following two conditions:

(T1) The vertices $y_0, \ldots, y_p$ are distinct, $e_1 = e$, and, for $i = 1, \ldots, p$, we have
$e_i \in E_G(\{y_0, \ldots, y_{i-1}\}, y_i)$.

Graph Edge Coloring: Vizing's Theorem and Goldberg's Conjecture,
First Edition. By M. Stiebitz, D. Scheide, B. Toft, and L. M. Favrholdt
Copyright © 2012 John Wiley & Sons, Inc.

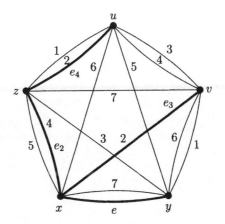

Figure 5.1   A Tashkinov tree $T = (x, e, y, e_2, z, e_3, v, e_4, u)$ with $p(T) = 4$.

(T2) For every edge e_i with $2 \leq i \leq p$, there is a vertex y_j with $0 \leq j < i$ such that $\varphi(e_i) \in \overline{\varphi}(y_j)$.

If T is a Tashkinov tree, then (T1) implies that the graph $(V(T), E(T))$ is indeed a tree. If $F = (e_1, y_1, \ldots, e_p, y_p)$ is a multi-fan at x with respect to e and φ, then $T = (x, F)$ is a Tashkinov tree with respect to e and φ, provided that the vertices $y_1, \ldots, y_p$ are distinct. Furthermore, every Kierstead path with respect to e and φ is a Tashkinov tree with respect to e and φ.

Let $T = (y_0, e_1, y_1, \ldots, e_p, y_p)$ be a Tashkinov tree with respect to e and a coloring of $\varphi \in \mathcal{C}^k(G - e)$. Clearly, $Ty_i = (y_0, e_1, \ldots, e_i, y_i)$ with $1 \leq i \leq p$ is a Tashkinov tree with respect to e and φ, too. Furthermore, $y_pT = (y_p)$ is a path of G of length 0. Hence there is a smallest integer $i \in \{0, \ldots, p\}$ such that the sequence $y_iT = (y_i, e_{i+1}, \ldots, e_p, y_p)$ corresponds to a path of G, that is, $e_j \in E_G(y_{j-1}, y_j)$ for $j = i+1, \ldots, p$. We refer to this number i as the **path number** of T and write $p(T) = i$. Clearly, if $p(T) = 0$, then T is a Kierstead path with respect to e and φ.

We say that a color α is **used** on T with respect to φ if $\varphi(e') = \alpha$ for some edge $e' \in E(T)$. Otherwise, we say that α is **unused** on T with respect to φ. Furthermore, we say that (u, v) is a (γ, δ)-**pair** with respect φ if $u, v \in V(G)$, $\gamma \in \overline{\varphi}(u)$ and $\delta \in \overline{\varphi}(v)$.

The following result is due to Tashkinov [291]. In case of $\chi'(G) \geq \Delta(G) + 2$ this result generalizes both Theorem 2.1 and Theorem 3.1. The proof of this result is similar to the proof of Theorem 3.1, but much more sophisticated.

Theorem 5.1 (Tashkinov [291] 2000) *Let G be a graph with $\chi'(G) = k + 1$ for an integer $k \geq \Delta(G) + 1$, and let $e \in E(G)$ be a critical edge of G. If T is a Tashkinov tree with respect to e and a coloring $\varphi \in \mathcal{C}^k(G - e)$, then $V(T)$ is elementary with respect to φ.*

Proof: By a *bad pair* we mean a pair (T, φ) consisting of a coloring $\varphi \in \mathcal{C}^k(G - e)$ and a Tashkinov tree T with respect to e and φ such that $V(T)$ is not elementary with respect to φ. For the proof of the theorem we have to show that there is no bad pair. Suppose, on the contrary, that there is a bad pair. Then we choose a bad pair (T, φ_0) such that

(a) $p(T)$ is minimum, and

(b) $|V(T)|$ is minimum subject to (a).

Let $T = (y_0, e_1, y_1, \ldots, y_{p-1}, e_p, y_p)$. First, we claim that $p(T) \geq 3$. To prove the claim, assume that $p(T) \leq 2$. Since $e_1 \in E_G(y_1, y_0)$, this implies that $p(T) = 0$ or $p(T) = 2$. In the first case T is a Kierstead path with respect to e and φ_0. In the second case $e_2 \in E_G(y_2, y_0)$ and $T' = (y_1, e_1, y_0, e_2, y_2, \ldots, e_p, y_p)$ is a Kierstead path with respect to e and φ_0, where $V(T') = V(T)$. Hence, in both cases, Theorem 3.1 implies that $V(T)$ is elementary with respect to φ_0, a contradiction. This contradiction shows that $p(T) \geq 3$.

Let $\mathcal{P}$ denote the set of all colorings $\varphi \in \mathcal{C}^k(G - e)$ such that (T, φ) is a bad pair. Clearly, $\varphi_0 \in \mathcal{P}$. Now consider an arbitrary coloring $\varphi \in \mathcal{P}$. Since T is a Tashkinov tree with respect to e and φ, we conclude that, for all $y_i \in V(T)$ with $i \geq 1$,

$$Ty_i = (y_0, e_1, y_1, e_2, y_2, \ldots, e_i, y_i)$$

is a Tashkinov tree with respect to e and φ, too. By the choice of T, this implies, in particular, that $V(Ty_i)$ is elementary with respect to φ, provided that $i < p$. Since $V(T)$ is not elementary with respect to φ, there is at least one index $i \in \{0, 1, \ldots, p-1\}$ such that $\overline{\varphi}(y_i) \cap \overline{\varphi}(y_p) \neq \emptyset$. We denote the set of all these indices by I_φ.

Before we continue the proof of Theorem 5.1, we shall first prove the following two claims.

(c) *Suppose that (y_i, y_j) is a (γ, δ)-pair with respect to $\varphi \in \mathcal{P}$ such that $0 \leq i < j < p$ and γ is unused on Ty_j with respect to φ. Then $\gamma \neq \delta$ and the (γ, δ)-chain $P = P_{y_j}(\gamma, \delta, \varphi)$ is a path having endvertices y_i and y_j. Moreover, for the coloring $\varphi' = \varphi/P$ we have $\varphi' \in \mathcal{P}$ and*

$$\overline{\varphi}'(y) = \begin{cases} (\overline{\varphi}(y_i) \setminus \{\gamma\}) \cup \{\delta\} & \text{if } y = y_i \\ (\overline{\varphi}(y_j) \setminus \{\delta\}) \cup \{\gamma\} & \text{if } y = y_j \\ \overline{\varphi}(y) & \text{otherwise.} \end{cases}$$

In the sequel, we briefly write $\varphi' = \varphi(i, j, \gamma, \delta)$.

(d) *If $1 \leq j < p$, then there are at least four colors in $\overline{\varphi}(V(Ty_j))$ that are unused on Ty_j with respect to φ.*

Proof of (c): Let $T' = Ty_j$. Since $\delta \in \overline{\varphi}(y_j)$ and $V(Ty_{p-1})$ is elementary with respect to φ, it follows from (T2) that δ is unused on T'. Since also γ is unused on T', it follows then that $\delta \in \overline{\varphi}(y_\ell)$ implies $\ell \in \{j, p\}$ and $\gamma \in \overline{\varphi}(y_\ell)$ implies $\ell \in \{i, p\}$. In particular, we have $\gamma \neq \delta$ and $\gamma \in \varphi(y_j)$. Consequently, the (γ, δ)-chain $P = P_{y_j}(\gamma, \delta, \varphi)$ is a path, where one endvertex is y_j and the other endvertex is some vertex $z \neq y_j$. Now, let φ' be the coloring obtained from φ by recoloring P. If $z \neq y_i$, then, because γ and δ are unused on T', we conclude that T' is a Tashkinov tree with respect to e and φ', where $\gamma \in \overline{\varphi}'(y_i) \cap \overline{\varphi}'(y_j)$. Hence (T', φ') is a bad pair with $p(T') \leq p(T)$ and $|V(T')| < |V(T)|$, a contradiction to the choice of (T, φ). This contradiction shows that $z = y_i$. Then $\overline{\varphi}'(y_i) = (\overline{\varphi}(y_i) \setminus \{\gamma\}) \cup \{\delta\}$, $\overline{\varphi}'(y_j) = (\overline{\varphi}(y_j) \setminus \{\delta\}) \cup \{\gamma\}$ and $\overline{\varphi}'(y_\ell) = \overline{\varphi}(y_\ell)$ for all $\ell \neq i, j$. Since both γ and δ are unused on T' with respect to φ, it follows that T is a Tashkinov tree with respect to e and φ' and, moreover, that $V(T)$ is not elementary with respect to φ'. Hence $\varphi' \in \mathcal{P}$. This proves statement (c). ∎

Proof of (d): Let $T' = Ty_j$. Clearly, T' has $j + 1$ vertices and j edges. For the set $\overline{\varphi}(v)$ of missing colors at $v \in V(T)$, we have $|\overline{\varphi}(v)| = k - d_G(v) + 1 \geq 2$ if $v = y_0, y_1$ and $|\overline{\varphi}(v)| = k - d_G(v) \geq 1$ otherwise. Since $V(T')$ is elementary with respect to φ and $y_0, y_1 \in V(T')$, it follows that $|\overline{\varphi}(V(T'))| \geq j + 3$. Since $|E(T')| = j$ and $e_1 = e \in E(T')$ is uncolored, at least 4 colors in $\overline{\varphi}(V(T'))$ are unused on T' with respect to φ. This proves (d). ∎

We now continue the proof of Theorem 5.1. To arrive at a contradiction we distinguish three cases.

Case 1: $p(T) = p$. This implies that $e_p \in E_G(y_h, y_p)$ for some index $h \leq p - 2$. Note that $p = p(T) \geq 3$. We claim that there is a coloring $\varphi \in \mathcal{P}$ such that

$$I_\varphi \text{ contains an index } i \neq p - 1. \tag{5.1}$$

To this end, we choose an arbitrary coloring $\varphi \in \mathcal{P}$. If (5.1) is not satisfied, then there is a color $\alpha \in \overline{\varphi}(y_p) \cap \overline{\varphi}(y_{p-1})$. By (d), at least 4 colors in $\overline{\varphi}(V(Ty_{p-2}))$ are unused on Ty_{p-2}. Consequently, there is a color $\beta \in \overline{\varphi}(V(Ty_{p-2}))$, say $\beta \in \overline{\varphi}(y_i)$ with $i \leq p - 2$, such that β is unused on T. Since $V(Ty_{p-1})$ is elementary with respect to φ, it follows that β is distinct from α. Then, by (c), the coloring $\varphi' = \varphi(i, p-1, \beta, \alpha)$ belongs to $\mathcal{P}$ and satisfies (5.1), since $\alpha \in \overline{\varphi}'(y_i) \cap \overline{\varphi}'(y_p)$.

Next we claim that there is even a coloring $\varphi \in \mathcal{P}$ satisfying the following:

$$I_\varphi \text{ contains an index } i \neq p - 1 \text{ such that some color} \\ \alpha \in \overline{\varphi}(y_p) \cap \overline{\varphi}(y_i) \text{ is unused on } T \text{ with respect to } \varphi. \tag{5.2}$$

To prove this, we choose an arbitrary coloring $\varphi \in \mathcal{P}$ that satisfies (5.1). Then there is an index $i < p - 1$ and a color $\alpha \in \overline{\varphi}(y_p) \cap \overline{\varphi}(y_i)$. If α is unused on T with respect to φ, we are done. Otherwise, α is used on T with respect to φ and we proceed as follows. By (d), at least 4 colors in $\overline{\varphi}(V(Ty_{p-2}))$ are unused on Ty_{p-2}. Consequently, there is a color $\gamma \in \overline{\varphi}(V(Ty_{p-2}))$, say $\gamma \in \overline{\varphi}(y_j)$ with $j \leq p - 2$, such that γ is unused on T with respect to φ. Since α is used on T, it follows that γ is distinct from α. If $\gamma \in \overline{\varphi}(y_p)$, then (5.2) is already satisfied for φ and we are done. Otherwise, we have $\gamma \in \varphi(y_p)$. Consequently, the chain $P = P_{y_p}(\alpha, \gamma, \varphi)$ is a path, where one endvertex is y_p and the other endvertex is some vertex distinct from y_p. If $V(P) \cap V(Ty_{p-1}) \neq \emptyset$, then there is an index $i_0 \in \{0, \ldots, p-1\}$ such that y_{i_0} belongs to P, and the subpath P' of P joining y_p and y_{i_0} does not contain any other vertex of $V(Ty_{p-1})$. Let $P' = \text{Path}(y_{i_0}, f_1, z_1, \ldots, f_m, z_m)$, where $z_m = y_p$. If

$i_0 < p - 1$, then, since $\alpha \in \overline{\varphi}(y_i)$, where $i \le p - 2$, and $\gamma \in \overline{\varphi}(y_j)$, where $j \le p - 2$,

$$T' = (y_0, e_1, y_1, \ldots, e_{p-2}, y_{p-2}, f_1, z_1, \ldots, f_m, z_m)$$

is a Tashkinov tree with respect to φ satisfying $\alpha \in \overline{\varphi}(y_i) \cap \overline{\varphi}(z_m)$. Hence (T', φ) is a bad pair with $p(T') < p = p(T)$, contradicting (a). Otherwise, $i_0 = p - 1$ and

$$T' = (y_0, e_1, y_1, \ldots, e_{p-1}, y_{p-1}, f_1, z_1, \ldots, f_m, z_m)$$

is a Tashkinov tree with respect to φ satisfying $\alpha \in \overline{\varphi}(y_i) \cap \overline{\varphi}(z_m)$, and, therefore, (T', φ) is a bad pair with $p(T') < p = p(T)$, again contradicting (a). Consequently, we have $V(P) \cap V(Ty_{p-1}) = \emptyset$. Let φ' be the coloring obtained from φ by recoloring the (α, γ)-chain P. Since P does not meet Ty_{p-1}, it follows that T is a Tashkinov tree with respect to e and φ' such that $\gamma \in \overline{\varphi}'(y_j) \cap \overline{\varphi}'(y_p)$ and γ remains unused on T with respect to φ'. Hence, the coloring φ' belongs to $\mathcal{P}$ and satisfies (5.2). This proves the claim.

Now we claim that there is a coloring $\varphi \in \mathcal{P}$ such that

$$\varphi(e_p) \in \varphi(y_{p-1}). \tag{5.3}$$

To show this, let $\varphi \in \mathcal{P}$ be an arbitrary coloring satisfying (5.2). Then there is a color $\alpha \in \overline{\varphi}(y_p) \cap \overline{\varphi}(y_i)$, where $i \le p - 2$, and α is unused on T with respect to φ. If $\beta = \varphi(e_p)$ belongs to $\varphi(y_{p-1})$, we are done. Otherwise, we have $\beta \in \overline{\varphi}(y_{p-1})$. From (c) we conclude that there is an (α, β)-chain P with respect to φ joining y_{p-1} and y_i and, moreover, the coloring $\varphi' = \varphi(i, p - 1, \alpha, \beta)$ belongs to $\mathcal{P}$. Since $\alpha \in \overline{\varphi}(y_p)$, we also deduce that P does not contain the vertex y_p and, therefore, we have $\varphi'(e_p) = \varphi(e_p) = \beta \in \varphi'(y_{p-1})$. This proves the claim.

Eventually, let $\varphi \in \mathcal{P}$ be a coloring satisfying (5.3). If φ also satisfies (5.1), then $T' = (y_0, e_1, y_1, \ldots, e_{p-2}, y_{p-2}, e_p, y_p)$ is a Tashkinov tree with respect to e and φ such that $V(T')$ is not elementary with respect to φ. Hence (T', φ) is a bad pair with $p(T') < p = p(T)$, contradicting (a).

If φ does not satisfy (5.1), then there is a color $\alpha \in \overline{\varphi}(y_p) \cap \overline{\varphi}(y_{p-1})$. By (d), at least 4 colors in $\overline{\varphi}(V(Ty_{p-2}))$ are unused on Ty_{p-2}. Consequently, there is a color $\gamma \in \overline{\varphi}(V(Ty_{p-2}))$, say $\gamma \in \overline{\varphi}(y_i)$ with $i \le p - 2$, such that γ is unused on T. Then (c) implies that there is an (α, γ)-chain P with respect to φ having endvertices y_i and y_{p-1}. Since φ does not satisfy (5.1), the set $\{y_0, \ldots, y_{p-2}, y_p\}$ is elementary with respect to φ. Since $\varphi(e_p) \in \varphi(y_{p-1})$, we conclude that $T' = (y_0, e_1, y_1, \ldots, e_{p-2}, y_{p-2}, e_p, y_p, e_{p-1}, y_{p-1})$ is a Tashkinov tree with respect to e and φ. Furthermore, (T', φ) is a bad pair such that $p(T') = p(T) = p$, $|V(T')| = |V(T)|$, $V(T'y_p)$ is elementary with respect to φ, and γ is unused on T' with respect to φ. Then (c) may be applied to T' and the coloring φ, implying that there is an (α, γ)-chain P' with respect to φ joining y_i and y_p. Then we have two distinct (α, γ)-chains P and P' with respect to φ, but $y_i \in V(P) \cap V(P')$, a contradiction. This settles Case 1.

Case 2: $p(T) \le p - 1$ and $\max(I_\varphi) \ge p(T)$ for some $\varphi \in \mathcal{P}$. Then we choose a coloring $\varphi \in \mathcal{P}$ such that $i = \max(I_\varphi)$ is maximum. We claim that $i = p - 1$. To prove the claim, assume that $i < p - 1$. Then there is a color $\alpha \in \overline{\varphi}(y_i) \cap \overline{\varphi}(y_p)$. Since $k \geq \Delta(G) + 1$, there is a color $\beta \in \overline{\varphi}(y_{i+1})$. By the hypothesis of Case 2, $i \geq p(T)$ and, therefore, $e_{i+1} \in E_G(y_i, y_{i+1})$. Since $V(Ty_{i+1})$ is elementary with respect to φ, it follows from (T2) that α is unused on Ty_{i+1}. Then (c) implies that the coloring $\varphi' = \varphi(i, i + 1, \alpha, \beta)$ belongs to $\mathcal{P}$ and $\max(I_{\varphi'}) > \max(I_{\varphi})$, a contradiction to the choice of φ. This proves the claim that $i = p - 1$.

Consequently, for the bad pair (T, φ), there is a color $\alpha \in \overline{\varphi}(y_{p-1}) \cap \overline{\varphi}(y_p)$. By (T2), for the color $\beta = \varphi(e_p)$, there is a vertex y_j with $j < p$ such that $\beta \in \overline{\varphi}(y_j)$. Since $e_p \in E_G(y_p, y_{p-1})$, we have $j < p - 1$. Recolor e_p with α. This results in a coloring $\varphi' \in \mathcal{C}^k(G - e)$ such that $T' = Ty_{p-1}$ is a Tashkinov tree with respect to e and φ' where $\beta \in \overline{\varphi}'(y_{p-1}) \cap \overline{\varphi}'(y_j)$. Hence (T', φ') is a bad pair with $p(T') = p(T)$ and $|V(T')| < |V(T)|$, contradicting (b). This contradiction completes the proof in Case 2.

Case 3: $p(T) \leq p - 1$ and $\max(I_{\varphi}) < p(T)$ for every $\varphi \in \mathcal{P}$. Let $j = p(T)$. Then $j \geq 3$ and $e_j \notin E_G(y_{j-1}, y_j)$. Now we choose a coloring $\varphi \in \mathcal{P}$ such that $i = \min(I_{\varphi})$ is minimum. Then $i \leq j - 1$ and we claim that $i < j - 1$. Otherwise, $i = j - 1$ and, since $i \in I_{\varphi}$, there would be a color $\alpha \in \overline{\varphi}(y_i) \cap \overline{\varphi}(y_p)$. It follows from (d) that there is a color $\gamma \in \overline{\varphi}(V(Ty_{j-2}))$ unused on Ty_{j-1}, say $\gamma \in \overline{\varphi}(y_h)$, where $h \leq j - 2$. Then, by (c), the coloring $\varphi' = \varphi(h, j - 1, \gamma, \alpha)$ would belong to $\mathcal{P}$ and $\min(I_{\varphi'}) < i = \min(I_{\varphi})$, a contradiction to the choice of φ. This proves the claim that $i < j - 1$.

Since $i \in I_{\varphi}$, there is a color $\alpha \in \overline{\varphi}(y_i) \cap \overline{\varphi}(y_p)$. Since $k \geq \Delta(G) + 1$, there is a color $\delta \in \overline{\varphi}(y_j)$. It follows from (d) that there is a color $\gamma \in \overline{\varphi}(V(Ty_{j-2})) \setminus \{\alpha\}$ that is unused on Ty_j. Then $\gamma \in \overline{\varphi}(y_h)$, where $h \leq j - 2$. Since $V(Ty_{p-1})$ is elementary with respect to φ, the color γ is distinct from δ. Moreover, $\gamma \in \varphi(y_p)$. Otherwise, by (c), $\varphi' = \varphi(h, j, \gamma, \delta)$ would be a coloring in $\mathcal{P}$ with $\max(I_{\varphi'}) \geq j = p(T)$, a contradiction to the hypothesis of Case 3. Consequently, the (α, γ)-chain P with respect to φ containing y_p is a path, where one endvertex is y_p and the other endvertex is some vertex distinct from y_p.

If $V(P) \cap V(Ty_{j-1}) \neq \emptyset$, then there is an index $i_0 \in \{0, \ldots, j - 1\}$ such that y_{i_0} belongs to P and the subpath P' of P joining y_p and y_{i_0} does not contain any other vertex of $V(Ty_{j-1})$. Let $P' = \text{Path}(y_{i_0}, f_1, z_1, \ldots, f_m, z_m)$, where $z_m = y_p$. If $i_0 < j - 1$, then, since $\alpha \in \overline{\varphi}(y_i)$, where $i \leq j - 2$, and $\gamma \in \overline{\varphi}(y_h)$, where $h \leq j - 2$,

$$T' = (y_0, e_1, y_1, \ldots, e_{j-2}, y_{j-2}, f_1, z_1, \ldots, f_m, z_m)$$

is a Tashkinov tree with respect to φ satisfying $\alpha \in \overline{\varphi}(y_i) \cap \overline{\varphi}(z_m)$. Hence (T', φ) is a bad pair with $p(T') < j = p(T)$, contradicting (a). Otherwise $i_0 = j - 1$ and

$$T' = (y_0, e_1, y_1, \ldots, e_{j-1}, y_{j-1}, f_1, z_1, \ldots, f_m, z_m)$$

is a Tashkinov tree with respect to φ satisfying $\alpha \in \overline{\varphi}(y_i) \cap \overline{\varphi}(z_m)$, and, therefore, (T', φ) is a bad pair with $p(T') < j = p(T)$, again contradicting (a).

If $V(P) \cap V(Ty_{j-1}) = \emptyset$, then we argue as follows. Let z be the endvertex of P distinct from y_p. Since $V(Ty_{p-1})$ is elementary with respect to φ, it follows that

$\alpha, \gamma \in \varphi(y)$ whenever $y \in \{y_j, y_{j+1}, \ldots, y_{p-1}\}$. Consequently, $z \notin V(T)$. Now, let φ' be the coloring obtained from φ by recoloring the (α, γ)-chain P. Then

$$\overline{\varphi}'(y_\ell) = \begin{cases} (\overline{\varphi}(y_p) \setminus \{\alpha\}) \cup \{\gamma\} & \text{if } \ell = p, \\ \overline{\varphi}(y_\ell) & \text{otherwise.} \end{cases}$$

Since P does not meet Ty_{j-1}, it follows that T is a Tashkinov tree with respect to φ' and, moreover, $\gamma \in \overline{\varphi}(y_h) = \overline{\varphi}'(y_h)$ remains unused on Ty_j with respect to φ', and $\delta \in \overline{\varphi}(y_j) = \overline{\varphi}'(y_j)$. Then, by (c), the coloring $\varphi'' = \varphi'(h, j, \gamma, \delta)$ belongs to $\mathcal{P}$. Since $\gamma \in \overline{\varphi}''(y_j) \cap \overline{\varphi}''(y_p)$, we have $\max(I_{\varphi''}) \geq j = p(T)$, contradicting the hypothesis of Case 3. This completes the proof in Case 3.

Therefore, in all three cases we arrive at a contradiction. Hence Theorem 5.1 is proved. ∎

Theorem 5.1 and its proof implies that Tashkinov trees are suitable test objects for the edge coloring problem. During the coloring procedure the path number of the Tashkinov tree never increases, but the order of the Tashkinov tree may increase. Therefore, the number of Kempe changes for one call of the corresponding subroutine EXT is not only dependent on the order of the Tashkinov tree, but also on the order of the graph.

Theorem 5.2 (Tashkinov [291] 2000) *Let G be a graph, let $e \in E(G)$ be an edge, and let $\varphi \in \mathcal{C}^k(G - e)$ be a coloring with $k \geq \Delta(G) + 1$. If T is Tashkinov tree with respect to e and φ such that $V(T)$ is not elementary with respect to φ, then after $O(|V(T)||V(G)|^2)$ Kempe changes a coloring $\varphi' \in \mathcal{C}^k(G)$ is obtainable.*

Up to now, Tashkinov trees seem to be the most general tool for obtaining an elementary set. All other known results about elementary sets in a given critical graph G are corollaries of Theorem 5.1, provided that $\chi'(G) \geq \Delta(G) + 2$.

Let G be a graph, let $e \in E(G)$ be an edge, and let $\varphi \in \mathcal{C}^k(G - e)$ be a coloring. Recall that a set $X \subseteq V(G)$ is closed with respect to φ if, for every colored edge $f \in \partial_G(X)$, the color $\varphi(f)$ is present at every vertex of X, i.e., $\varphi(f) \in \varphi(v)$ for every $v \in X$. Furthermore, the set X is strongly closed with respect to φ if X is closed and if $\varphi(f) \neq \varphi(f')$ for every two distinct colored edges $f, f' \in \partial_G(X)$.

Let G be a graph and let H be its **underlying simple graph**, that is, $H \subseteq G$ is a simple graph such that $V(H) = V(G)$ and $\mu_H(x, y) = 1$ iff $\mu_G(x, y) \geq 1$ for all $x, y \in V(G)$. As usual, the **length** of a path or cycle is the number of its edges, and a path or cycle is **odd** or **even** depending on whether its length is odd or even. The **girth** $g(G)$ of G (respectively, the **odd girth** $g_o(G)$ of G) is defined to be the minimum length of a cycle (respectively, an odd cycle) contained in H. If H does not contain such a cycle, we put $g(G) = \infty$ (respectively, $g_o(G) = \infty$).

Theorem 5.3 *Let G be a graph with $\chi'(G) = k + 1$ for an integer $k \geq \Delta(G) + 1$, let $e \in E(G)$ be a critical edge of G, and let $\varphi \in \mathcal{C}^k(G - e)$ be a coloring. Moreover, let T be a maximal Tashkinov tree with respect to e and φ, and let $T' = (y_0, e_1, y_1, \ldots, e_p, y_p)$ be an arbitrary Tashkinov tree with respect to e and φ. Then the following statements hold:*

(a) $V(T)$ *is elementary and closed, both with respect to* φ.

(b) $|V(T)| \equiv 1 \bmod 2$ *and* $\delta(G[V(T)]) \geq (|V(T)| - 1)(\chi'(G) - \Delta(G) - 1) + 2$.

(c) $V(T') \subseteq V(T)$.

(d) *There is a Tashkinov tree* $\tilde{T}$ *with respect to* e *and* φ *satisfying* $V(\tilde{T}) = V(T)$ *and* $\tilde{T}y_p = T'$.

(e) *Suppose that* (y_i, y_j) *is a* (γ, δ)-*pair with respect to* φ, *where* $1 \leq i < j \leq p$. *Then* $\gamma \neq \delta$ *and there is a* (γ, δ)-*chain* P *with respect to* φ *satisfying the following conditions:*

  (1) P *is a path with endvertices* y_i *and* y_j.

  (2) $|E(P)|$ *is even.*

  (3) $V(P) \subseteq V(T)$.

  (4) *If* γ *is unused on* $T'y_j$, *then* T' *is a Tashkinov tree with respect to the edge* e *and the coloring* $\varphi' \in \mathcal{C}^k(G - e)$ *obtained from* φ *by recoloring the* (γ, δ)-*chain* P.

(f) $G[V(T)]$ *contains an odd cycle.*

Proof: Let $T = (z_0, f_1, z_1, \ldots, f_m, z_m)$. Theorem 5.1 implies that $V(T)$ is elementary with respect to φ. Now we consider an arbitrary edge $f \in \partial_G(V(T))$ and denote by z the endvertex of f belonging to $V(G) \setminus V(T)$. If $\varphi(f) \in \overline{\varphi}(V(T))$, then $F' = (z_0, f_1, z_1, \ldots, f_m, e_m, f, z)$ is a Tashkinov tree with respect to e and φ, contradicting the maximality of T. Hence, $V(T)$ is closed with respect φ. This completes the proof of (a).

For the proof of (b), let $H = G[V(T)]$. If $\alpha \in \overline{\varphi}(u)$ for some vertex u of H, then (a) implies that every other vertex of H is incident with an edge colored with α, where this edge belongs to H. This implies, in particular, that $|V(T)|$ is odd. Furthermore, if u is a vertex of H, then $|\overline{\varphi}(u)| = k - d_G(u) + 1 \geq \chi'(G) - \Delta(G)$ if $e \in E_G(u)$ and $|\overline{\varphi}(u)| = k - d_G(u) \geq \chi'(G) - \Delta(G) - 1$ otherwise. Since the uncolored edge e belongs to H and since $V(H) = V(T)$, this implies that

$$d_H(z) \geq a + \sum_{u \in V(H), u \neq z} |\overline{\varphi}(u)| \geq (|V(T)| - 1)(\chi'(G) - \Delta(G) - 1) + 2$$

for every vertex z of H, where $a = 1$ if $e \in E_G(z)$ and $a = 0$ otherwise. This completes the proof of (b).

For the proof of (c), suppose that $V(T')$ is not contained in $V(T)$. Since $y_0, y_1 \in V(T)$, this implies that there is an index $i \geq 1$ such that $y_0, \ldots, y_i \in V(T)$ and $y_{i+1} \notin V(T)$. Since $e_{i+1} \in E_G(y_{i+1}, y_j)$ for some $j \leq i$ and $\varphi(e_{i+1}) \in \overline{\varphi}(y_\ell)$ for some $\ell \leq i$, this implies that $T'' = (z_0, f_1, z_1, \ldots, f_m, z_m, e_{i+1}, y_{i+1})$ is a Tashkinov tree with respect to e and φ, contradicting the assumption that T is a maximal Tashkinov tree. Hence, (c) is proved.

The proof of (d) is similar to the proof of (c). Since T' is a Tashkinov tree with respect to e and φ, there is a Tashkinov tree $\tilde{T}$ with respect to e and φ such that $\tilde{T}y_p = T'$ and such that $\tilde{T}$ is maximal with respect to this condition. By (c), we then have $V(\tilde{T}) \subseteq V(T)$. Hence, if $V(\tilde{T}) \neq V(T)$, then there is an index $i \geq 1$, such that $z_0, \ldots, z_i \in V(\tilde{T})$ and $z_{i+1} \notin V(\tilde{T})$. But then $\tilde{T}' = (\tilde{T}, f_{i+1}, z_{i+1})$ is a Tashkinov tree with respect to e and φ, contradicting the choice of $\tilde{T}$. This contradiction proves (d).

Next, we prove (e). By (a) and (c), $V(T)$ is elementary with respect to φ and $V(T') \subseteq V(T)$. Therefore, by the hypothesis of (e), y_i is the only vertex of T with $\gamma \in \overline{\varphi}(y_i)$, and y_j is the only vertex of T with $\delta \in \overline{\varphi}(y_j)$. Consequently, $\gamma \neq \delta$ and the (γ, δ)-chain P with respect to φ containing y_i is a path, where one endvertex is y_i and the other endvertex is some vertex $z \neq y_i$. Since, by (a), $V(T)$ is closed with respect to φ, it follows that every edge of P belongs to $G[V(T)]$. This implies that $V(P) \subseteq V(T)$. Since $\overline{\varphi}(z)$ contains γ or δ, we deduce that $z = y_j$ and, therefore, $|E(P)|$ is even. Thus (1), (2) and (3) are proved.

For the proof of (4), let φ' be the coloring obtained from φ by recoloring P and let $T_1 = T'y_j$. It follows from (1) that $\overline{\varphi}'(y_i) = (\overline{\varphi}(y_i) \setminus \{\gamma\}) \cup \{\delta\}$, $\overline{\varphi}'(y_j) = (\overline{\varphi}(y_j) \setminus \{\delta\}) \cup \{\gamma\}$, and $\overline{\varphi}'(y_\ell) = \overline{\varphi}(y_\ell)$ for all $\ell \neq i, j$. Since $\delta \in \overline{\varphi}(y_j)$ and $V(T')$ is elementary with respect to e and φ, it follows from (T2) that δ is unused on T_1 with respect to φ. Since, by hypothesis, γ is also unused on T_1 with respect to φ, we then deduce that T' is a Tashkinov tree with respect to e and φ'. This proves (4). Hence the proof of (e) is complete.

Finally, we prove (f). Obviously, (y_0, e, y_1) is a Tashkinov tree with respect to e and φ, and there is a pair (α, β) of colors such that (y_0, y_1) is an (α, β)-pair with respect to φ. Then, by (e), there is a path P in $G - e$ such that P has endvertices y_0 and y_1, $|E(P)|$ is even, and $V(P) \subseteq V(T)$. Consequently, $P + e$ is an odd cycle of $G[V(T)]$. Hence (e) is proved. ∎

Let G be a graph with $\chi'(G) = k+1$ for an integer $k \geq \Delta(G)+1$, let $e \in E_G(x, y)$ be a critical edge of G, and let $\varphi \in \mathcal{C}^k(G - e)$ be a coloring. Obviously, a Tashkinov tree with respect to e and φ is a maximal Tashkinov tree with respect to e and φ if and only if it is closed with respect to φ. Furthermore, Theorem 5.3 tells us that the vertex set X of a maximal Tashkinov tree T with respect to e and φ is a unique set completely determined by e and φ; it does not depend on the structure of T. The set X is the smallest set that contains x, y and is closed with respect to φ. Obviously, for each multi-fan F at x with respect to e and φ, we have $V(F) \cup \{x\} \subseteq X$. Moreover, X contains the vertex set of each Kierstead path with respect to e and φ. Hence a coloring algorithm based on Tashkinov trees as test objects uses at most $\min\{\Delta(G) + \mu(G), 3\Delta(G)/2\}$ colors. This also follows from statement (b) in Theorem 5.3. To get a better estimate for the number of colors, we shall extend Tashkinov's method, see Sect. 5.2. It will also prove useful to construct maximal Tashkinov trees where the number of used colors is relatively small.

Theorem 5.4 *Let G be a graph with $\chi'(G) = k+1$ for an integer $k \geq \Delta(G)+1$, let $e \in E(G)$ be a critical edge of G, and let $\varphi \in \mathcal{C}^k(G - e)$ be a coloring. Then there* *is a maximal Tashkinov tree T with respect to e and φ such that at most $\frac{|V(T)|-1}{2}$ colors are used on T with respect to φ.*

Proof: Obviously, there exists a maximal Tashkinov tree T' with respect to e and φ. By Theorem 5.3, we have $|V(T')| = 2\ell + 1$ for some integer $\ell \geq 1$. To obtain T, we construct, for $i = 1, \ldots, \ell$, a Tashkinov tree $T_i = (y_0, e_1, y_1, \ldots, e_{2i}, y_{2i})$ with respect to e and φ such that at most i colors are used on T_i with respect to φ. Since $|V(T_\ell)| = |V(T')|$, it follows then from Theorem 5.3 that $T = T_\ell$ is a maximal Tashkinov tree with respect to φ. Since at most ℓ colors are used on T with respect to φ, we are done,

To construct T_1, we consider the Tashkinov tree T'. Since $|V(T')| \geq 3$, there is a vertex $y \in V(T)$ such that $T'y$ has exactly three vertices. Then $T_1 = T'y$ is a Tashkinov tree with respect to e and φ such that exactly one color is used on T with respect to φ.

Now assume that, for $1 \leq i < \ell$, a Tashkinov tree $T_i = (y_0, e_1, y_1, \ldots, e_{2i}, y_{2i})$ with respect to e and φ have already been constructed such that at most i colors are used on T_i with respect to φ. Since $|V(T_i)| < |V(T')|$, by Theorem 5.3, there is a maximal Tashkinov tree $\tilde{T}$ with respect to e and φ such that $V(\tilde{T}) = V(T')$ and $\tilde{T}y_{2i} = T_i$. Hence there is an edge $f \in \partial_G(V(T_i))$ such that $\varphi(f) \in \overline{\varphi}(V(T_i))$. Let z be the endvertex of f belonging to $V(G) \setminus V(T_i)$. Then, $T_i' = (T_i, f, z)$ is a Tashkinov tree with respect to e and φ with $|V(T_i')| = 2i + 2$. Since $i \geq 1$, there exists a color α used on T_i. Since, by Theorem 5.1, $V(T_i')$ is elementary with respect to φ, this implies that there is a unique vertex $u \in V(T_i')$ such that $\alpha \in \overline{\varphi}(u)$. Since $|V(T_i')|$ is even and the edges of color α form an matching, this implies that there is an edge $f' \in \partial_G(V(T_i'))$ such that $\varphi(f') = \alpha$. Hence, if z' is the endvertex of f' contained in $V(G) \setminus V(T_i')$, then $T_{i+1} = (T_i, f, z, f', z')$ is a Tashkinov tree with respect to e and φ such that at most $i + 1$ colors are used on T_{i+1} with respect to φ. This completes the construction. ∎

Let G be a graph with $\chi'(G) = k+1$ for an integer $k \geq \Delta(G)+1$, let $e \in E_G(x, y)$ be a critical edge of G, and let $\varphi \in C^k(G - e)$ be an arbitrary coloring. Then $T = (x, e, y)$ is a Tashkinov tree with respect to e and φ. Hence, Theorem 5.3 implies that if (x, y) is an (α, β)-pair with respect to φ, then $\alpha \neq \beta$ and $G - e$ has an (α, β)-chain P with respect to φ, where P is a path with endvertices x and y. Furthermore, $V(P)$ belongs to the vertex set of the maximal Tashkinov tree with respect to e and φ. Usually, such a chain is called a **critical chain** with respect to e and φ. As an immediate consequence of Theorem 5.3 we obtain the following result. This result was proved by Goldberg [114] and, independently, by Andersen [5].

Corollary 5.5 (Goldberg [111] 1974, Andersen [5] 1977) *Let G be a graph such that $\chi'(G) = k + 1$ for an integer $k \geq \Delta(G) + 1$. Suppose that $e \in E_G(x, y)$ is a critical edge of G and $\varphi \in C^k(G - e)$ is an arbitrary coloring. If (α, β) is a pair of colors satisfying $\alpha \in \overline{\varphi}(x)$ and $\beta \in \overline{\varphi}(y)$, then $\alpha \neq \beta$ and $G - e$ has an (α, β)-chain P with respect to φ such that P has endvertices x and y. Moreover, if X is the vertex set of all these chains, then X is elementary with respect to φ.*

The following theorem that can be deduced easily from Theorem 5.3 improves a result of Kierstead [166] (see also [168, 169]). It also provides an affirmative answer to a problem posed by Kierstead [167, Problem 8].

Theorem 5.6 *Let G be a graph with $\chi'(G) \geq \Delta(G) + s$ for an integer $s \geq 2$. Then the following statements hold:*

(a) *G contains a subgraph H such that H contains an odd cycle, $m = |V(H)|$ is odd, $m \geq g_o(G) \geq 3$, and $\delta(H) \geq (m-1)(s-1) + 2$.*

(b) *G contains a triple $T = \{x, y, z\}$ of vertices such that $\mu_G(x, y) \geq s$ and $\mu_G(x, z) + \mu_G(y, z) \geq 2s - 1$.*

Proof: By Proposition 1.1(a), G contains a critical subgraph G' with $\chi'(G') = \chi'(G)$. From the hypothesis, $\chi'(G') = \chi'(G) \geq \Delta(G) + s = \Delta(G') + s'$, where $s' \geq s \geq 2$. Then G' contains an edge e and, therefore, G' contains a maximal Tashkinov tree T with respect to e and a coloring $\varphi \in C^k(G' - e)$, where $k = \chi'(G') - 1$. We then deduce from Theorem 5.3 that $H = G'[V(T)]$ is a subgraph of G such that H contains an odd cycle, $m = |V(H)|$ is odd, and $\delta(H) \geq (m-1)(\chi'(G') - \Delta(G') - 1) + 2 \geq (m-1)(s'-1) + 2 \geq (m-1)(s-1) + 2$. Clearly, $m \geq g_o(G) \geq 3$. This proves statement (a).

In order to prove (b), we first deduce from (a) that G contains a subgraph H such that $m = |V(H)| \geq 3$ is odd and $\delta(H) \geq (m-1)(s-1) + 2$. Among all such subgraphs H, we choose one for which m is as small as possible. Since $\delta(H) \geq (m-1)(s-1)+2$, there are two vertices $x, y \in V(H)$ such that $\mu(x, y) \geq s$. Let $H' = H - x - y$ and $m' = |V(H')|$. If $m' = 1$, then $m = 3$ and $V(H')$ consists of a single vertex z and, therefore, $\mu_H(x, z) + \mu_H(y, z) = d_H(z) \geq \delta(H) \geq 2s$. Hence $T = \{x, y, z\}$ is a desired triple. Otherwise, $m' = m - 2 \geq 3$ is odd. By the choice of H, this implies that $\delta(H') \leq (m'-1)(s-1) + 1 = (m-3)(s-1) + 1$. Consequently, there is a vertex z in H' such that $d_{H'}(z) \leq (m-3)(s-1) + 1$. Since $d_H(z) \geq (m-1)(s-1) + 2$, we conclude that $\mu_H(x, z) + \mu_H(y, z) = d_H(z) - d_{H'}(z) \geq 2s - 1$. Hence $T = \{x, y, z\}$ is a desired triple. This proves (b). ∎

Theorem 5.6 can be used to prove the following generalization of Vizing's theorem, due to Steffen [287].

Theorem 5.7 (Steffen [287] 2000) *Let G be a graph with girth $g \geq 3$. Then*

$$\chi'(G) \leq \Delta(G) + \left\lceil \frac{\mu(G)}{\lfloor g/2 \rfloor} \right\rceil.$$

Proof: Suppose that the statement is false, i.e., there is a graph G with girth $g \geq 3$ and $\chi'(G) \geq \Delta(G) + t + 1$, where

$$t = \left\lceil \frac{\mu(G)}{\lfloor g/2 \rfloor} \right\rceil.$$

By Vizing's bound, we have $g \geq 4$. Furthermore, Theorem 5.6(a) implies that G contains a subgraph H such that H contains an odd cycle, $m = |V(H)|$ is odd, and $\delta(H) \geq (m - 1)t + 2$. Hence, $2h \leq g \leq 2h + 1 \leq m$ for an integer $h \geq 2$. Since $\mu(H) \leq \mu(G)$, we deduce that $t \geq \frac{\mu(H)}{h}$ and $\delta(H) \geq 2\mu(H) + 2$. The last inequality implies, in particular, that $|N_H(x)| \geq 3$ for each vertex $x \in V(H)$.

Now, let us choose a shortest odd cycle C in H. Then we have $|V(C)| \geq 2h + 1 \geq 5$. Consequently, each vertex $x \in V(C)$ has exactly two neighbors in C and each vertex $x \in V(H) \setminus V(C)$ has at most one neighbor in C. This implies that each vertex of C has a neighbor in $V(H) \setminus V(C)$.

Next, let us choose a longest path P in H such that $|V(C) \cap V(P)| = 1$. Then P joins a vertex $x \in V(C)$ with a vertex $y \in V(H) \setminus V(C)$. Clearly, we have $N_H(y) \subseteq V(P) \cup V(C)$. Since $|N_H(y)| \geq 3$ and $|N_H(y) \cap V(C)| \leq 1$, we then conclude that $p = |V(P)| - 1$ satisfies $p \geq 3$. Hence, there are unique integers $N, r \geq 0$ such that $p - 2 = N(g - 2) + r$ and $r \leq g - 3$. Then we conclude that y has at most $N + 1$ neighbors in $V(P - x)$ and at most one neighbor in $V(C)$. This implies that $d_H(y) \leq (N + 2)\mu(H)$. On the other hand, we have $|V(H)| \geq |V(C)| + p$ and, therefore, $\delta(H) \geq (|V(C)| + p - 1)t + 2 \geq (2h + p)(\mu(H)/h)$. Hence, we obtain $(N + 2)\mu(H) \geq (2h + p)(\mu(H)/h) + 2$ and, therefore, $hN > p$. Since $p = N(g - 2) + r + 2$ and $g \geq 2h$, this gives $hN > N(2h - 2) + r + 2$ and, therefore, $N(2 - h) - 2 > r$. Since $N, r \geq 0$ and $h \geq 2$, this yields a contradiction. ∎

As a simple consequence of Theorem 5.3(b)(f), respectively Theorem 5.6(a), we obtain the following result due to Goldberg [109, 114].

Theorem 5.8 (Goldberg [109] 1972) *Let G be a graph with odd girth $g_o \geq 3$. Then*

$$\chi'(G) \leq \Delta(G) + 1 + \frac{\Delta(G) - 2}{g_o - 1}.$$

Proof: Let G be a graph with maximum degree Δ and odd girth g_o. If $\chi'(G) \leq \Delta + 1$, then $\chi'(G) \leq \Delta + 1 + (\Delta - 2)/(g_o - 1)$ and we are done. If $\chi'(G) \geq \Delta + 2$, then Theorem 5.3(b)(f) implies that $\Delta \geq (g_o - 1)(\chi'(G') - \Delta - 1) + 2$. This yields $\chi'(G) \leq \Delta + 1 + (\Delta - 2)/(g_o - 1)$ and we are done, too. ∎

By Theorem 5.8, every graph G with $\chi'(G) \geq \Delta(G) + 2$ has odd girth at most $\lfloor (\chi'(G) - 3)/(\chi'(G) - \Delta(G) - 1) \rfloor$. A similar result for graphs G with $\chi'(G) \geq \Delta(G) + 1$ does not hold. The odd girth of such graphs can be arbitrarily large, but not infinite. The last statement is König's theorem [174] and follows immediately from statements (a) and (b) in Theorem 2.1.

Theorem 5.9 (König [174] 1916) *Every bipartite graph G satisfies $\chi'(G) = \Delta(G)$.*

5.2 EXTENDED TASHKINOV TREES

One advantage of Tashkinov's method is that the vertex set of a maximal Tashkinov tree is not only elementary, but also closed. However, as we have shown in Theorem 1.4, in order to prove Goldberg's conjecture, we need an elementary set that is not only closed, but also strongly closed. Nevertheless, even the vertex set X of a maximum Tashkinov tree need not be strongly closed. Hence it is natural to search for a possibility to extend X to a larger elementary set, provided that X is not strongly closed. One such possibility will be discussed in this section.

In the sequel, let G be a critical graph such that $\chi'(G) = k + 1$ for a given integer $k \geq \Delta(G) + 1$. Since G is critical, for every edge $e \in E(G)$ and every coloring $\varphi \in C^k(G - e)$, there is a Tashkinov tree T with respect to e and φ. Hence there is a largest number p such that $p = |V(T)|$ for such a Tashkinov tree T. We call p the **Tashkinov order** of G and write $t(G) = p$. Furthermore, we denote by $\mathcal{T}(G)$ the set of all triples (T, e, φ) such that $e \in E(G)$, $\varphi \in C^k(G - e)$, and T is a Tashkinov tree on p vertices with respect to e and φ. Evidently, $\mathcal{T}(G) \neq \emptyset$.

For a triple $(T, e, \varphi) \in \mathcal{T}(G)$ we introduce the following notation. For a color $\alpha \in \{1, \ldots, k\}$, let $E_{\varphi,\alpha} = \{e' \in E(G) \setminus \{e\} \mid \varphi(e') = \alpha\}$; furthermore, let

$$E_\alpha(T, e, \varphi) = E_{\varphi,\alpha} \cap \partial_G(V(T)).$$

We call α a **defective color** with respect to (T, e, φ) if $|E_\alpha(T, e, \varphi)| \geq 2$. The set of all defective colors with respect to (T, e, φ) is denoted by $\Gamma^d(T, e, \varphi)$. Furthermore, α is said to be a **free color** with respect to (T, e, φ) if $\alpha \in \overline{\varphi}(V(T))$ and α is unused on T with respect to φ. The set of all free colors with respect to (T, e, φ) is denoted by $\Gamma^f(T, e, \varphi)$.

Proposition 5.10 *Let G be a critical graph with $\chi'(G) = k + 1$ for an integer $k \geq \Delta(G) + 1$ and let $(T, e, \varphi) \in \mathcal{T}(G)$. Then the following statements hold:*

(a) $|V(T)| = t(G)$, $t(G)$ *is odd, and* $t(G) \geq 3$.

(b) $V(T)$ *is elementary and closed both with respect to* φ.

(c) $V(T)$ *is strongly closed with respect to* φ *iff* $\Gamma^d(T, e, \varphi) = \emptyset$.

(d) *If* $\alpha \in \overline{\varphi}(V(T))$, *then* $E_\alpha(T, e, \varphi) = \emptyset$.

(e) *If* $\alpha \in \Gamma^d(T, e, \varphi)$, *then* $|E_\alpha(T, e, \varphi)|$ *is odd and* ≥ 3.

(f) *For a vertex* $x \in V(T)$, *we have* $|\overline{\varphi}(x)| = k - d_G(x) + 1 \geq 2$ *if* $e \in E_G(x)$ *and* $|\overline{\varphi}(x)| = k - d_G(x) \geq 1$ *otherwise. Moreover,* $|\Gamma^f(T, e, \varphi)| \geq 4$.

(g) *Every color in* $\Gamma^d(T, e, \varphi) \cup \Gamma^f(T, e, \varphi)$ *is unused on* T *with respect to* φ.

Proof: Both statements (a) and (b) are simple consequences of Theorem 5.3 and the fact that T is, in particular, a maximal Tashkinov tree with respect to e and φ. Evidently, (b) implies both statements (c) and (d).

For the proof of (e), consider a defective color $\alpha \in \Gamma^d(T, e, \varphi)$ and let $E_\alpha = E_\alpha(T, e, \varphi)$. Then, by definition, $|E_\alpha| \geq 2$. By (d), we know that $\alpha \in \varphi(v)$ for every $v \in V(T)$ and, by (a), we know that $|V(T)|$ is odd. Since $E_{\varphi,\alpha}$ is a matching of G, this implies that $|E_\alpha|$ is odd and, therefore, $|E_\alpha| \geq 3$. Thus (e) is proved.

The first part of statement (f) follows simply from the fact that $\varphi \in \mathcal{C}^k(G - e)$ and $k \geq \Delta(G) + 1$. Since, by (b), $V(T)$ is elementary with respect to φ, this implies that $|\overline{\varphi}(V(T))| \geq p + 2$, where $p = |V(T)|$. Since T has $p - 1$ edges and the uncolored edge e belongs to T, we easily conclude that at least 4 colors in $\overline{\varphi}(V(T))$ are unused on T with respect to φ, i.e. $|\Gamma^f(T, e, \varphi)| \geq 4$.

It remains to prove (g). By definition, every free color is unused on T. So let $\alpha \in \Gamma^d(T, e, \varphi)$ be a defective color. Then $E_\alpha(T, e, \varphi) \neq \emptyset$ and, by (d), $\alpha \in \varphi(v)$ for every vertex v of T. By (T2), this implies that α is unused on T, too. This completes the proof of Proposition 5.10. ∎

Let P be a path, and let u, v be two vertices of P. Then there is a unique subpath P' of P having u and v as endvertices. We denote this subpath by uPv or vPu. If we fix an endvertex of P, say u, then we obtain a linear order $\preceq_{(u,P)}$ of the vertex set of P in a natural way, where $x \preceq_{(u,P)} y$ if the vertex x belongs to the subpath uPy. The next crucial proposition we learned from Tashkinov's paper [291].

Proposition 5.11 *Let G be a critical graph with $\chi'(G) = k + 1$ for an integer $k \geq \Delta(G) + 1$ and let $(T, e, \varphi) \in \mathcal{T}(G)$. Let $\alpha \in \Gamma^d(T, e, \varphi)$ be a defective color, and let u be a vertex of T such that $\overline{\varphi}(u)$ contains a free color $\gamma \in \Gamma^f(T, e, \varphi)$. Then, for the (α, γ)-chain $P = P_u(\alpha, \gamma, \varphi)$, the following statements hold:*

(a) *P is a path, where one endvertex is u and the other endvertex is some vertex $z \in V(G) \setminus V(T)$.*

(b) *$E_\alpha(T, e, \varphi) = E(P) \cap \partial_G(V(T))$.*

(c) *In the linear order $\preceq_{(u,P)}$ there is a first vertex v^1 such that $v^1 \in V(G) \setminus V(T)$ and there is a last vertex v^2 such that $v^2 \in V(T)$, where $v^1 \preceq_{(u,P)} v^2$. In the same order, v^1 has a successor w^1 and $w^1 \in V(G) \setminus V(T)$.*

(d) *$\overline{\varphi}(v^2) \cap \Gamma^f(T, e, \varphi) = \emptyset$.*

(e) *$V(T) \cup \{v^1, w^1\}$ is elementary with respect to φ.*

Proof: Let $E_\alpha = E_\alpha(T, e, \varphi)$. From Proposition 5.10 it follows that $\alpha \in \varphi(v)$ for all $v \in V(T)$, $|E_\alpha| \geq 3$ is odd, $E_\gamma(T, e, \varphi) = \emptyset$, and, moreover, $V(T)$ is elementary with respect to φ. Hence, u is the only vertex in T such that $\gamma \in \overline{\varphi}(u)$ and P is a path, where one endvertex is u and the other endvertex is some vertex $z \in V(G) \setminus V(T)$. Thus (a) is proved.

Since $E_\gamma(T, e, \varphi) = \emptyset$, for the proof of (b) it is sufficient to show that every edge of E_α belongs to P. Suppose this is false. Then there is an edge $e' \in E_\alpha \setminus E(P)$. Now, consider the coloring $\varphi' = \varphi/P$ obtained from φ by recoloring $P = P_u(\alpha, \gamma)$. Since, by Proposition 5.10(g), both colors α and γ are unused on T with respect to

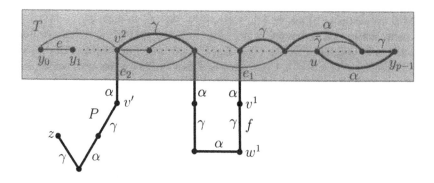

Figure 5.2   A triple $(T, e, \varphi) \in \mathcal{T}(G)$ with a defective color α and a free color γ.

φ, we conclude that T is a Tashkinov tree with respect to e and φ'. Consequently, $(T, e, \varphi') \in \mathcal{T}$. Furthermore, $\alpha \in \overline{\varphi}'(u)$ and, since $e' \notin E(P)$, $\varphi'(e') = \varphi(e') = \alpha$ implying $E_\alpha(T, e, \varphi') \neq \emptyset$. However, this is a contradiction to Proposition 5.10(d). Hence the proof of (b) is complete.

Since $E_\gamma(T, e, \varphi) = \emptyset$ and $|E_\alpha| \geq 3$ is odd, the existence of v^1 and v^2 is a simple consequence of (a) and (b). Clearly, $v^1 \preceq_{(u,P)} v^2$. Since v^1 is the first vertex in the linear order $\preceq_{(u,P)}$ that belongs to $V(G) \setminus V(T)$, (b) implies that there is an edge $e_1 \in E_\alpha \cap E(P)$ that is incident with v^1. Moreover, since v^2 is the last vertex in the linear order $\preceq_{(u,P)}$ that belongs to $V(T)$, we also conclude that there is an edge $e_2 \in E_\alpha \cap E(P)$ that is incident with v^2. If v' is the second endvertex of e_2, then $v' \in V(P)$ and the subpath $v'Pz$ is completely contained in $G - V(T)$ (see Fig. 5.2). Obviously, v^1 is not an endvertex of P and, therefore, v^1 has a successor w^1 in the linear order $\preceq_{(u,P)}$. Then there is an edge $f \in E_G(v^1, w^1) \cap E(P)$ with $\varphi(f) = \gamma$. Since $E_\gamma(T, e, \varphi) = \emptyset$, we have $w^1 \in V(G) \setminus V(T)$. Thus (c) is proved.

In order to prove (d), assume that it is false. Then there is a color $\delta \in \overline{\varphi}(v^2)$ that is unused on T with respect to φ. From (c) it follows that $v^2 \neq u$. Since $\gamma \in \overline{\varphi}(u)$ is also unused on T with respect to φ, we then deduce from Theorem 5.3(e) that $\gamma \neq \delta$ and there is a (γ, δ)-chain P_1 with respect to φ having endvertices u and v^2. Furthermore, we obtain from Theorem 5.3(e) that $V(P_1) \subseteq V(T)$ and T is a Tashkinov tree with respect to e and $\varphi_1 = \varphi/P_1$. Hence, $(T, e, \varphi_1) \in \mathcal{T}(G)$ and we have $E_\alpha(T, e, \varphi_1) = E_\alpha$, $\delta \in \overline{\varphi}_1(u)$, and $\gamma \in \overline{\varphi}_1(v^2)$. Obviously, neither γ, δ nor α is used on T with respect to φ_1. Since the subpath $v'Pz$ of $P = P_u(\alpha, \gamma, \varphi)$ is contained in $G - V(T)$ and $V(P_1) \subseteq V(T)$, it follows that the subpath $P_2 = v^2Pz$ is now an (α, γ)-chain with respect to φ_1. Since both α and γ are unused on T with respect to φ_1, it follows that T remains a Tashkinov tree with respect to e and $\varphi_2 = \varphi_1/P_2$, and, therefore, $(T, e, \varphi_2) \in \mathcal{T}(G)$. But then $\alpha \in \overline{\varphi}_2(v^2)$ and $e_1 \in E_\alpha(T, e, \varphi_2)$, contradicting Proposition 5.10(d). This proves (d).

In order to prove (e), we first claim that $V(T) \cup \{v^1\}$ is elementary with respect to φ. Suppose this is false. Since $V(T)$ is elementary with respect to φ, this implies that there is a color $\beta \in \overline{\varphi}(V(T)) \cap \overline{\varphi}(v^1)$. Since v^1 is incident with $e_1 \in E_\alpha$, the colors

α and β are distinct. Since, by (c), $\gamma \in \varphi(v^1)$, the (β, γ)-chain $P_3 = P_{v^1}(\beta, \gamma)$ is a path, where one endvertex is v^1. Since $\beta, \gamma \in \overline{\varphi}(V(T))$, it follows from Proposition 5.10(d) that $E_\gamma(T, e, \varphi) = E_\beta(T, e, \varphi) = \emptyset$. This implies that the path P_3 is contained in $G - V(T)$. Consequently, T is a Tashkinov tree with respect to e and $\varphi_3 = \varphi/P_3$ and, therefore, $(T, e, \varphi_3) \in \mathcal{T}(G)$. Then $\gamma \in \overline{\varphi}_3(u) \cap \Gamma^f(T, e, \varphi_3)$, $\gamma \in \overline{\varphi}_3(v^1)$, $E_\alpha(T, e, \varphi_3) = E_\alpha$, and $\alpha \in \Gamma^d(T, e, \varphi_3)$ is a defective color. Since v^1 is the first vertex in the linear order $\preceq_{(u,P)}$ that belongs to $V(G) \setminus V(T)$ and P_3 is contained in $G - V(T)$, we then obtain that $P_u(\alpha, \gamma, \varphi_3) = uPv^1$. Consequently, $P_u(\alpha, \gamma, \varphi_3)$ does not contain all edges from E_α, a contradiction to (b). Hence, the claim that $V(T) \cup \{v^1\}$ is elementary with respect to φ is proved.

Next, we claim that $\{v^1, w^1\}$ is elementary with respect to φ. Otherwise, there is a color $\beta \in \overline{\varphi}(v^1) \cap \overline{\varphi}(w^1)$ and we obtain a new coloring φ' by recoloring the edge $f \in E_G(v^1, w^1) \cap E(P)$ with β. Observe that $\beta \notin \{\alpha, \gamma\}$. Then we have $(T, e, \varphi') \in \mathcal{T}(G)$, $\gamma \in \overline{\varphi}'(u) \cap \Gamma^f(T, e, \varphi')$, $P_u(\alpha, \gamma, \varphi') = uPv^1$, $E_\alpha(T, e, \varphi') = E_\alpha$, and $\alpha \in \Gamma^d(T, e, \varphi')$. Consequently, $P_u(\alpha, \gamma, \varphi')$ does not contain all edges from E_α, a contradiction to (b). This proves the claim.

Eventually, we claim that $\overline{\varphi}(V(T)) \cap \overline{\varphi}(w^1) = \emptyset$. Suppose this is false. Then there is a color $\beta' \in \overline{\varphi}(V(T)) \cap \overline{\varphi}(w^1)$. By Proposition 5.10(f), there is a free color $\beta \in \Gamma^f(T, e, \varphi) \setminus \{\beta', \gamma\}$. Since $\beta, \beta' \in \overline{\varphi}(V(T))$, Proposition 5.10(d) implies that $E_\beta(T, e, \varphi) = E_{\beta'}(T, e, \varphi) = \emptyset$. Hence, $\alpha \notin \{\beta, \beta'\}$ and $P_1 = P_{w^1}(\beta, \beta', \varphi)$ is a path such that w^1 is an endvertex of P_1 and $V(P_1) \cap V(T) = \emptyset$. Consequently, for the coloring $\varphi_1 = \varphi/P_1$, we have $(T, e, \varphi_1) \in \mathcal{T}(G)$, $E_\alpha(T, e, \varphi_1) = E_\alpha$, $\alpha \in \Gamma^d(T, e, \varphi_1)$, $\gamma \in \overline{\varphi}_1(u)$, $P_u(\alpha, \gamma, \varphi_1) = P$, $\beta \in \overline{\varphi}_1(V(T)) \cap \overline{\varphi}_1(w^1)$, and $\beta, \gamma \in \Gamma^f(T, e, \varphi_1)$.

By Proposition 5.10(f), there is a color $\delta \in \overline{\varphi}_1(v^1)$. Since $\{v^1, w^1\}$ is elementary with respect to φ_1, we have $\delta \neq \beta$. Since $V(T) \cup \{v^1\}$ is elementary with respect to φ_1, we have $\delta \notin \overline{\varphi}_1(V(T))$ and, therefore, δ is unused on T with respect to φ_1. Clearly, $\delta \notin \{\alpha, \gamma\}$. Now, let $P' = P_{v^1}(\beta, \delta, \varphi_1)$ and $\varphi' = \varphi/P_1$. Then P' is a path, where one endvertex is v^1 and the other endvertex is some vertex $z^1 \neq v^1$. Since neither β nor δ is used on T with respect to φ_1, T remains a Tashkinov tree with respect to φ' and, moreover, $(T, e, \varphi') \in \mathcal{T}(G)$. Furthermore, we have $E_\alpha(T, e, \varphi') = E_\alpha$, $\alpha \in \Gamma^d(T, e, \varphi')$, $\gamma \in \overline{\varphi}'(u) \cap \Gamma^f(T, e, \varphi')$, and $P_u(\alpha, \gamma, \varphi_1) = P$. This implies that $V(T) \cup \{v^1\}$ as well as $\{v^1, w^1\}$ is elementary with respect to φ'. If $z^1 \neq w^1$, then we have $\beta \in \overline{\varphi}'(v^1) \cap \overline{\varphi}'(w^1)$, a contradiction. If $z^1 = w^1$, then we have $\beta \in \overline{\varphi}'(V(T)) \cap \overline{\varphi}(v^1)$, a contradiction, too. This completes the proof of (e). ∎

Let G be a critical graph with $\chi'(G) = k + 1$ for an integer $k \geq \Delta(G) + 1$ and let $(T, e, \varphi) \in \mathcal{T}(G)$. A vertex v in $V(G) \setminus V(T)$ is called **absorbing** with respect to (T, e, φ) if, for every color $\delta \in \overline{\varphi}(v)$ and every free color $\gamma \in \Gamma^f(T, e, \varphi)$ with $\gamma \neq \delta$, the (γ, δ)-chain $P_v(\gamma, \delta)$ contains a vertex $u \in V(T)$ such that $\gamma \in \overline{\varphi}(u)$. Since, by Proposition 5.10(b), $V(T)$ is elementary with respect to φ, this vertex u is then the unique vertex in T such that $\gamma \in \overline{\varphi}(u)$ and, moreover, $P_v(\gamma, \delta)$ is a path whose endvertices are u and v. Clearly, u belongs to $P_v(\gamma, \delta)$ if and only if v belongs to $P_u(\gamma, \delta)$. Let $A(T, e, \varphi)$ denote the set of all vertices in $V(G) \setminus V(T)$ that are absorbing with respect to (T, e, φ).

Proposition 5.12 *Let G be a critical graph with $\chi'(G) = k + 1$ for an integer $k \geq \Delta(G) + 1$ and let $(T, e, \varphi) \in \mathcal{T}(G)$. Then the vertex set $V(T) \cup A(T, e, \varphi)$ is elementary with respect to φ.*

Proof: Suppose, on the contrary, that $Z = V(T) \cup A(T, e, \varphi)$ is not elementary with respect to φ. Then there are two distinct vertices $v, v' \in Z$ and a color $\delta \in \overline{\varphi}(v) \cap \overline{\varphi}(v')$. Since, by Proposition 5.10(b), $V(T)$ is elementary with respect to φ, at least one of the these two vertices belongs to $A = A(T, e, \varphi)$, say $v \in A$. From Proposition 5.10(f) it follows that there is a free color $\gamma \in \Gamma^{\mathrm{f}}(T, e, \varphi)$ distinct from δ. Then, since $V(T)$ is elementary with respect to φ, there is a unique vertex u in T such that $\gamma \in \overline{\varphi}(u)$. Since $v \in A$, it follows that v belongs to the chain $P_u = P_u(\gamma, \delta)$. Hence, P_u is a path and u, v are the endvertices of P_u, where $u \neq v$. Since $\delta \in \overline{\varphi}(v')$ and $v' \neq v$, we then conclude that $v' \notin V(P_u)$ and, therefore, $v' \neq u$. Clearly, this implies that $v' \notin A$. Consequently, $v' \in V(T) \setminus \{u\}$. Since $\gamma \in \overline{\varphi}(u)$ and $\delta \in \overline{\varphi}(v')$, it follows then from Theorem 5.3(e) that the vertex v' belongs to P_u, a contradiction. This completes the proof. ∎

Proposition 5.13 *Let G be a critical graph with $\chi'(G) = k + 1$ for an integer $k \geq \Delta(G) + 1$ and let $(T, e, \varphi) \in \mathcal{T}(G)$. Let u be a vertex of T such that $\overline{\varphi}(u)$ contains a free color $\gamma \in \Gamma^{\mathrm{f}}(T, e, \varphi)$, and let δ be an arbitrary color distinct from γ. Then the following statements hold:*

(a) $E_\gamma(T, e, \varphi) = \emptyset$.

(b) $E_\delta(T, e, \varphi) \subseteq E(P_u(\gamma, \delta))$.

(c) *If $\delta \in \overline{\varphi}(v)$ for some vertex $v \in V(T) \cup A(T, e, \varphi)$, then v belongs to $P_u(\gamma, \delta)$.*

(d) *If P is a (γ, δ)-chain with respect to φ distinct from $P_u(\gamma, \delta)$, then every endvertex of P belongs to $V(G) \setminus (V(T) \cup A(T, e, \varphi))$.*

(e) *If P is a (γ, δ)-chain with respect to φ such that P is a path and every endvertex of P belongs to $V(G) \setminus V(T)$, then P and T are vertex disjoint.*

Proof: Since $\gamma \in \overline{\varphi}(V(T))$, statement (a) follows from Proposition 5.10(d). For the proof of (b), we need only consider the case that the set $E_\delta = E_\delta(T, e, \varphi)$ is nonempty. If $\delta \in \Gamma^{\mathrm{d}}(T, e, \varphi)$ is a defective color, then Proposition 5.11(b) implies that every edge of E_δ belongs to $P_u = P_u(\gamma, \delta)$. Otherwise, $|E_\delta| = 1$ and we argue as follows. By Proposition 5.10(d), $\delta \in \varphi(v)$ for every vertex v of T. Since, by Proposition 5.10(b), $V(T)$ is elementary with respect to φ, we then conclude that P_u is a path, where u is one endvertex of P_u and the other endvertex is some vertex in $V(G) \setminus V(T)$. Since, by (a), $E_\gamma(T, e, \varphi) = \emptyset$, this implies that P_u contains the only edge of E_δ. This completes the proof of (b).

For the proof of (c), let $v \in V(T) \cup A(T, e, \varphi)$ be a vertex such that $\delta \in \overline{\varphi}(v)$. If $v = u$, then, evidently, v belongs to P_u. If $v \in V(T) \setminus \{u\}$, then Theorem 5.3(e) implies that v belongs to P_u. Otherwise, $v \in A(T, e, \varphi)$ and v belongs to P_u by definition of absorbing. This proves (c).

For the proof of (d), let P be a (γ, δ)-chain distinct from $P_u = P_u(\gamma, \delta)$. Clearly, P and P_u are vertex disjoint. Now suppose that P has an endvertex v belonging to $V(T) \cup A(T, e, \varphi)$. Then $\overline{\varphi}(v)$ contains γ or δ. Clearly, $v \neq u$ and, by Proposition 5.12, $V(T) \cup A(T, e, \varphi)$ is elementary with respect to φ. Since $\gamma \in \overline{\varphi}(u)$, it follows that $\delta \in \overline{\varphi}(v)$. Then (c) implies that v belongs to P_u, contradicting the fact that P and P_u are vertex disjoint. This proves (d).

For the proof of (e), let P be a (γ, δ)-chain with respect to φ such that P is a path whose endvertices belong to $V(G) \setminus V(T)$. If P and T have a vertex in common, then at least one edge of P, say e', belongs to $\partial_G(V(T))$. Since, by (a), $E_\gamma(T, e, \varphi) = \emptyset$, this implies $e' \in E_\delta(T, e, \varphi)$. Then, by (b), e' belongs to $P_u = P_u(\gamma, \delta)$, too. Consequently, $P = P_u$. This contradicts the hypothesis of (e), since u is an endvertex of $P = P_u$ contained in T. Thus (e) is proved. ∎

We call $v \in V(G)$ a **defective vertex** with respect to $(T, e, \varphi) \in \mathcal{T}(G)$ if there are two distinct colors α and γ such that $\alpha \in \Gamma^d(T, e, \varphi)$ is a defective color, $\gamma \in \Gamma^f(T, e, \varphi)$ is a free color, and v is the first vertex in the linear order $\preceq_{(u,P)}$ belonging to $V(G) \setminus V(T)$, where u is the unique vertex in T with $\gamma \in \overline{\varphi}(u)$ and $P = P_u(\alpha, \gamma)$. The set of all defective vertices with respect to (T, e, φ) is denoted by $D(T, e, \varphi)$.

Proposition 5.14 *Let G be a critical graph with $\chi'(G) = k + 1$ for an integer $k \geq \Delta(G) + 1$ and let $(T, e, \varphi) \in \mathcal{T}(G)$. Then $D(T, e, \varphi) \subseteq A(T, e, \varphi)$.*

Proof: Let v be an arbitrary vertex of $D(T, e, \varphi)$. Then there are two distinct colors α and γ such that $\alpha \in \Gamma^d(T, e, \varphi)$ is a defective color, $\gamma \in \Gamma^f(T, e, \varphi)$ is a free color, and v is the first vertex in the linear order $\preceq_{(u,P)}$ belonging to $V(G) \setminus V(T)$, where u is the unique vertex in T with $\gamma \in \overline{\varphi}(u)$ and $P = P_u(\alpha, \gamma)$.

Now consider an arbitrary color $\delta \in \overline{\varphi}(v)$ and a free color $\gamma' \in \Gamma^f(T, e, \varphi)$. By Proposition 5.11(e), the set $V(T) \cup \{v\}$ is elementary with respect to φ. Consequently, there is a unique vertex u' in T such that $\gamma' \in \overline{\varphi}(u')$ and, moreover, $\gamma' \neq \delta$. Clearly, $\alpha \neq \delta$ and, by Proposition 5.10(d), $\alpha \neq \gamma'$.

In order to prove that $v \in A(T, e, \varphi)$, we have to show that u' belongs to the (γ', δ)-chain $P' = P_v(\gamma', \delta)$. Suppose, on the contrary, that $u' \notin V(P')$. Since $V(T) \cup \{v\}$ is elementary with respect to φ, this implies, in particular, that no endvertex of P' belongs to T. Since P' is a path and γ' is a free color, it follows from Proposition 5.13(e) that P' and T are vertex disjoint. Consequently, T is a Tashkinov tree with respect to $\varphi' = \varphi/P'$ and, therefore, $(T, e, \varphi') \in \mathcal{T}(G)$. For the coloring φ', we then obtain that $\gamma' \in \overline{\varphi}'(u') \cap \overline{\varphi}'(v)$, $E_\alpha(T, e, \varphi) = E_\alpha(T, e, \varphi')$ and, moreover, $u P_u(\alpha, \gamma, \varphi) v = u P_u(\alpha, \gamma, \varphi') v$. Hence, $\alpha \in \Gamma^d(T, e, \varphi')$ and v remains the first vertex in the linear order $\preceq_{(u, P_u(\alpha, \gamma, \varphi'))}$ that belongs to $V(G) \setminus V(T)$. From Proposition 5.11(e) it follows then that $V(T) \cup \{v\}$ is elementary with respect to φ'. Since $\gamma' \in \overline{\varphi}'(u') \cap \overline{\varphi}'(v)$, this is a contradiction. Consequently, $v \in A(T, e, \varphi)$ and we are done. ∎

Proposition 5.15 *Let G be a critical graph with $\chi'(G) = k + 1$ for an integer $k \geq \Delta(G) + 1$ and let $(T, e, \varphi) \in \mathcal{T}(G)$. Furthermore, let $\gamma, \delta \in \{1, \ldots, k\}$ be* *two distinct colors, and let P be an (γ, δ)-chain with respect to φ. Suppose that $V(P) \cap V(T) = \emptyset$. Then the coloring $\varphi_1 = \varphi/P$ satisfies the following:*

(a) $(T, e, \varphi_1) \in \mathcal{T}(G)$.

(b) $\overline{\varphi}_1(V(T)) = \overline{\varphi}(V(T))$ and $\Gamma^f(T, e, \varphi_1) = \Gamma^f(T, e, \varphi)$.

(c) $\Gamma^d(T, e, \varphi_1) = \Gamma^d(T, e, \varphi)$ and $E_\alpha(T, e, \varphi_1) = E_\alpha(T, e, \varphi)$ for every defective color $\alpha \in \Gamma^d(T, e, \varphi)$.

(d) $D(T, e, \varphi_1) = D(T, e, \varphi)$.

Proof: Since $V(P) \cap V(T) = \emptyset$, we conclude that $\varphi_1(f) = \varphi(f)$ for every edge $f \in E_G(V(T), V(G)) \setminus \{e\}$ and $\overline{\varphi}_1(v) = \overline{\varphi}(v)$ for every vertex $v \in V(T)$. Consequently, T is a Tashkinov tree with respect to e end φ_1, and, therefore, $(T, e, \varphi_1) \in \mathcal{T}(G)$. Furthermore, $\overline{\varphi}_1(V(T)) = \overline{\varphi}(V(T))$ and $\Gamma^f(T, e, \varphi_1) = \Gamma^f(T, e, \varphi)$. This proves (a) and (b). Clearly, $E_\alpha(T, e, \varphi_1) = E_\alpha(T, e, \varphi)$ for every color α. This implies (c).

For the proof of (d), consider an arbitrary vertex $v \in D(T, e, \varphi)$. Then there is a defective color $\alpha \in \Gamma^d(T, e, \varphi)$ and a free color $\gamma' \in \Gamma^f(T, e, \varphi)$ such that v is the first vertex in the linear order $\preceq_{(u, P')}$ belonging to $V(G) \setminus V(T)$, where u is the unique vertex of T such that $\gamma' \in \overline{\varphi}(u)$ and $P' = P_u(\gamma', \alpha, \varphi)$. The subpath $P^* = uP'v$ then satisfies $E(P^*) \subseteq E_G(V(T), V(G))$. This implies $\varphi_1(f) = \varphi(f)$ for every edge $f \in E(P^*)$. By (b) and (c), we also have $\alpha \in \Gamma^d(T, e, \varphi')$, $\gamma' \in \Gamma^f(T, e, \varphi')$ and $\gamma' \in \overline{\varphi}_1(u)$. Hence if $P_1 = P_u(\gamma', \alpha, \varphi_1)$, then v remains the first vertex in the linear order $\preceq_{(u, P_1)}$ belonging to $V(G) \setminus V(T)$. Consequently, $v \in D(T, e, \varphi_1)$ is a defective vertex with respect to (T, e, φ_1). This implies that $D(T, e, \varphi) \subseteq D(T, e, \varphi_1)$. Since P remains an (γ, δ)-chain with respect to φ_1 and $\varphi = \varphi_1/P$, we conclude in a similar way that $D(T, e, \varphi_1) \subseteq D(T, e, \varphi)$. This completes the proof of (d). ∎

Proposition 5.16 *Let G be a critical graph with $\chi'(G) = k + 1$ for an integer $k \geq \Delta(G) + 1$ and let $(T, e, \varphi) \in \mathcal{T}(G)$. Let P be a (γ, δ)-chain with respect to φ, where $\gamma \in \Gamma^f(T, e, \varphi)$ is a free color and δ is an arbitrary color distinct from γ. Suppose that P is a path and every endvertex of P belongs to $V(G) \setminus V(T)$. Then the coloring $\varphi_1 = \varphi/P$ satisfies the following:*

(a) $(T, e, \varphi_1) \in \mathcal{T}(G)$.

(b) $\overline{\varphi}_1(V(T)) = \overline{\varphi}(V(T))$ and $\Gamma^f(T, e, \varphi_1) = \Gamma^f(T, e, \varphi)$.

(c) $\Gamma^d(T, e, \varphi_1) = \Gamma^d(T, e, \varphi)$ and $E_\alpha(T, e, \varphi_1) = E_\alpha(T, e, \varphi)$ for every defective color $\alpha \in \Gamma^d(T, e, \varphi)$.

(d) $D(T, e, \varphi_1) = D(T, e, \varphi)$.

Proof: From Proposition 5.13(e) it follows that P and T are vertex disjoint. Hence, Proposition 5.16 is an immediate consequence of Proposition 5.15. ∎

Proposition 5.17 *Let G be a critical graph with $\chi'(G) = k + 1$ for an integer $k \geq \Delta(G) + 1$ and let $(T, e, \varphi) \in \mathcal{T}(G)$. Furthermore, let $Y \subseteq D(T, e, \varphi)$ and $Z = V(T) \cup Y$. Suppose that $e' \in E_G(z, v)$ is an edge such that $z \in Z$, $v \in V(G) \setminus Z$, and $\overline{\varphi}(e') \in \overline{\varphi}(Z)$. Then the vertex set $Z \cup \{v\}$ is elementary with respect to φ.*

Proof: For the proof we shall analyze a possible counterexample. By Proposition 5.14, $Y \subseteq A(T, e, \varphi)$ and, therefore, $Z = V(T) \cup Y$ is elementary with respect to φ by Proposition 5.12. Hence, a possible counterexample can be described as a tuple $(T, e, \varphi, Y, e', z, v, u, \delta)$ such that $(T, e, \varphi) \in \mathcal{T}(G), Y \subseteq D(T, e, \varphi), e' \in E_G(z, v)$, $z \in Z, v \in V(G) \setminus Z, \varphi(e') \in \overline{\varphi}(Z), u \in Z$, and $\delta \in \overline{\varphi}(v) \cap \overline{\varphi}(u)$. We call δ the *bad color* of the corresponding counterexample.

First, we claim that there is a counterexample such that the bad color belongs to $\Gamma^{\mathrm{f}}(T, e, \varphi)$. To this end, consider a counterexample $(T, e, \varphi, Y, e', z, v, u, \delta)$. By Proposition 5.10(f), there is a free color $\gamma \in \Gamma^{\mathrm{f}}(T, e, \varphi)$ such that γ is distinct from δ as well as from $\varphi(e')$. Then there is a vertex u' in T such that $\gamma \in \overline{\varphi}(u')$. Since γ is a free color and $\delta \in \overline{\varphi}(u)$, it follows from Proposition 5.13(c) that u belongs to $P' = P_{u'}(\gamma, \delta)$. Consequently, u and u' are the endvertices of P', possibly $u = u'$. Since $\delta \in \overline{\varphi}(v)$ and since v is distinct from both u and u', this implies that $P = P_v(\gamma, \delta)$ is a path that is distinct from P'. Then, by Proposition 5.13(d), every endvertex of P belongs to $V(G) \setminus (V(T) \cup A(T, e, \varphi))$ and, therefore, to $V(G) \setminus Z$. Then the coloring $\varphi_1 = \varphi/P$ satisfies $\overline{\varphi}_1(Z) = \overline{\varphi}(Z)$ and $\gamma \in \overline{\varphi}_1(v)$. From Proposition 5.16 it follows then that $(T, e, \varphi_1, Y, e', z, v, u', \gamma)$ is a counterexample having the desired property. This proves the claim

Now, let us consider a counterexample $(T, e, \varphi, Y, e', z, v, u, \gamma)$ such that the bad color $\gamma \in \Gamma^{\mathrm{f}}(T, e, \varphi)$ is a free color. Since Z is elementary with respect to φ, this implies that u is the unique vertex of T such that $\gamma \in \overline{\varphi}(u)$. Let $\beta = \varphi(e')$. Since $\gamma \in \overline{\varphi}(v)$ and $e' \in E_G(z, v)$, we have $\gamma \neq \beta$. By hypothesis, there is a vertex $z' \in Z$ such that $\beta \in \overline{\varphi}(z')$. Clearly, $z' \neq z, v$. To arrive at a contradiction, we distinguish two cases.

Case 1: $z \in Y$. By Proposition 5.10(f), there is a color $\delta \in \overline{\varphi}(z)$. Clearly, $\delta \neq \beta = \varphi(e')$ and $\delta \neq \gamma$. From Proposition 5.13(c) it follows that z belongs to $P_u = P_u(\gamma, \delta)$. Consequently, z and u are the endvertices of P_u. Since $\gamma \in \overline{\varphi}(v)$ and $v \notin \{u, z\}$, this implies that $P = P_v(\gamma, \delta)$ is a path that is distinct from P_u. Then, by Proposition 5.13(d), every endvertex of P belongs to $V(G) \setminus (V(T) \cup A(T, e, \varphi))$ and so to $V(G) \setminus Z$. Then the coloring $\varphi_1 = \varphi/P$ satisfies $\overline{\varphi}_1(Z) = \overline{\varphi}(Z)$ and $\delta \in \overline{\varphi}_1(v)$. From Proposition 5.16 it now follows that $(T, e, \varphi_1, Y, e', z, v, z, \delta)$ is a counterexample, too. Furthermore, we have $\beta = \varphi(e') = \varphi_1(e') \in \overline{\varphi}(z') = \overline{\varphi}_1(z')$, where $\beta \notin \{\gamma, \delta\}$ and $z' \neq z$.

Now consider the coloring φ_2 such that $\varphi_2(f) = \varphi_1(f)$ for all edges $f \in E(G) \setminus \{e, e'\}$ and $\varphi_2(e') = \delta$. Since $e' \in E_G(z, v)$ and $\delta \in \overline{\varphi}_1(z) \cap \overline{\varphi}_1(v)$, we have $\varphi_2 \in \mathcal{C}^k(G - e)$. Furthermore, we have $\beta \in \overline{\varphi}_2(z)$ and $\beta \in \overline{\varphi}_1(z') = \overline{\varphi}_2(z')$. Since e' is not contained in $E(T) \cup \partial_G(V(T))$, we easily conclude that $(T, e, \varphi_2) \in \mathcal{T}(G)$ and $D(T, e, \varphi_2) = D(T, e, \varphi_1)$. Since $Y \subseteq D(T, e, \varphi_1)$, it follows then from Proposition 5.14 that $Y \subseteq A(T, e, \varphi_2)$ Then, by Proposition 5.11, the vertex set

$Z = V(T) \cup Y$ is elementary with respect to φ_2. Since $\beta \in \overline{\varphi}_2(z) \cap \overline{\varphi}_2(z')$, this is a contradiction.

Case 2: $z \in V(T)$. Clearly, this implies $e' \in E_\beta(T, e, \varphi)$. Since u is a vertex of T and $\gamma \in \overline{\varphi}(u)$ is a free color, it follows from Proposition 5.13(b) that every edge of $E_\beta(T, e, \varphi)$ belongs to $P_u = P_u(\gamma, \beta)$. Since $e' \in E_G(v)$ and $\gamma \in \overline{\varphi}(v)$, this implies that v belongs to P_u, too. Consequently, u and v are the endvertices of P_u. From Proposition 5.10(d) it follows that $\beta \in \varphi(x)$ for every vertex x of T. Since $\beta \in \overline{\varphi}(z')$ and $z' \in Z$, this implies $z' \in Y$ and, therefore, $z' \neq u$. Since z' is also distinct from v, this implies that z' is not contained in P_u. By Proposition 5.14, the vertex z' belongs to $A(T, e, \varphi)$. But then Proposition 5.13(c) implies that z' belongs to P_u, a contradiction.

Therefore, in both cases we arrive at a contradiction. Hence the proof of Proposition 5.17 is complete. ∎

Consider a triple $(T, e, \varphi) \in \mathcal{T}(G)$ and a set $Z \subseteq V(G)$ containing all vertices of T. Furthermore, let $F = (e_1, u_1, \ldots, e_p, u_p)$ be a sequence consisting of distinct edges $e_1, \ldots, e_p \in E(G)$ and distinct vertices $u_1, \ldots, u_p \in V(G)$. The sequence F is called a **fan** at Z with respect to φ if for every $i \in \{1, \ldots, p\}$ there are two vertices z, z' satisfying $z \in Z$, $z' \in Z \cup \{u_1, \ldots, u_{i-1}\}$, $e_i \in E_G(z, u_i)$, and $\varphi(e_i) \in \overline{\varphi}(z')$.

Theorem 5.18 *Let G be a critical graph with $\chi'(G) = k + 1$ for an integer $k \geq \Delta(G) + 1$ and let $(T, e, \varphi) \in \mathcal{T}(G)$. Furthermore, let $Y \subseteq D(T, e, \varphi)$ and $Z = V(T) \cup Y$. If F is a fan at Z with respect to φ, then $Z \cup V(F)$ is elementary with respect to φ.*

Proof: By a *bad tuple* we mean a tuple (T, e, φ, Y, F) such that $(T, e, \varphi) \in \mathcal{T}(G)$, $Y \subseteq D(T, e, \varphi)$, F is a fan at $Z = V(T) \cup Y$ with respect to φ, and $Z \cup V(F)$ is not elementary with respect to φ. For the proof of the theorem, we have to show that there is no bad tuple. Suppose, on the contrary, that there is a bad tuple. Then we choose a bad tuple (T, e, φ_0, Y, F) such that

(a) $|V(F)|$ is minimum.

Let $Z = V(T) \cup Y$ and $F = (e_1, u_1, \ldots, e_p, u_p)$. Furthermore, let $\mathcal{P}$ denote the set of all colorings $\varphi \in \mathcal{C}^k(G - e)$ such that (T, e, φ, Y, F) is a bad tuple. Clearly, $\varphi_0 \in \mathcal{P}$.

Now let us consider an arbitrary coloring $\varphi \in \mathcal{P}$. Then (T, e, φ, Y, F) is a bad tuple and Proposition 5.17 implies that $p \geq 2$. Furthermore, since the sequence $Fu_{p-1} = (e_1, u_1, \ldots, e_{p-1}, u_{p-1})$ is also a fan at Z with respect to φ, it follows from (a) that

(b) $Z \cup \{u_1, \ldots, u_{p-1}\}$ is elementary with respect to φ.

Proposition 5.14 implies that

(c) $Y \subseteq A(T, e, \varphi)$.

Let $i \in \{1, \ldots, p\}$. Since $F = (e_1, u_1, \ldots, e_p, u_p)$ is a fan at Z with respect to φ, there is a sequence $i_1, \ldots, i_q$ of integers such that $1 \leq i_1 < i_2 < \cdots < i_q = i$,

$\varphi(e_{i_j}) \in \overline{\varphi}(u_{i_{j-1}})$ for $2 \leq j \leq q$, and $\varphi(e_{i_1}) \in \overline{\varphi}(Z)$. Then, clearly, the sequence $F' = (e_{i_1}, u_{i_1}, \ldots, e_{i_q}, u_{i_q})$ is a fan at Z with respect to φ, where $q = |V(F')| \leq i$. From (b) it follows that F' is uniquely determined by the edge e_i and the coloring φ. The fan F' is called the subfan of F induced by e_i and φ, and we write $F' = F[e_i, \varphi]$. We need the following statement.

(d) *Let $i \in \{1, \ldots, p\}$ and let $F' = (e_{i_1}, u_{i_1}, \ldots, e_{i_q}, u_{i_q})$ be the subfan of F induced by e_i and φ such that $q < p$. Furthermore, let u be a vertex of T such that $\overline{\varphi}(u)$ contains a free color $\gamma \in \Gamma^f(T, e, \varphi)$ that is distinct from the color $\varphi(e_{i_1})$. If $\delta \in \overline{\varphi}(u_i)$, then the vertex u_i belongs to $P_u(\gamma, \delta)$.*

Proof of (d): Since F' is a fan at Z with respect to φ and $q = |V(F')| < p$, it follows from (a) that $Z \cup V(F')$ is elementary with respect to φ. Since, by hypothesis, $\gamma \in \overline{\varphi}(u)$, $\gamma \neq \varphi(e_{i_1})$, and $\varphi(e_{i_j}) \in \overline{\varphi}(u_{i_{j-1}})$ for $2 \leq j \leq q$, this implies that γ is not contained in $\Gamma = \{\varphi(e_{i_1}), \ldots, \varphi(e_{i_q})\}$. Since $\delta \in \overline{\varphi}(u_i)$ and $u_i = u_{i_q}$, we also obtain $\delta \notin \Gamma$ and $\delta \neq \gamma$.

Now suppose that u_i is not contained in $P_u = P_u(\gamma, \delta)$. Then $P = P_{u_i}(\gamma, \delta)$ is distinct from P_u. Since $\delta \in \overline{\varphi}(u_i)$, P is a path and from Proposition 5.13(d) we deduce that every endvertex of P belongs to $W = V(G) \setminus (V(T) \cup A(T, e, \varphi))$. By (c), $W \subseteq V(G) \setminus Z$. Then the coloring $\varphi_1 = \varphi/P$ satisfies $\overline{\varphi}_1(Z) = \overline{\varphi}(Z)$ and $\gamma \in \overline{\varphi}_1(u_i)$. Furthermore, it follows from Proposition 5.16 that $(T, e, \varphi_1) \in \mathcal{T}(G)$ and $Y \subseteq D(T, e, \varphi) = D(T, e, \varphi_1)$. Since $\gamma, \delta \notin \Gamma$, we then conclude that F' is a fan at Z with respect to φ_1. The color γ belongs to $\overline{\varphi}_1(u_i)$ as well as to $\overline{\varphi}_1(u) = \overline{\varphi}(u)$. Consequently, (T, e, φ_1, Y, F') is a bad tuple with $|V(F')| < |V(F)|$, a contradiction to (a). This proves (d).   ∎

We continue the proof of Theorem 5.18. Since (T, e, φ, Y, F) is a bad tuple, the set $Z \cup V(F) = Z \cup \{u_1, \ldots, u_p\}$ is not elementary with respect to φ. On the other hand, by (b), the set $Z \cup \{u_1, \ldots, u_{p-1}\}$ is elementary with respect to φ. Consequently, there is a pair (z, δ) consisting of a vertex $z \in Z \cup \{u_1, \ldots, u_{p-1}\}$ and a color $\delta \in \overline{\varphi}(z) \cap \overline{\varphi}(u_p)$. In what follows, let J denote the set of all triples (φ, z, δ) such that $\varphi \in \mathcal{P}$, $z \in Z$, and $\delta \in \overline{\varphi}(z) \cap \overline{\varphi}(u_p)$. To arrive at a contradiction, we consider two cases.

Case 1: $J = \emptyset$. Let $\varphi \in \mathcal{P}$ be an arbitrary coloring. Then (T, e, φ, Y, F) is a bad tuple and, since $J = \emptyset$, there is an index $j \in \{1, \ldots, p-1\}$ and a color δ such that $\delta \in \overline{\varphi}(u_p) \cap \overline{\varphi}(u_j)$.

Let $F_1 = F[e_p, \varphi]$ be the subfan of F induced by e_p and φ. First, we consider the case when $|V(F_1)| < p = |V(F)|$. Then $F_1 = (e_{i_1}, u_{i_1}, \ldots, e_{i_q}, u_{i_q})$ with $e_{i_q} = e_p$ and $u_{i_q} = u_p$. For the subfan $F_2 = F[e_j, \varphi]$, we have $|V(F_2)| \leq j < p$ and $F_2 = (e_{j_1}, u_{j_1}, \ldots, e_{j_r}, u_{j_r})$ with $e_{j_r} = e_j$ and $u_{j_r} = u_j$. By Proposition 5.10(f), there is a free color $\gamma \in \Gamma^f(T, e, \varphi)$ that is distinct from both $\varphi(e_{i_1})$ and $\varphi(e_{j_1})$. Then there is a unique vertex u in T such that $\gamma \in \overline{\varphi}(u)$. From (d) it then follows that $P_u(\gamma, \delta)$ contains both vertices u_p and u_j. This, however, is a contradiction.

It remains to consider the case when $|V(F_1)| = p$. Then $\varphi(e_i) \in \overline{\varphi}(u_{i-1})$ for $2 \leq i \leq p$ and $\varphi(e_1) \in \overline{\varphi}(Z)$. By Proposition 5.10(f), there is a free color $\gamma \in \Gamma^f(T, e, \varphi)$ that is distinct from $\varphi(e_1)$. Then there is a unique vertex u in T such that $\gamma \in \overline{\varphi}(u)$.

For the subfan $F_2 = F[e_j, \varphi]$, we clearly have $F_2 = (e_1, u_1, \ldots, e_j, u_j)$. Since $\delta \in \overline{\varphi}(u_j)$, it follows from (d) that u_j belongs to $P_u = P_u(\gamma, \delta)$. Consequently, P_u is a path whose endvertices are u and u_j. Since $\delta \in \overline{\varphi}(u_p)$ and $u_p \notin \{u, u_j\}$, this implies that $P = P_{u_p}(\gamma, \delta)$ is a path that is distinct from P_u. Then, by Proposition 5.13(d), every endvertex of P belongs to $W = V(G) \setminus (V(T) \cup A(T, e, \varphi))$. By (c), $W \subseteq V(G) \setminus Z$. Then, for the coloring $\varphi_1 = \varphi/P$, we have $\overline{\varphi}_1(Z) = \overline{\varphi}(Z)$ and $\gamma \in \overline{\varphi}_1(u_p)$. Furthermore, it follows from Proposition 5.16 that $(T, e, \varphi_1) \in \mathcal{T}(G)$, $Y \subseteq D(T, e, \varphi) = D(T, e, \varphi_1)$, and $\Gamma^f(T, e, \varphi) = \Gamma^f(T, e, \varphi_1)$.

Eventually, we claim that F is a fan at Z with respect to φ_1. By (b), the set $Z \cup \{u_1, \ldots, u_{p-1}\}$ is elementary with respect to φ. Furthermore, we have $\varphi(e_i) \in \overline{\varphi}(u_{i-1})$ for $2 \leq i \leq p$, $\varphi(e_1) \in \overline{\varphi}(Z)$, $\gamma \neq \varphi(e_1)$, $\gamma \in \overline{\varphi}(Z)$, and $\delta \in \overline{\varphi}(u_p) \cap \overline{\varphi}(u_j)$. Consequently, $\gamma \notin \{\varphi(e_1), \ldots, \varphi(e_p)\}$. Furthermore, since u_j is not contained in $P = P_{u_p}(\gamma, \delta)$, we conclude that no endvertex of P belongs to $\{u_1, \ldots, u_{p-1}\}$. Thus the coloring φ_1 satisfies $\gamma \in \overline{\varphi}_1(Z) = \overline{\varphi}(Z)$ and $\overline{\varphi}_1(u_i) = \overline{\varphi}(u_i)$ for $1 \leq i \leq p - 1$. Now, consider an edge e_i of the fan F with $1 \leq i \leq p$. If $e_i \notin E(P)$, then we obtain $\varphi_1(e_i) = \varphi(e_i)$ and, therefore, $\varphi_1(e_i) \in \overline{\varphi}(u_{i-1}) = \overline{\varphi}_1(u_{i-1})$ if $2 \leq i \leq p$ or $\varphi_1(e_i) \in \overline{\varphi}(Z) = \overline{\varphi}_1(Z)$ if $i = 1$. If $e_i \in E(P)$, then $\gamma \neq \varphi(e_i)$ implies $\varphi(e_i) = \delta$ and, therefore, $\varphi_1(e_i) = \gamma \in \overline{\varphi}(Z) = \overline{\varphi}_1(Z)$. This proves the claim that F is a fan at Z with respect to φ_1.

Since $\gamma \in \overline{\varphi}_1(u) \cap \overline{\varphi}_1(u_p)$, it follows then that (T, e, φ_1, Y, F) is a bad tuple. Consequently, we have $\varphi_1 \in \mathcal{P}$ and $(\varphi_1, u, \gamma) \in J$, contradicting the hypothesis of Case 1.

Case 2: $J \neq \emptyset$. First, we establish some basic facts for triples in J and explain how we can move from one triple to another triple.

(e) *Let $(\varphi, z, \delta) \in J$ be an arbitrary triple. Then the following statements hold:*

(1) $\varphi(e_i) \in \varphi(u_{i-1})$ *for* $2 \leq i \leq p$ *and, moreover,* $\varphi(e_1) \in \overline{\varphi}(Z)$.

(2) $\varphi(e_i) \neq \varphi(e_j)$ *for* $1 \leq i < j \leq p$ *and* $\delta \notin \{\varphi(e_2), \ldots, \varphi(e_p)\}$.

(3) *Let $u \in V(T)$ and let $\gamma \in \overline{\varphi}(u)$ be a color with $\gamma \neq \delta$. Assume that γ or δ belongs to $\Gamma^f(T, e, \varphi)$. Then $P_u(\gamma, \delta)$ is a path whose endvertices are u and z. Furthermore, $P = P_{u_p}(\gamma, \delta)$ is a path whose endvertices belong to $V(G) \setminus (Z \cup \{u_1, \ldots, u_{p-1}\})$ and, for the coloring $\varphi_1 = \varphi/P$, we have $\varphi_1 \in \mathcal{P}$, $\Gamma^f(T, e, \varphi_1) = \Gamma^f(T, e, \varphi)$, and $(\varphi_1, u, \gamma) \in J$.*

Proof of (e): In order to prove (1), assume that it is false. Since the sequence $F = (e_1, u_1, \ldots, e_p, u_p)$ is a fan at Z with respect to φ, we have $\varphi(e_1) \in \overline{\varphi}(Z)$. Therefore, for some $i \in \{2, \ldots, p\}$, there is a vertex $v \in Z \cup \{u_1, \ldots, u_{i-2}\}$ such that $\varphi(e_i) \in \overline{\varphi}(v)$. This, however, implies that the subfan $F' = F[e_p, \varphi]$ satisfies $|V(F')| < |V(F)|$. On the other hand, since $z \in Z$, $u_p \in V(F')$ and $\delta \in \overline{\varphi}(z) \cap \overline{\varphi}(u_p)$, we easily conclude that (T, e, φ, Y, F') is a bad tuple, a contradiction to (a). This proves (1).

Since the set $Z \cup \{u_1, \ldots, u_{p-1}\}$ is elementary with respect to φ (by (b)), statement (2) is an immediate consequence of (1).

For the proof of (3), let u be a vertex of T and let $\gamma \in \overline{\varphi}(u)$ be a color with $\gamma \neq \delta$. Assume that γ or δ belongs to $\Gamma^f(T, e, \varphi)$.

The vertex z belongs to $Z = V(T) \cup Y$ and, by (c), $Y \subseteq A(T, e, \varphi)$. Since, by hypothesis, $\delta \in \overline{\varphi}(z)$, $\gamma \in \overline{\varphi}(u)$, and γ or δ belongs to $\Gamma^f(T, e, \varphi)$, it follows from Proposition 5.13(c) that z belongs to $P_u = P_u(\gamma, \delta)$. Consequently, z and u are the endvertices of P_u. Since $\delta \in \overline{\varphi}(u_p)$ and $u_p \notin \{u, z\}$, this implies that $P = P_{u_p}(\gamma, \delta)$ is a path that is distinct from P_u. Then, by Proposition 5.13(d), every endvertex of P belongs to $W = V(G) \setminus (V(T) \cup A(T, e, \varphi))$. By (c), $W \subseteq V(G) \setminus Z$. Then the coloring $\varphi_1 = \varphi/P$ satisfies $\overline{\varphi}_1(Z) = \overline{\varphi}(Z)$ and $\gamma \in \overline{\varphi}_1(u_p)$. Furthermore, we deduce from Proposition 5.16 that $(T, e, \varphi_1) \in \mathcal{T}(G)$, $Y \subseteq D(T, e, \varphi) = D(T, e, \varphi_1)$, and $\Gamma^f(T, e, \varphi) = \Gamma^f(T, e, \varphi_1)$.

Eventually, we claim that F is a fan at Z with respect to φ_1. Since, by (b), the set $Z \cup \{u_1, \ldots, u_{p-1}\}$ is elementary with respect to φ, it follows from (e1) that $\gamma, \delta \notin \{\varphi(e_2), \ldots, \varphi(e_p)\}$. Since $\gamma, \delta \in \overline{\varphi}(Z)$, (b) implies that no endvertex of $P = P_{u_p}(\gamma, \delta)$ belongs to the set $\{u_1, \ldots, u_{p-1}\}$. Consequently, for $2 \leq i \leq p$, we have $\varphi_1(e_i) = \varphi(e_i)$ as well as $\overline{\varphi}_1(u_{i-1}) = \overline{\varphi}(u_{i-1})$. By (e1), this implies that $\varphi_1(e_i) \in \overline{\varphi}_1(u_{i-1})$ for $2 \leq i \leq p$. Furthermore, since $\gamma, \delta \in \overline{\varphi}(Z) = \varphi_1(Z)$, we obtain from (e1) that $\varphi_1(e_1) \in \overline{\varphi}_1(Z)$. This proves the claim that F is a fan at Z with respect to φ_1

Since $\gamma \in \overline{\varphi}_1(u) \cap \overline{\varphi}_1(u_p)$, it follows then that (T, e, φ_1, Y, F) is a bad tuple, and, therefore, $\varphi_1 \in \mathcal{P}$ and $(\varphi_1, u, \gamma) \in J$. This completes the proof of statement (3). ∎

Next, we claim that there is a triple $(\varphi, z, \delta) \in J$ such that $\delta \neq \varphi(e_1)$. Suppose this is false. To arrive at a contradiction, we first choose an arbitrary triple $(\varphi, z, \delta) \in J$. By Proposition 5.10(f), there is a free color $\gamma \in \Gamma^f(T, e, \varphi)$ that is distinct from $\delta = \varphi(e_1)$. Then there is a unique vertex u in T such that $\gamma \in \overline{\varphi}(u)$. From (e3) it follows that there is a coloring $\varphi_1 \in \mathcal{P}$ such that $(\varphi_1, u, \gamma) \in J$, where $\gamma \in \Gamma^f(T, e, \varphi) = \Gamma^f(T, e, \varphi_1)$ is a free color. If $\gamma \neq \varphi_1(e_1)$, we have a contradiction. Otherwise, $\gamma = \varphi_1(e_1)$ and we argue as follows.

Since $\varphi_1 \in \mathcal{P}$, the sequence $F = (e_1, u_1, \ldots, e_p, u_p)$ is a fan at Z with respect to φ_1 and, therefore, the edge e_1 has an endvertex $v \in Z$. By Proposition 5.10 (f), there is a color $\delta' \in \overline{\varphi}(v)$. Since $\gamma \in \overline{\varphi}_1(u)$ is a free color, it follows from (e3) that $P'_u = P_u(\gamma, \delta', \varphi_1)$ is a path whose endvertices are u and v. Since $\varphi_1(e_1) = \gamma$ and $e_1 \in E_G(v)$, the edge e_1 belongs to P'_u, too. Furthermore, by (e3), $P' = P_{u_p}(\gamma, \delta', \varphi_1)$ is a path whose endvertices belong to $V(G) \setminus (Z \cup \{u_1, \ldots, u_{p-1}\})$ and, for the coloring $\varphi_2 = \varphi_1/P'$, we have $\varphi_2 \in \mathcal{P}$, $\Gamma^f(T, e, \varphi_2) = \Gamma^f(T, e, \varphi_1)$, and $(\varphi_2, v, \delta') \in J$. Since the edge e_1 belongs to P'_u, this implies that $\varphi_2(e_1) = \varphi_1(e_1) = \gamma$ and, therefore, $\delta' \neq \varphi_2(e_1)$, a contradiction. This proves the claim that J contains a triple (φ, z, δ) such that $\delta \neq \varphi(e_1)$.

Now, let J' denote the set of all triples (φ, z, δ) such that $\delta \neq \varphi(e_1)$. Clearly, $J' \neq \emptyset$ and we claim that there is a triple $(\varphi, u, \gamma) \in J'$ such that $\gamma \in \Gamma^f(T, e, \varphi)$ is a free color. For the proof, consider an arbitrary triple $(\varphi, z, \delta) \in J'$. If $\delta \in \Gamma^f(T, e, \varphi)$ we are done. Otherwise, by Proposition 5.10 (f), there is a color $\gamma \in \Gamma^f(T, e, \varphi)$ such that γ is distinct from both δ and $\varphi(e_1)$. Then, by (e3), for the coloring $\varphi_1 = \varphi/P_{u_p}(\gamma, \delta)$, we have $\varphi_1 \in \mathcal{P}, \Gamma^f(T, e, \varphi_1) = \Gamma^f(T, e, \varphi)$, and $(\varphi_1, u, \gamma) \in J$. Since both γ and δ are distinct from $\varphi(e_1)$, this implies that $\varphi_1(e_1) = \varphi(e_1)$ and, therefore, $\gamma \neq \varphi_1(e_1)$. Hence $(\varphi_1, u, \gamma) \in J'$ and $\gamma \in \Gamma^f(T, e, \varphi_1)$. This proves the claim.

Finally, to reach a contradiction, we choose a triple $(\varphi, u, \gamma) \in J'$ such that $\gamma \in \Gamma^f(T, e, \varphi)$ is a free color. Let $\beta = \varphi(e_p)$. Then, by (e1), we have $\beta \in \overline{\varphi}(u_{p-1})$. From (e1) and (d) it follows that u_{p-1} belongs to $P_u = P_u(\gamma, \beta)$. Consequently, u_{p-1} and u are the endvertices of P_u. Since $\gamma \in \overline{\varphi}(u_p)$ and $u_p \notin \{u_{p-1}, u\}$, this implies that $P = P_{u_p}(\gamma, \beta)$ is a path that is distinct from P_u. Then, by Proposition 5.13(d), every endvertex of P belongs to $V(G) \backslash (V(T) \cup A(T, e, \varphi))$, and, therefore, to $V(G) \backslash Z$. Then the coloring $\varphi_1 = \varphi/P$ satisfies $\overline{\varphi}_1(Z) = \overline{\varphi}(Z)$ and $\beta \in \overline{\varphi}_1(u_p) \cap \overline{\varphi}_1(u_{p-1})$. Furthermore, it follows from Proposition 5.16 that $(T, e, \varphi_1) \in \mathcal{T}(G)$, $Y \subseteq D(T, e, \varphi) = D(T, e, \varphi_1)$, and $\Gamma^f(T, e, \varphi) = \Gamma^f(T, e, \varphi_1)$.

Now we claim that F is a fan at Z with respect to φ_1. Since $(\varphi, u, \gamma) \in J'$, we have $\gamma \neq \varphi(e_1)$. From (e2) it follows then that $\gamma \notin \{\varphi(e_1), \ldots, \varphi(e_p)\}$ and $\beta = \varphi(e_p) \notin \{\varphi(e_1), \ldots, \varphi(e_{p-1})\}$. Since u_{p-1} is an endvertex of P_u and $\beta \in \overline{\varphi}(u_{p-1})$, it follows from (b) that no endvertex of P belongs to the set $\{u_1, \ldots, u_{p-1}\}$. Consequently, for $1 \leq i \leq p-1$, we have $\varphi_1(e_i) = \varphi(e_i)$ as well as $\overline{\varphi}_1(u_i) = \overline{\varphi}(u_i)$. Furthermore, $\varphi_1(e_p) = \gamma$ and $\gamma \in \overline{\varphi}(Z) = \overline{\varphi}_1(Z)$. Then (e1) implies that F is a fan at Z with respect to φ_1.

Since $\beta \in \overline{\varphi}_1(u_p) \cap \overline{\varphi}_1(u_{p-1})$, it now follows that (T, e, φ_1, Y, F) is a bad tuple and, therefore, $\varphi_1 \in \mathcal{P}$. Moreover, on one hand, $F_1 = (e_1, u_1, \ldots, e_{p-1}, u_{p-1})$ is a fan at Z with respect to φ_1 such that $\varphi_1(e_i) \in \overline{\varphi}_1(u_{i-1})$ for $2 \leq i \leq p-1$ and $\varphi_1(e_1) \in \overline{\varphi}_1(Z)$. On the other hand, the sequence $F_2 = (e_p, u_p)$ is a fan at Z with respect to φ_1, where $\gamma = \varphi_1(e_p) \in \overline{\varphi}_1(Z)$. By Proposition 5.10(f), there is a free color $\gamma' \in \Gamma^f(T, e, \varphi_1)$ that is distinct from both γ and $\varphi_1(e_1)$. Then there is a unique vertex u' in T such that $\gamma' \in \overline{\varphi}_1(u')$. Since $\beta \in \overline{\varphi}_1(u_p) \cap \overline{\varphi}_1(u_{p-1})$, it follows from (d) that $P_{u'}(\gamma', \beta, \varphi_1)$ contains both vertices u_p and u_{p-1}. However, this is a contradiction.

Therefore, in both cases we arrive at a contradiction. Hence the proof of Theorem 5.18 is complete. ∎

5.3 ASYMPTOTIC BOUNDS

That the number of colors used by an algorithm based on Tashkinov trees as test objects is bounded from above by $\Delta + \mu$ follows from statement (b) in Theorem 5.3. However, if we use some of the methods developed in Sect. 5.2, a better bound for the number of colors can be obtained. This follows from the following result.

Theorem 5.19 *Let G be a critical graph with $\chi'(G) = k + 1$ for an integer $k \geq \Delta(G) + 1$ and let $(T^*, e, \varphi) \in \mathcal{T}(G)$. If G is not elementary, then the Tashkinov order of G satisfies*

$$2(k - \Delta(G)) + 3 \leq t(G) \leq \frac{\Delta(G) - 3}{k - \Delta(G)} - 1 \tag{5.4}$$

and, for every maximal Tashkinov tree T with respect to e and φ, the following statements hold:

(a) $V(T) = V(T^*)$, $t(G) = |V(T)|$ *is odd and* ≥ 3.

(b) $V(T)$ *is elementary and closed with respect to* φ, *but not strongly closed.*

(c) *There is a vertex* $v \in V(T)$ *such that each color in* $\overline{\varphi}(v)$ *is used on* T *with respect to* φ, *and at least* $k - \Delta(G) + 1$ *colors are used on* T *with respect to* φ.

Proof: From Proposition 5.10 it follows that $|V(T^*)| = t(G) \geq 3$, $t(G)$ is odd, and $V(T^*)$ is elementary and closed with respect to φ. Let $T = (y_0, e_1, \ldots, e_p, y_p)$ be an arbitrary maximal Tashkinov tree with respect to e and φ. From Theorem 5.3 we conclude that $V(T) = V(T^*)$ and, therefore, $(T, e, \varphi) \in \mathcal{T}(G)$. Since G is not elementary, Theorem 1.4 implies that $V(T)$ is not strongly closed with respect to φ. Hence (a) and (b) are proved.

By Proposition 5.10, there is a defective color $\alpha \in \Gamma^d(T, e, \varphi)$ as well as a free color $\gamma \in \Gamma^f(T, e, \varphi)$. Since $V(T)$ is elementary with respect to φ, there is a unique vertex $u \in V(T)$ such that $\gamma \in \overline{\varphi}(u)$.

Consider the (α, γ)-chain $P = P_u(\alpha, \gamma, \varphi)$. By Proposition 5.11, we have $E_\alpha(T, e, \varphi) = E(P) \cap \partial_G(V(T))$ and, moreover, in the linear order $\preceq_{(u,P)}$ there is a first vertex v^1 that belongs to $V(G) \setminus V(T)$ and there is a last vertex v^2 that belongs to V(T), where $v^1 \preceq_{(u,P)} v^2$.

First, Proposition 5.11 implies $\overline{\varphi}(v^2) \cap \Gamma^f(T, e, \varphi) = \emptyset$. Hence every color of $\overline{\varphi}(v^2)$ is used on T with respect to φ. If $v^2 \in \{y_0, y_1\}$ we have $|\overline{\varphi}(v^2)| = k - d_G(v^2) + 1 \geq k - \Delta(G) + 1$, and so at least $k - \Delta(G) + 1$ colors are used on T with respect to φ. Otherwise, we have $v^2 = y_j$ for some $j \in \{2, \ldots, p\}$ and hence $|\overline{\varphi}(v^2)| = k - d_G(v^2) \geq k - \Delta(G)$. Then the color $\beta = \varphi(e_j)$ is used on T with respect to φ. Since $e_j \in E_G(v^2)$, $\beta \notin \overline{\varphi}(v^2)$. Hence at least $k - \Delta(G) + 1$ colors are used on T with respect to φ. Consequently, in both cases, at least $k - \Delta(G) + 1$ colors are used on T with respect to φ. This proves (c).

Second, Proposition 5.11 implies that in the linear order $\preceq_{(u,P)}$ the vertex v^1 has a successor w^1, where $w^1 \in V(G) \setminus V(T)$. Furthermore, it follows that $X = V(T) \cup \{v^1, w^1\}$ is elementary with respect to φ. Since $\alpha \in \Gamma^d(T, e, \varphi)$, Proposition 5.10 implies $\alpha \notin \overline{\varphi}(V(T))$. By Proposition 5.11, w^1 is not an endvertex of P and, therefore, $\alpha \notin \overline{\varphi}(v^1, w^1)$. Hence α is present at any vertex of X with respect to φ and so $|\overline{\varphi}(X)| \leq k - 1$. Since X is elementary with respect to φ and since $y_0, y_1 \in X$, we conclude from Proposition 5.10(f) that

$$|\overline{\varphi}(X)| = 2 + \sum_{y \in X} (k - d_G(y)) \geq 2 + |X|(k - \Delta(G)).$$

Since $|X| = |V(T)| + 2$, we then obtain

$$t(G) = |V(T)| \leq \frac{k - 3}{k - \Delta(G)} - 2 = \frac{\Delta(G) - 3}{k - \Delta(G)} - 1.$$

From Theorem 5.4 it follows that we may choose a maximal Tashkinov tree T with respect to e and φ such that at most $\frac{|V(T)| - 1}{2}$ colors are used on T with respect to φ. By (c), this implies

$$k - \Delta + 1 \leq \frac{|V(T)| - 1}{2}$$

and hence

$$2(k - \Delta(G)) + 3 \leq |V(T)| = t(G).$$

Thus the proof of the theorem is complete. ∎

From Theorem 5.19 we can deduce the following two upper bounds for the chromatic index. The first upper bound is a graph parameter τ defined by

$$\tau(G) = \max\{\Delta(G) + \sqrt{(\Delta(G) - 1)/2}, \, w(G)\}$$

if $G \neq \emptyset$ and $\tau(G) = 0$ otherwise. The second upper bound is an ε-dependent graph parameter τ_ε with $\varepsilon > 0$ defined by

$$\tau_\varepsilon(G) = \max\left\{(1 + \varepsilon)\Delta(G) + 1 - 3\varepsilon, \, \Delta(G) - 1 + \frac{1}{2\varepsilon}, \, w(G)\right\}$$

if $G \neq \emptyset$ and $\tau_\varepsilon(G) = 0$ otherwise.

Corollary 5.20 *Every graph G satisfies $\chi'(G) \leq \tau(G)$.*

Proof: Since both parameters are monotone, it suffices to show that every critical graph G satisfies $\chi'(G) \leq \tau(G)$ (Proposition 1.1). If $\Delta(G) \leq 2$, then $\chi'(G) = w(G) \leq \tau(G)$ and we are done. Now, assume that $\Delta(G) \geq 3$. If $\chi'(G) \leq \Delta(G) + 1$, then $\chi'(G) \leq \tau(G)$ and we are done, too. Otherwise, $\chi'(G) = k + 1$ for some integer $k \geq \Delta(G) + 1$ and we argue as follows. Since G is critical, Theorem 5.19 implies $\chi'(G) = w(G)$ or

$$2(k - \Delta(G)) + 3 \leq \frac{\Delta(G) - 3}{k - \Delta(G)} - 1.$$

The last inequality is equivalent to

$$(k - \Delta(G))^2 + 2(k - \Delta(G)) \leq \frac{1}{2}(\Delta(G) - 3),$$

which implies that

$$\chi'(G) = k + 1 \leq \Delta(G) + \sqrt{(\Delta(G) - 1)/2}.$$

Hence, in both cases we obtain $\chi'(G) \leq \tau(G)$. This completes the proof. ∎

Corollary 5.21 *For every $\varepsilon > 0$ any graph G satisfies $\chi'(G) \leq \tau_\varepsilon(G)$.*

Proof: Let $\varepsilon > 0$ be a real number. Since both parameters are monotone, it suffices to show that every critical graph G satisfies $\chi'(G) \leq \tau_\varepsilon(G)$ (Proposition 1.1).

If $\Delta(G) \leq 2$, then $\chi'(G) = w(G) \leq \tau_\varepsilon(G)$ and we are done. So assume $\Delta(G) \geq 3$. Then $\tau_\varepsilon(G) \geq \Delta(G) + 1$ and we need only to consider the case that

$\chi'(G) = k + 1$ for some integer $k \geq \Delta(G) + 1$. From Theorem 5.19 it then follows that $\chi'(G) = w(G)$ or

$$2(k - \Delta(G)) + 3 \leq \frac{\Delta(G) - 3}{k - \Delta(G)} - 1.$$

In the former case, we are obviously done. In the latter case, we distinguish two subcases. If $2(k-\Delta(G))+3 \leq \frac{1}{\varepsilon}-1$, then $\chi'(G) = k+1 \leq \Delta(G)-1+\frac{1}{2\varepsilon} \leq \tau_\varepsilon(G)$ and we are done. Otherwise, we have

$$\frac{\Delta(G) - 3}{k - \Delta(G)} > \frac{1}{\varepsilon}.$$

Since $\Delta(G) \geq 3$, this implies that $\chi'(G) = k + 1 \leq (1 + \varepsilon)\Delta(G) + 1 - 3\varepsilon$ and we are done, too. Thus the proof is complete. ∎

Haxell and McDonald [132] established an upper bound for $\chi'(G)$ in terms of the maximum degree $\Delta(G)$, the density $w(G)$, and the maximum multiplicity $\mu(G)$. First, we need a result similar to Theorem 5.19.

Theorem 5.22 (Haxell and McDonald [132] 2011) *Let G be a critical graph with* $\chi'(G) = k + 1$ *for an integer* $k \geq \Delta(G) + 1$. *If G is not elementary, then*

$$t(G) > \left(1 + \frac{k - \Delta(G)}{2\mu(G)}\right)^{k-\Delta(G)} + 1. \tag{5.5}$$

Proof: There is certainly a triple $(T^*, e, \varphi) \in \mathcal{T}(G)$. By Theorem 5.19, $t(G) = |V(T^*)| \geq 3$, $t(G)$ is odd, and $V(T^*)$ is elementary and closed with respect to φ, but not strongly closed with respect to φ. Furthermore, every Tashkinov tree T with respect to e and φ satisfies $V(T) \subseteq V(T^*)$ with equality if and only if T is a maximal Tashkinov tree. In order to show that (5.5) holds, we introduce the following notation.

If T is a Tashkinov tree with respect to e and φ, then let $\Gamma^u(T)$ denote the set of all colors used on T with respect to φ, and let

$$m(T) = \max\{|\overline\varphi(v) \cap \Gamma^u(T)| \mid v \in V(T)\}.$$

Furthermore, we define $V^m(T) = \{v \in V(T) \mid |\overline\varphi(v) \cap \Gamma^u(T)| = m(T)\}$ and $W^m(T) = V(T) \setminus V^m(T)$.

If T is a maximal Tashkinov tree with respect to e and φ, then Theorem 5.19(c) implies that there is a vertex $v \in V(T)$ such that $\overline\varphi(v) \subseteq \Gamma^u(T)$ and, therefore, $m(T) \geq |\overline\varphi(v)| \geq k - \Delta(G)$.

Let T be a Tashkinov tree with respect to e and φ, and let $e' \in \partial_G(V(T))$. Then we denote by $v(e', T)$ the endvertex of e' belonging to $V(G) \setminus V(T)$. Furthermore, we denote the sequence $T' = (T, e', v(e', T))$ by $T + e'$. If $\varphi(e') \in \overline\varphi(V(T))$, then $T + e'$ is a Tashkinov tree with respect to e and φ and, moreover, $\Gamma^u(T + e') = \Gamma^u(T) \cup \{\varphi(e')\}$. Furthermore, $V(T + e')$ is elementary with respect to φ and from

(T2) it follows that $\overline{\varphi}(v(e')) \cap \Gamma^u(T + e') = \emptyset$. Hence, either $m(T + e') = m(T)$, or $m(T + e') = m(T) + 1$ and $\varphi(e') \in \overline{\varphi}(V^m(T)) \cap (\{1, \ldots, k\} \setminus \Gamma^u(T))$.

By an ℓ-*tree* we mean a Tashkinov tree T with respect to e and φ such that $m(T) = \ell$ and $|\Gamma^u(T)| \leq \lceil|V(T)| - 1\rceil/2$. First, we claim that any maximal ℓ-tree T with $\ell \geq 1$ has an odd number of vertices. Otherwise, $|V(T)| \geq 4$ is even and there is a color $\alpha \in \Gamma^u(T)$. Since $V(T)$ is elementary with respect to φ, there is an edge $e' \in \partial_G(V(T))$ such that $\varphi(e') = \alpha$. Then $T + e'$ is an ℓ-tree, contradicting the choice of T. This proves the claim.

Now, we prove by induction on ℓ ($1 \leq \ell \leq k - \Delta(G)$) that there exists an ℓ-tree with at least n_ℓ vertices, where

$$n_\ell = 3 \left(1 + \frac{k - \Delta(G)}{2\mu(G)}\right)^{\ell-1}.$$

If $T^* = (y_0, e_1, y_1, \ldots, e_{p-1}, y_{p-1})$, then $p = t(G) \geq 3$ and, obviously, $T = (y_0, e_1, y_1, e_2, y_2)$ is a 1-tree and $|V(T)| = 3 = n_1$.

So suppose $1 \leq \ell \leq k - \Delta(G) - 1$ and there exists an ℓ-tree with at least n_ℓ vertices. Then there is a maximal ℓ-tree T with $|V(T)| \geq n_\ell$. This implies that $|V(T)|$ is odd. Since any maximal Tashkinov tree T' with respect to e and φ satisfies $m(T') \geq k - \Delta(G) > \ell = m(T)$, we conclude that T is not a maximal Tashkinov tree with respect to e and φ. Hence, $V(T)$ is not closed with respect to φ and, therefore, there is an edge $e_1 \in \partial_G(V(T))$ such that $\varphi(e_1) \in \overline{\varphi}(V(T))$. Consequently, $T_1 = T + e_1$ is a Tashkinov tree with respect to e and φ. Since $|V(T)|$ is odd, we obtain that $|\Gamma^u(T_1)| \leq \lceil|V(T_1)|-1\rceil/2$. Then $m(T_1) = m(T)+1 = \ell+1$, since otherwise T_1 would be a larger ℓ-tree. Consequently, T_1 is an $(\ell + 1)$-tree and we can extend this tree to a maximal $(\ell + 1)$-tree T'. This implies that $|V(T')|$ is odd. To complete the induction step, it suffices to show that $|V(T')| \geq n_{\ell+1}$.

Let $v_1 = v(e_1, T)$ and let α be an arbitrary color of the set $\Gamma = \overline{\varphi}(W^m(T))$. Since $V(T_1)$ is elementary with respect to φ, there is an edge $e_\alpha \in E_G(v_1)$ and $e_\alpha \notin \partial_G(V(T))$, since otherwise $T_2 = T + e_\alpha$ would be a larger ℓ-tree. Hence, there is a vertex $v_\alpha \in V(G) \setminus V(T_1)$ such that $e_\alpha \in E_G(v_1, v_\alpha)$. Next, we claim that $v_\alpha \in V(T')$. Suppose this is false. Since T' is an $(\ell + 1)$-tree and $|V(T')|$ is odd, this implies that $T' + e_\alpha$ is also an $(\ell + 1)$-tree, contradicting the choice of T'. This proves the claim. Consequently, each vertex of the set $X = \{v_\alpha \mid \alpha \in \Gamma\}$ is contained in $V(T')$ and, therefore, $|V(T')| \geq |V(T)| + |X| \geq n_\ell + |X|$. Since $V(T)$ is elementary with respect to φ, we obtain that

$$|X| \geq \frac{|\Gamma|}{\mu(G)} \geq \frac{|W^m(T)|(k - \Delta(G))|}{\mu(G)} = \frac{(|V(T)| - |V^m(T)|)(k - \Delta(G))}{\mu(G)}.$$

Furthermore, since each vertex $v \in V^m(T)$ satisfies $|\overline{\varphi}(v) \cap \Gamma^u(T)| = m(T) = \ell \geq 1$, we conclude that $|V^m(T)| \leq |\Gamma^u(T)|/\ell \leq |\Gamma^u(T)|$. Since $|\Gamma^u(T)| \leq \lceil|V(T)| - 1\rceil/2 \leq |V(T)|/2$, we eventually obtain from the above inequality that

$$|X| \geq \frac{|V(T)|(k - \Delta(G))}{2\mu(G)} \geq \frac{n_\ell(k - \Delta(G))}{2\mu(G)},$$

which implies that

$$|V(T')| \geq n_\ell + |X| \geq n_\ell \left(1 + \frac{k - \Delta(G)}{2\mu(G)}\right) = n_{\ell+1}.$$

This completes the proof of the induction step.

As a consequence, we obtain that $t(G) \geq n_{k-\Delta(G)}$. From Vizing's bound we deduce $1 \leq k - \Delta(G) = \chi'(G) - 1 - \Delta(G) \leq \mu(G) - 1$, and hence

$$1 < \left(1 + \frac{k - \Delta(G)}{2\mu(G)}\right) < 2.$$

Consequently, we obtain that

$$t(G) \geq 3 \left(1 + \frac{k - \Delta(G)}{2\mu(G)}\right)^{k-\Delta(G)-1} > \left(1 + \frac{k - \Delta(G)}{2\mu(G)}\right)^{k-\Delta(G)} + 1.$$

This completes the proof.   ∎

Corollary 5.23 (Haxell and McDonald [132] 2011) *Every nonempty graph G satisfies*

$$\chi'(G) \leq \max\{w(G), \Delta(G) + 2\sqrt{\mu(G) \ln \Delta(G)}\}.$$

Proof: By Proposition 1.1, it suffices to establish the statement for any critical graph $G \neq \emptyset$. If $\chi'(G) = w(G)$ or $\chi'(G) \leq \Delta(G) + 1$, then the statement is true. So we may assume that G is not elementary and $\chi'(G) = k + 1$ for some integer $k \geq \Delta(G) + 1$. From Theorem 5.19 and Theorem 5.22 we then deduce

$$\left(1 + \frac{k - \Delta(G)}{2\mu(G)}\right)^{k-\Delta(G)} + 1 < \frac{\Delta(G) - 3}{k - \Delta(G)} - 1.$$

Since Vizing's bound implies that $t = k - \Delta(G) = \chi'(G) - 1 - \Delta(G)$ satisfies $0 < t/(2\mu(G)) < 1/2$, from the above inequality we obtain

$$e^{\frac{t^2}{4\mu(G)}} < \left(1 + \frac{t}{2\mu(G)}\right)^t < \frac{\Delta(G)}{t}.$$

Since $t \geq 1$, this yields

$$\frac{t^2}{4\mu(G)} < \ln(\Delta(G)) - \ln(t) \leq \ln(\Delta(G)),$$

and hence $t < 2\sqrt{\mu(G) \ln(\Delta(G))}$ as required.   ∎

5.4 TASHKINOV'S COLORING ALGORITHM

The proof that both parameters τ and τ_ε are upper bounds for the chromatic index uses critical graphs as explained in Proposition 1.1 and Theorem 5.19. The proof of Theorem 5.19 is based on Theorem 5.3, Proposition 5.10, and Proposition 5.11. However, the proofs of all these results are constructive and can be transformed into an algorithm, showing that both parameters are efficiently realizable upper bounds for the chromatic index χ'. A proof of the next result was first given by Scheide [269] in his doctoral thesis.

Theorem 5.24 *Let $\varepsilon > 0$ be a real number. Both parameters τ and τ_ε are efficiently realizable upper bounds for the chromatic index χ'.*

Sketch of Proof: For the proof, we design a coloring algorithm **TasCol**. The input of this algorithm is a graph G and its output is a pair (k, φ) such that $\varphi \in C^k(G)$.

TasCol(G):

1. **If** $\Delta(G) \leq 2$ **then** compute an optimal coloring φ of G
   and **return** (k, φ) with $k = |\varphi(E(G))|$.

2. Let G' be the edgeless graph with $V(G') = V(G)$,
   let φ be the empty coloring of G', and let $k = \Delta(G) + 1$.

3. **For** every edge $e \in E(G)$ **do**

    (a) Let x, y be the two endvertices of e.

    (b) $E(G') \leftarrow E(G') \cup \{e\}$.

    (c) $(k, \varphi) \leftarrow$ **TasExt**$(G', e, x, y, k, \varphi)$.

4. **Return** (k, φ).

End

The core of the algorithm **TasCol** is a subroutine **TasExt** that extends a given partial coloring of the input graph G. The input of **TasExt** is a tuple $(G', e, x, y, k, \varphi)$, where G' is a graph, $e \in E_{G'}(x, y)$, and $\varphi \in C^k(G' - e)$. The graph G' is a subgraph of the input graph G of **TasCol** consisting of all edges that are already colored and the next uncolored edge e. Obviously, we have $k \geq \Delta(G') + 1$. The output of **TasExt** is a pair (k', φ'), where $k' \in \{k, k + 1\}$ and $\varphi' \in C^{k'}(G')$. For the input $(G', e, x, y, k, \varphi)$, the subroutine computes a Tashkinov tree T with respect to the edge e and the coloring $\varphi \in C^k(G' - e)$. If $V(T)$ is elementary and closed both with respect to φ, we write $(T, e, \varphi) \in \mathcal{T}^*$. For a triple $(T, e, \varphi) \in \mathcal{T}^*$, we then adopt the notation introduced at the beginning of Sect. 5.2 for the triples in $\mathcal{T}(G)$: defective colors, free colors, and so on.

TasExt$(G', e, x, y, k, \varphi)$:

1. $p \leftarrow 1, e_p \leftarrow e, y_p \leftarrow y, y_0 \leftarrow x, T \leftarrow (y_0, e_p, y_p)$.

2. **If** $\overline{\varphi}(V(Ty_{p-1})) \cap \overline{\varphi}(y_p) \neq \emptyset$ **then**

    (a) Compute $\varphi' \in C^k(G')$ as in Theorem 5.1.

    (b) **Return** (k, φ').

3. **If** $\exists e_{p+1} \in \overline{\varphi}(V(T)) : \varphi(e_{p+1}) \in \varphi(E(T))$ **then**

   (a) Let y_{p+1} be the endvertex of e_{p+1} that is not in $V(T)$.

   (b) $T \leftarrow (T, e_{p+1}, y_{p+1}), p \leftarrow p + 1$.

   (c) **Goto 2**

4. **If** $\exists e_{p+1} \in \partial_G(V(T)) : \varphi(e_{p+1}) \in \overline{\varphi}(V(T)) \setminus \varphi(E(T))$ **then**

   (a) Let y_{p+1} be the endvertex of e_{p+1} that is not in $V(T)$.

   (b) $T \leftarrow (T, e_{p+1}, y_{p+1}), p \leftarrow p + 1$.

   (c) **Goto 2.**

5. **If** $\Gamma^d(T, e, \varphi) = \emptyset$ **then**

   (a) $\varphi' \leftarrow \varphi, \varphi'(e) \leftarrow k + 1$.

   (b) **Return** $(k + 1, \varphi')$.

6. Choose $\alpha \in \Gamma^d(T, e, \varphi)$ and $\gamma \in \Gamma^f(T, e, \varphi)$.
   Let $u \in V(T)$ with $\gamma \in \overline{\varphi}(u)$ and $P \leftarrow P_u(\alpha, \gamma, \varphi)$.

7. **If** $E_\alpha(T, e, \varphi) \nsubseteq E(P)$ **then**

   (a) Compute $\varphi' \in \mathcal{C}^k(G' - e)$ according to Proposition 5.11(b),
   i.e., $\varphi' = \varphi/P$.

   (b) $\varphi \leftarrow \varphi'$.

   (c) **Goto 3.**

8. Set v^1, v^2 according to Proposition 5.11(c) and let w^1 be the
   neighbor of v^1 in P that belongs to $V(G) \setminus V(T)$ as in
   the proof of Theorem 5.19.

9. **If** $\overline{\varphi}(v^2) \cap \Gamma^f(T, e, \varphi) \neq \emptyset$ **then**

   (a) Compute $\varphi' \in \mathcal{C}^k(G' - e)$ according to Proposition 5.11(d).

   (b) $\varphi \leftarrow \varphi'$.

   (c) **Goto 3.**

10. **If** $V(T) \cup \{v^1, w^1\}$ is not elementary with respect to φ **then**

    (a) Compute $\varphi' \in \mathcal{C}^k(G' - e)$ according to Proposition 5.11(e).

    (b) $\varphi \leftarrow \varphi'$.

    (c) **Goto 3.**

11. $\varphi' \leftarrow \varphi, \varphi'(e) \leftarrow k + 1$.

12. **Return** $(k + 1, \varphi')$.

End

First, let us analyze the subroutine **TasExt**. We will refer to all variables as they are valued in the current state of the algorithm, not as they are valued in the input $(G', e, x, y, k, \varphi)$. It is not difficult to check that the triple (T, e, φ) satisfy the following conditions:

- When entering step 2, we have $p \geq 1$ and $T = (y_0, e_1, y_1, \ldots, e_p, y_p)$ is a Tashkinov tree with respect to e and $\varphi \in \mathcal{C}^k(G' - e)$ such that $V(Ty_{p-1})$ is elementary with respect to φ. Then we check whether $V(T)$ is elementary with respect to φ or not. If $V(T)$ is not elementary with respect to φ, then according to the proof of Theorem 5.1 we can compute a coloring $\varphi' \in \mathcal{C}^k(G')$ and **TasExt** returns (k, φ'). Otherwise, we go to step 3.

- When entering step 3 or 4, T is a Tashkinov tree with respect to e and φ such that $V(T)$ is elementary with respect to φ.

- When entering step 5, T is a Tashkinov tree with respect to e and φ such that $V(T)$ is elementary and closed both with respect to φ, i.e., $(T, e, \varphi) \in \mathcal{T}^*$. Then $|V(T)| \geq 3$ is odd and (see the proof of Theorem 5.4) at most $\frac{|V(T)|-1}{2}$ colors are used on T with respect to φ. If the set of defective colors $\Gamma^{\mathrm{d}}(T, e, \varphi)$ is empty, then $V(T)$ is also strongly closed with respect to φ and **TasExt** returns $(k + 1, \varphi')$, where φ' is the coloring obtained from φ by assigning the new color $k + 1$ to φ.

- When entering step 7, $(T, e, \varphi) \in \mathcal{T}^*$, α is a defective color, γ is a free color, u is the unique vertex in T with $\gamma \in \overline{\varphi}(u)$, and $P = P_u(\alpha, \gamma, \varphi)$. If $E_\alpha(T, e, \varphi) \not\subseteq E(P)$, then, see the proof Proposition 5.11(b), T remains a Tashkinov tree with respect to e and the coloring $\varphi' = \varphi/P$ and, moreover, $V(T)$ is elementary with respect to φ', but not closed with respect to φ'. Furthermore, $\varphi \leftarrow \varphi'$ and we go to step 3.

- When entering step 9 or step 10, $(T, e, \varphi) \in \mathcal{T}^*$, α is a defective color γ is a free color, u is the unique vertex in T with $\gamma \in \overline{\varphi}(u)$, and $P = P_u(\alpha, \gamma, \varphi)$. Furthermore, $E_\alpha(T, e, \varphi) \subseteq E(P)$, vertices v^1, v^2 are chosen according to Proposition 5.11(c), and w^1 is the neighbor of v^1 in P that belongs to $V(G) \setminus V(T)$ as in the proof of Theorem 5.19.

- In step 9, we check whether the set $\varphi(v^2)$ contains a free color with respect to (T, e, φ) or not. If there exists such a free color, then according to the proof of Proposition 5.11(d) we compute a coloring $\varphi' \in \mathcal{C}^k(G' - e)$ such that T remains a Tashkinov tree with respect to e and φ and, moreover, $V(T)$ is elementary with respect to φ', but not closed with respect to φ'. Furthermore, $\varphi \leftarrow \varphi'$ and we go to step 3. If there is no such free color, we go to step 10 and check whether $X = V(T) \cup \{v^1, w^1\}$ is elementary with respect to φ or not. If X is not elementary, then according to the proof of Proposition 5.11(e), we compute a coloring $\varphi' \in \mathcal{C}^k(G' - e)$ such that T remains a Tashkinov tree with respect to e and φ and, moreover, $V(T)$ is elementary with respect to φ', but not closed with respect to φ'. Furthermore, $\varphi \leftarrow \varphi'$ and we go to step 3. If X is elementary with respect to φ, then **TasExt** returns $(k + 1, \varphi')$, where φ' is the coloring obtained from φ by assigning the new color $k + 1$ to φ.

Hence, the subroutine **TasExt** works correctly provided it terminates. Apart from the fact that the subroutine may jump to step 2 or step 3, there are no other loops. If the subroutine jumps to step 2, then it came either from step 3c or from step 4c. In both cases the value of p was increased before. Since p is never decreased and $p + 1 = |V(T)| \leq |V(G')|$, this implies that there is only a finite number of jumps to step 2. If **TasExt** jumps to step 3, then it came from step 7c, 9c, or 10c. In all these cases, see above, the set $V(T)$ is elementary with respect to φ, but not closed with respect to φ and, therefore, the checks in step 3 and 4 cannot both fail. Then the algorithm has to jump back to step 2 again, but this can happen only a finite number of times. Consequently, the algorithm **TasExt** has to terminate at some point and the output is a pair (k', φ') such that $k' \in \{k, k+1\}$ and $\varphi' \in \mathcal{C}_{k'}(G')$.

If $k' = k + 1$, then the subroutine terminates either in step 5b or in step 12. In the former case, $V(T)$ is elementary and strongly closed with respect to φ and, therefore, $k + 1 = w(G')$. In the latter case, we have $(T, e, \varphi) \in \mathcal{T}^*$, $\overline{\varphi}(v^2) \cap \Gamma^f(T, e, \varphi) = \emptyset$, $X = V(T) \cup \{v^1, w^1\}$ is elementary with respect to φ, and at most $\frac{|V(T)|-1}{2}$ colors are used on T with respect to φ. Then (see the proof of Theorem 5.19) we obtain

$$2(k - \Delta(G')) + 3 \leq |V(T)| \leq \frac{\Delta(G') - 3}{k - \Delta(G')} - 1$$

Since $k \geq \Delta(G') + 1$, this implies that $\Delta(G') \geq 3$. Then, see the proofs of Corollary 5.20 and Corollary 5.21, we obtain $k + 1 \leq \tau(G')$ as well as $k + 1 \leq \tau_\varepsilon(G')$.

Now, let us analyze the coloring algorithm **TasCol**. The input is a graph G and the output is an edge coloring φ of G using $A(G)$ colors. We claim that $A(G) \leq \tau(G)$ and $A(G) \leq \tau_\varepsilon(G)$. If $\Delta(G) \leq 2$, then **TasCol** computes an optimal edge coloring of G and, therefore, $A(G) = \chi'(G) = w(G)$. Hence we are done. If $\Delta(G) \geq 3$, then we start the coloring algorithm with $k = \Delta(G) + 1$ colors. If the algorithm never uses a new color, then $A(G) = \Delta(G) + 1$ and, since $\Delta(G) \geq 3$, we have $\Delta(G) + 1 \leq \tau(G)$ and $\Delta(G) + 1 \leq \tau_\varepsilon(G)$. Hence we are done. Otherwise, let us consider the last call of **Ext** where a new color is used. The input is a tuple $(G', e, x, y, k, \varphi)$ where G' is a subgraph of G, $e \in E_{G'}(x, y)$, and $\varphi \in \mathcal{C}^k(G' - e)$. The output is a pair (k', φ') such that $k' = k+1$ and $\varphi' \in \mathcal{C}^{k'}(G')$. Then $A(G) = k+1$ and, see above, $k + 1 \leq \tau(G') \leq \tau(G)$ and $k + 1 \leq \tau_\varepsilon(G') \leq \tau_\varepsilon(G)$. Hence we are done, too.

If we take into account Theorem 5.2 and the implementation details discussed at the end of Sect. 1.5, it is not difficult to show that the time complexity of **TasCol** is bounded from above by a polynomial in $|E(G)|$ and $|V(G)|$; the time complexity is $O(|E(G)|(|V(G)| + \Delta(G))|V(G)|^3)$. ∎

5.5 POLYNOMIAL TIME ALGORITHMS

All previously considered edge coloring algorithms have an execution time that is bounded from above by a polynomial depending on the number of vertices and the number of edges of the input graph. But a graph G is, up to isomorphisms, completely determined by its vertex set V and its edge multiplicities $\mu_G(x, y) = |E_G(x, y)|$ for all vertex pairs (x, y). Hence the size of the graph representation may be of order

$|V(G)|^2 \log \mu(G)$ and, therefore, the previous coloring algorithms are only pseudo-polynomial. The running time of a real polynomial time algorithm has to be bounded by a polynomial in $|V(G)|$ and $\log \mu(G)$. This also implies that we cannot color the edges one-by-one as was done before. Using two concepts from Sanders and Steurer [262], we will develop a polynomial time edge coloring algorithm based on the algorithm TasExt. The first concept is an elegant data storing method such that the running time of the standard recoloring operations is bounded by a polynomial depending only of the order of the input graph, provided that the size of the stored data is sufficiently small. The second concept is a divide-and-conquer strategy that provides a partial edge coloring of a graph G with at most $\max\{\Delta(G), w(G)\}$ colors such that the number of uncolored edges is sufficiently small and such that the storage space needed is sufficiently small.

For a set X, let $X^{(2)}$ be the set of all subsets of X with two elements. A graph $G = (V, \mu)$ is then defined to be a pair consisting of a vertex set V and a multiplicity function $\mu : V^{(2)} \to \mathbb{N}_0$. Then the representation of a graph G has bounded size, from above as well as from below, by polynomials in $|V(G)|$ and $\log \mu(G)$.

Let $G = (V, \mu)$ be a graph. A **partial k-edge-coloring** φ of G is a k-edge-coloring of a subgraph of G, that is, $\varphi : V^{(2)} \to 2^{\{1,\dots,k\}}$ is a map satisfying $|\varphi(e)| \leq \mu(e)$ for all $e \in V^{(2)}$ and $\varphi(e) \cap \varphi(e') = \emptyset$ for all $e, e' \in V^{(2)}$ with $|e \cap e'| = 1$. So the map φ may be viewed as a k-edge-coloring of the subgraph $G' = (V, \mu')$ with $\mu'(e) = |\varphi(e)|$ for all $e \in V^{(2)}$.

Let $G = (V, \mu)$ be a graph, and let φ be a k-edge-coloring of the graph G' obtained from G by deleting an edge from $\{x, y\}$. We can now easily adapt all concepts, like multi-fans, Kierstead paths, or Tashkinov trees to the new graph model. Clearly, we may still think of a Tashkinov tree T with respect to an edge $e \in E_G(x, y)$ and φ as sequence $T = (y_0, e_1, y_1, \dots, e_p, y_p)$ of vertices $y_0, \dots, y_p$ and edges $e_1, \dots, e_p$ satisfying the conditions (T1) and (T2).

To obtain a better performance of our coloring algorithms it is not enough to consider a graph with multiple edges as a pair consisting of a vertex set and a multiplicity function. This graph model, however, gives us the possibility to introduce more efficient data structures. Consider a partial k-edge coloring φ of G. For each color $\alpha \in \{1, \dots, k\}$, we have the color class $E_\alpha(\varphi) = \{e \in V^{(2)} \mid \alpha \in \varphi(e)\}$. Following Sanders and Steurer [262], we can contract consecutive colors with the same color class to **color intervals**. For such a color interval I, let $E_I(\varphi) = E_\alpha(\varphi)$ for some $\alpha \in I$. For every color interval I, a **same-color list** is stored, containing all edges of $E_I(\varphi)$. Every element of a same-color list also contains a pointer to the corresponding color interval. Since there is a one-to-one correspondence between the color classes and the matchings of G, we have $|E_I(\varphi)| \leq \frac{n}{2}$, where $n = |V(G)|$. Note that all this is similar to the same-color list described in Sect. 1.5, now only for color intervals instead of single colors. Since the color intervals are pairwise disjoint, we may order them and store them as a sorted list $L(\varphi)$. In order to store a partial edge coloring φ of G we need space $O(n \cdot \ell)$, where $n = |V(G)|$ and $\ell = |L(\varphi)|$ is the number of color intervals.

The coloring algorithm is now working on color intervals rather than on single colors. If such an algorithm uses a set of colors (e.g., missing colors, free colors, defective colors), this set is represented as an ordered list of color intervals, which can be computed from the ordered list $L(\varphi)$ in time $O(\ell)$.

If an algorithm performs a typical recoloring operation, like a Kempe change or coloring an edge, one or more colors from given intervals are needed. If the algorithm chooses a color from a given color interval, this interval must be split, where the splitting operation also includes copying the same-color list of the corresponding interval, which can be done time $O(n)$. Observe that each such a recoloring operation increases the number of color intervals only by a constant. After splitting the involved color intervals, the time complexity of such a recoloring operation remains the same as in the usual implementation. Although an algorithm works on color intervals, we will still use the notation of special colors when formulating the algorithm.

Let **TasExtI** be the version of **TasExt** using the new data structures and techniques. That is, **TasExtI** works on an input (G', x, y, k, φ) that is valid if $G' = (V, \mu)$ is a graph, $\mu(x, y) > 0$, $k \geq \Delta(G') + 1$, and φ is a k-edge coloring of the graph obtained from G' by deleting one edge between x and y. On such a valid input, **TasExtI** returns a pair (k', φ') where $k' \in \{k, k+1\}$ and $\varphi' \in C^{k'}(G')$. Moreover, if φ is contracted to $\ell = |L(\varphi)|$ color intervals then the running time is bounded by a polynomial in $|V|$ and ℓ, and the number $|L(\varphi')| - \ell$ of new color intervals is bounded by a polynomial in $|V|$. This simply follows from the running time of **TasExt** and the properties of the new data structures.

Lemma 5.25 *Let (G', x, y, k, φ) be a valid input for **TasExtI** satisfying $k \geq \Delta(G') + \lfloor |V|/2 \rfloor$, and let φ be contracted to $\ell = |L(\varphi)|$ color intervals. Then the output (φ', k') of **TasExtI** satisfies $k' \in \{k, w(G')\}$.*

Proof: Clearly, we have $k' \in \{k, k+1\}$. Moreover, see the analysis of the algorithm **TasExt**, if $k' = k + 1$ then either $k' = w(G')$ or inequality (5.4) can be applied. In the second case, we especially have $|V| - 1 \geq 2(k - \Delta(G') + 1)$, implying $k + 1 \leq \Delta(G') + \lfloor |V|/2 \rfloor$, which contradicts the hypothesis. Hence $k' \in \{k, w(G')\}$. ∎

Based on an idea from Sanders and Steurer [262], we use the algorithm **TasExtI** to construct the following divide-and-conquer algorithm, that produces a partial edge coloring of an input graph.

ParCol$(G = (V, \mu))$:

1. If $\Delta(G) \leq \left\lfloor \frac{|V|}{2} \right\rfloor$ then return $((V, \emptyset), 0, \emptyset)$.

2. Initialize $G' = (V, \mu')$ with G.
   Delete edges from G' until $\Delta(G') = \Delta(G) - \left\lfloor \frac{|V|}{2} \right\rfloor$.

3. $G_0 \leftarrow \left\lfloor \frac{G'}{2} \right\rfloor$, i.e., $G_0 = (V, \mu_0)$ and $\forall e \in V^{(2)}: \mu_0(e) \leftarrow \left\lfloor \frac{\mu'(e)}{2} \right\rfloor$.

4. $(G_0', k, \varphi) \leftarrow$ **ParCol**(G_0).

5. $G_1' \leftarrow 2G_0'$, i.e., $\forall e \in V^{(2)}: \mu_1'(e) \leftarrow 2\mu_0'(e)$.

6. $\varphi \leftarrow 2\varphi$, i.e., $k \leftarrow 2k$ and $I \leftarrow [2a - 1, 2b]$ for every color interval $I = [a, b]$ of φ.

7. **If** $k < \Delta(G)$ **then** $k \leftarrow \Delta(G)$.

8. Extend φ to a coloring of G' by successively coloring the remaining edges using **TasExtI**.

9. **Return** (G', k, φ).

End

Proposition 5.26 *Let $G = (V, \mu)$ be a graph. On the input G, the algorithm **ParCol** returns a tuple (G', k, φ) satisfying the following conditions:*

(a) $G' \subseteq G$ and $\Delta(G') = \max\{0, \Delta(G) - \lfloor \frac{|V|}{2} \rfloor\}$.

(b) $|E(G')| \geq |E(G)| - \lfloor \frac{|V|^2}{2} \rfloor$.

(c) $\varphi \in C^k(G')$ and $k \leq \max\{\Delta(G), w(G)\}$.

(d) *The number $\ell = |L(\varphi)|$ of color intervals satisfies $\ell \leq p(|V|) \cdot \log \mu(G)$, where p is a polynomial.*

(e) *The running time of **ParCol** is bounded by a polynomial in $|V|$ and $\log \mu(G)$.*

Proof: Statement (a) is an immediate consequence of steps 1 and 2, in which G' is established. Observe that in step 2 at most $|V| \cdot \lfloor |V|/2 \rfloor \leq \lfloor |V|^2/2 \rfloor$ edges are deleted from G'. This proves (b).

To prove (c), we proceed by induction on $\Delta(G)$. If $\Delta(G) \leq \lfloor |V|/2 \rfloor$ statement (c) is evident, since the algorithm returns the edgeless graph on $n = |V|$ vertices and the empty edge coloring (see step 1). This proves the basic case.

Now suppose $\Delta(G) > \lfloor |V|/2 \rfloor$. Then, at the end of step 3, we have $\Delta(G_0) = \lfloor \Delta(G')/2 \rfloor < \Delta(G)$. Hence the induction hypothesis implies that after step 4 we have $\varphi \in C^k(G_0)$ with $k \leq \max\{\Delta(G_0), w(G_0)\}$. After step 6 we then obtain $\varphi \in C^k(G_1')$, where $k \leq 2\max\{\Delta(G_0), w(G_0)\} \leq \max\{\Delta(G'), w(G')\}$. Therefore, after step 7 we get $\Delta(G') + \lfloor |V|/2 \rfloor = \Delta(G) \leq k \leq \max\{\Delta(G), w(G)\}$. It follows from Lemma 5.25 that the value of k is increased in step 8 only if $k + 1 \leq w(G)$. This completes the proof of (c).

The recursion depth of the algorithm is at most $\log(\Delta(G)) \leq \log(|V|\mu(G))$. The number of color intervals is increased only in the recursion step and in step 8, in which the subroutine **TasColI** is called at most $|E(G')| - 2|E(G_0')| \leq |E(G')| - 2|E(G_0)| + |V|^2 \leq 2|V|^2$ times. This implies that step 8 increases the number of color intervals by at most a polynomial in $|V|$ (see the discussion before Lemma 5.25). Thus (d) is proved. Statement (e) is a consequence of (d) and Lemma 5.25. ∎

The algorithm **ParCol** provides a partial coloring of a graph $G = (V, \mu)$ contracted to ℓ color intervals, where at most $\lfloor |V|^2/2 \rfloor$ edges remain uncolored. These uncolored edges may be colored one-by-one using any subroutine **Ext** that runs in polynomial time with respect to $|V|$ and ℓ, and increases ℓ by at most a polynomial in $|V|$. This way we maintain an overall polynomial time. Moreover, since **ParCol** uses at most $\max\{\Delta(G), w(G)\} \le \chi'(G)$ colors, we still maintain all approximation guarantees of the routine **Ext**. A proper candidate for the subroutine **Ext** would be the algorithm **TasExtI**, resulting in an algorithm **TasColI** that has the same approximation guarantees as **TasCol**.

TasColI$(G = (V, \mu))$:

1. $(G', k, \varphi) \leftarrow$ **ParCol**(G).
2. **If** $k < \Delta(G) + 1$ **then** $k \leftarrow \Delta(G) + 1$.
3. Extend φ to G by successively coloring the remaining edges using **TasExtI**.
4. **Return** (k, φ).

End

Corollary 5.27 *The algorithm* **TasColI** *is polynomial and colors the edges of a graph G with at most $\tau(G)$ colors.*

5.6 NOTES

The fundamental paper of Tashkinov [291] from 2000 contains two theorems and their proofs, but no separate auxiliary results. The first theorem shows that the new test objects, which have become known as Tashkinov trees, are suitable test objects for the edge coloring problem. Tashkinov formulated this theorem as a statement about the existence of an algorithm extending a partial coloring (Theorem 5.2), but not as a statement about critical graphs (Theorem 5.1). Tashkinov's proof describes a coloring algorithm and shows its correctness without estimating the running time. Nevertheless, our proof of Theorem 5.1 is essentially Tashkinov's original proof transformed into the language of critical graphs. The second theorem in Tashkinov's paper [291] asserts that every graph G satisfies

$$\chi(G) \le \max\{\frac{11}{10}\Delta(G) + \frac{8}{10}, w(G)\}. \tag{5.6}$$

Again, the proof of this theorem given in Tashkinov [291] describes an algorithm showing that the right hand side of the above inequality defines an efficiently realizable upper bound for the chromatic index. That this is the case was first proved by Nishizeki and Kashiwagi [237] in 1990. Their proof is based on the critical chain method; it uses this method to its limits. Tashkinov's proof is not only much shorter, but it also demonstrates the great advantage Tashkinov trees have compared to other types of test objects for the edge coloring problem.

From Tashkinov's proof of (5.6) one can deduce several very useful results about Tashkinov trees, in particular, Theorems 5.3 and 5.4 as well as Propositions 5.10

and 5.11. Tashkinov himself did not formulate these results in explicit form; in fact he did not consider critical graphs at all. This might be the reason why Tashkinov did not recognize that these results immediately imply Theorem 5.19 and hence the bound in Corollary 5.20 as well as the bound in Corollary 5.21. That this is indeed the case was first pointed out by Scheide [268] in 2007, see also Scheide [269, 271]. [2] Tashkinov's analysis of his coloring algorithm is slightly different from the one given in Sect. 5.4 (see Theorem 5.19). Translated into the language of critical graphs, he proved that if G is a critical graph with $\chi'(G) \geq \Delta(G) + 2$, then G is elementary or $t(G) \geq 9$; see Lemma 6.17 for an extension of this result. Combined with the upper bound for $t(G)$ in (5.4) this yields (5.6), first for critical graphs and therefore, because of Proposition 1.1, for graphs in general. One can even replace the additive constant $\frac{8}{10}$ in (5.6) by $\frac{7}{10}$, see also Corollary 6.21.

The results in Sect. 5.2 were first published in a preprint by Favrholdt, Stiebitz and Toft [91] from 2006. Theorem 5.18 is the main result of this section and describes a possibility to extend Tashkinov trees to a larger class of test objects. In what follows, we shall discuss another such possibility. First, we need some new definitions.

Let G be a graph, let $e \in E_G(x, y)$, and let $\varphi \in C^k(G - e)$ be a coloring for an integer $k \geq 0$. By a *tree sequence* of G with respect to e we mean a sequence $T = (y_0, e_1, y_1, \ldots, e_p, y_p)$ consisting of edges $e_1, \ldots, e_p$ and vertices $y_0, \ldots, y_p$ satisfying (T1). Now consider an arbitrary tree-sequence T of G with respect to e. For a color $\alpha \in \{1, \ldots, k\}$, let $E_\alpha(T, \varphi) = \{e' \in \partial_G(V(T)) \mid \varphi(e') = \alpha\}$. The color α is said to be a *defective color* with respect to (T, φ) if $|E_\alpha(T, \varphi)| \geq 2$. We say that α is *used* on T with respect to φ if there is an edge $e' \in E(T)$ such that $\varphi(e') = \alpha$. If $e' \in \partial_G(V(T))$, then we denote by $T + e'$ the sequence (T, e', z), where z is the endvertex of e' belonging to $V(G) \setminus V(T)$. Clearly, $T + e'$ is again a tree-sequence of G with respect to e.

We now define the class $\mathcal{T}^{\text{ext}}(G, e, \varphi)$ of tree sequences of G with respect to e and φ recursively by the following three rules:

(E1) $T = (x, e, y) \in \mathcal{T}^{\text{ext}}(G, e, \varphi)$.

(E2) Let $T \in \mathcal{T}^{\text{ext}}(G, e, \varphi)$, $e' \in \partial_G(V(T))$, and $\varphi(e') \in \overline{\varphi}(V(T))$. Then $T + e' \in \mathcal{T}^{\text{ext}}(G, e, \varphi)$.

(E3) Let $T \in \mathcal{T}^{\text{ext}}(T, e, \varphi)$ such that $V(T)$ is closed with respect to φ, let α be a defective color with respect to (T, φ), and let $e' \in E_\alpha(T, \varphi)$. Then $T + e' \in \mathcal{T}^{\text{ext}}(G, e, \varphi)$ if there is a vertex $u \in V(T)$ and a color $\gamma \in \overline{\varphi}(u)$ such that γ is not used on T with respect to φ and e' is the first edge of the chain $P = P_u(\alpha, \gamma, \varphi)$ in the linear order $\preceq_{(u,P)}$ which belongs to the coboundary $\partial_G(V(T))$. We then call e' a *defective edge* of $T + e'$.

The condition (E3) is motivated by Proposition 5.11. Clearly, a tree sequence T of G with respect to e is a Tashkinov tree with respect to e and φ if and only if T belongs to $\mathcal{T}^{\text{ext}}(G, e, \varphi)$ and T has no defective edge. To use the tree sequences of

---

[2]Paper [271] is a revised version of paper [268] with a more appropriate title.

$\mathcal{T}^{\text{ext}}(G, e, \varphi)$ as test objects for the edge coloring problem, we need a result analogue to Theorem 5.1, namely

(O) Let G be a graph with $\chi'(G) = k + 1$ for an integer $k \geq \Delta(G) + 1$, let $e \in E(G)$ be a critical edge of G, and let $\varphi \in C^k(G - e)$ be a coloring. Then the vertex set of each tree sequence $T \in \mathcal{T}^{\text{ext}}(G, e, \varphi)$ is elementary with respect to φ.

Unfortunately, we are not able to prove or disprove (O); all that we can prove in this direction is Theorem 5.18. Chen, Yu and Zang [44] introduced another family of possible test objects, the so-called *VKT-trees* with respect to (G, e, φ). The definition of VKT-trees is very much similar to the definition of $\mathcal{T}^{\text{ext}}(G, e, \varphi)$, one need only to modify the condition (E3) slightly. However, also Chen, Yu, and Zang were not able to prove (O) for VKT-trees instead of trees in $\mathcal{T}^{\text{ext}}(G, e, \varphi)$. They only succeeded to prove this statement under the additional assumption that $k \geq \lfloor \Delta(G) + \sqrt{\Delta(G)/2} \rfloor$. As shown in Chen, Yu, and Zang [44], this result gives another proof of Corollary 5.20 (with $\sqrt{\Delta/2}$ instead of $\sqrt{(\Delta - 1)/2}$), but no improvement of this bound. Compared to Scheide's proof, given in this chapter, the proof of Chen, Yu and Zang seems harder. Surprisingly, the proof of Chen, Yu, and Zhang does not yield a polynomial coloring algorithm; the number of necessary Kempe changes is bounded from above by $O(n^{O(\sqrt{\Delta}+3)}m\Delta)$. Kurt [187] announced that he found a new class of test object, similar to $\mathcal{T}^{\text{ext}}(G, e, \varphi)$, for which one can prove statement (O) under the additional assumption that $k \geq \lfloor \Delta(G) + \sqrt[3]{\Delta(G)/2} \rfloor$. This gives an improvement of the result in Corollary 5.20.

The edge coloring algorithm **TasCol** in Sect. 5.4 is an extension of Tashkinov's original algorithm implied by Tashkinov's theorem (Theorem 5.2 respectively Theorem 5.1). For a detailed discussion of Tashkinov's original algorithm the reader is referred to the thesis of Scheide [269], respectively to the thesis of McDonald [214]. As pointed out by McDonald [214] one can modify the algorithm slightly so that the number of recolorings required only depends on Δ rather than on the order n, as stated in Theorem 5.2. This follows from the fact that any Tashkinov tree with at least Δ vertices is not elementary, because of Proposition 1.7. Hence we can always truncate a too large Tashkinov tree. Note that, although the number of recolorings in one call of the subroutine **TasExt** depends only on Δ, many of these recolorings involve swapping an alternating path, which could involve all the vertices of the graph. Hence, overall, Tashkinov's algorithm is polynomial, depending on $|E|$, $|V|$, and Δ, namely $O(|E|(|V| + \Delta)\Delta^3)$

The results in Sect. 5.5 are taken from Scheide's thesis [269] (see also Scheide's paper [270]); he developed further ideas due to Sanders and Steurer [262].

The proof of Theorem 5.22 is due to Haxell and McDonald [132]. As remarked in reference [132], the constants are not best possible. In particular, using results from the next chapter, one can deduce that $n_3 \geq 7$.

Plantholt [245] proved that every graph G with even order $n \geq 572$ satisfies $\chi'(G) \leq \max\{\Delta(G), w(G)\} + 1 + \sqrt{n \ln n / 6}$.

CHAPTER 6

GOLDBERG'S CONJECTURE

The density $w(G)$ of a graph G is closely related to its fractional chromatic index $\chi'^*(G)$, and Goldberg's conjecture implies that every graph G satisfies $\chi'^*(G) \leq \chi'(G) \leq \chi'^*(G) + 1$. In this chapter we shall give a purely combinatorial proof of Kahn's result [161] that $\chi'(G)$ is asymptotically equal to $\chi'^*(G)$, even with a better asymptotic value (Theorem 6.2). We shall also prove the "next step" of a parameterized version of Goldberg's conjecture (Theorem 6.5).

6.1 DENSITY AND FRACTIONAL CHROMATIC INDEX

The density $w(G)$ of a graph G is related to the so-called fractional edge chromatic number or fractional chromatic index of G. A **fractional edge coloring** of a graph G is an assignment of a nonnegative weight w_M to each matching M of G such that every edge $e \in E(G)$ satisfies

$$\sum_{M : e \in M} w_M = 1.$$

The **fractional chromatic index** $\chi'^*(G)$ of G is the minimum value of $\sum_M w_M$, where the sum is over all matchings M of G and the minimum is over all fractional edge colorings w of G. In the case $|E(G)| = 0$ we have $\chi'^*(G) = 0$. From the definition of the fractional chromatic index it follows that $\Delta(G) \leq \chi'^*(G) \leq \chi'(G)$ for every graph G. The computation of the chromatic index is NP-hard, but with matching techniques one can compute the fractional chromatic index in polynomial time, see the books [276, 277].

From Edmonds' matching polytope theorem [76] the following characterization of the fractional edge chromatic number of an arbitrary graph G can be obtained (see Appendix B):

$$\chi'^*(G) = \max \left\{ \Delta(G), \max_{H \subseteq G, |V(H)| \geq 2} \frac{|E(H)|}{\left\lfloor \frac{1}{2}|V(H)| \right\rfloor} \right\}. \tag{6.1}$$

As an immediate consequence of this characterization we obtain the following result.

Proposition 6.1 *Let G be an arbitrary graph. Then the following statements hold:*

(a) *If $\chi'^*(G) = \Delta(G)$, then $w(G) \leq \Delta(G)$.*

(b) *If $\chi'^*(G) > \Delta(G)$, then $w(G) = \lceil \chi'^*(G) \rceil$ and $w(G)$ can be computed in polynomial time.*

Proposition 6.1 implies that Goldberg's conjecture (Conjecture 1.2) is equivalent to the claim that $\chi'(G) = \lceil \chi'^*(G) \rceil$ for every graph G with $\chi'(G) \geq \Delta(G) + 2$. Another equivalent formulation of Goldberg's conjecture is the claim that every graph G satisfies the inequality

$$\chi'(G) \leq \max\{\Delta(G) + 1, w(G)\} = \max\{\Delta(G) + 1, \lceil \chi'^*(G) \rceil\}. \tag{6.2}$$

Hence in case (a) of Proposition 6.1 $\chi'(G)$ will be either $\Delta(G)$ or $\Delta(G) + 1$, whereas in case (b) of Proposition 6.1 $\chi'(G)$ will be $\lceil \chi'^*(G) \rceil$. Thus Goldberg's conjecture implies that the difficulty in determining $\chi'(G)$ is only to distinguish between the two cases $\chi'(G) = \Delta(G)$ and $\chi'(G) = \Delta(G) + 1$.

Using probabilistic methods, Kahn [161] proved in 1996 that every graph G satisfies $\chi'(G) \sim \chi'^*(G)$ as $\chi'^*(G) \to \infty$. Sanders and Steurer [262] extended this result to the statement that every graph G satisfies the inequality

$$\chi'(G) \leq \chi'^*(G) + \sqrt{9\chi'^*(G)/2}.$$

As a simple consequence of Corollary 5.20 we obtain the following strengthening due to Scheide [268, 269, 271] of these results.

Theorem 6.2 *Every graph G satisfies $\chi'(G) \leq \chi'^*(G) + \sqrt{\chi'^*(G)/2}$.*

Proof: The inequality is certainly true if $\Delta(G) \leq 1$, since then $\chi'(G) = \Delta(G) = w(G) = \chi'^*(G)$. So assume $\Delta(G) \geq 2$. By Corollary 5.20, $\chi'(G) \leq \tau(G) = \max\{\Delta(G) + \sqrt{(\Delta(G) - 1)/2}, w(G)\}$. Furthermore, $\chi'^*(G) \geq \Delta(G) \geq 2$ and

$\lceil \chi'^*(G) \rceil \geq w(G)$ (Proposition 6.1), which gives $\chi'^*(G) + \sqrt{\chi'^*(G)/2} \geq \chi'^*(G) + 1 \geq w(G)$, and hence

$$\chi'^*(G) + \sqrt{\chi'^*(G)/2} \geq \tau(G) \geq \chi'(G).$$

This completes the proof. ∎

Recall that a graph H is defined to be an **inflation graph** of a given graph G if both graphs have the same vertex set and if every pair (x, y) of distinct vertices satisfies $\mu_H(x, y) = 0$ if $\mu_G(x, y) = 0$ and $\mu_H(x, y) \geq \mu_G(x, y)$ otherwise. An inflation graph of a cycle with $n \geq 3$ vertices is also called a **ring graph**.

The following result about the chromatic index of ring graphs seems to be mentioned first in the book of Ore [238]. In Ore's book this result was attributed to B. Rothschild and J. Stemple, and the proof for ring graphs with an odd number of vertices was said to be a little involved and it was left to the reader.

Theorem 6.3 *If G is a subgraph of a ring graph satisfying $n = |V(G)| \geq 2$ and $m = |E(G)| \geq 0$, then $\chi'(G) = \max\{\Delta(G), \lceil \frac{m}{\lfloor n/2 \rfloor} \rceil\}$.*

Proof: Since $\chi'(G) \geq \max\{\Delta(G), \lceil \frac{m}{\lfloor n/2 \rfloor} \rceil\}$, it suffices to show that $\chi'(G) \leq \max\{\Delta(G), \lceil \frac{m}{\lfloor n/2 \rfloor} \rceil\}$. We prove this inequality by induction on $m \geq 0$. Clearly, the statement is evident if $\chi'(G) \leq \Delta(G)$. Now suppose that $\chi'(G) \geq \Delta(G) + 1$. This implies that G is not bipartite and, therefore, G is an inflation graph of an odd cycle with $n \geq 3$ vertices and so $m \geq n$.

First consider the case that there is a vertex $x \in V(G)$ such that $d_G(x) < \Delta(G)$. Since G is an inflation graph of an odd cycle, there is a perfect matching M of $G - x$ and, therefore, the graph $H = G - M$ satisfies $m' = |E(H)| = m - \lfloor n/2 \rfloor$ and $\Delta(H) \leq \Delta(G) - 1$. It follows from the induction hypothesis that $\chi'(H) \leq \max\{\Delta(H), \lceil \frac{m'}{\lfloor n/2 \rfloor} \rceil\} = \max\{\Delta(G), \lceil \frac{m}{\lfloor n/2 \rfloor} \rceil\} - 1$. Consequently, we have $\chi'(G) \leq \chi'(H) + 1 \leq \max\{\Delta(G), \lceil \frac{m}{\lfloor n/2 \rfloor} \rceil\}$.

Now consider the case that $d_G(x) = \Delta(G)$ for all $x \in V(G)$. Since G is an inflation graph of an odd cycle C with $n \geq 3$ vertices, it follows that there is an integer $\ell \geq 1$ such that $\mu_G(x, y) = \ell$ provided that $\mu_C(x, y) = 1$. Consequently, $m = \ell n$ and $\Delta(G) = 2\ell$. Again, there is a perfect matching M of $G - x$, where x is some vertex of G. For the graph $H = G - M$, we have $m' = |E(H)| = m - \lfloor n/2 \rfloor$ and $\Delta(H) = \Delta(G)$. Then the induction hypothesis implies that $\chi'(H) \leq \max\{\Delta(H), \lceil \frac{m'}{\lfloor n/2 \rfloor} \rceil\} = \max\{\Delta(G), \lceil \frac{m}{\lfloor n/2 \rfloor} \rceil - 1\}$. Since $\Delta(G) + 1 = 2\ell + 1 \leq \lceil \frac{2\ell n}{n-1} \rceil = \lceil \frac{m}{\lfloor n/2 \rfloor} \rceil$, this implies that $\chi'(G) \leq \chi'(H) + 1 \leq \max\{\Delta(G), \lceil \frac{m}{\lfloor n/2 \rfloor} \rceil\}$. This completes the proof. ∎

Theorem 6.3 was independently obtained by T. Gallai (personal communication in 1969). However, Gallai formulated the theorem as a statement about the vertex-criticality (with respect to χ) of line graphs of ring graphs (see Krusenstjerna-Hafstrøm and Toft [185] for an account of Gallai's result). Based on Theorem 6.3 it is not difficult to show that a ring graph G satisfying $|V(G)| = 2p + 1$ and $\chi'(G) = k + 1$ is critical if and only if $\Delta(G) \leq k$ and $|E(G)| = kp + 1$. The study of critical ring graphs led I. T. Jakobsen to the following conjecture, posed at the graph theory conference in Keszthely (Hungary) in 1973 and published in 1975 in the proceedings of the conference.

Conjecture 6.4 (Jakobsen [156] 1975) *Let G be a critical graph, and let*

$$\chi'(G) > \frac{m}{m-1} \Delta(G) + \frac{m-3}{m-1}$$

for an odd integer $m \geq 3$. Then $|V(G)| \leq m - 2$.

For an integer $m \geq 3$, let $\mathcal{J}_m$ denote the class of all graphs G satisfying

$$\chi'(G) > \frac{m}{m-1} \Delta(G) + \frac{m-3}{m-1}.$$

Shannon's theorem (Theorem 2.4) implies that $\mathcal{J}_3$ is empty. Furthermore, for every integer $m \geq 3$, we have $\mathcal{J}_m \subseteq \mathcal{J}_{m+1}$ and the class

$$\mathcal{J} = \bigcup_{m=3}^{\infty} \mathcal{J}_m$$

consists of all graphs G such that $\chi'(G) \geq \Delta(G) + 2$. In order to show that, for an odd integer $m \geq 5$, every critical graph in $\mathcal{J}_m$ has at most $m-2$ vertices it is sufficient to show that every critical graph in $\mathcal{J}_m$ is elementary. This follows immediately from Theorem 1.4, Proposition 1.7, and Proposition 1.3. Hence Jakobsen's conjecture is a weaker version of Goldberg's conjecture.

Up to now, Jakobsen's conjecture is known to be true only for odd m with $5 \leq m \leq 15$. In all these cases the proof of the statement that every critical graph in $\mathcal{J}_m$ has at most $m - 2$ vertices is based on a proof of the seemingly more general statement that every graph in $\mathcal{J}_m$ is elementary. This was proved, for $m = 5$ independently by B. Aa. Sørensen (unpublished), Andersen [5], and Goldberg [110], for $m = 7$ independently by B. Aa. Sørensen (unpublished) and Andersen [5], for $m = 9$ by Goldberg [112, 114], and for $m = 11$ by Nishizeki and Kashiwagi [237]. All proofs presented by these authors rely on the critical chain method described in Corollary 5.5. The proof given by Nishizeki and Kashiwagi in 1990 for the case $m = 11$ is very long and exploits the critical chain method to its limit. Ten years later Tashkinov [291] used his new method to give a much shorter proof for the case $m = 11$. Based on an extension of Tashkinov methods Favrholdt et al. [91] established the case $m = 13$ in 2006. The next step $m = 15$ was obtained in 2007 by Scheide [268, 269, 271].

Theorem 6.5 [271] *Every graph G such that*

$$\chi'(G) > \frac{15}{14} \Delta(G) + \frac{12}{14}$$

is an elementary graph.

The proof of Theorem 6.5 will be presented in Sect. 6.3. In order to prove this result it is sufficient to show that every critical graph in $\mathcal{J}_{15}$ is elementary. This follows from the next result.

Proposition 6.6 *Let $m \geq 3$ be an odd integer. If every critical graph in $\mathcal{J}_m$ is elementary, then every graph in $\mathcal{J}_m$ is elementary.*

Proof: Consider an arbitrary graph $G \in \mathcal{J}_m$. By Proposition 1.1, G contains a subgraph H such that H is critical and $\chi'(H) = \chi'(G)$. Then $\Delta(H) \leq \Delta(G)$ and, therefore, $H \in \mathcal{J}_m$. From the hypothesis it follows that H is an elementary graph. Thus $w(G) \leq \chi'(G) = \chi'(H) = w(H) \leq w(G)$ and, therefore, $\chi'(G) = w(G)$. This proves that G is an elementary graph, too. ∎

As an immediate consequence of Theorem 6.5 and Proposition 6.1 we obtain the following upper bound for the chromatic index.

Corollary 6.7 *Every graph G satisfies*

$$\chi'(G) \leq \max \left\{ \frac{15}{14} \Delta(G) + \frac{12}{14}, w(G) \right\} = \max \left\{ \frac{15}{14} \Delta(G) + \frac{12}{14}, \lceil \chi'^*(G) \rceil \right\}.$$

Corollary 6.8 *Let m be an odd integer satisfying $3 \leq m \leq 15$. Then every critical graph $G \in \mathcal{J}_m$ has at most $m - 2$ vertices.*

We conclude this section with a simple consequence of Corollary 5.21. This statement generalizes an earlier result by Favrholdt et al. [91, Theorem 10.8].

Corollary 6.9 *Let $m \geq 3$ be an odd integer, and let G be a graph such that $\Delta(G) \geq \frac{1}{2}(m - 3)^2$. Then*

$$\chi'(G) \leq \max\{ \frac{m}{m - 1} \Delta(G) + \frac{m - 3}{m - 1}, w(G) \}.$$

Proof: Let $\Delta = \Delta(G)$ and $\varepsilon = \frac{1}{m-1}$. By hypothesis,

$$\Delta \geq \frac{1}{2}(m - 3)^2 = \frac{1}{2}(\frac{1}{\varepsilon} - 2)^2 = \frac{1}{2\varepsilon^2} - \frac{2}{\varepsilon} + 2,$$

and, therefore,

$$(1 + \varepsilon)\Delta + 1 - 2\varepsilon \geq \Delta + 1 + \varepsilon(\frac{1}{2\varepsilon} - \frac{2}{\varepsilon}) = \Delta - 1 + \frac{1}{2\varepsilon}.$$

By Corollary 5.21, we then obtain

$$\begin{aligned}
\chi'(G) &\leq \max\{(1 + \varepsilon)\Delta + 1 - 3\varepsilon, \Delta - 1 + \frac{1}{2\varepsilon}, w(G)\} \\
&\leq \max\{(1 + \varepsilon)\Delta + 1 - 2\varepsilon, w(G)\} \\
&= \max\{\frac{m}{m - 1}\Delta + \frac{m - 3}{m - 1}, w(G)\}.
\end{aligned}$$

This completes the proof. ∎

6.2   BALANCED TASHKINOV TREES

Using Tashkinov trees as test objects for coloring algorithms, it has proved advantageous to choose the Tashkinov tree in an appropriate way, to obtain a Tashkinov tree, where the number of used colors is small and the number of free colors is large.

Let G be a critical graph with $\chi'(G) = k + 1$ for an integer $k \geq \Delta(G) + 1$ and let $(T, e, \varphi) \in \mathcal{T}(G)$. Then the Tashkinov tree T has the form

$$T = (y_0, e_1, y_1, \ldots, e_{p-1}, y_{p-1}),$$

where $p = t(G)$. Then T is called a **normal Tashkinov tree** with respect to e and φ if there are two colors $\alpha \in \overline{\varphi}(y_0)$ and $\beta \in \overline{\varphi}(y_1)$, an integer $2 \leq h \leq p-1$, and an edge $f \in E_G(y_0, y_{h-1})$ such that the path $\mathrm{Path}(y_1, e_2, y_2, \ldots, e_{h-1}, y_{h-1}, f, y_0)$ is an (α, β)-chain with respect to φ. In this case $T y_{h-1}$ is called the (α, β)-**trunk** of T and the number h is called the **height** of T written $h(T) = h$. Furthermore, let $\mathcal{T}^N(G)$ denote the set of all triples $(T, e, \varphi) \in \mathcal{T}(G)$ for which T is a normal Tashkinov tree, and let $h(G)$ denote the greatest number h such that there is a triple $(T, e, \varphi) \in \mathcal{T}^N(G)$ with $h(T) = h$. The following lemma shows that normal Tashkinov trees can be generated from arbitrary ones, which also implies that $\mathcal{T}^N(G) \neq \emptyset$.

Proposition 6.10 *Let G be a critical graph with $\chi'(G) = k + 1$ for an integer $k \geq \Delta(G) + 1$ and let $(T, e, \varphi) \in \mathcal{T}(G)$ with $e \in E_G(x, y)$. Then there are two colors α, β and a Tashkinov tree T' with respect to e and φ satisfying $\alpha \in \overline{\varphi}(x)$, $\beta \in \overline{\varphi}(y)$, $V(T') = V(T)$, $(T', e, \varphi) \in \mathcal{T}^N(G)$, and $h(T') = |V(P)|$ where $P = P_x(\alpha, \beta, \varphi)$.*

Proof: Since $k \geq \Delta(G) + 1$, there are two colors $\alpha \in \overline{\varphi}(x)$ and $\beta \in \overline{\varphi}(y)$. By Theorem 2.1, we have $\alpha \neq \beta$ and $P = P_x(\alpha, \beta, \varphi)$ is a path having endvertices x and y. This means P is a path of the form

$$P = \mathrm{Path}(v_1, f_2, v_2, \ldots, f_h, v_h)$$

with $v_1 = y$ and $v_h = x$. Evidently, $f_j \in E_G(v_{j-1}, v_j)$ and $\varphi(f_j) \in \{\alpha, \beta\} \subseteq \overline{\varphi}(x, y)$ for all $j \in \{2, \ldots, h\}$. Hence $T_1 = (x, e, y, f_2, v_2, \ldots, f_{h-1}, v_{h-1})$ is a Tashkinov tree with respect to e and φ. By Theorem 5.3, there is a Tashkinov tree T' with respect to e and φ such that $V(T') = V(T)$ and $T' v_{h-1} = T_1$. Then T' is normal with respect to e and φ and, therefore, $(T', e, \varphi) \in \mathcal{T}^N(G)$. Moreover, T_1 is the (α, β)-trunk of T' and $h(T') = |V(T_1)| = |V(P)|$, which completes the proof. ∎

Consider an arbitrary triple $(T, e, \varphi) \in \mathcal{T}^N(G)$, where T has the form $T = (y_0, e_1, y_1, \ldots, e_{p-1}, y_{p-1})$. Then (T, e, φ) is called a **balanced triple** with respect to e and φ if $h(T) = h(G)$ and $\varphi(e_{2j}) = \varphi(e_{2j-1})$ for $h(T) < 2j < p$. Let $\mathcal{T}^B(G)$ denote the set of all balanced triples $(T, e, \varphi) \in \mathcal{T}(G)$. The following proposition shows that $\mathcal{T}^B(G) \neq \emptyset$.

Proposition 6.11 *Let G be a critical graph with $\chi'(G) = k + 1$ for an integer $k \geq \Delta(G) + 1$. Then the following statements hold:*

(a) $h(G)$ is odd and $h(G) \geq 3$.

(b) If $(T, e, \varphi) \in \mathcal{T}^N(G)$ with $h(T) = h(G)$, then there is a Tashkinov tree T' with respect to e and φ such that $V(T') = V(T)$, $(T', e, \varphi) \in \mathcal{T}^B(G)$ and, moreover, all colors used on T' are used on T.

(c) $\mathcal{T}^B(G) \neq \emptyset$.

Proof: Let $(T, e, \varphi) \in \mathcal{T}^N(G)$ with $h(T) = h(G) = h$. Then T has the form $T = (y_0, e_1, y_1, \ldots, e_{p-1}, y_{p-1})$, where $p = t(G)$ is odd (Proposition 5.10) and Ty_{h-1} is the (α, β)-trunk of T. Consequently, we have $\alpha \in \overline{\varphi}(y_0)$, $\beta \in \overline{\varphi}(y_1)$, and there is an edge $f \in E_G(y_0, y_{h-1})$ satisfying $P = P_{y_0}(\alpha, \beta, \varphi) = \text{Path}(y_1, e_2, \ldots, e_{h-1}, y_{h-1}, f, y_0)$. Hence P is a path with two distinct endvertices whose edges are alternately colored with α and β with respect to φ. Since each of these two colors is missing at one of the two endvertices of P, the length $|E(P)|$ is even. Hence $h = h(G)$ is odd and $h \geq 3$. Thus (a) is proved.

Let $i \leq p$ be the greatest odd integer for which there exists a Tashkinov tree $T' = (y_0', e_1', y_1', \ldots, e_{i-1}', y_{i-1}')$ with respect to e and φ satisfying $T'y_{h-1}' = Ty_{h-1}$, $\varphi(E(T')) \subseteq \varphi(E(T))$, and $\varphi(e_{2j-1}') = \varphi(e_{2j}')$ whenever $h < 2j < i$. Evidently, we have $i \geq h$, since Ty_{i-1} fulfils these requirements for $i = h$.

Now suppose that $i < p$. Then there is a smallest integer r satisfying $y_r \in V(T) \setminus V(T')$. Let $y_i' = y_r$ and $e_i' = e_r$. Hence $e_i' \in E_G(V(T'), y_i')$ and $\varphi(e_i') \in \overline{\varphi}(V(T'))$, implying that $T_1 = (T', e_i', y_i')$ is a Tashkinov tree with respect to e and φ. Let $\gamma = \varphi(e_i')$. Clearly, $|V(T_1)|$ is even, $\gamma \in \overline{\varphi}(V(T_1))$, and $V(T_1)$ is elementary with respect to φ (Theorem 5.1). Hence there is an edge $e_{i+1}' \in E_G(V(T_1), y_{i+1}')$ satisfying $y_{i+1}' \in V(G) \setminus V(T_1)$ and $\varphi(e_{i+1}') = \gamma$. Evidently, $T_2 = (T_1, e_{i+1}', y_{i+1}')$ is a Tashkinov tree with respect to e and φ satisfying $T_2y_{h-1}' = Ty_{h-1}$, $\varphi(E(T_2)) \subseteq \varphi(E(T))$, and $\varphi(e_{2j-1}') = \varphi(e_{2j}')$ whenever $h < 2j < i + 2$. This contradicts the maximality of i. Consequently, we have $i = p$ and so $V(T') = V(T)$ (Theorem 5.3(d)). Hence $(T', e, \varphi) \in \mathcal{T}^B(G)$ with $\varphi(E(T')) \subseteq \varphi(E(T))$ and (b) is proved. Furthermore, (c) follows simply from (b) and the fact that $\mathcal{T}^N(G) \neq \emptyset$. This completes the proof. ∎

Consider a critical graph G and a balanced triple $(T, e, \varphi) \in \mathcal{T}^B(G)$. Then T has the form

$$T = (y_0, e_1, y_1, \ldots, e_{p-1}, y_{p-1})$$

and Ty_h is the (α, β)-trunk of T, where $h = h(G)$, $\alpha \in \overline{\varphi}(y_0)$ and $\beta \in \overline{\varphi}(y_1)$. Moreover, there is an edge $f_h \in E_G(y_0, y_{h-1})$ with $\varphi(f_h) = \beta$. For $1 \leq i \leq h - 1$, let $f_i = e_i$. Clearly, the edges $f_1, \ldots, f_h$ form a cycle in G. Furthermore, f_1 is the uncolored e and the edges $f_2, \ldots, f_h$ are colored alternately with α and β with respect to φ. Now, choose an index $j \in \{1, \ldots, h-1\}$. Since (y_0, y_1) is a (α, β)-pair with respect to φ, there is a coloring $\varphi' \in \mathcal{C}^k(G - f_{j+1})$ such that $\varphi'(e') = \varphi(e')$ for all edges $e' \in E(G) \setminus \{f_1, \ldots, f_h\}$ and the edges $f_{j+2}, \ldots, f_h, f_1, \ldots, f_j$ are colored alternately with α and β with respect to φ'. Then

$$T' = (y_j Ty_{h-1}, f_h, y_0 Ty_{j-1}, e_h, y_h, \ldots, e_{p-1}, y_{p-1})$$

is a normal Tashkinov tree, where $T'y_{j-1}$ is the (α, β)-trunk of T' and, moreover, the triple (T', f_{j+1}, φ') is balanced. In the sequel, we write $(T', f_{j+1}, \varphi') = (T, e, \varphi)(y_0 \to y_j)$.

6.3 OBSTRUCTIONS

For the proof of Theorem 6.5, it suffices to show that every critical graph $G \in \mathcal{J}_{15}$ is elementary. Let $G \in \mathcal{J}_m$ be a critical graph for an odd integer $m \geq 3$. Clearly, this implies that $\chi'(G) = k + 1$ for some integer $k \geq \Delta(G) + 1$. Hence we can use the results from Chap. 5. By Theorem 1.4, we know that G is elementary if and only if there is an edge $e \in E(G)$, a coloring $\varphi \in \mathcal{C}^k(G - e)$, and a set $X \subseteq V(G)$ containing both endvertices of e such that X is elementary as well as strongly closed with respect to φ. One candidate for such a set X is the vertex set of a maximum Tashkinov tree T with respect to e and φ, i.e., the triple (T, e, φ) belongs to $\mathcal{T}(G)$. Then $V(T)$ is elementary and closed with respect to φ. If $V(T)$ is also strongly closed we are done. Otherwise, by the results from Chap. 5, the triple (T, e, φ) possesses certain properties. In particular, by Proposition 5.10, there are defective colors, i.e., $\Gamma^d(T, e, \varphi) \neq \emptyset$. Furthermore, by Proposition 5.11, there is a vertex $v \in V(T)$ such that every color missing at v with respect to φ is used on T. From Proposition 1.7 it also follows that the cardinality of any elementary set with respect to φ containing both endvertices of e is at most $m - 1$. However, all these properties are not sufficient to show that a desired set X exists, even under the additional assumption that $m = 15$. Lemma 6.13 below describes a possibility how a desired set X can be obtained by extending the vertex set of T by means of the defective colors, provided that T satisfies several conditions. This lemma is one of the keys to the proof of Theorem 6.5. First, we introduce the concept of an exit vertex and establish a simple property of those vertices.

Let G be a critical graph with $\chi'(G) = k + 1$ for an integer $k \geq \Delta(G) + 1$ and let $(T, e, \varphi) \in \mathcal{T}(G)$. Let $\gamma \in \overline{\varphi}(u)$ for a vertex $u \in V(T)$ and let $\delta \in \Gamma^d(T, e, \varphi)$. Clearly, the (γ, δ)-chain $P = P_u(\gamma, \delta, \varphi)$ is a path, where u is one endvertex of P and, moreover, exactly one of the two colors γ or δ is missing at the second endvertex of P with respect to φ. Since $V(T)$ is elementary with respect to φ and the defective color δ is present at every vertex in $V(T)$, the second endvertex of P belongs to $V(G) \setminus V(T)$. Hence in the linear order $\preceq_{(u,P)}$ there is a last vertex v that belongs to $V(T)$. This vertex is said to be an **exit vertex** with respect to (T, e, φ). The set of all exit vertices with respect to (T, e, φ) is be denoted by $F(T, e, \varphi)$.

Proposition 6.12 [271] *Let G be a critical graph with $\chi'(G) = k + 1$ for an integer $k \geq \Delta(G) + 1$ and let $(T, e, \varphi) \in \mathcal{T}(G)$. Then $\overline{\varphi}(F(T, e, \varphi)) \cap \Gamma^f(T, e, \varphi) = \emptyset$.*

Proof: Let $v \in F(T, e, \varphi)$ be an exit vertex. Then there is a vertex $u \in V(T)$, a color $\gamma \in \overline{\varphi}(u)$, and a defective color $\delta \in \Gamma^d(T, e, \varphi)$ such that v is the last vertex in the linear order $\preceq_{(u,P)}$ that belongs to $V(T)$, where $P = P_u(\gamma, \delta, \varphi)$. Clearly, P is a path with one endvertex u and another endvertex $z \in V(G) \setminus V(T)$. Suppose there is a color $\alpha \in \overline{\varphi}(v) \cap \Gamma^f(T, e, \varphi)$. By Proposition 5.10, $V(T)$ is elementary and closed with respect to φ and, therefore, no edge in $\partial_G(V(T))$ is colored with α or γ with respect to φ. Hence a coloring $\varphi' \in \mathcal{C}^k(G-e)$ is obtained from φ by interchanging the colors α and γ on all edges in $E_G(V(G) \setminus V(T), V(G) \setminus V(T))$. Then, evidently, $(T, e, \varphi') \in \mathcal{T}(G)$, $\Gamma^f(T, e, \varphi') = \Gamma^f(T, e, \varphi)$ and $\Gamma^d(T, e, \varphi') = \Gamma^d(T, e, \varphi)$. In particular, we have $\alpha \in \overline{\varphi}'(v) \cap \Gamma^f(T, e, \varphi')$ and $\delta \in \Gamma^d(T, e, \varphi')$. Moreover, the chain $P' = P_v(\alpha, \delta, \varphi')$ satisfies $P' = vPz$ and, therefore, we have $|E(P') \cap \partial_G(V(T))| = 1$. On the other hand, Proposition 5.10 implies that $|E(P') \cap \partial_G(V(T))| = |E_\delta(T, e, \varphi')| > 1$, a contradiction. Hence we have $\overline{\varphi}(v) \cap \Gamma^f(T, e, \varphi) = \emptyset$. Clearly, this completes the proof.   ∎

Lemma 6.13 [271] *Let G be a critical graph with $\chi'(G) = k + 1$ for an integer $k \geq \Delta(G) + 1$ and let $(T, e, \varphi) \in \mathcal{T}(G)$. Suppose that T has the form*

$$T = (y_0, e_1, y_1, \ldots, e_{r-1}, y_{r-1}, f_{\gamma_1}^1, u_{\gamma_1}^1, f_{\gamma_1}^2, u_{\gamma_1}^2, \ldots, f_{\gamma_s}^1, u_{\gamma_s}^1, f_{\gamma_s}^2, u_{\gamma_s}^2)$$

where $\Gamma = \{\gamma_1, \ldots, \gamma_s\}$ is a set of s colors, $Y = \{y_0, \ldots, y_{r-1}\}$ and the following conditions hold:

(S1) *$f_\gamma^j \in E_G(Y, u_\gamma^j)$ for every $\gamma \in \Gamma$ and $j \in \{1, 2\}$.*

(S2) *$\varphi(f_\gamma^1) = \varphi(f_\gamma^2) = \gamma \in \overline{\varphi}(Y)$ for every $\gamma \in \Gamma$.*

(S3) *$\varphi(e_j) \notin \Gamma$ for every $j \in \{2, \ldots, r-1\}$.*

(S4) *For every $v \in F(T, e, \varphi)$ there is a color $\gamma \in \Gamma$ such that $\gamma \in \overline{\varphi}(v)$.*

Then G is an elementary graph.

Proof: Since $(T, e, \varphi) \in \mathcal{T}(G)$ and G is critical, it follows from Proposition 5.10 that $V(T)$ is elementary and closed both with respect to φ. If $V(T)$ is also strongly closed with respect to φ, then Theorem 1.4 implies that G is elementary and we are done. Hence, we need only to consider the case that $V(T)$ is not strongly closed with respect to φ. Then we construct a superset X of $V(T)$ such that X is elementary and strongly closed with respect to φ. Again, by Theorem 1.4, this implies that G is elementary and we are done, too.

Since $V(T)$ is not strongly closed with respect to φ, it follows from Proposition 5.10 that $\Gamma^d(T, e, \varphi) \neq \emptyset$. Clearly, this implies $F(T, e, \varphi) \neq \emptyset$ and, therefore, $s' = |F(T, e, \varphi)| \geq 1$. Since $V(T)$ is elementary with respect to φ, we conclude from (S4) that there is a set $\Gamma' \subseteq \Gamma$ of s' colors such that, for every $v \in F(T, e, \varphi)$, there is a unique color $\gamma \in \Gamma'$ satisfying $\gamma \in \overline{\varphi}(v)$. Hence, there is a one to one correspondence between $F(T, e, \varphi)$ and Γ'. In particular, $|\overline{\varphi}(v) \cap \Gamma'| = 1$ for every $v \in F(T, e, \varphi)$ and $\overline{\varphi}(v) \cap \Gamma' = \emptyset$ for every $v \in V(T) \setminus F(T, e, \varphi)$.

By Proposition 5.10(f), $|\overline{\varphi}(y_0)| \geq 2$. Hence there is a color $\alpha_0 \in \overline{\varphi}(y_0) \setminus \Gamma'$. Now consider an arbitrary defective color $\delta \in \Gamma^d(T, e, \varphi)$. In the sequel, let P_δ denote the chain defined by

$$P_\delta = P_{y_0}(\alpha_0, \delta, \varphi).$$

By Proposition 5.10, δ is present at any vertex of T. Since $\alpha_0 \in \overline{\varphi}(y_0)$ and $V(T)$ is elementary as well as closed with respect to φ, we conclude that P_δ is a path, where one endvertex is y_0 and the other endvertex, denoted by z_δ, belongs to $V(G) \setminus V(T)$. This implies that there is a last vertex in the linear order $\preceq_{(y_0, P_\delta)}$ that belongs to $V(T)$; we denote this vertex by v_δ^0. By definition, $v_\delta^0 \in F(T, e, \varphi)$ is an exit vertex. Hence there is a unique color $\gamma \in \Gamma'$ such that $\gamma \in \overline{\varphi}(v_\delta^0)$. We denote this color by $\gamma = \gamma(\delta)$. Furthermore, v_δ^0 is incident with an edge, denoted by f_δ^0, such that $\varphi(f_\delta^0) = \delta$. We denote the second endvertex of f_δ^0 by u_δ^0. Clearly, $f_\delta^0 \in E_\delta(T, e, \varphi)$ and $u_\delta^0 \in V(G) \setminus V(T)$. As we shall see later, $u_\delta^0 \neq z_\delta$ and $|E_\delta(T, e, \varphi)| = 3$. Hence there is a set $U_\delta \subseteq V(G) \setminus V(T)$ consisting of two vertices distinct from u_δ^0 that are incident with edges in $E_\delta(T, e, \varphi)$. Our goal is to show that the set

$$X = V(T) \cup \bigcup_{\delta \in \Gamma^d(T, e, \varphi)} U_\delta$$

is elementary and strongly closed with respect to φ. The proof is long and relies on several statements.

Claim 6.13.1 $F(T, e, \varphi) \subseteq Y$ and as a consequence $v_\delta^0 \in Y$ for all $\delta \in \Gamma^d(T, e, \varphi)$.

Proof: It follows from (S2) that all colors used on T with respect to φ are contained in $\overline{\varphi}(Y)$. Since $V(T)$ is elementary with respect to φ and since $\overline{\varphi}(v) \neq \emptyset$ for every $v \in V(G)$, this implies that $\overline{\varphi}(v) \cap \Gamma^f(T, e, \varphi) \neq \emptyset$ for every $v \in V(T) \setminus Y$. By Proposition 6.12, $\overline{\varphi}(v) \cap \Gamma^f(T, e, \varphi) = \emptyset$ for every exit vertex $v \in F(T, e, \varphi)$. Hence, we obtain $F(T, e, \varphi) \subseteq Y$. $\triangle$

For a color $\gamma \in \Gamma$, let $T^{(-\gamma)}$ denote the sequence obtained from T by deleting the edges f_γ^1, f_γ^2 as well as the vertices u_γ^1, u_γ^2. By (S1) and (S2), it follows that $T^{(-\gamma)}$ is a Tashkinov tree with respect to e and φ. In the sequel, let

$$U_\gamma = \{u_\gamma^1, u_\gamma^2\}$$

and

$$Z_\gamma = V(T^{(-\gamma)}) = V(T) \setminus U_\gamma.$$

Claim 6.13.2 $E_G(u_\gamma^1, u_\gamma^2) \cap E_{\varphi, \alpha} \neq \emptyset$ for every color $\gamma \in \Gamma'$ and every color $\alpha \in \overline{\varphi}(Z_\gamma) \setminus \{\gamma\}$.

Proof: Since $V(T)$ is elementary and closed with respect to φ, for every $\gamma \in \Gamma'$ and every $\alpha \in \overline{\varphi}(Z_\gamma) \setminus \{\gamma\}$, there is an edge $f \in E_G(u_\gamma^1, V(T))$ with $\varphi(f) = \alpha$. Now suppose $f \in E_G(u_\gamma^1, Z_\gamma)$. Then there is a second edge $f' \in E_G(u_\gamma^2, Z_\gamma)$ with $\varphi(f') = \alpha$. Hence, $T' = (T^{(-\gamma)}, f, u_\gamma^1, f', u_\gamma^2)$ is a Tashkinov tree with respect to e and φ satisfying $(T', e, \varphi) \in \mathcal{T}(G)$, $V(T') = V(T)$, and $\gamma \in \Gamma^f(T', e, \varphi)$. Moreover, $\Gamma^d(T', e, \varphi) = \Gamma^d(T, e, \varphi)$, which implies $F(T', e, \varphi) = F(T, e, \varphi)$ and, therefore, $\gamma \in \overline{\varphi}(F(T', e, \varphi)) \cap \Gamma^f(T', e, \varphi)$, a contradiction to Proposition 6.12. Consequently, $f \notin E_G(u_\gamma^1, Z_\gamma)$, but $f \in E_G(u_\gamma^1, u_\gamma^2)$. $\triangle$

Claim 6.13.3 $E_G(U_{\gamma(\delta)}, Z_{\gamma(\delta)}) \cap E_{\varphi, \delta} = \emptyset$ for every $\delta \in \Gamma^d(T, e, \varphi)$.

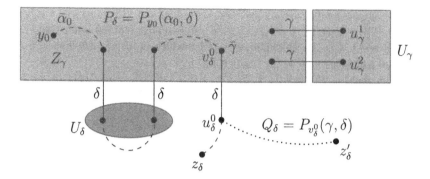

Figure 6.1 The Tashkinov tree T, a defective color δ and $\gamma = \gamma(\delta)$.

Proof: Suppose on the contrary that there is a color $\delta \in \Gamma^d(T, e, \varphi)$ and an edge $g_1 \in E_G(U_{\gamma(\delta)}, Z_{\gamma(\delta)})$ with $\varphi(g_1) = \delta$, say g_1 is incident to $u^1_{\gamma(\delta)}$. This implies that $|E_G(U_{\gamma(\delta)}, V(G) \setminus V(T)) \cap E_\delta(e, \varphi)| \leq 1$. Since $|E_\delta(T, e, \varphi)| \geq 3$ by Proposition 5.10(e), we then conclude that $|E_G(Z_{\gamma(\delta)}, V(G) \setminus V(T)) \cap E_\delta(e, \varphi)| \geq 2$. Hence, there is an edge $g_2 \in E_G(Z_{\gamma(\delta)}, V(G) \setminus V(T)) \setminus \{f^0_\delta\}$ with $\varphi(g_3) = \delta$. Let u_2 be the endvertex of g_2 that belongs to $V(G) \setminus V(T)$. From Claim 6.13.2 we know that there is an edge $g_3 \in E_G(u^1_{\gamma(\delta)}, u^2_{\gamma(\delta)})$ with $\varphi(g_3) = \alpha_0$.

Now consider the subpath $P_1 = v^0_\delta P_\delta z_\delta$. Then, clearly, $V(P_1) \cap V(T) = \{v^0_\delta\}$. Furthermore, since $\alpha_0, \gamma(\delta) \in \overline{\varphi}(V(T))$ and $V(T)$ is closed with respect to φ, a new coloring $\varphi_1 \in \mathcal{C}^k(G - e)$ is obtained from φ by interchanging the colors α_0 and $\gamma(\delta)$ on all edges in $E_G(V(G) \setminus V(T), V(G) \setminus V(T))$. Then we conclude that $P_1 = P_{v^0_\delta}(\gamma(\delta), \delta, \varphi_1)$. Hence the coloring $\varphi_2 = \varphi_1/P_1$ belongs to $\mathcal{C}^k(G - e)$, and $T_1 = T^{(-\gamma(\delta))}$ is a Tashkinov tree with respect to e and φ_2 satisfying $V(T_1) = Z_{\gamma(\delta)}$ and $\delta \in \overline{\varphi}_2(v^0_\delta) \subseteq \overline{\varphi}_2(V(T_1))$. Since g_1, g_2, g_3 neither belong to $E_G(V(G) \setminus V(T), V(G) \setminus V(T))$ nor to $E(P_1)$, their colors did not change and, therefore, we have $\varphi_2(g_1) = \varphi_2(g_2) = \delta$ and $\varphi_2(g_3) = \alpha_0$. Then, evidently, $T_2 = (T_1, g_1, u^1_{\gamma(\delta)}, g_3, u^2_{\gamma(\delta)}, g_2, u_2)$ is a Tashkinov tree with respect to e and φ_2 satisfying $|V(T_2)| > |V(T)| = t(G)$, a contradiction. $\triangle$

For a defective color $\delta \in \Gamma^d(T, e, \varphi)$, let Q_δ denote the chain defined by

$$Q_\delta = P_{v^0_\delta}(\gamma(\delta), \delta, \varphi)$$

and, moreover, let $\varphi_\delta \in \mathcal{C}^k(G - e)$ denote the coloring defined by

$$\varphi_\delta = \varphi/Q_\delta.$$

Evidently, Q_δ is a a path, where one endvertex is v^0_δ and the other endvertex, denoted by z'_δ, belongs to $V(G) \setminus V(T)$ (Fig. 6.1).

Claim 6.13.4 $V(Q_\delta) \cap V(T) = \{v^0_\delta\}$ *for every* $\delta \in \Gamma^d(T, e, \varphi)$.

Proof: Suppose on the contrary that there is a color $\delta \in \Gamma^d(T, e, \varphi)$ with $V(Q_\delta) \cap V(T) \neq \{v_\delta^0\}$. Since $v_\delta^0 \in V(Q_\delta) \cap V(T)$, this implies that the last vertex v_1 in the linear order $\preceq_{(v_\delta^0, Q_\delta)}$ belonging to $V(T)$ satisfies $v_1 \neq v_\delta^0$. Obviously, $v_1 \in F(T, e, \varphi)$ and hence $v_1 \in Y$ (by Claim 6.13.1).

Clearly, there is an edge $f_1 \in E_G(v_1, V(G) \setminus V(T))$ with $\varphi(f_1) = \delta$. Let $u_1 \in V(G) \setminus V(T)$ be the second endvertex of f_1. Note that $f_1 \neq f_\delta^0$. Furthermore, let $P_1 = v_\delta^0 P_\delta z_\delta$ and $P_1' = v_1 Q_\delta z_\delta'$. Then $V(P_1) \cap V(T) = \{v_\delta^0\}$ and $V(P_1') \cap V(T) = \{v_1\}$. Since $v_1 \in F(T, e, \varphi)$, there is a unique color $\gamma' \in \Gamma'$ with $\gamma' \in \overline{\varphi}(v_1)$. Moreover, $v_1 \neq v_\delta^0$ implies $\gamma(\delta) \neq \gamma'$. For convenience, let $\gamma = \gamma(\delta)$.

Since $V(T)$ is closed with respect to φ, no edge in $E_G(V(T), V(G) \setminus V(T))$ is colored with α_0, γ or γ' with respect to φ. Hence we can obtain two new colorings from φ, the first one $\varphi_1 \in C^k(G - e)$ by interchanging the colors α_0 and γ on all edges in $E_G(V(G) \setminus V(T), V(G) \setminus V(T))$, the second one $\varphi_1' \in C^k(G - e)$ by interchanging the colors γ and γ' on all edges in $E_G(V(G) \setminus V(T), V(G) \setminus V(T))$. Clearly, we have $(T, e, \varphi_1) \in \mathcal{T}(G)$ and $(T, e, \varphi_1') \in \mathcal{T}(G)$, $\Gamma^f(T, e, \varphi_1) = \Gamma^f(T, e, \varphi_1') = \Gamma^f(T, e, \varphi)$, $\Gamma^d(T, e, \varphi_1) = \Gamma^d(T, e, \varphi_1') = \Gamma^d(T, e, \varphi)$ and, moreover, $P_1 = P_{v_\delta^0}(\gamma, \delta, \varphi_1)$ and $P_1' = P_{v_1}(\gamma', \delta, \varphi_1')$.

Then $\varphi_2 = \varphi_1/P_1 \in C^k(G - e)$ and we deduce that $T_1 = T^{(-\gamma)}$ is a Tashkinov tree with respect to e and φ_2 satisfying $V(T_1) = Z_\gamma$ and $\delta \in \overline{\varphi}_2(v_\delta^0) \subseteq \overline{\varphi}(V(T_1))$. Since f_1 belongs neither to $E_G(V(G) \setminus V(T), V(G) \setminus V(T))$ nor to $E(P_1)$, its color did not change, so we have $\varphi_2(f_1) = \delta$. Moreover, $v_1 \in Y \subseteq V(T_1)$ and, therefore, $T_2 = (T_1, f_1, u_1)$ is a Tashkinov tree with respect to e and φ_2.

Similarly, the coloring $\varphi_2' = \varphi_1'/P_1' \in C^k(G - e)$ and $T_1' = T^{(-\gamma')}$ is a Tashkinov tree with respect to e and φ_2' satisfying $V(T_1') = Z_{\gamma'}$ and $\delta \in \overline{\varphi}_2'(v_1) \subseteq \overline{\varphi}(v(T_1'))$. Since f_δ^0 belongs neither to $E_G(V(G) \setminus V(T), V(G) \setminus V(T))$ nor to P_1', its color did not change, so we have $\varphi_2'(f_\delta^0) = \delta$. Moreover, $v_\delta^0 \in Y \subseteq V(T_1')$ and, therefore, $T_2' = (T_1', f_\delta^0, u_\delta^0)$ is a Tashkinov tree with respect to e and φ_2'.

Let $Z = V(T_1) \cap V(T_1') = V(T) \setminus U_\gamma \setminus U_{\gamma'}$. Since $\delta \notin \overline{\varphi}(Z)$ and $|Z|$ is odd, also $|E_G(Z, V(G) \setminus Z) \cap E_\delta(e, \varphi)|$ is odd. So, apart from f_δ^0 and f_1, there is another edge $f_2 \in E_G(Z, V(G) \setminus Z)$ with $\varphi(f_2) = \delta$. Since f_2 has an endvertex in $Z \subseteq V(T)$, but is distinct from f_δ^0 or f_1, it neither belongs to $E(P_1)$, to $E(P_1')$, or to $E_G(V(G) \setminus V(T), V(G) \setminus V(T))$. So none of the recolorings have an effect on f_2, which leads to $\varphi_2(f_2) = \varphi_2'(f_2) = \delta$.

Let u_2 be the endvertex of f_2 that belongs to $V(G) \setminus Z$. We claim that $u_2 \notin U_{\gamma'}$. Suppose, on the contrary, that $u_2 \in U_{\gamma'}$, say $u_2 = u_{\gamma'}^1$. From Claim 6.13.2 we then conclude that there is an edge $f' \in E_G(u_{\gamma'}^1, u_{\gamma'}^2)$ with $\varphi(f') = \alpha_0$. Obviously, we have $\varphi_2'(f') = \alpha_0 \in \overline{\varphi}_2'(V(T_2'))$ and, therefore, $T' = (T_2', f_2, u_{\gamma'}^1, f', u_{\gamma'}^2)$ is a Tashkinov tree with respect to e and φ_2' satisfying $|V(T')| > |V(T)| = t(G)$, a contradiction. This proves the claim that $u_2 \notin U_{\gamma'}$. Moreover, from Claim 6.13.3 we conclude that $u_2 \notin U_\gamma$ and, therefore, $u_2 \in V(G) \setminus V(T)$. Hence $T_3 = (T_2, f_2, u_2)$ is a Tashkinov tree with respect to e and φ_2, and $T_3' = (T_2', f_2, u_2)$ is a Tashkinov tree with respect to e and φ_2'. Since $|V(T_3)| = |V(T_3')| = |V(T)| = t(G)$, this implies $(T_3, e, \varphi_2), (T_3', e, \varphi_2') \in \mathcal{T}(G)$.

From Proposition 5.10(b,f) it follows that $|\overline{\varphi}(\{y_0, y_1\})| \geq 4$. So there is a color $\beta \in \overline{\varphi}(\{y_0, y_1\})$ with $\beta \notin \{\alpha_0, \gamma, \gamma'\}$. Obviously, we also have $\beta \neq \delta$ and, therefore, the color β does not matter in any of the mentioned recolorings, which leads to $E_\beta(e, \varphi) = E_\beta(e, \varphi_2) = E_\beta(e, \varphi_2')$. Then, evidently, $\beta \in \overline{\varphi}_2(V(T_3))$. By Proposition 5.10(b), $V(T_3)$ is elementary and closed both with respect to φ_2. Hence there is an edge $f_3 \in E_G(u_2, V(T_3))$ with $\varphi_2(f_3) = \beta$. Clearly, we also have $\varphi(f_3) = \beta$, but since $V(T)$ is closed with respect to φ, the edge f_3 cannot have an endvertex in $V(T)$. Therefore, we conclude that $f_3 \in E_G(u_2, u_1)$. Moreover, we have $\varphi_2'(f_3) = \beta \in \overline{\varphi}_2'(V(T_3'))$ and hence $T_4' = (T_3, f_3, u_1)$ is a Tashkinov tree with respect to e and φ_2' satisfying $|V(T_4')| > |V(T)| = t(G)$, a contradiction. This proves Claim 6.13.4. $\triangle$

As an immediate consequence of Claim 6.13.4 we obtain the following properties of the coloring φ_δ.

Claim 6.13.5 *For every $\delta \in \Gamma^d(T, e, \varphi)$ the coloring $\varphi_\delta \in \mathcal{C}^k(G - e)$ satisfies:*

1. $\varphi_\delta(f_\delta^0) = \gamma(\delta)$,

2. $\varphi_\delta(f) = \varphi(f)$ *for every edge $f \in E_{G-e}(V(T), V(G)) \setminus \{f_\delta^0\}$,*

3. $\varphi_\delta(f) = \varphi(f)$ *for every edge $f \in E(G - e) \setminus E(Q_\delta)$,*

4. $\overline{\varphi}_\delta(v_\delta^0) = \overline{\varphi}(v_\delta^0) \setminus \{\gamma(\delta)\} \cup \{\delta\}$,

5. $\overline{\varphi}_\delta(v) = \overline{\varphi}(v)$ *for every vertex $v \in V(T) \setminus \{v_\delta^0\}$, and*

6. $\overline{\varphi}_\delta(v) = \overline{\varphi}(v)$ *for every vertex $v \in V(G) \setminus V(Q_\delta)$.*

To continue the proof, we establish the following property of the vertex set $Z_{\gamma(\delta)}$.

Claim 6.13.6 $|E_G(Z_{\gamma(\delta)}, V(G) \setminus V(T)) \cap E_{\varphi, \delta}| = 3$ *for every $\delta \in \Gamma^d(T, e, \varphi)$.*

Proof: Suppose on the contrary that there is a color $\delta \in \Gamma^d(T, e, \varphi)$ such that $m = |E_G(Z_{\gamma(\delta)}, V(G) \setminus V(T)) \cap E_{\varphi, \delta}|$ satisfies $m \neq 3$.

First consider the case that $m > 3$. Then apart from f_δ^0 there are another three edges $g_1, g_2, g_3 \in E_G(Z_{\gamma(\delta)}, V(G) \setminus V(T))$ having color δ with respect to φ and hence also with respect to φ_δ (by Claim 6.13.5). Let z_1, z_2, z_3 denote the endvertices of g_1, g_2, g_3 belonging to $V(G) \setminus V(T)$. Then $T_1 = (T^{(-\gamma(\delta))}, g_1, z_1, g_2, z_2, g_3, z_3)$ is a Tashkinov tree with respect to e and φ_δ satisfying $|V(T_1)| > |V(T)| = t(G)$, a contradiction.

It remains to consider the case that $m < 3$. From Claim 6.13.3 we know that $E_G(Z_{\gamma(\delta)}, U_{\gamma(\delta)}) \cap E_{\varphi, \delta} = \emptyset$. Therefore, we have $|E_G(Z_{\gamma(\delta)}, V(G) \setminus Z_{\gamma(\delta)}) \cap E_{\varphi, \delta}| = |E_G(Z_{\gamma(\delta)}, V(G) \setminus V(T)) \cap E_{\varphi, \delta}| < 3$. Since $\delta \notin \overline{\varphi}(Z_{\gamma(\delta)})$ and $|Z_{\gamma(\delta)}|$ is odd, this implies that $|E_G(Z_{\gamma(\delta)}, V(G) \setminus Z_{\gamma(\delta)}) \cap E_{\varphi, \delta}|$ is odd, too, which leads to $|E_G(Z_{\gamma(\delta)}, V(G) \setminus Z_{\gamma(\delta)}) \cap E_{\varphi, \delta}| = 1$. Consequently, the only edge in $E_G(Z_{\gamma(\delta)}, V(G) \setminus Z_{\gamma(\delta)}) \cap E_{\varphi, \delta}$ is f_δ^0.

By Claim 6.13.2, we have $E_G(Z_{\gamma(\delta)}, U_{\gamma(\delta)}) \cap E_{\varphi,\alpha_0} = \emptyset$. Since $\alpha_0 \in \overline{\varphi}(V(T))$ and $V(T)$ is closed with respect to φ, this implies $E_G(Z_{\gamma(\delta)}, V(G) \setminus Z_{\gamma(\delta)}) \cap E_{\varphi,\alpha_0} = \emptyset$. Hence we obtain $E_G(Z_{\gamma(\delta)}, V(G) \setminus Z_{\gamma(\delta)}) \cap E(P_\delta) = \{f_\delta^0\}$. For the subpath $P_1 = y_0 P_\delta v_\delta^0$, this means $V(P_1) \subseteq Z_{\gamma(\delta)}$. Since v_δ^0 is the last vertex in the linear order $\preceq_{(y_0, P_\delta)}$ that belongs to $V(T)$, we conclude that $V(P_\delta) \cap V(T) \subseteq Z_{\gamma(\delta)}$.

Then especially the vertices $u_{\gamma(\delta)}^1, u_{\gamma(\delta)}^2$ do not belong to P_δ. Hence the chain $P_2 = P_{u_{\gamma(\delta)}^1}(\alpha_0, \delta, \varphi)$ is vertex disjoint to P_δ and, moreover, $V(P_2) \cap V(T) \subseteq U_{\gamma(\delta)}$. Then, evidently, $E(P_2) \cap E(T) = \emptyset$ and hence T is a Tashkinov tree with respect to e and the coloring $\varphi_2 = \varphi/P_2$. Since $|E_G(Z_{\gamma(\delta)}, V(G) \setminus V(T)) \cap E_{\varphi,\delta}| = 1$, Proposition 5.10(e) implies that there are two edges $g_4 \in E_G(u_\delta^1, V(G) \setminus V(T))$ and $g_5 \in E_G(u_\delta^2, V(G) \setminus V(T))$ with $\varphi(g_4) = \varphi(g_5) = \delta$. Evidently, $g_4 \in E(P_2)$ and $\varphi_2(g_4) = \alpha_0 \in \overline{\varphi}_2(y_0)$. If u_4 is the endvertex of g_4 belonging to $V(G) \setminus V(T)$, then $T_2 = (T, g_4, u_4)$ is a Tashkinov tree with respect to e and φ_2 satisfying $|V(T_2)| > |V(T)| = t(G)$, a contradiction. This proves the claim. $\triangle$

For every $\delta \in \Gamma^d(T, e, \varphi)$, we know from Claim 6.13.6 that besides f_δ^0 there are two other edges $f_\delta^1, f_\delta^2 \in E_G(Z_{\gamma(\delta)}, V(G) \setminus V(T))$ with $\varphi(f_\delta^1) = \varphi(f_\delta^2) = \delta$. For $j = 1, 2$, let $f_\delta^j \in E_G(v_\delta^j, u_\delta^j)$ where $v_\delta^1, v_\delta^2 \in Z_{\gamma(\delta)}$ and $u_\delta^1, u_\delta^2 \in V(G) \setminus V(T)$. Furthermore, let

$$U_\delta = \{u_\delta^1, u_\delta^2\}.$$

By Claim 6.13.4, we have $f_\delta^1, f_\delta^2 \notin E(Q_\delta)$ and, therefore, $u_\delta^1, u_\delta^2 \notin V(Q_\delta)$. Hence 6.13.5 implies the following result.

Claim 6.13.7 $\varphi_\delta(f) = \varphi(f)$ *for every color* $\delta \in \Gamma^d(T, e, \varphi)$ *and every edge* $f \in E_G(U_\delta, V(G))$.

In particular, for every $\delta \in \Gamma^d(T, e, \varphi)$, this leads to $\varphi_\delta(f_\delta^1) = \varphi_\delta(f_\delta^2) = \delta$. From $\delta \in \overline{\varphi}_\delta(v_\delta^0)$ it then follows that the sequence T_δ, defined by

$$T_\delta = (T^{(-\gamma(\delta))}, f_\delta^1, u_\delta^1, f_\delta^2, u_\delta^2),$$

is a Tashkinov tree with respect to e and φ_δ satisfying $|V(T_\delta)| = |V(T)|$. Therefore, we obtain the following result.

Claim 6.13.8 $(T_\delta, e, \varphi_\delta) \in \mathcal{T}(G)$ *for every color* $\delta \in \Gamma^d(T, e, \varphi)$.

Since $V(T)$ is closed with respect to φ, for every $\delta \in \Gamma^d(T, e, \varphi)$ and every $\alpha \in \overline{\varphi}(Z_{\gamma(\delta)}) \setminus \{\gamma(\delta)\}$, we have $E_G(Z_{\gamma(\delta)}, V(G) \setminus V(T)) \cap E_{\varphi,\alpha} = \emptyset$. This implies $E_G(Z_{\gamma(\delta)}, U_\delta) \cap E_{\varphi,\alpha} = \emptyset$ and, moreover, $E_G(Z_{\gamma(\delta)}, U_\delta) \cap E_{\varphi_\delta,\alpha} = \emptyset$ (Claim 6.13.7). Since $\alpha \in \overline{\varphi}_\delta(Z_{\gamma(\delta)}) \subseteq \overline{\varphi}_\delta(V(T_\delta))$ and since $V(T_\delta)$ is elementary and closed with respect to φ_δ (Proposition 5.10(b)), there must be an edge between u_δ^1 and u_δ^2 colored with α with respect to φ_δ and also with respect to φ (Claim 6.13.7). Therefore, we obtain the following result.

Claim 6.13.9 $E_G(u_\delta^1, u_\delta^2) \cap E_{\varphi,\alpha} \neq \emptyset$ *for every* $\delta \in \Gamma^d(T, e, \varphi)$ *and every* $\alpha \in \overline{\varphi}(Z_{\gamma(\delta)}) \setminus \{\gamma(\delta)\}$.

Claim 6.13.10 $E_G(u^1_{\gamma(\delta)}, u^2_{\gamma(\delta)}) \cap E_{\varphi,\delta} \neq \emptyset$ *for every* $\delta \in \Gamma^d(T, e, \varphi)$.

Proof: Let $\delta \in \Gamma^d(T, e, \varphi)$. Obviously, $|Z_{\gamma(\delta)}| = r + 2s - 2$ and, therefore, $|\overline{\varphi}(Z_{\gamma(\delta)})| \geq r + 2s$ (Proposition 5.10(b,f)). Since there are at most $r - 2 + s$ colors used on T with respect to φ, there is a color $\beta \in \overline{\varphi}(Z_{\gamma(\delta)}) \cap \Gamma^f(T, e, \varphi)$. Let $v \in Z_{\gamma(\delta)}$ be the unique vertex with $\beta \in \overline{\varphi}(v)$ and let $P = P_v(\beta, \delta, \varphi)$. From Claim 6.13.6 and Proposition 5.11 we conclude that P is a path having one endvertex v and another endvertex $z \in V(G) \setminus V(T)$, and that P satisfies

$$E(P) \cap E_G(Z_{\gamma(\delta)}, V(G) \setminus V(T)) = \{f^0_\delta, f^1_\delta, f^2_\delta\}.$$

By Claim 6.13.1, we have $F(T, e, \varphi) \subseteq Y \subseteq Z_{\gamma(\delta)}$, and so, for the last vertex v' in the linear order $\preceq_{(v,P)}$ belonging to $V(T)$, we deduce that $v' \in \{v^0_\delta, v^1_\delta, v^2_\delta\}$.

From Claim 6.13.9 it follows that there is an edge $g \in E_G(u^1_\delta, u^2_\delta)$ with $\varphi(g) = \beta$. Hence the path

$$P_1 = \text{Path}(v^1_\delta, f^1_\delta, u^1_\delta, g, u^2_\delta, f^2_\delta, v^2_\delta)$$

is a subpath of P and, therefore, $v' \notin \{v^1_\delta, v^2_\delta\}$, but $v' = v^0_\delta$.

We have $E_G(U_{\gamma(\delta)}, Z_{\gamma(\delta)}) \cap E_{\varphi,\beta} = \emptyset$ (Claim 6.13.2) and $E_G(U_{\gamma(\delta)}, Z_{\gamma(\delta)}) \cap E_{\varphi,\delta} = \emptyset$ (Claim 6.13.3). Hence, for the subpath $P_2 = vPv^0_\delta$, we conclude that $V(P_2) \subseteq Z_{\gamma(\delta)} \cup U_\delta$. Further, for $P_3 = v^0_\delta Pz$, we clearly have $V(P_3) \cap V(T) = \{v^0_\delta\}$ and hence $V(P) \cap U_{\gamma(\delta)} = \emptyset$. From Proposition 5.11(b) it then follows that $E_G(U_{\gamma(\delta)}, V(G) \setminus V(T)) \cap E_{\varphi,\delta} = \emptyset$. Since $E_G(U_{\gamma(\delta)}, Z_{\gamma(\delta)}) \cap E_{\varphi,\delta} = \emptyset$ and $\delta \notin \overline{\varphi}(V(T))$, we conclude that $E_G(u^1_{\gamma(\delta)}, u^2_{\gamma(\delta)}) \cap E_{\varphi,\delta} \neq \emptyset$, as required. $\triangle$

Claim 6.13.11 $E_G(u^1_\delta, u^2_\delta) \cap E_{\varphi,\gamma(\delta)} \neq \emptyset$ *for every* $\delta \in \Gamma^d(T, e, \varphi)$.

Proof: Let $\delta \in \Gamma^d(T, e, \varphi)$. Obviously, $|Z_{\gamma(\delta)}| = r + 2s - 2$ and, therefore, $|\overline{\varphi}(Z_{\gamma(\delta)})| \geq r + 2s$ (Proposition 5.10(b,f)). Since there are at most $r - 2 + s$ colors used on T_δ with respect to φ_δ, there is a color $\beta \in \overline{\varphi}_\delta(Z_{\gamma(\delta)}) \cap \Gamma^f(T, e, \varphi_\delta)$. Let $v \in Z_{\gamma(\delta)}$ be the unique vertex with $\beta \in \overline{\varphi}_\delta(v)$.

All three edges $f^0_\delta, f^1_{\gamma(\delta)}, f^1_{\gamma(\delta)}$ are contained in $F = E_G(Z_{\gamma(\delta)}, V(G) \setminus V(T_\delta))$ and from Claim 6.13.5 it follows that $\varphi_\delta(f^0_\delta) = \varphi_\delta(f^1_{\gamma(\delta)}) = \varphi_\delta(f^2_{\gamma(\delta)}) = \gamma(\delta)$. Furthermore, these three edges are the only edges in F having color $\gamma(\delta)$ with respect to the coloring φ_δ. Otherwise, such an edge f would, by Claim 6.13.5, satisfy $\varphi(f) = \gamma(\delta)$, and f would belong to $E_G(V(T), V(G) \setminus V(T))$. Since $\gamma(\delta) \in \overline{\varphi}(V(T))$, this would contradict the fact that $V(T)$ is closed with respect to φ. Hence we have $E_G(Z_{\gamma(\delta)}, V(G) \setminus V(T_\delta)) \cap E_{\varphi_\delta,\gamma(\delta)} = \{f^0_\delta, f^1_{\gamma(\delta)}, f^2_{\gamma(\delta)}\}$. From this and Proposition 5.11 we conclude that $P = P_v(\beta, \gamma(\delta), \varphi_\delta)$ is a path, where one endvertex is v and the other endvertex is some vertex $z \in V(G) \setminus V(T_\delta)$, and that P satisfies $E(P) \cap E_G(Z_{\gamma(\delta)}, V(G) \setminus V(T_\delta)) = \{f^0_\delta, f^1_{\gamma(\delta)}, f^2_{\gamma(\delta)}\}$.

By Proposition 5.10(f), we have $\overline{\varphi}_\delta(u^1_\delta) \neq \emptyset$ and $\overline{\varphi}_\delta(u^2_\delta) \neq \emptyset$. Since no color in $\overline{\varphi}_\delta(U_\delta)$ is used on T_δ with respect to φ_δ, Proposition 6.12 implies $F(T_\delta, e, \varphi_\delta) \subseteq Z_{\gamma(\delta)}$. Hence, for the last vertex v' in the linear order $\preceq_{(v,P)}$ belonging to $V(T)$, we conclude that $v' \in \{v^0_\delta, v_1, v_2\}$, where v_1, v_2 are the two endvertices of $f^1_{\gamma(\delta)}, f^2_{\gamma(\delta)}$ belonging to $Z_{\gamma(\delta)}$.

From Claim 6.13.2 it follows that there is an edge $g \in E_G(u^1_{\gamma(\delta)}, u^2_{\gamma(\delta)})$ with $\varphi(g) = \beta$ and, by Claim 6.13.5, also with $\varphi_\delta(g) = \beta$. Hence

$$P_1 = \text{Path}(v_1, f^1_{\gamma(\delta)}, u^1_{\gamma(\delta)}, g, u^2_{\gamma(\delta)}, f^2_{\gamma(\delta)}, v_2)$$

is a subpath of P and, therefore, $v' \notin \{v_1, v_2\}$, but $v' = v^0_\delta$.

Since $V(T)$ is closed with respect to φ, we clearly have $E_G(U_\delta, Z_{\gamma(\delta)}) \cap E_{\varphi, \beta} = \emptyset$ and $E_G(U_\delta, Z_{\gamma(\delta)}) \cap E_{\varphi, \gamma(\delta)} = \emptyset$. Therefore, by Claim 6.13.5, $E_G(U_\delta, Z_{\gamma(\delta)}) \cap E_{\varphi_\delta, \beta}) = \emptyset$ and $E_G(U_\delta, Z_{\gamma(\delta)}) \cap E_{\varphi_\delta, \gamma(\delta)} = \emptyset$. Hence, for the subpath $P_2 = vPv^0_\delta$, we conclude that $V(P_2) \subseteq Z_{\gamma(\delta)} \cup U_{\gamma(\delta)}$. Furthermore, for $P_3 = v^0_\delta Pz$, we clearly have $V(P_3) \cap V(T_\delta) = \{v^0_\delta\}$ and, therefore, $V(P) \cap U_\delta = \emptyset$. From Proposition 5.11(b) it then follows that $E_G(U_\delta, V(G) \setminus V(T_\delta)) \cap E_{\varphi_\delta, \gamma(\delta)} = \emptyset$. Since we have $E_G(U_\delta, Z_{\gamma(\delta)}) \cap E_{\varphi_\delta, \gamma(\delta)} = \emptyset$ and $\gamma(\delta) \notin \overline{\varphi}_\delta(V(T_\delta))$, we conclude that $E_G(u^1_\delta, u^2_\delta) \cap E_{\varphi_\delta, \gamma(\delta)} \neq \emptyset$. From Claim 6.13.7 it then follows that $E_G(u^1_\delta, u^2_\delta) \cap E_{\varphi, \gamma(\delta)} \neq \emptyset$. This proves the claim. $\triangle$

Claim 6.13.12 $E_G(u^1_{\gamma(\delta)}, u^2_{\gamma(\delta)}) \cap E_{\varphi, \alpha} \neq \emptyset$ *for every* $\delta \in \Gamma^d(T, e, \varphi)$ *and every* $\alpha \in \overline{\varphi}(U_\delta)$.

Proof: Let $\delta \in \Gamma^d(T, e, \varphi)$ and let $\alpha \in \overline{\varphi}(U_\delta)$. Then, clearly, $\alpha \neq \delta$ and, by Claim 6.13.11, also $\alpha \neq \gamma(\delta)$. Hence we have $E_{\varphi, \alpha} = E_{\varphi_\delta, \alpha}$ and, therefore, $\alpha \in \overline{\varphi}_\delta(U_\delta)$. Moreover, $\alpha_0 \in \overline{\varphi}_\delta(y_0)$ and $V(T_\delta)$ is elementary with respect to φ_δ. Hence we have $\alpha \neq \alpha_0$. Since $V(T_\delta)$ is also closed with respect to φ_δ, for $P = P_{u^1_{\gamma(\delta)}}(\alpha_0, \alpha, \varphi) = P_{u^1_{\gamma(\delta)}}(\alpha_0, \alpha, \varphi_\delta)$, we have $V(P) \subseteq V(G) \setminus V(T_\delta)$. This implies $V(P) \cap V(T) \subseteq U_{\gamma(\delta)}$ and, therefore, $E(P) \cap E(T) = \emptyset$. Then T is a Tashkinov tree with respect to e and the coloring $\varphi' = \varphi/P$, and from $y_0 \notin V(P)$ we conclude that $\alpha_0 \in \overline{\varphi}'(V(T))$.

From Claim 6.13.2 we know that $E_G(u^1_{\gamma(\delta)}, u^2_{\gamma(\delta)}) \cap E_{\varphi, \alpha_0} \neq \emptyset$ and, therefore, we have $u^2_{\gamma(\delta)} \in V(P)$. If there is an edge $g \in E_G(U_{\gamma(\delta)}, z)$ with $z \in V(G) \setminus V(T)$ and $\varphi(g) = \alpha$, then we would have $\varphi'(g) = \alpha_0 \in \overline{\varphi}'(V(T))$, and $T' = (T, g, z)$ would be a Tashkinov tree with respect to e and φ' satisfying $|V(T')| > |V(T)| = t(G)$, a contradiction. If there is an edge $g \in E_G(U_{\gamma(\delta)}, Z_{\gamma(\delta)})$ with $\varphi(g) = \alpha$, then we would have $\varphi_\delta(g) = \alpha \in \overline{\varphi}_\delta(V(T_\delta))$, a contradiction, too, because $V(T_\delta)$ is closed with respect to φ_δ. Consequently, we have $E_G(U_{\gamma(\delta)}, V(G) \setminus U_{\gamma(\delta)}) \cap E_{\varphi, \alpha} = \emptyset$. Since $V(T)$ is elementary with respect to φ, we conclude that there is an edge $g \in E_G(u^1_{\gamma(\delta)}, u^2_{\gamma(\delta)})$ with $\varphi(g) = \alpha$. This proves the claim. $\triangle$

Claim 6.13.13 $E_G(u^1_\delta, u^2_\delta) \cap E_{\varphi, \alpha} \neq \emptyset$ *for every* $\delta \in \Gamma^d(T, e, \varphi)$ *and every* $\alpha \in \overline{\varphi}(U_{\gamma(\delta)})$.

Proof: Let $\delta \in \Gamma^d(T, e, \varphi)$ and let $\alpha \in \overline{\varphi}(U_{\gamma(\delta)})$. Then, clearly, we have $\alpha \neq \gamma(\delta)$ and, by Claim 6.13.10, also $\alpha \neq \delta$. Hence we have $E_{\varphi, \alpha} = E_{\varphi_\delta, \alpha}$. Moreover, $\alpha_0 \in \overline{\varphi}_\delta(y_0)$ and $V(T)$ is elementary with respect to φ implying $\alpha \neq \alpha_0$. Since $V(T)$ is also closed with respect to φ, for $P = P_{u^1_\delta}(\alpha_0, \alpha, \varphi) = P_{u^1_\delta}(\alpha_0, \alpha, \varphi_\delta)$, we have $V(P) \subseteq V(G) \setminus V(T)$. This implies $V(P) \cap V(T_\delta) \subseteq U_\delta$ and, therefore, $E(P) \cap E(T_\delta) = \emptyset$. Then T_δ is a Tashkinov tree with respect to e and the coloring $\varphi' = \varphi_\delta/P$, and from $y_0 \notin V(P)$ we conclude that $\alpha_0 \in \overline{\varphi}'(V(T_\delta))$.

From Claim 6.13.9 we know that $E_G(u_\delta^1, u_\delta^2) \cap E_{\varphi,\alpha_0} \neq \emptyset$ and, therefore, we have $u_\delta^2 \in V(P)$. If there is an edge $g \in E_G(U_\delta, z)$ with $z \in V(G) \setminus V(T_\delta)$ and $\varphi_\delta(g) = \alpha$, then we would have $\varphi'(g) = \alpha_0 \in \overline{\varphi}'(V(T_\delta))$, and $T' = (T_\delta, g, z)$ would be a Tashkinov tree with respect to e and φ' satisfying $|V(T')| > |V(T_\delta)| = t(G)$, a contradiction. If there is an edge $g \in E_G(U_\delta, Z_{\gamma(\delta)})$ with $\varphi_\delta(g) = \alpha$, then we would have $\varphi(g) = \alpha \in \overline{\varphi}(V(T))$, a contradiction, too, because $V(T)$ is closed with respect to φ. Consequently, we have $E_G(U_\delta, V(G) \setminus U_\delta) \cap E_{\varphi_\delta, \alpha} = \emptyset$. Since $V(T_\delta)$ is elementary with respect to φ_δ, we conclude that there is an edge $g \in E_G(u_\delta^1, u_\delta^2)$ with $\varphi_\delta(g) = \varphi(g) = \alpha$. This proves the claim. $\triangle$

Claim 6.13.14 $U_\delta \subseteq A(T, e, \varphi)$ *for every* $\delta \in \Gamma^d(T, e, \varphi)$.

Proof: Let $\delta \in \Gamma^d(T, e, \varphi)$ and let $u \in U_\delta$. Our aim is to show that $u \in A(T, e, \varphi)$. To this end, choose an arbitrary color $\alpha \in \overline{\varphi}(u)$ and a free color $\beta \in \Gamma^f(T, e, \varphi) \setminus \{\alpha\}$. Since $\delta \in \varphi(u)$ and $\gamma(\delta) \in \varphi(u)$ (Claim 6.13.11), we have $\alpha \notin \{\delta, \gamma(\delta)\}$. Since neither δ nor $\gamma(\delta)$ is a free color with respect to (T, e, φ), we also have $\beta \notin \{\delta, \gamma(\delta)\}$. Hence $E_{\varphi,\alpha} = E_{\varphi_\delta,\alpha}$ and $E_{\varphi,\beta} = E_{\varphi_\delta,\beta}$. Since $V(T)$ is elementary with respect to φ, there is a unique vertex v in T such that $\beta \in \overline{\varphi}(v)$. Now let $P = P_u(\alpha, \beta, \varphi)$. Then $P = P_u(\alpha, \beta, \varphi_\delta)$ and P is a path having u as an endvertex. To complete the proof that $u \in A(T, e, \varphi)$, we have to show that v is the other endvertex of P. For the proof we shall use the following facts: $(T_\delta, e, \varphi_\delta) \in \mathcal{T}(G)$ (Claim 6.13.8), $U_\delta = \{u_\delta^1, u_\delta^2\}$ consists of two vertices, $\overline{\varphi}(w) = \overline{\varphi}_\delta(w)$ for $w \in U_\delta$ (Claim 6.13.7), and $\beta \in \overline{\varphi}_\delta(v)$ (Claim 6.13.5).

Consider first the case that $v \in Z_{\gamma(\delta)}$. Then $u, v \in V(T_\delta)$ and (u, v) is an (α, β)-pair with respect to φ_δ. Since $P = P_u(\alpha, \beta, \varphi_\delta)$, we conclude from Theorem 5.3(e) that u and v are the endvertices of P.

It remains to consider the case that $v \in U_{\gamma(\delta)}$. By Claim 6.13.3, we have $E_G(u_\delta^1, u_\delta^2) \cap E_{\varphi,\beta} \neq \emptyset$ and, therefore, $U_\delta \subseteq V(P)$. Since $\alpha \in \overline{\varphi}_\delta(u)$ and $V(T_\delta)$ is elementary and closed with respect to φ_δ, there is an edge $f_2 \in E_G(U_\delta, Z_{\gamma(\delta)})$ with $\varphi_\delta(f_2) = \alpha$ and, moreover, $E_G(Z_{\gamma(\delta)}, V(G) \setminus Z_{\gamma(\delta)}) \cap E_{\varphi_\delta, \alpha} = \{f_2\}$. Therefore, $E_G(Z_{\gamma(\delta)}, V(G) \setminus Z_{\gamma(\delta)}) \cap E_{\varphi,\alpha} = \{f_2\}$ and $f_2 \in E(P)$. Since $\beta \in \overline{\varphi}(U_{\gamma(\delta)})$ and $V(T)$ is elementary and closed with respect to φ, there is an edge $f_3 \in E_G(U_{\gamma(\delta)}, Z_{\gamma(\delta)})$ with $\varphi(f_3) = \beta$ and, moreover, $E_G(Z_{\gamma(\delta)}, V(G) \setminus Z_{\gamma(\delta)}) \cap E_{\varphi,\beta} = \{f_3\}$. Since we have $f_2 \in E(P)$ and $\alpha, \beta \notin \overline{\varphi}(Z_{\gamma(\delta)})$, and since f_2, f_3 are the only two edges in $E_G(Z_{\gamma(\delta)}, V(G) \setminus Z_{\gamma(\delta)})$ colored with α or β with respect to φ, we conclude that $f_3 \in E(P)$. By Claim 6.13.12, there is an edge $f_4 \in E_G(u_{\gamma(\delta)}^1, u_{\gamma(\delta)}^2)$ with $\varphi(f_4) = \alpha$. This implies $f_4 \in E(P)$ and, therefore, $U_{\gamma(\delta)} \subseteq V(P)$. Hence we have $v \in V(P)$, so v must be the second endvertex of P.

In both cases P is a path with endvertices u and v. Hence, by definition, we have $u \in A(T, e, \varphi)$ and the claim is proved. $\triangle$

Now, consider the set

$$X = V(T) \cup \bigcup_{\delta \in \Gamma^{\mathrm{d}}(T,e,\varphi)} U_\delta.$$

Then, by Claim 6.13.14, we have $X \subseteq V(T) \cup A(T, e, \varphi)$. Hence Proposition 5.12 implies the following result.

Claim 6.13.15 *X is elementary with respect to φ.*

Claim 6.13.16 *The set $X_\delta = V(T) \cup U_\delta$ is closed with respect to φ for every $\delta \in \Gamma^{\mathrm{d}}(T, e, \varphi)$.*

Proof: Let $\delta \in \Gamma^{\mathrm{d}}(T, e, \varphi)$ and let $\alpha \in \overline{\varphi}(X_\delta)$. Now we have to show that $\partial_G(X_\delta) \cap E_{\varphi,\alpha} = \emptyset$.

If $\alpha \in \overline{\varphi}(V(T))$, then based on Claim 6.13.9, Claim 6.13.11, and Claim 6.13.13 we deduce that $E_G(u_\delta^1, u_\delta^2) \cap E_{\varphi,\alpha} \neq \emptyset$ and, therefore, $E_G(U_\delta, V(G) \setminus X_\delta) \cap E_{\varphi,\alpha} = \emptyset$. Since $V(T)$ is closed with respect to φ, we have $E_G(V(T), V(G) \setminus X_\delta) \cap E_{\varphi,\alpha} = \emptyset$. This implies that $\partial_G(X_\delta) \cap E_{\varphi,\alpha} = \emptyset$.

If $\alpha \in \overline{\varphi}(U_\delta)$, then we have $\alpha \neq \delta$ and also $\alpha \neq \gamma(\delta)$ (Claim 6.13.11). Hence $E_{\varphi,a} = E_{\varphi_\delta,\alpha}$ and, therefore, $\alpha \in \overline{\varphi}_\delta(V(T_\delta))$. Since $V(T_\delta)$ is closed with respect to φ_δ, $E_G(V(T_\delta), V(G) \setminus X_\delta) \cap E_{\varphi,\alpha} = E_G(V(T_\delta), V(G) \setminus X_\delta) \cap E_{\varphi_\delta,\alpha} = \emptyset$. From Claim 6.13.12 we know that $E_G(u_{\gamma(\delta)}^1, u_{\gamma(\delta)}^2) \cap E_{\varphi,\alpha} \neq \emptyset$ and, therefore, $E_G(U_{\gamma(\delta)}, V(G) \setminus X_\delta) \cap E_{\varphi,\alpha} = \emptyset$. This implies that $\partial_G(X_\delta) \cap E_{\varphi,\alpha} = \emptyset$.   △

By Claim 6.13.9, $E_G(u_\delta^1, u_\delta^2) \cap E_{\varphi,\alpha_0} \neq \emptyset$ for all $\delta \in \Gamma^{\mathrm{d}}(T, e, \varphi)$. Hence we obtain the following result.

Claim 6.13.17 *For any $\delta, \delta' \in \Gamma^{\mathrm{d}}(T, e, \varphi)$ the sets U_δ and $U_{\delta'}$ are either equal or disjoint.*

Claim 6.13.18 *The set X is closed with respect to φ.*

Proof: Suppose on the contrary that X is not closed with respect to φ, that is, there exists a color $\alpha \in \overline{\varphi}(X)$ and an edge $f \in \partial_G(X)$ with $\varphi(f) = \alpha$. Then, clearly, there is a color $\delta \in \Gamma^{\mathrm{d}}(T, e, \varphi)$ satisfying $f \in E_G(V(T) \cup U_\delta, V(G) \setminus X)$. Since, by Claim 6.13.16, $V(T) \cup U_\delta$ is closed with respect to φ, we conclude that $\alpha \in \overline{\varphi}(X \setminus V(T) \setminus U_\delta)$ and, therefore, by Claim 6.13.17, $\alpha \in \overline{\varphi}(U_{\delta'})$ for a color $\delta' \in \Gamma^{\mathrm{d}}(T, e, \varphi)$ with $U_\delta \cap U_{\delta'} = \emptyset$.

Since, by Claim 6.13.16, also $V(T) \cup U_{\delta'}$ is closed with respect to φ, we have $f \notin E_G(V(T) \cup U_{\delta'}, V(G) \setminus X)$. In particular, this means $f \notin E_G(V(T), V(G) \setminus X)$ and, therefore, we conclude that $f \in E_G(u, v)$ for two vertices $u \in U_\delta$ and $v \in V(G) \setminus X$. Now let $P = P_u(\alpha_0, \alpha, \varphi)$. Since $\alpha_0, \alpha \in \overline{\varphi}(V(T) \cup U_{\delta'})$ and $V(T) \cup U_{\delta'}$ is closed with respect to φ, this implies $V(P) \cap V(T) = \emptyset$. Hence we have $E(P) \cap E(T_\delta) = \emptyset$.

By Claim 6.13.15, X is elementary with respect to φ. Since $\alpha \in \overline{\varphi}(U_{\delta'})$ and $\Gamma' \subseteq \overline{\varphi}(V(T))$, we obtain $\alpha \notin \Gamma'$. Moreover, since $V(T) \cup U_{\delta'}$ is closed with respect to φ (Claim 6.13.16) and since $f_\delta^1 \in E_G(V(T), V(G) \setminus V(T) \setminus U_{\delta'})$, we conclude that $\alpha \neq \delta$. Moreover, we also have $\alpha_0 \notin \Gamma'$ and $\alpha_0 \neq \delta$. Consequently, we have $E_{\varphi,\alpha} = E_{\varphi_\delta,\alpha}$ and $E_{\varphi,\alpha_0} = E_{\varphi_\delta,\alpha_0}$, which especially implies $P = P_u(\alpha_0, \alpha, \varphi_\delta)$. From $E(P) \cap E(T_\delta) = \emptyset$ it then follows that T_δ is a Tashkinov tree with respect to e and $\varphi' = \varphi_\delta/P$. Since $f \in E(P)$, we have $\varphi'(f) = \alpha_0 \in \overline{\varphi}'(V(T_\delta))$. Hence $T' = (T_\delta, f, v)$ is a Tashkinov tree with respect to e and φ' satisfying $|V(T')| > |V(T_\delta)| = t(G)$, a contradiction. This proves the claim. △

Claim 6.13.19 *If $\alpha \notin \overline{\varphi}(X)$ and $P = P_{y_0}(\alpha_0, \alpha, \varphi)$, then $|E(P) \cap \partial_G(X)| = 1$.*

Proof: By Claim 6.13.15, X is elementary with respect to φ and, since $\alpha_0 \in \overline{\varphi}(y_0)$, we know that P is a path, where one endvertex is y_0 and the another endvertex is some vertex $z \in V(G) \setminus X$. Evidently, there is a last vertex v in the linear order $\preceq_{(y_0, P)}$ that belongs to X, and there is an edge in $g \in E_G(v, V(G) \setminus X)$ with $\varphi(g) = \alpha$. For the subpath $P_1 = y_0 P v$ of P, we have to show that $V(P_1) \subseteq X$, this would complete the proof of the claim. To this end, we distinguish three cases.

Case 1: $v \in V(T)$ *and* $\alpha \notin \Gamma^d(T, e, \varphi)$. Then we have $\partial_G(V(T)) \cap E_{\varphi,\alpha} = \{g\}$. Since $\alpha_0 \in \overline{\varphi}(V(T))$ and since $V(T)$ is closed with respect to φ, we conclude that $E(P) \cap \partial_G(V(T)) = \{g\}$ and, therefore, $V(P_1) \subseteq V(T) \subseteq X$.

Case 2: $v \in V(T)$ *and* $\alpha \in \Gamma^d(T, e, \varphi)$. Then from Claim 6.13.6 and Claim 6.13.10 we conclude that $\partial_G(V(T)) \cap E_{\varphi,\alpha} = \{f_\alpha^0, f_\alpha^1, f_\alpha^2\}$. Hence we have $\partial_G(X_\alpha) = \{f_\alpha^0\}$, which implies $g = f_\alpha^0$. Since $\alpha_0 \in \overline{\varphi}(X_\alpha)$ and since X_α is closed with respect to φ (Claim 6.13.16), it follows that $V(P_1) \subseteq X_\alpha \subseteq X$.

Case 3: $v \notin V(T)$. Then, evidently, $v \in U_\delta$ for some $\delta \in \Gamma^d(T, e, \varphi)$. Since $\varphi(g) = \alpha$, we conclude that $\alpha \neq \delta$. Clearly, we also have $\alpha \neq \gamma(\delta)$ and, therefore, we deduce that $E_{\varphi,\alpha} = E_{\varphi_\delta,\alpha}$. Moreover, we also have $\alpha_0 \notin \{\delta, \gamma(\delta)\}$ and, therefore, it follows that $P = P_{y_0}(\alpha_0, \alpha, \varphi_\delta)$. Clearly, v is the last vertex in $\preceq_{(y_0, P)}$ that belongs to $V(T_\delta)$. Since no color from $\overline{\varphi}_\delta(v)$ is used on T_δ with respect to φ_δ, we deduce that $\overline{\varphi}_\delta(v) \cap \Gamma^f(T_\delta, e, \varphi_\delta) \neq \emptyset$. From Proposition 6.12 it then follows that $v \notin F(T_\delta, e, \varphi_\delta)$ and, therefore, $\alpha \notin \Gamma^d(T_\delta, e, \varphi_\delta)$. Hence we conclude that $E(P) \cap \partial_G(V(T_\delta)) = \{g\}$, which implies $V(P_1) \subseteq V(T_\delta) \subseteq X$.

In any of the three cases we have $V(P_1) \subseteq X$, which implies $E(P) \cap \partial_G(X) = \{g\}$. Hence the proof is finished. △

Claim 6.13.20 *X is strongly closed with respect to φ*

Proof: Suppose on the contrary that X is not strongly closed with respect to φ. Since, by Claim 6.13.18, X is closed with respect to φ, this implies that there is a color α satisfying $\alpha \notin \overline{\varphi}(X)$ and $|\partial_G(X) \cap E_{\varphi,\alpha}| \geq 2$. Obviously, this implies $|\partial_G(X) \cap E_{\varphi,\alpha}| \geq 3$, because $|X|$ is odd.

If $|E_G(V(T), V(G) \setminus X) \cap E_{\varphi,\alpha}| \geq 2$ then we would have $\alpha \in \Gamma^d(T, e, \varphi)$, but then Claim 6.13.6 and Claim 6.13.10 would imply $\partial_G(V(T)) \cap E_{\varphi,\alpha} = \{f_\alpha^0, f_\alpha^1, f_\alpha^2\}$ and, therefore, $E_G(V(T), V(G) \setminus X) \cap E_{\varphi,\alpha} = \{f_\alpha^0\}$, a contradiction. Consequently, $|E_G(V(T), V(G) \setminus X) \cap E_{\varphi,\alpha}| \leq 1$, which leads to $|E_G(X \setminus V(T), V(G) \setminus X) \cap E_{\varphi,\alpha}| \geq 2$.

The path $P = P_{y_0}(\alpha_0, \alpha, \varphi)$ satisfies $|E(P) \cap \partial_G(X)| = 1$ (Claim 6.13.19). Hence there is a color $\delta \in \Gamma^{\mathrm{d}}(T, e, \varphi)$ and an edge $f \in E_G(U_\delta, V(G) \setminus X)$ satisfying $\varphi(f) = \alpha$ and $f \notin E(P)$. Let u be the endvertex of f that belongs to $V(G) \setminus X$ and let $P' = P_u(\alpha_0, \alpha, \varphi)$. Since $f \in E(P') \setminus E(P)$, we deduce that P and P' are vertex disjoint. Furthermore, we claim that $V(P') \cap V(T) = \emptyset$. To prove this, we consider two cases.

Case 1: $\alpha \in \Gamma^{\mathrm{d}}(T, e, \varphi)$. From Claim 6.13.6 and Claim 6.13.10 we conclude that $\partial_G(V(T)) \cap E_{\varphi,\alpha} = \{f_\alpha^0, f_\alpha^1, f_\alpha^2\}$ and, therefore, we have $|\partial_G(X_\alpha) \cap E_{\varphi,\alpha}| = 1$. Since $\alpha_0 \in \overline{\varphi}(X_\alpha)$ and since X_α is closed with respect to φ (Claim 6.13.16), we also have $\partial_G(X_\alpha) \cap E_{\varphi,\alpha_0} = \emptyset$. Since the only edge in $\partial_G(X_\alpha) \cap E_{\varphi,\alpha}$ must belong to $E(P)$, we conclude that $V(P') \cap X_\alpha = \emptyset$ and, therefore, also $V(P') \cap V(T) = \emptyset$.

Case 2: $\alpha \notin \Gamma^{\mathrm{d}}(T, e, \varphi)$. Then the only edge in $\partial_G(V(T)) \cap E_{\varphi,\alpha}$ belongs to $E(P)$. Since $\alpha_0 \in \overline{\varphi}(V(T))$ and since $V(T)$ is closed with respect to φ, we conclude that $V(P') \cap V(T) = \emptyset$.

In any case we have $V(P') \cap V(T) = \emptyset$, which implies $E(P') \cap E(T_\delta) = \emptyset$. Moreover, from $\alpha, \alpha_0 \notin \{\delta, \gamma(\delta)\}$ we conclude that $P' = P_u(\alpha_0, \alpha, \varphi_\delta)$. Then, evidently, T_δ is a Tashkinov tree with respect to e and $\varphi' = \varphi_\delta/P'$. From $\varphi'(f) = \alpha_0 \in \overline{\varphi}'(V(T_\delta))$ it then follows that $T' = (T_\delta, f, u)$ is also a Tashkinov tree with respect to e and φ' satisfying $|V(T')| > |V(T_\delta)| = t(G)$, a contradiction. This proves Claim 6.13.20 $\triangle$

Now, by Claim 6.13.15 and Claim 6.13.20, X is elementary and strongly closed with respect to φ. Hence Theorem 1.4 implies that G is an elementary graph, which completes the proof of Lemma 6.13.   ∎

Lemma 6.14 [271] *Let G be a critical graph with $\chi'(G) = k + 1$ for an integer $k \geq \Delta(G) + 1$. If $h(G) > t(G) - 4$, then G is an elementary graph.*

Proof: Both $h(G)$ and $t(G)$ are odd (Proposition 5.10(a), Proposition 6.11(a)) and $h(G) \leq t(G)$. By hypothesis, this gives either $h(G) = t(G)$ or $h(G) = t(G) - 2$.

By Proposition 6.11(c), $\mathcal{T}^B(G) \neq \emptyset$. Thus there is a triple $(T, e, \varphi) \in \mathcal{T}^B(G)$, where $e \in E_G(x, y)$ is an edge, $\varphi \in \mathcal{C}^k(G - e)$ is a coloring, and T is a Tashkinov tree with respect to e and φ. Clearly, $h(T) = h(G)$.

If $h(G) = t(G)$, then T consists only of its trunk and so only two colors, say $\alpha \in \overline{\varphi}(x)$ and $\beta \in \overline{\varphi}(y)$ are used on T with respect to φ. By Proposition 5.10(f), this implies that $\overline{\varphi}(v) \cap \Gamma^{\mathrm{f}}(T, e, \varphi) \neq \emptyset$ for every vertex $v \in V(T)$. From Proposition 6.12 we then conclude that $F(T, e, \varphi) = \emptyset$ and hence $\Gamma^{\mathrm{d}}(T, e, \varphi) = \emptyset$. Then $V(T)$ is elementary and strongly closed with respect to φ (Proposition 5.10(b,c)) and, therefore, G is an elementary graph (Theorem 1.4).

It remains to consider the case $h(G) = t(G) - 2$. Then T has the form

$$T = (y_0, e_1, y_1, \ldots, e_{r-1}, y_{r-1}, f_1, u_1, f_2, u_2),$$

where $y_0 = x$, $y_1 = y$, and $r = h(T)$. Clearly, exactly two colors $\alpha \in \overline{\varphi}(x)$ and $\beta \in \overline{\varphi}(y)$ are used on Ty_{r-1} with respect to φ. Moreover, $\varphi(f_1) = \varphi(f_2) = \gamma \in \overline{\varphi}(y_0, \ldots, y_{p-1}) \setminus \{\alpha, \beta\}$ and $f_j \in E_G(\{y_0, \ldots, y_{r-1}\}, u_j)$ for $j = 1, 2$.

Since $V(T)$ is elementary with respect to φ (Proposition 5.10(b)), there is a unique vertex $y_s \in \{y_0, \ldots, y_{r-1}\}$ with $\gamma \in \overline{\varphi}(y_s)$. Since there are exactly three colors used on T with respect to φ, namely $\alpha \in \overline{\varphi}(x)$, $\beta \in \overline{\varphi}(y)$ and $\gamma \in \overline{\varphi}(y_s)$, we conclude from Proposition 5.10(f) that $\overline{\varphi}(v) \cap \Gamma^f(T, e, \varphi) \neq \emptyset$ for every vertex $v \in V(T) \setminus \{y_s\}$. Then Proposition 6.12 implies $F(T, e, \varphi) \subseteq \{y_s\}$. Hence (T, e, φ) fulfils the structural conditions of Lemma 6.13 and, therefore, G is an elementary graph. This completes the proof. ∎

Lemma 6.15 [271] *Let G be a critical graph with $\chi'(G) = k + 1$ for an integer $k \geq \Delta(G) + 1$. If $h(G) < 5$, then G is an elementary graph.*

Proof: Since $h(G) < 5$, Proposition 6.11(a) implies that $h(G) = 3$. By Proposition 6.11(c), there is a triple $(T, e, \varphi) \in \mathcal{T}^B(G)$. Hence T has the form

$$T = (y_0, e_1, y_1, \ldots, e_{p-1}, y_{p-1}),$$

where $p = t(G)$, and $T_1 = (y_0, e_1, y_1, e_2, y_2)$ is the (α, β)-trunk of T, where $\alpha \in \overline{\varphi}(y_0)$ and $\beta \in \overline{\varphi}(y_1)$. Furthermore, there is an edge $f \in E_G(y_2, y_0)$ with $\varphi(f) = \beta$.

Suppose that G is not an elementary graph. Then, by Lemma 6.14, we have $p = t(G) \geq h(G) + 4 = 7$. Hence there is a color $\gamma = \varphi(e_3) = \varphi(e_4) \in \overline{\varphi}(y_j)$ for some $j \in \{1, 2, 3\}$. Without loss of generality, we may assume that $j = 0$, otherwise we could replace the triple (T, e, φ) by the balanced triple $(T, e, \varphi)(y_0 \to y_j)$. Therefore, we have $e_3, e_4 \in E_G(\{y_1, y_2\}, \{y_3, y_4\})$ and, moreover, (y_0, y_1) is a (γ, β)-pair with respect to φ. From Theorem 5.3(e) we then conclude that there is a (γ, β)-chain P with respect to φ having endvertices y_1 and y_0 and satisfying $V(P) \subseteq V(T)$. Evidently, $h = |V(P)|$ is odd, $f, e_3, e_4 \in E(P)$, and $y_0, y_1, y_2, y_3, y_4 \in V(P)$. Therefore, we have $h \geq 5$. Then, by Proposition 6.10, there is a Tashkinov tree T' with respect to e and φ satisfying $(T', e, \varphi) \in \mathcal{T}^N(G)$ and $h(T') = h \geq 5 > h(G)$, a contradiction. Hence G is an elementary graph. ∎

Lemma 6.16 [271] *Let G be a critical graph with $\chi'(G) = k + 1$ for an integer $k \geq \Delta(G) + 1$ such that $h(G) = 5$. Furthermore, let $(T, e, \varphi) \in \mathcal{T}^N(G)$ and let $T' = (y_0, e_1, y_1, e_2, y_2, e_3, y_3, e_4, y_4)$ be the (α, β)-trunk of T. If $\gamma \in \overline{\varphi}(y_0)$ is a color such that $E_G(V(T'), V(T) \setminus V(T')) \cap E_\gamma(e, \varphi) \neq \emptyset$, then the following statements hold:*

(a) *There are three edges $f_1 \in E_G(y_1, V(T) \setminus V(T'))$, $f_2 \in E_G(y_4, V(T) \setminus V(T'))$ and $f_3 \in E_G(y_2, y_3)$ with $\varphi(f_1) = \varphi(f_2) = \varphi(f_3) = \gamma$.*

(b) *For the two endvertices v_1, v_2 of the two edges f_1, f_2 that belong to $V(T) \setminus V(T')$, we have $E_G(v_1, v_2) \cap E_\alpha \neq \emptyset$ and $E_G(v_1, v_2) \cap E_\beta \neq \emptyset$.*

Proof: By hypothesis, $e_1 = e$, $\varphi(e_2) = \varphi(e_4) = \alpha \in \overline{\varphi}(y_0)$, $\varphi(e_3) = \beta \in \overline{\varphi}(y_1)$, and there is an edge $e_0 \in E_G(y_4, y_0)$ with $\varphi(e_0) = \beta$.

Let $P_1 = P_{y_0}(\gamma, \beta, \varphi)$. By Theorem 5.3(e), P_1 is a path of even length having endvertices y_0 and y_1. Then, by Proposition 6.10, there is a Tashkinov tree T_1 with respect to e and φ satisfying $(T_1, e, \varphi) \in \mathcal{T}^N(G)$ and $h(T_1) = |V(P_1)|$. Since $h(G) = 5$, we conclude that $|V(P_1)| \leq 5$. Since y_0, y_1 are the endvertices of P_1 and $\varphi(e_0) = \beta$, we have $e_0 \in E(P_1)$ and $y_0, y_1, y_4 \in V(P_1)$.

Now we claim that $E_G(\{y_2, y_3\}, V(T) \setminus V(T')) \cap E_{\varphi, \gamma} = \emptyset$. Suppose this is not true. Then there is an edge $g \in E_G(\{y_2, y_3\}, V(T) \setminus V(T'))$ with $\varphi(g) = \gamma$. Let $v \in V(T) \setminus V(T')$ be the second endvertex of g. We conclude that none of the three vertices y_2, y_3, v belongs to $V(P_1)$, otherwise all three would belong to $V(P_1)$ and, therefore, we would have $|V(P_1)| \geq 6$, a contradiction. Hence, by Theorem 5.3(e), T is a Tashkinov tree with respect to e and $\varphi_1 = \varphi/P_1$ satisfying $\alpha \in \overline{\varphi}_1(y_0)$, $\gamma \in \overline{\varphi}_1(y_1)$, $\varphi_1(e_0) = \varphi_1(g) = \gamma$, and $\varphi_1(e_2) = \varphi_1(e_4) = \alpha$. Evidently, $P_2 = P_{y_0}(\alpha, \gamma, \varphi_1)$ contains the subpaths $\text{Path}(y_0, e_0, y_4, e_4, y_3)$ and $\text{Path}(y_1, e_2, y_2)$, which implies $g \in E(P_2)$ and $v \in V(P_2)$. Hence we have $|V(P_2)| \geq 6$ and, by Proposition 6.10, there is a Tashkinov tree T_2 with respect to e and φ_1 satisfying $(T_2, e, \varphi_1) \in \mathcal{T}^N(G)$ and $h(T_2) = |V(P_2)| \geq 6 > 5 = h(G)$, a contradiction. This proves the claim.

Since $\gamma \in \overline{\varphi}(y_0)$ and since $V(T)$ is elementary with respect to φ (Proposition 5.10(b)), there are edges $f_3 \in E_G(y_2, V(T'))$ and $f_3' \in E_G(y_3, V(T'))$ with $\varphi(f_3) = \varphi(f_3') = \gamma$. Then $f_3 \neq f_3'$ would imply $f_3, f_3' \in E_G(\{y_2, y_3\}, \{y_1, y_4\})$ and, therefore, $E_G(V(T'), V(T) \setminus V(T')) \cap E_{\varphi, \gamma} = \emptyset$, a contradiction. Consequently, we have $f_3 = f_3' \in E_G(y_2, y_3)$ with $\varphi(f_3) = \gamma$, exactly as claimed in (a).

Since $V(T)$ is elementary with respect to φ, there are edges $f_1 \in E_G(y_1)$ and $f_2 \in E_G(y_4)$ with $\varphi(f_1) = \varphi(f_2) = \gamma$. If $f_1 = f_2$, then we would have $E_G(V(T'), V(T) \setminus V(T')) \cap E_{\varphi, \gamma} = \emptyset$, contradicting the hypothesis. Consequently, we have $f_1 \neq f_2$ and, therefore, $f_1 \in E_G(y_1, V(T) \setminus V(T'))$ and $f_2 \in E_G(y_4, V(T) \setminus V(T'))$, which completes the proof of (a).

Let v_1, v_2 be the two endvertices of f_1, f_2 belonging to $V(T) \setminus V(T')$. Then P_1 contains the two subpaths $\text{Path}(y_0, e_0, y_4, f_2, v_2)$ and $\text{Path}(y_1, f_1, v_1)$. Since $|V(P_1)| \leq 5$, there is an edge $g_1 \in E_G(v_1, v_2)$ with $\varphi(g_1) = \beta$ and, therefore, we have $P_1 = \text{Path}(y_1, e_5, y_5, g_1, y_6, e_6, y_4, e_0, y_0)$, which proves part of (b).

Let $P_3 = P_{y_0}(\alpha, \beta, \varphi) = \text{Path}(y_1, e_2, y_2, e_3, y_3, e_4, y_4, e_0, y_0)$ and let $\varphi_2 = \varphi/P_3$. Then, by Theorem 5.3(e), T is a Tashkinov tree with respect to e and φ_2 satisfying $\alpha \in \overline{\varphi}_2(y_1)$, $\gamma \in \overline{\varphi}_2(y_0)$, $\varphi_2(e_0) = \alpha$, and $\varphi_2(f_1) = \varphi_2(f_2) = \gamma$. Hence $P_4 = P_{y_0}(\gamma, \alpha, \varphi_2)$ contains the two subpaths $\text{Path}(y_0, e_0, y_4, f_2, v_2)$ and $\text{Path}(y_1, f_1, v_1)$ and, therefore, we have $|V(P_4)| \geq 5$. If we would have $|V(P_4)| > 5$ then, by Proposition 6.10, there would be a Tashkinov tree T_3 with respect to e and φ_2 satisfying $(T_3, e, \varphi_2) \in \mathcal{T}^N(G)$ and $h(T_3) = |V(P_4)| > 5 = h(G)$, a contradiction. Hence we have $|V(P_4)| = 5$, and there is an edge $g_2 \in E_G(v_1, v_2)$ with $\varphi_2(g_2) = \alpha$. Since $g_2 \notin E(P_3)$, we also have $\varphi(g_2) = \alpha$, which eventually proves (b). ∎

Lemma 6.17 [271] *Let G be a critical graph with $\chi'(G) = k + 1$ for an integer $k \geq \Delta(G) + 1$. If $t(G) < 11$, then G is an elementary graph.*

Proof: Suppose on the contrary that there exists a critical graph G such that $\chi'(G) = k+1 \geq \Delta(G)+2$ and $t(G) < 11$, but G is not elementary. By Proposition 5.10(a), $t(G)$ is odd and, therefore, we have $t(G) \leq 9$. Moreover, from Lemma 6.14

and Lemma 6.15 we conclude that $5 \leq h(G) \leq t(G) - 4 \leq 5$ and, therefore, we have $h(G) = 5$ and $t(G) = 9$.

Clearly, there is a triple $(T, e, \varphi) \in \mathcal{T}^B(G)$. Then T has the form

$$T = (y_0, e_1, y_1, e_2, y_2, e_3, y_3, e_4, y_4, e_5, y_5, e_6, y_6, e_7, y_7, e_8, y_8),$$

where $e_1 = e$, $\varphi(e_2) = \varphi(e_4) = \alpha \in \overline{\varphi}(y_0)$, $\varphi(e_3) = \beta \in \overline{\varphi}(y_1)$, $\varphi(e_5) = \varphi(e_6) = \gamma_1 \in \overline{\varphi}(\{y_0, \ldots, y_4\})$, and $\varphi(e_7) = \varphi(e_8) = \gamma_2 \in \overline{\varphi}(\{y_0, \ldots, y_6\})$. Moreover, $T_1 = Ty_4$ is the (α, β)-trunk of T, and there is an edge $e_0 \in E_G(y_4, y_0)$ with $\varphi(e_0) = \beta$.

Clearly, $\gamma_1 \in \overline{\varphi}(y_i)$ for some $i \in \{0, \ldots, 4\}$. Without loss of generality, we may assume that $i = 0$, otherwise we could replace the triple (T, e, φ) by the balanced triple $(T, e, \varphi)(y_0 \to y_i)$. Since $e_5 \in E_G(\{y_0, \ldots, y_4\}, y_5)$, we conclude from Lemma 6.16 that there are five edges f_1, f_2, f_3, g_1, g_2 satisfying $f_1 \in E_G(y_1, v_1)$ for a vertex $v_1 \in \{y_5, \ldots, y_8\}$, $f_2 \in E_G(y_4, v_2)$ for a vertex $v_2 \in \{y_5, \ldots, y_8\}$, $\varphi(f_1) = \varphi(f_2) = \gamma_1$, $f_3 \in E_G(y_2, y_3)$, $\varphi(f_3) = \gamma_1$, $g_1, g_2 \in E_G(v_1, v_2)$, $\varphi(g_1) = \alpha$, and $\varphi(g_2) = \beta$. In particular, this implies $\{e_5, e_6\} = \{f_1, f_2\}$ and $\{y_5, y_6\} = \{v_1, v_2\}$. By symmetry, we may assume that $v_1 = y_5$ and $v_2 = y_6$, so that $f_1 = e_5$ and $f_2 = e_6$.

Now we obtain $\varphi(e_2) = \varphi(e_4) = \varphi(g_1) = \alpha \in \overline{\varphi}(y_0)$, $\varphi(e_0) = \varphi(e_3) = \varphi(g_2) = \beta \in \overline{\varphi}(y_1)$, $\varphi(f_1) = \varphi(f_2) = \varphi(f_3) = \gamma_1 \in \overline{\varphi}(y_0)$, and $e_7, e_8 \in E_G(\{y_0, \ldots, y_6\}, \{y_7, y_8\})$ and, therefore, $\gamma_2 \notin \{\alpha, \beta, \gamma_1\}$. Since $V(T)$ is elementary and closed with respect to φ (Proposition 5.10(b)), there are three edges $f_4, g_3, g_4 \in E_G(y_7, y_8)$ satisfying $\varphi(f_4) = \gamma_1$, $\varphi(g_3) = \alpha$, and $\varphi(g_4) = \beta$.

We may assume that $\gamma_2 \in \overline{\varphi}(y_0, \ldots, y_4)$, otherwise we could simply replace T by

$$T_1 = (y_0, e_1, y_1, f_1, v_1, g_2, v_2, f_2, y_4, e_2, y_2, e_4, y_3, e_7, y_7, e_8, y_8).$$

Obviously, we have $(T_1, e, \varphi) \in \mathcal{T}^B(G)$, and T_1y_4 is the (γ_1, β)-trunk of T. Hence T_1 has the same structure as T, just the two colors α and γ_1 changed their role.

Now we claim that $E_G(\{y_0, \ldots, y_4\}, \{y_7, y_8\}) \cap E_{\varphi, \gamma_2} \neq \emptyset$. Suppose this is not true. Then we have $e_7, e_8 \in E_G(\{y_5, y_6\}, \{y_7, y_8\})$ and, by symmetry, we may assume $e_7 \in E_G(y_5, y_7)$ and $e_8 \in E_G(y_6, y_8)$. Evidently, the chain $P_1 = P_{y_7}(\gamma_2, \beta, \varphi)$ is a cycle of the form $(y_7, e_7, y_5, g_2, y_6, e_8, y_8, g_4, y_7)$, and T is a Tashkinov tree with respect to e and $\varphi_1 = \varphi/P_1$. Then, clearly, the chain $P_2 = P_{y_0}(\gamma_1, \beta, \varphi_1)$ satisfies $P_2 = \text{Path}(y_1, f_1, v_1, e_7, y_7, f_4, y_8, e_8, v_2, f_2, y_4, e_0, y_0)$. Therefore, by Proposition 6.10, there is a Tashkinov tree T_2 with respect to e and φ_1 satisfying $(T_2, e, \varphi_1) \in \mathcal{T}^N(G)$ and $h(T_2) = |V(P_2)| = 7 > h(G)$, a contradiction. This proves the claim.

Since we have $\gamma_2 \in \overline{\varphi}(y_j)$ for some $j \in \{0, \ldots, 4\}$, we can construct a new Tashkinov tree as follows. In the case $j = 0$ let $(T', e', \varphi') = (T, e, \varphi)$, otherwise let $(T', e', \varphi') = (T, e, \varphi)(y_0 \to y_j)$. In any case we have $(T', e', \varphi') \in \mathcal{T}^B(G)$, and T' has the form

$$T' = (y_0', e_1', y_1', e_1', y_2', e_3', y_3', e_4', y_4', e_5, y_5, e_6, y_6, e_7, y_7, e_8, y_8)$$

such that $\{y_0', \ldots, y_4'\} = \{y_0, \ldots, y_4\}$, $y_0' = y_j$, $\alpha, \gamma_2 \in \overline{\varphi}'(y_0')$, $\beta \in \overline{\varphi}'(y_1')$, and $T'y_4'$ is the (α, β)-trunk of T'. Thus, by Lemma 6.16, there are two vertices

$v'_1, v'_2 \in \{y_5, \ldots, y_8\}$ and five edges $f'_1, f'_2, f'_3, g'_1, g'_2$ satisfying $f'_1 \in E_G(y'_1, v'_1)$, $f'_2 \in E_G(y'_4, v'_2)$, $\varphi'(f_1) = \varphi'(f_2) = \gamma_2$, $f'_3 \in E_G(y'_2, y'_3)$, $\varphi'(f'_3) = \gamma_2$, $g'_1, g'_2 \in E_G(v'_1, v'_2)$, $\varphi'(g'_1) = \alpha$, and $\varphi'(g'_2) = \beta$. Consequently, we have

$$f'_1, f'_2 \in E_G(\{y_0, \ldots, y_4\}, \{y_5, \ldots, y_8\}), f'_3 \in E_G(\{y_0, \ldots, y_4\}, \{y_0, \ldots, y_4\}),$$

$\varphi(f'_1) = \varphi(f'_2) = \varphi(f'_3) = \gamma_2$, and $\varphi(g'_1) = \alpha$. This implies

$$|E_G(\{y_0, \ldots, y_4\}, \{y_5, \ldots, y_8\}) \cap E_{\gamma_2}(e, \varphi)| = 2.$$

We also have $E_G(\{y_0, \ldots, y_4\}, \{y_7, y_8\}) \cap E_{\gamma_2}(e, \varphi) \neq \emptyset$ and, therefore, we obtain $\{y_7, y_8\} \cap \{v'_1, v'_2\} \neq \emptyset$. Since $\varphi(g'_1) = \varphi(g_1) = \alpha$, we then deduce that $\{y_7, y_8\} = \{v'_1, v'_2\}$.

Now we have $\gamma_1 \in \overline{\varphi}(y_0)$ and $\gamma_2 \in \overline{\varphi}(y_j)$ for some $j \in \{0, \ldots, 4\}$. Moreover, Proposition 5.10(f) implies $|\overline{\varphi}(v) \setminus \{\alpha, \beta\}| \geq 1$ for every $v \in V(T)$. Since no colors apart from α, β, γ_1 and γ_2 are used on T with respect to φ, we conclude that, for every vertex $v \in V(T) \setminus \{y_0, y_j\}$, the set $\overline{\varphi}(v)$ contains at least one free color with respect to (T, e, φ). From Proposition 6.12 it then follows that $F(T, e, \varphi) \subseteq \{y_0, y_j\}$. Since $\gamma_1 \neq \gamma_2$ and $e_5, e_6, e_7, e_8 \in E_G(\{y_0, \ldots, y_4\})$, the triple (T, e, φ) fulfils the structural conditions from Lemma 6.13 and, therefore, G is an elementary graph. This completes the proof. ∎

Lemma 6.18 [271] *Let $G \in \mathcal{J}_m$ be a critical graph for an odd integer $m \geq 3$, let $(T, e, \varphi) \in \mathcal{T}(G)$, and let $Z = V(T) \cup D(T, e, \varphi)$. Then the following statements hold:*

(a) $|Z| \leq m - 2$.

(b) *If $|Z| = m - 2$, then G is an elementary graph.*

Proof: By the hypothesis, $\chi'(G) > \Delta(G)$ and hence $\Delta(G) \geq 2$. Furthermore, we have $\chi'(G) > \frac{m}{m-1}\Delta(G) + \frac{m-3}{m-1} \geq \Delta(G) + 1$ and, therefore, $\chi'(G) \geq \Delta(G) + 2$. Hence, for $k = \chi'(G) - 1$, we have $k \geq \Delta(G) + 1$ and $\varphi \in \mathcal{C}^k(G - e)$.

We now deduce from Proposition 5.12 and Proposition 5.14 that Z is elementary with respect to φ. Then Proposition 1.7(c) implies $|Z| \leq m - 1$.

Now suppose $|Z| = m - 1$. Since $k \geq \Delta(G) + 1$, there is a color $\alpha \in \overline{\varphi}(Z)$. Moreover, Z is elementary with respect to φ and $|Z|$ is even, so there is an edge $g \in \partial_G(Z)$ having an endvertex $z \in V(G) \setminus Z$ and satisfying $\varphi(g) = \alpha$. Then $F = (g, z)$ is a fan at Z with respect to φ, and Theorem 5.18 implies that $Z \cup \{z\}$ is elementary with respect to φ. Since $|Z \cup \{z\}| = m$, this contradicts Proposition 1.7(c). Consequently, we have $|Z| \leq m - 2$ and (a) is proved.

For the proof of (b), assume that $|Z| = m - 2$. In the sequel, we will use the following abbreviation. For a coloring $\varphi' \in \mathcal{C}^k(G - e)$, a color $\alpha \in \{1, \ldots, k\}$ and a set $X \subseteq V(G)$, let

$$E_\alpha(X, e, \varphi') = \partial_G(X) \cap E_{\varphi', \alpha}.$$

First, we claim that Z is closed with respect to φ. Suppose this is not true. Then there is a color $\alpha' \in \overline{\varphi}(Z)$ satisfying $E_{\alpha'}(Z, e, \varphi) \neq \emptyset$. Since Z is elementary with respect to φ and since $|Z|$ is odd, we conclude that $|E_{\alpha'}(Z,e,\varphi)| \geq 2$. Then there are two distinct edges $g_1, g_2 \in E_{\alpha'}(Z,e,\varphi)$. For $j = 1,2$, let z_j denote the endvertex of g_j that belongs to $V(G) \setminus Z$. Clearly, $z_1 \neq z_2$ and, therefore, $F' = (g_1, z_1, g_2, z_2)$ is a fan at Z with respect to φ. Hence $Z \cup \{z_1, z_2\}$ is elementary with respect to φ (Theorem 5.18), but then $|Z \cup \{z_1, z_2\}| = m$ contradicts Proposition 1.7(c). This proves the claim that Z is closed with respect to φ.

Now we want to show that Z is also strongly closed with respect to φ. Suppose this is not true. Then there is a color $\delta \in \{1, \ldots, k\}$ satisfying $\delta \notin \overline{\varphi}(Z)$ and $|E_\delta(Z,e,\varphi)| \geq 2$. Since $|Z|$ is odd, we then conclude that $|E_\delta(Z,e,\varphi)| \geq 3$. Moreover, by Proposition 5.10(g), there is a color $\gamma \in \Gamma^f(T,e,\varphi)$, and there is a unique vertex $v \in V(T)$ satisfying $\gamma \in \overline{\varphi}(v)$. Let $P = P_v(\gamma, \delta, \varphi)$. Then P is a path, and v is an endvertex of P. Since $\delta \notin \overline{\varphi}(Z)$ and Z is elementary with respect to φ, the other endvertex of P belongs to $V(G) \setminus Z$. Hence in the linear order $\preceq_{(v,P)}$ there is a first vertex u that belongs to $V(G) \setminus Z$.

Let Φ denote the set of all colorings $\varphi' \in \mathcal{C}^k(G - e)$ such that $(T, e, \varphi') \in \mathcal{T}(G)$, $\Gamma^f(T,e,\varphi') = \Gamma^f(T,e,\varphi)$, $D(T,e,\varphi') = D(T,e,\varphi)$, $E_{\varphi',\delta} = E_{\varphi,\delta}$ and $E_G(Z,Z) \cap E_\gamma(e,\varphi') = E_G(Z,Z) \cap E_\gamma(e,\varphi)$. Clearly, $\varphi \in \Phi$.

Consider an arbitrary coloring $\varphi' \in \Phi$. Then $Z = V(T) \cup D(T,e,\varphi')$ and, by Proposition 5.12 and Proposition 5.14, it follows that Z is elementary with respect to φ'. Since $|Z| = m - 2$, it follows from the above proof that Z is closed with respect to φ'. Since $E_{\varphi',\delta} = E_{\varphi,\delta}$, it follows that $\delta \notin \overline{\varphi}'(Z)$ and $E_\delta(Z,e,\varphi') = E_\delta(Z,e,\varphi)$. Consequently, Z is not strongly closed with respect to φ'. Furthermore, $\gamma \in \Gamma^f(T,e,\varphi) = \Gamma^f(T,e,\varphi')$ and, since $E_G(Z,Z) \cap E_{\varphi',\gamma} = E_G(Z,Z) \cap E_{\varphi,\gamma}$, we have $\gamma \in \overline{\varphi}'(v)$. Moreover, since $E(vPu) \subseteq E_{\varphi,\delta} \cup (E_G(Z,Z) \cap E_{\varphi,\gamma})$, it follows that $\varphi'(f') = \varphi(f')$ for all edges $f' \in E(vPu)$. Hence we have $vP'u = vPu$, where $P' = P_v(\gamma, \delta, \varphi')$. In particular, this implies that u is the first vertex in the linear order $\preceq_{(v,P')}$ that belongs to $V(G) \setminus Z$.

We claim that there is a coloring $\varphi' \in \Phi$ such that $\gamma \in \overline{\varphi}'(u)$. Clearly, this implies that $P_v(\gamma, \delta, \varphi') = vPu$. For the proof of this claim, we consider the following two cases.

Case 1: $\overline{\varphi}(u) \cap \overline{\varphi}(Z) \neq \emptyset$. Then there is a color $\beta \in \overline{\varphi}(u) \cap \overline{\varphi}(Z)$. If $\beta = \gamma$ we are done, because $\varphi \in \Phi$. If $\beta \neq \gamma$, then let $P_1 = P_u(\gamma, \beta, \varphi)$ and $\varphi' = \varphi/P$. Since Z is closed with respect to φ, we obtain $E_\gamma(Z,e,\varphi) = E_\beta(Z,e,\varphi) = \emptyset$. Hence we have $V(P_1) \cap Z = \emptyset$. By Proposition 5.15, it follows that $(T,e,\varphi') \in \mathcal{T}(G)$, $\Gamma^f(T,e,\varphi') = \Gamma^f(T,e,\varphi)$, and $D(T,e,\varphi') = D(T,e,\varphi)$. Since the recoloring involves neither edges of $E_G(Z,Z)$ nor edges colored with δ, this implies that $\varphi' \in \Phi$. Moreover, $\gamma \in \overline{\varphi}'(u)$ and, therefore, φ' has the desired properties.

Case 2: $\overline{\varphi}(u) \cap \overline{\varphi}(Z) = \emptyset$. Since Z is elementary with respect to φ, this implies that $Z \cup \{u\}$ is elementary with respect to φ, too. Since $\gamma \in \overline{\varphi}(Z)$, there is a vertex $u' \in V(G)$ and an edge $f \in E_G(u,u')$ with $\varphi(f) = \gamma$. Since Z is closed with respect to φ, it follows that $u' \in V(G) \setminus Z$. Then $|Z \cup \{u,u'\}| = m$ and hence, by Proposition 1.7(c), the set $Z \cup \{u,u'\}$ is not elementary with respect to φ. Since $Z \cup \{u\}$ is elementary with respect to φ, this implies that $\overline{\varphi}(u') \cap \overline{\varphi}(Z \cup \{u\}) \neq \emptyset$. We consider three subcases.

Case 2a: *There is a color* $\gamma_1 \in \overline{\varphi}(u') \cap \overline{\varphi}(u)$. Then we can simply obtain the desired coloring $\varphi' \in \Phi$ from φ by recoloring the edge f with the color γ_1.

Case 2b: *There is a color* $\gamma_1 \in \overline{\varphi}(u') \cap \overline{\varphi}(Z)$ *satisfying* $\gamma_1 \in \Gamma^f(T, e, \varphi)$. Then there is a unique vertex $v' \in V(T)$ with $\gamma_1 \in \overline{\varphi}(v')$. Moreover, there is a color $\gamma_2 \in \overline{\varphi}(u)$ (Proposition 5.10(g)). Since $Z \cup \{u\}$ is elementary with respect to φ, we clearly have $\gamma_1 \neq \gamma_2$ and $\gamma_2 \notin \overline{\varphi}(Z)$. Furthermore, from $\varphi(f) = \gamma$ it follows that $\gamma \notin \{\gamma_1, \gamma_2\}$. Since $\delta \notin \overline{\varphi}(Z)$ and $\delta \in \varphi(u)$, we also obtain $\delta \notin \{\gamma_1, \gamma_2\}$.

Now let $P_2 = P_{v'}(\gamma_1, \gamma_2, \varphi)$. Then P_2 is a path, where one endvertex is v' and the other endvertex is some vertex $w' \in V(G) \setminus Z$. Since $\gamma_1 \in \overline{\varphi}(V(T))$, Proposition 5.10(d) implies $E_{\gamma_1}(T, e, \varphi) = \emptyset$. Hence we obtain $\emptyset \neq \partial_G(V(T)) \cap E(P_2) \subseteq E_{\varphi, \gamma_2}$. Then we conclude that $E_{\gamma_2}(T, e, \varphi) \subseteq E(P_2)$. This is clear if $|E_{\gamma_2}(T, e, \varphi)| = 1$, otherwise it follows from Proposition 5.11(b) and the fact that $\gamma_1 \in \Gamma^f(T, e, \varphi)$.

If $w' = u$ then, evidently, u' is not an endvertex of P_2 and, therefore, u' does not belong to $V(P_2)$ at all. Let $P_3 = P_{u'}(\gamma_1, \gamma_2, \varphi)$ and $\varphi_3 = \varphi/P_3$. Clearly, P_3 and P_2 are disjoint. Since $E_{\gamma_1}(T, e, \varphi) = \emptyset$ and $E_{\gamma_2}(T, e, \varphi) \subseteq E(P_2)$, this implies that $V(P_3) \cap V(T) = \emptyset$. By Proposition 5.15, it follows that $(T, e, \varphi_3) \in \mathcal{T}(G)$, $\Gamma^f(T, e, \varphi_3) = \Gamma^f(T, e, \varphi)$ and $D(T, e, \varphi_3) = D(T, e, \varphi)$. Since the recoloring does not involve edges colored with γ or δ, this implies that $\varphi_3 \in \Phi$. Moreover, since $u \notin V(P_3)$, we have $\gamma_2 \in \overline{\varphi}_3(u) \cap \overline{\varphi}_3(u')$ and $\varphi_3(f) = \gamma$. Hence we can obtain the desired coloring $\varphi' \in \Phi$ from φ_3 by recoloring the edge f with the color γ_2.

If otherwise $w' \neq u$ then, evidently, u does not belong to $V(P_2)$ at all. Let $P_4 = P_u(\gamma_1, \gamma_2, \varphi)$ and $\varphi_4 = \varphi/P_4$. Clearly, P_4 and P_2 are disjoint. Since $E_{\gamma_1}(T, e, \varphi) = \emptyset$ and $E_{\gamma_2}(T, e, \varphi) \subseteq E(P_2)$, this implies that $V(P_4) \cap V(T) = \emptyset$. By Proposition 5.15, it follows that $(T, e, \varphi_4) \in \mathcal{T}(G)$, $\Gamma^f(T, e, \varphi_4) = \Gamma^f(T, e, \varphi)$ and $D(T, e, \varphi_4) = D(T, e, \varphi)$. Since the recoloring does not involve edges colored with γ or δ, this implies that $\varphi_4 \in \Phi$. Moreover, we have $\gamma_1 \in \overline{\varphi}_4(u) \cap \overline{\varphi}_4(Z)$ and $\varphi_4(f) = \gamma$. Hence we are in the same situation as in Case 1, just with the coloring φ_4 instead of φ, and the color γ_1 instead of β. Then we can obtain the desired coloring $\varphi' \in \Phi$ from φ_4 analogously to Case 1, that is, $\varphi' = \varphi_4/P_u(\gamma, \gamma_1, \varphi_4)$.

Case 2c: *There is a color* $\gamma_1 \in \overline{\varphi}(u') \cap \overline{\varphi}(Z)$ *satisfying* $\gamma_1 \notin \Gamma^f(T, e, \varphi)$. By Proposition 5.10(g), there is a color $\gamma_3 \in \Gamma^f(T, e, \varphi) \setminus \{\gamma\}$. Evidently, $\gamma_1 \neq \gamma_3$. Let $P_5 = P_{u'}(\gamma_1, \gamma_3, \varphi)$ and $\varphi_5 = \varphi/P_5$. Since $\gamma_1, \gamma_3 \in \overline{\varphi}(Z)$ and Z is closed with respect to φ, we obtain $E_{\gamma_1}(Z, e, \varphi) = E_{\gamma_3}(Z, e, \varphi) = \emptyset$. Consequently, P_5 is a path satisfying $V(P_5) \cap Z = \emptyset$. From Proposition 5.15 it then follows that $(T, e, \varphi_5) \in \mathcal{T}(G)$, $\Gamma^f(T, e, \varphi_5) = \Gamma^f(T, e, \varphi)$ and $D(T, e, \varphi_5) = D(T, e, \varphi)$. Since the recoloring does not involve edges colored with γ or δ, this implies that $\varphi_5 \in \Phi$. Moreover, we have $\gamma_3 \in \overline{\varphi}_5(u') \cap \Gamma^f(T, e, \varphi_5)$. Since $\gamma_3 \neq \gamma$, we also have $\varphi_5(f) = \gamma$. Hence we are in the same situation as in Case 2b, just with the coloring φ_5 instead of φ, and the color γ_3 instead of γ_1. Then we can obtain the desired coloring $\varphi' \in \Phi$ from φ_4 analogously to Case 2b.

Thus the claim is proved, that is, there is a coloring $\varphi' \in \Phi$ such that $\gamma \in \overline{\varphi}'(u)$. Then $P' = P_v(\gamma, \delta, \varphi') = vPu$. Evidently, we have $|E(P') \cap \partial_G(Z)| = 1$. Since $|E_\delta(Z, e, \varphi)| \geq 3$, there must be two edges $f_1, f_2 \in \partial_G(Z) \setminus E(P')$ with $\varphi(f_1) =$

$\varphi(f_2) = \delta$. Since $\varphi' \in \Phi$, this implies $\varphi'(f_1) = \varphi'(f_2) = \delta$. For $j = 1, 2$, let $v_j \in Z$ and $u_j \in V(G) \setminus Z$ denote the endvertices of f_j.

Now let $P_1' = P_{u_1}(\gamma, \delta, \varphi')$ and $P_2' = P_{u_2}(\gamma, \delta, \varphi')$. Note that P_1' and P_2' may be equal. Since $E(P') \cap \{f_1, f_2\} = \emptyset$, both chains P_1' and P_2' are vertex disjoint to P'. Since $\gamma \in \overline{\varphi}'(V(T))$, Proposition 5.10(d) implies $E_\gamma(T, e, \varphi') = \emptyset$. Moreover, by Proposition 5.11(b), $E_\delta(T, e, \varphi) \subseteq P'$. Consequently, we obtain $V(P_1') \cap V(T) = V(P_2') \cap V(T) = \emptyset$. Define $\varphi_2' = \varphi'/P_1'$ if $P_1' = P_2'$, otherwise define $\varphi_2' = (\varphi'/P_1')/P_2'$. From Proposition 5.15 we then conclude that $(T, e, \varphi_2') \in \mathcal{T}(G)$ and $D(T, e, \varphi_2') = D(T, e, \varphi)$. Moreover, we have $\varphi_2'(f_1) = \varphi_2'(f_2) = \gamma \in \overline{\varphi}_2'(V(T))$ and, therefore, $F = (f_1, u_1, f_2, u_2)$ is a fan at Z with respect to φ_2'. From Theorem 5.18 it then follows that $Z \cup \{u_1, u_2\}$ is elementary with respect to φ_2'. Since $|Z \cup \{u_1, u_2\}| = m$, this contradicts Proposition 1.7(c). This proves that Z is strongly closed with respect to φ. Then, by Theorem 1.4, G is an elementary graph. This completes the proof of statement (b). ∎

Lemma 6.19 [271] *Let $G \in \mathcal{J}_m$ be a critical for an odd integer $m \geq 3$. Then the following statements hold:*

(a) *If $t(G) > m - 4$, then G is an elementary graph.*

(b) *If $t(G) = m - 4$ and $h(G) > t(G) - 8$, then G is an elementary graph.*

Proof: As in the proof of Lemma 6.18, we conclude from the hypothesis that $\chi'(G) \geq \Delta(G) + 2$. Graph G being critical, there is a triple $(T, e, \varphi) \in \mathcal{T}^B(G)$ (Proposition 6.11). Clearly, $e \in E_G(x, y)$ for two vertices $x, y \in V(T)$ and $\varphi \in \mathcal{C}^k(G - e)$, where $k = \chi'(G) - 1$.

To prove (a), assume that $t(G) > m - 4$. Since m is odd (by hypothesis) and $t(G)$ is odd (by Proposition 5.10(a)), this implies $t(G) \geq m - 2$. Evidently, $|V(T) \cup D(T, e, \varphi)| \geq m - 2$. Then Lemma 6.18 implies that $|V(T) \cup D(T, e, \varphi)| = m - 2$ and G is elementary. This proves (a).

To prove (b), assume that $t(G) = m - 4$ and $h(G) > t(G) - 8$. If the triple (T, e, φ) satisfies $\Gamma^d(T, e, \varphi) = \emptyset$, then Proposition 5.10 implies that $V(T)$ is elementary as well as strongly closed with respect to φ and, therefore, G is an elementary graph (Theorem 1.4). So, for the rest of the proof, assume that $\Gamma^d(T, e, \varphi) \neq \emptyset$. In particular, this implies $D = D(T, e, \varphi) \neq \emptyset$. If $|D| \geq 2$, then $|V(T) \cup D| \geq m - 2$ implying that $|V(T) \cup D| = m - 2$ and G is elementary (Lemma 6.18). So from now on we assume that $|D| = 1$.

Let $\delta \in \Gamma^d(T, e, \varphi)$ and $E' = E_\delta(T, e, \varphi)$. By Proposition 5.10(e), $|E'|$ is odd and $|E'| \geq 3$. Let s be the number of vertices $v \in V(T)$ such that $\overline{\varphi}(v)$ contains no free color with respect to φ. Then, obviously, we have $s \geq |E'| - |D| \geq |E'| - 1 \geq 2$.

Let $\alpha_1 \in \overline{\varphi}(x)$ and $\alpha_2 \in \overline{\varphi}(y)$ be the two colors used on the trunk of T with respect to φ. Clearly, we have $|\overline{\varphi}(v)| \geq k - \Delta(G) + 1$ for $v \in \{x, y\}$ and $|\overline{\varphi}(v)| \geq k - \Delta(G)$ for $v \in V(T) \setminus \{x, y\}$. Since $V(T)$ is elementary with respect to φ, this implies that $|\overline{\varphi}(v) \setminus \{\alpha_1, \alpha_2\}| \geq k - \Delta(G)$ for every $v \in V(T)$. Since (T, e, φ) is a balanced triple and $h(T) > |V(T)| - 8$, we conclude that, apart from α_1 and α_2, there are at most 3 other colors used on T with respect to φ, which leads to $s \leq \frac{3}{k - \Delta(G)}$.

Now we have $2 \leq |E'| - 1 \leq s \leq \frac{3}{k-\Delta(G)}$, which implies $k = \Delta(G) + 1$ and, therefore, $2 \leq |E'| - 1 \leq 3$. Since $|E'|$ is odd, this gives $|E'| = 3$.

Let $E' = \{f_1, f_2, f_3\}$ and, for $i = 1, 2, 3$, let u_i be the endvertex of f_i belonging to $V(G) \setminus V(T)$. Clearly, one of these vertices belongs to D, say $u_1 \in D$. Since $|D| = 1$, we then have $D = \{u_1\}$. From Proposition 5.12 and Proposition 5.14 we conclude that $Z = V(T) \cup D = V(T) \cup \{u_1\}$ is elementary with respect to φ. Since, $V(T)$ is closed with respect to φ (Proposition 5.10(b)), there is a vertex $u \in V(G) \setminus Z$ and an edge $f \in E_G(u_1, u)$ with $\varphi(f) \in \overline{\varphi}(V(T))$. Then (f, u) is a fan at Z with respect to φ and, therefore, $X = Z \cup \{u\}$ is elementary with respect to φ (Theorem 5.18).

Now we claim that $\delta \notin \overline{\varphi}(X)$. Suppose this is not true. Since $\delta \notin \overline{\varphi}(Z)$, this implies that $\delta \in \overline{\varphi}(u)$. Then $(f, u, f_2, u_2, f_3, u_3)$ is a fan at Z with respect to φ and, therefore, $X_1 = X \cup \{u_2, u_3\}$ is elementary with respect to φ (Theorem 5.18). Since $|X_1| = m$, this contradicts Proposition 1.7(c). This proves the claim.

Consequently, we obtain $k \geq |\overline{\varphi}(X)| + 1$. Set X being elementary with respect to φ, we obtain $|\overline{\varphi}(X)| \geq |X| + 2 = m$ (Proposition 5.10(b,f)). Hence $k \geq m + 1$. On the other hand, we have $k + 1 = \chi'(G) > \frac{m}{m-1}\Delta(G) + \frac{m-3}{m-1} = k + \frac{k-3}{m-1}$, which leads to $k < m + 2$. Both k and m being integers, we conclude that $k = m + 1 = |\overline{\varphi}(X)| + 1$.

Now we claim that X is closed with respect to φ. Suppose this is not true. Then there is a color $\alpha \in \overline{\varphi}(X)$ satisfying $E_1 = \partial_G(X) \cap E_{\varphi,\alpha} \neq \emptyset$. Since X is elementary with respect to φ, there is a unique vertex in X where the color α is missing with respect to φ. Moreover, $|X| = m - 2$ is odd and, therefore, $|E_1|$ is even and $|E_1| \geq 2$. Hence there is at least one edge $f' \in E_1$ having an endvertex in Z. Let $u' \in V(G) \setminus X$ be the other endvertex of f'. Then (f, u, f', u') is a fan at Z with respect to φ and, by Theorem 5.18, $X_2 = X \cup \{u'\}$ is elementary with respect to φ. From $k = m + 1$ and $|X_2| = m - 1$ we obtain $k = |X_2| + 2 \leq |\overline{\varphi}(X_2)| \leq k$ and, therefore, $|\overline{\varphi}(X_2)| = k$. This implies $\delta \in \overline{\varphi}(X_2)$, and from $\delta \notin \overline{\varphi}(X)$ we then conclude that $\delta \in \overline{\varphi}(u')$. Consequently, we have $u' \notin \{u_2, u_3\}$ and, therefore, at least one of the vertices u_2, u_3 does not belong to X_2, say $u_2 \notin X_2$. Then, evidently, (f, u, f', u', f_2, u_2) is a fan at Z with respect to φ and, therefore, by Theorem 5.18, $X_3 = X_2 \cup \{u_2\}$ is elementary with respect to φ, but $|X_3| = m$ contradicts Proposition 1.7(c). This proves the claim.

Let $E'' = \partial_G(X) \cap E_{\varphi,\delta}$. Since $\delta \notin \overline{\varphi}(X)$ and $|X|$ is odd, we conclude that $|E''| \geq 1$ is odd, too. We claim that $|E''| = 1$. Suppose, on the contrary, that $|E''| > 1$. Since $|E''|$ is odd, this implies $|E''| \geq 3$. From $E' = \{f_1, f_2, f_3\}$ and $f_1 \in E_G(X, X)$ it then follows that $|E''| = 3$ and $E'' = \{f_2, f_3, g\}$ for an edge $g \in E_G(u, v)$, where $v \in V(G) \setminus X$. Let $\beta \in \overline{\varphi}(v)$. Evidently, we have $\beta \neq \delta$ and, since $k = |\overline{\varphi}(X)| + 1$ and $\delta \notin \overline{\varphi}(X)$, we also have $\beta \in \overline{\varphi}(X)$. Moreover, by Proposition 5.10(g), there is a color $\gamma \in \Gamma^f(T, e, \varphi) \setminus \{\beta\}$. Now let $P = P_v(\beta, \gamma, \varphi)$ and $\varphi' = \varphi/P$. Since X is closed with respect to φ, we conclude that $V(P) \cap X = \emptyset$. By Proposition 5.15, we then have $(T, e, \varphi') \in \mathcal{T}(G)$, $\gamma \in \Gamma^f(T, e, \varphi')$, $\delta \in \Gamma^d(T, e, \varphi')$, and $D(T, e, \varphi') = \{u_1\}$. Moreover, we have $\varphi'(f_1) = \varphi'(f_2) = \varphi'(f_3) = \varphi'(g) = \delta$, and $\gamma \in \overline{\varphi}(v)$. Since both $V(T)$ and X are closed with respect to φ, there is an edge $g' \in E_G(u_1, u)$ satisfying $\varphi'(g') = \varphi(g') = \gamma$. Let $v' \in V(T)$ be the unique vertex with $\gamma \in \overline{\varphi}'(v)$ and let

$P' = P_{v'}(\gamma, \delta, \varphi')$. Since $D(T, e, \varphi') = \{u_1\}$, we conclude that u_1 is the first vertex in the linear order $\preceq_{(P', v')}$ that belongs to $V(G) \setminus V(T)$. Then, evidently, u, v are the next vertices in the linear order $\preceq_{(P', v')}$ and, moreover, v is the second endvertex of P'. Consequently, we have $f_2, f_3 \notin E(P')$, a contradiction to Proposition 5.11(b). Hence the claim is proved.

Since $k = |\overline{\varphi}(X)| + 1$ and $\delta \notin \overline{\varphi}(X)$, and since X is closed with respect to φ, we then conclude that every edge in $\partial_G(X)$ is colored with δ with respect to φ. From $|E''| = 1$ it then follows that X is strongly closed with respect to φ. Since X is also elementary with respect to φ, we deduce from Theorem 1.4 that G is elementary. This, finally, proves (b). ∎

Proof of Theorem 6.5 :   For the proof it is sufficient to show that every critical graph in $\mathcal{J}_{15}$ is elementary. Let $G \in \mathcal{J}_{15}$ be a critical graph. Then $\chi'(G) > \frac{15}{14}\Delta(G) + \frac{12}{14}$ and, therefore, $\Delta(G) \geq 2$ and $\chi'(G) \geq \Delta(G) + 2$.

If $t(G) < 11$, then Lemma 6.17 implies that G is an elementary graph. If $t(G) > 11$, then Lemma 6.19(a) implies that G is an elementary graph. If $t(G) = 11$ and $h(G) > 3$, then Lemma 6.19(b) implies that G is an elementary graph. If $t(G) = 11$ and $h(G) \leq 3$, then Lemma 6.15 implies that G is an elementary graph. This completes the proof. ∎

6.4 APPROXIMATION ALGORITHMS

Theorem 6.5 provides a new upper bound for the chromatic index of an arbitrary graph G, namely $\chi'(G) \leq \tau'(G)$ where $\tau'(G) = \max\{\lfloor(15\Delta(G) + 12)/14\rfloor, w(G)\}$. The above proof of Theorem 6.5 can be transformed into an algorithm that colors the edges of an arbitrary graph $G = (V, E)$ using at most $\tau'(G)$ colors, where the algorithm has time complexity bounded from above by a polynomial in $|V|$ and $|E|$. If the approach discussed in Sect. 5.5 is used, we may even design a polynomial time algorithm. The algorithm is mainly based on Tashkinov trees as test objects and has a similar structure as Tashkinov's algorithm presented in Sect. 5.4. On one hand, the new algorithm for the bound τ' is much more sophisticated than Tashkinov's algorithm for the bound τ. On the other hand, we have $\tau(G) < \tau'(G)$, provided that $\Delta(G)$ is large enough.

Clearly, $\tau'(G) \leq \lfloor(15\chi'(G) + 12)/14\rfloor$, since both $\Delta(G)$ and $w(G)$ are lower bounds for chromatic index $\chi'(G)$. We conclude this section with a variation due to Caprara and Rizzi [36].

Theorem 6.20 (Caprara and Rizzi [36] 1998) *Suppose there exists a polynomial-time algorithm A which finds an edge coloring of an arbitrary graph G using at most* $\max\{\lfloor a\Delta(G) + b\rfloor, \chi'(G)\}$ *colors, where $1 \leq a$. Then there exists a polynomial-time algorithm I which finds an edge coloring of an arbitrary graph G using at most* $\lfloor a\chi'(G) + c\rfloor$ *colors, where $c = \max\{0, b + 1 - a, 4 - 3a\}$.*

Proof: If $\Delta(G) \leq 2$, the ECP can be easily solved in polynomial time and we are done, since $\chi'(G) \leq \lfloor a\chi'(G) + c\rfloor$. If $\Delta(G) \geq 3$, then the algorithm I works as follows.

First, consider the case that there exists a Δ-**matching** in G, that is, a matching M such that every vertex of maximum degree in G is an endvertex of some edge in M. Let $G' = G - M$ and let $\varphi \in \mathcal{C}^k(G')$ be the coloring of G' returned by the algorithm A. Then the algorithm I returns the coloring $\varphi' \in \mathcal{C}^{k+1}(G)$ obtained from φ by coloring all edges in M with the new color $k + 1$. Clearly, we have $\Delta(G') = \Delta(G) - 1$ and $k \leq \max\{\lfloor a\Delta(G') + b\rfloor, \chi'(G')\}$. If $k \leq \lfloor a\Delta(G') + b\rfloor$, then $k+1 \leq \lfloor a\Delta(G')+b+1\rfloor = \lfloor a\Delta(G)+b+1-a\rfloor \leq \lfloor a\chi'(G)+c\rfloor$. Otherwise, $k \leq \chi'(G') \leq \chi'(G)$ and, therefore, $k + 1 \leq \chi'(G) + 1$. Since $\chi'(G) \geq \Delta(G) \geq 3$ and $a \geq 1$, it then follows by a simple calculation that $k + 1 \leq \chi'(G) + 1 \leq \lfloor a\chi'(G) + c\rfloor$.

It remains to consider the case that there exists no Δ-matching in G. Then the algorithm I returns the coloring $\varphi \in \mathcal{C}^k(G)$ returned by the algorithm A. Clearly, $\chi'(G) \geq \Delta(G) + 1$, since otherwise there would exists a Δ-matching in G. Hence, for the number k of colors, we have

$$
\begin{aligned}
k &\leq \max\{\lfloor a\Delta(G) + b\rfloor, \chi'(G)\} \\
&\leq \max\{\lfloor a\chi'(G) + b - a\rfloor, \chi'(G)\} \\
&< \lfloor a\chi'(G) + c\rfloor.
\end{aligned}
$$

The determination of a Δ-matching in a graph G, if any, can be carried out by finding a maximum matching in the graph H obtained from G as follows. We take two copies G_1 and G_2 of G and, for every vertex $v \in V(G)$ with $d_G(v) < \Delta(G)$, we add an edge between v_1 and v_2, where v_1 and v_2 are the counterparts of v in G_1 and G_2, respectively. The resulting graph is H. Then it is easy to prove that G has a Δ-matching if and only if H has a perfect matching. Since the matching problem can be solved in polynomial time, it follows that I is a polynomial-time algorithm. ∎

Corollary 6.21 (Caprara and Rizzi [36] 1998) *Suppose there exists a polynomial-time algorithm A which finds an edge coloring of an arbitrary graph G using at most*

$$
\max\left\{\left\lfloor\frac{(2p + 1)\Delta(G) + (2p - 2)}{2p}\right\rfloor, \chi'(G)\right\}
$$

colors, where $p \geq 3/2$ is a real number. Then there exists a polynomial-time algorithm I which finds an edge coloring of an arbitrary graph G using at most

$$
\left\lfloor\frac{(2p + 1)\chi'(G) + (2p - 3)}{2p}\right\rfloor = \chi'(G) + 1 + \left\lfloor\frac{\chi'(G) - 3}{2p}\right\rfloor
$$

colors. Unless P = NP, *the additive term $(2p - 3)/2p$ is best possible.*

Proof: Clearly, $w(G) \leq \chi'(G)$. Furthermore, if $a = (2p + 1)/2p$ and $b = (2p - 2)/2p$, then $c = \max\{0, b + 1 - a, 4 - 3a\} = (2p - 3)/2p$. Hence, the first part of the statement follows from Theorem 6.20. The second part follows from Holyer's result [150] that it is NP-complete to decide whether a simple 3-regular graph G has chromatic index 3. ∎

6.5 GOLDBERG'S CONJECTURE FOR SMALL GRAPHS

As proved by Scheide [269], Theorem 6.5 and its proof implies that Goldberg's conjecture holds for graphs with small order or with small maximum degree.

Theorem 6.22 *If a graph G satisfies $\Delta(G) \leq 15$ or $|V(G)| \leq 15$, then $\chi'(G) \leq \max\{\Delta(G) + 1, w(G)\}$.*

Proof: By Proposition 1.1, it suffices to prove Theorem 6.22 for the class of critical graphs. Hence, in the sequel, let G be a critical graph with $\Delta(G) \leq 15$ or $|V(G)| \leq 15$. First, assume that $\Delta(G) \leq 15$. Then

$$\left\lfloor \frac{15}{14}\Delta(G) + \frac{12}{14} \right\rfloor = \left\lfloor \Delta(G) + \frac{\Delta(G) + 12}{14} \right\rfloor \leq \Delta(G) + 1$$

and Corollary 6.7 yields $\chi'(G) \leq \max\{\Delta(G) + 1, w(G)\}$.

Now, assume that $|V(G)| \leq 15$. If $\chi'(G) \leq \Delta(G) + 1$, then we are done. Otherwise, $\chi'(G) = k + 1$ for an integer $k \geq \Delta(G) + 1$ and we have to show that G is elementary. Since G is critical and $\chi'(G) \geq \Delta(G) + 2$, there is a triple $(T, e, \varphi) \in \mathcal{T}(G)$, that is, $e \in E(G)$, $\varphi \in \mathcal{C}^k(G - e)$, and T is a Tashkinov tree with respect to e and φ such that $|V(T)| = t(G)$. If $t(G) < 11$, then Lemma 6.17 implies that G is elementary.

It remains to consider the case that $t(G) \geq 11$. By Proposition 5.10, $|V(T)| = t(G)$ is odd and, moreover, $V(T)$ is elementary and closed with respect to φ. If $V(T)$ is also strongly closed, then G is elementary (Theorem 1.4) and we are done. If $V(T)$ is not strongly closed, then $\Gamma^{\mathrm{d}}(T, e, \varphi) \neq \emptyset$ (Proposition 5.10(c)) and, therefore, $D(T, e, \varphi) \neq \emptyset$. Hence, there is a vertex $u \in D(T, e, \varphi)$.

From Proposition 5.12 and Proposition 5.14 it follows that $Z = V(T) \cup \{u\}$ is elementary with respect to φ. Clearly, $\overline{\varphi}(Z) \neq \emptyset$ and $|Z| = t(G) + 1$ is even. Hence, there is a color $\gamma \in \overline{\varphi}(Z)$ and an edge $f_1 \in \partial_G(Z)$ with $\varphi(f_1) = \gamma$. Let u_1 denote the endvertex of f_1 that belongs to $V(G) \setminus Z$. Then (f_1, u_1) is a fan at Z with respect to φ and, therefore, $X_1 = Z \cup \{u_1\}$ is elementary with respect to φ (Theorem 5.18). Observe that $|X_1|$ is odd, $|X_1| \geq 13$, and $|V(G) \setminus X_1| \leq 2$. This implies that $|\partial_G(X_1) \cap E_{\varphi,\beta}| = 1$ for every color $\beta \notin \overline{\varphi}(X_1)$. Consequently, if X_1 is closed with respect to φ, then X_1 is also strongly closed with respect to φ, and so G is elementary (Theorem 1.4). So it remains to consider the case that X_1 is not closed with respect to φ. Since $|X_1|$ is odd, this implies that there are two edges $f_2, f_3 \in \partial_G(X_1)$ such that $\varphi(f_2) = \varphi(f_3) = \gamma'$ and $\gamma' \in \overline{\varphi}(X_1)$. For $j = 2, 3$, let u_j be the endvertex of f_j that belongs to $V(G) \setminus X_1$. Clearly, at least one of these two edges, say f_2, has an endvertex in Z. Then (f_1, u_1, f_2, u_2) is a fan at Z with respect to φ and hence $X_2 = Z \cup \{u_1, u_2\}$ is elementary with respect to φ (Theorem 5.18). Note that $|X_2|$ is even, $|X_2| \geq 14$ and $V(G) = X_2 \cup \{u_3\}$. Then, for every color $\alpha \in \overline{\varphi}(X_2)$, there is an edge $f_\alpha \in E_G(u_3, X_2)$ such that $\varphi(f_\alpha) = \alpha$ implying that $\alpha \in \varphi(u_3)$. Consequently, $V(G) = X_2 \cup \{u_3\}$ is elementary with respect to φ. Now Theorem 1.4 can be used to deduce that G is an elementary graph. This completes the proof of the theorem. ∎

6.6   ANOTHER CLASSIFICATION PROBLEM FOR GRAPHS

Every graph G satisfies $\chi'(G) \geq \max\{\Delta(G), w(G)\}$, and Goldberg's conjecture implies that the chromatic index is either equal to this trivial lower bound or it is just 1 more. This proposes a classification problem for graphs, similar to the one for simple graphs. A graph G is said to be of the **first class** if $\chi'(G) = \max\{\Delta(G), w(G)\}$, otherwise G is said to be a of the **second class**. Goldberg's conjecture that $\chi'(G) \leq \max\{\Delta(G) + 1, w(G)\}$ implies that a graph G is of the second class if and only if $\chi'(G) = \max\{\Delta(G), w(G)\} + 1$. It follows from the characterization (6.1) of the fractional chromatic index that a graph G is of the first class if $\chi'(G) = \lceil \chi'^*(G) \rceil$ and vice versa. On the other hand, every graph of the second class satisfies $\chi'(G) = \lceil \chi'^*(G) \rceil + 1$ and $w(G) \leq \Delta(G) = \lceil \chi'^*(G) \rceil$, provided that Goldberg's conjecture is true. Observe that simple graphs G satisfying $\chi'(G) = \Delta(G) + 1 = w(G)$ are of the first class in this new classification, but of class two in the standard classification. This may indicate why second class graphs are so rare. However, graphs of the second class do exist and our favorite candidate, the Petersen graph, is indeed such a graph. On the other hand, it seems most likely that "almost all" graphs belong to the first class. One class of graphs for which it has been proved that all its members belong to the first class are the ring graphs (Theorem 6.3). There is a much more interesting class of graphs for which it is conjectured that all of them are of the first class, namely planar graphs. This conjecture was posed by Seymour [281]. In this paper Seymour also posed, independently of Goldberg, the conjecture that $\chi' \leq \max\{\Delta + 1, w\}$.

Conjecture 6.23 (Seymour's Exact Conjecture) *Every planar graph G is of the first class, i.e., $\chi'(G) = \max\{\Delta(G), w(G)\}$.*

Let us first show that Seymour's exact conjecture implies the Four-Color Theorem. This may indicate why a proof of the conjecture is not easily available. A set X is called an **odd set**, respectively an **even set**, depending on whether $|X|$ is odd or even. For a graph G and an integer $k \geq 0$, we define the k-**deficiency** of G by

$$\text{def}_k(G) = \sum_{v \in V(G)} (k - d_G(v)).$$

Hence, $\text{def}_k(G) = k|V(G)| - 2|E(G)|$ and, for any odd set $X \subseteq V(G)$, we have

$$\text{def}_k(G[X]) \geq k \Leftrightarrow k|X| - 2|E(G[X])| \geq k \Leftrightarrow k \geq \frac{2|E(G[X])|}{|X| - 1} = \frac{|E(G[X])|}{\lfloor |X|/2 \rfloor}.$$

By (1.6), this implies that every graph G, having order at least 3, satisfies:

$$w(G) \leq k \text{ if and only if } \text{def}_k(G[X]) \geq k \text{ for every odd } X \subseteq V(G). \qquad (6.3)$$

Note that every graph G with at most two vertices satisfies $\chi'(G) = w(G) = \Delta(G)$. If G is an r-regular graph, then $|\partial_G(X)| = \text{def}_r(G[X])$ for all $X \subseteq V(G)$. For a bridgeless cubic graph G, a simple parity argument shows that $|\partial_G(X)| \geq 3$ for each odd set $X \subseteq V(G)$. By (6.3), this implies that every bridgeless cubic graph G satisfies $w(G) \leq 3$ and so $\max\{\Delta(G), w(G)\} = 3$. Hence Seymour's exact conjecture would imply that each bridgeless cubic planar graph has $\chi'(G) = 3$, which is equivalent to the Four-Color Theorem, as observed by Tait [290].

While a **subdivision** of a graph G is obtained from G by subdividing edges, a **minor** of G is obtained from G by contracting edges, deleting edges and deleting vertices, where to **contract** an edge $e \in E_G(x, y)$ is to contract the set $\{x, y\}$ (see Sect. 4.7). For example, the complete graph K_5 and the complete bipartite graph $K_{3,3}$ are both minors of the Petersen graph, and the Petersen graph contains a subdivision of $K_{3,3}$, but not of K_5. The class of planar graphs is obviously closed under taking subgraphs, subdivisions or minors; and one of the most well known theorems in graph theory, published by Kuratowski [186] in 1930, says that a graph is planar if and only if it does not contain a subdivision of K_5 or $K_{3,3}$ as a subgraph. Kuratowski's theorem was refined by K. Wagner in his 1935 thesis. Wagner [303] proved that a graph G is planar if and only if K_5 and $K_{3,3}$ are not minors of G. So the following result due to Marcotte [210] provides an affirmative solution to the exact conjecture for a subclass of planar graphs.

Theorem 6.24 (Marcotte [210] 2001) *Every graph G not containing K_5^- or $K_{3,3}$ as a minor is of the first class, i.e., $\chi'(G) = \max\{\Delta(G), w(G)\}$.*

To present a proof of Marcotte's theorem is beyond the scope of this book and we shall only outline the general approach. From Wagner's work [302, 303, 304] it follows that a connected graph without a K_5^- or $K_{3,3}$ minor contains a k-cut with $k \in \{1, 2\}$, or its underlying simple graph is isomorphic to K_n for $1 \leq n \leq 3$, the prism or the wheel. A k-**cut** of a graph G is a set $X \subseteq V(G)$ such that $|X| = k$ and $G - X$ is the disjoint union of two nonempty graphs G_1 and G_2. If G_1 and G_2 can be chosen in such a way that $|V(G_i)| \geq 2$ for $i = 1, 2$, then the k-cut X is called an **essential** k-**cut**.

For convenience, let $\kappa(G) = \max\{\Delta(G), w(G)\}$. Marcotte's proof of Theorem 6.24 is based on an investigation of a smallest counterexample G. Then $\chi'(G) \geq \kappa(G) + 1$ and, since the class of graphs under consideration is closed under taking subgraphs, G is a critical graph. This implies, in particular, that G is connected and has no 1-cut. If G has an essential 2-cut, a decomposition theorem from Marcotte [209] can be applied to get a contradiction. Let us mention two further decomposition results used in Marcotte's proof.

Proposition 6.25 (Marcotte [208] 1990) *Let G be a graph such that $w(G) \leq \Delta(G)$ and $\Delta(G) \geq 1$. Then the following statements hold:*

(a) *G contains a matching M such that $\Delta(G - M) = \Delta(G) - 1$.*

(b) *G contains $p \geq 1$ matchings $M_1, \ldots, M_p$ such that $G' = G - \bigcup_{i=1}^{p} M_i$ satisfies either $(E(G') = \emptyset$ and $\Delta(G) = p)$ or $(\Delta(G) = \Delta(G') + p$ and $w(G') = \Delta(G') + 1)$. Furthermore, $\chi'(G) \leq \kappa(G)$, provided that $\chi'(G') \leq \kappa(G')$.*

Proof: Statement (a) is a simple consequence of Edmond's matching polytope theorem and the hypothesis (see Theorem B.7). For the proof of (b), suppose that G is a graph with $w(G) \leq \Delta(G)$ and $\Delta(G) \geq 1$. The existence of the desired matchings follows from (a) by a simple inductive argument. If $E(G') = \emptyset$ and $\Delta(G) = p$, then $\chi'(G) \leq p \leq \kappa(G)$. Otherwise, $\chi'(G) \leq p + \chi'(G') \leq p + \kappa(G') = p + \Delta(G') + 1 = \Delta(G) + 1 \leq w(G) = \kappa(G)$, provided that $\chi'(G') \leq \kappa(G')$. ∎

Proposition 6.26 (Marcotte [210] 2001) *Let G be a graph such that $\Delta(G) \leq w(G)$, and let $X \subseteq V(G)$ be an odd set such that $3 \leq |X| < |V(G)|$ and $\mathrm{def}_k(G[X]) = k$, where $k = w(G)$. Then $\kappa(G/X) \leq \kappa(G)$ and, moreover, $\chi'(G) \leq \kappa(G)$, provided that $\chi'(G/X) \leq \kappa(G)$ and $\chi'(G[X]) \leq \kappa(G)$.*

Proof: Let z denote the new vertex of $G' = G/X$ obtained by contracting the set X. Note that $k = \kappa(G) = w(G)$. The degree of a vertex $x \neq z$ is the same in G' as in G. The degree of z in G' is given by $d_{G'}(z) = |\partial_G(X)| \leq \sum_{v \in X}(k - d_{G[X]}(v)) = k$ and so the graph $G' = G/X$ satisfies $\Delta(G') \leq k$. Next we show that $w(G') \leq k = w(G)$. If $|V(G')| \leq 2$ we have $w(G') = \Delta(G') \leq k$. So assume that $|V(G')| \geq 3$. Then we can use (1.6). Let $U \subseteq V(G')$ be an odd set. If $z \notin U$, then $|E(G'[U])| = |E(G[U])| \leq k\lfloor|U|/2\rfloor$. If $z \in U$, then $|E(G[X])| = k\lfloor|X|/2\rfloor$ (by hypothesis) and, therefore,

$$
\begin{aligned}
|E(G'[U])| &= |E(G[U \setminus \{z\} \cup X])| - |E(G[X])| \\
&\leq k\lfloor(|U| + |X| - 1)/2\rfloor - k\lfloor|X|/2\rfloor \\
&\leq k\lfloor|U|/2\rfloor.
\end{aligned}
$$

This shows that $w(G') \leq k$ and so $\kappa(G') \leq \kappa(G)$. To complete the proof suppose that $\chi'(G[X]) \leq k$ and $\chi'(G/X) \leq k$. Observe that X is a k-contractible set of G. It then follows from the parity argument (see the proof of Proposition 4.49) that $\chi'(G/X^c) \leq k$. Then $\chi'(G) \leq \max\{\chi'(G/X), \chi'(G/X^c)\} \leq k$, as required. ∎

The most elaborate part of Marcotte's proof, however, is concerned with graphs that cannot be further reduced by applying various decomposition results like Proposition 6.26. These graphs are so-called triangle-subdivisions of certain basic graphs, including the triangle, the prism or the wheels. Any triangle-subdivision of K_3 is a graph without a K_4 minor. Graphs without a K_4 minor are also referred to as **series-parallel** graphs. That series-parallel graphs are of the first class was first proved by Seymour [283] in 1990. Ten years later, a different proof was given by Fernandes and Thomas [94]. The proof of Fernandes and Thomas exploits the fact that each simple series-parallel graph has minimum degree at most 2 (by Dirac [73], respectively Duffin [74]), and so it has coloring number at most 3. However, for being a first class graph it is not sufficient to have coloring number at most 3. An example is the subgraph obtained from the Petersen graph by deleting any vertex. However, to prove that each series-parallel graph is of the first class, the following result due to Juvan, Mohar, and Thomas [159] can be successfully applied.

Lemma 6.27 (Juvan, Mohar, and Thomas [159] 1999) *Every nonempty simple series-parallel graph has one of the following:*

(a) *a vertex of degree at most 1,*

(b) *two distinct vertices of degree 2 with the same neighbors,*

(c) *two adjacent vertices of degree 2,*

(d) *a triangle with one vertex of degree 2 and one vertex of degree 3,*

(e) *two triangles u_1, v_1, w and u_2, v_2, w sharing exactly one vertex w of degree 4 such that both v_1 and v_2 have degree 2.*

Sketch of Proof: We proceed by induction on the order n of G. If $n \leq 3$ we are through. So let $n \geq 3$. It is easy to see that it suffices to proof the result under the assumption that G is edge-maximal without a K_4 minor. By a well known result (see for instance Diestel's book [69, Proposition 7.3.1]), it follows then that G contains a vertex v of degree 2 such that the two neighbors of v are adjacent. Since $G - v$ contains one of the five configurations (by induction), it is then easy to check that G contains one of these configurations, too. ∎

Lemma 6.27 provides us with a scheme for proving that each series-parallel graph is of the first class. The proof is by contradiction, where a smallest counterexample G is chosen first. Then, according to the lemma, there are five cases to look at. While the first two cases are easy to handle using Lemma 6.29, the remaining three cases require more sophisticated arguments. In each of these cases, we replace the corresponding configuration by a suitable smaller configuration and show that an edge coloring of the resulting graph G' can be extended to an edge coloring of the original graph G, which gives the required contradiction. In order to show that $\kappa(G') \leq \kappa(G)$, we use a simple observation due to Fernandes and Thomas [94]. First we need a new concept. For a graph G and a vertex x of G, the degree of x in the underlying simple graph of G is called the **simple degree** of x in G. So the simple degree of x is the number of neighbors of x in G.

Lemma 6.28 (Fernandes and Thomas [94] 2000) *Let k be an integer, and let G be a graph with $\Delta(G) \leq k$. Then $w(G) \leq k$ if and only if $2|E(G[U])| \leq k(|U| - 1)$ for every odd set $U \subseteq V(G)$ such that each vertex of U has simple degree at least 2 in $G[U]$.*

Proof: If G has order at most 2, then $w(G) = \Delta(G)$ and there is nothing to prove. So assume that the order of G is at least 3. The "only if" part is evident. To prove the "if" part, we must show that $2|E(G[U])| \leq k(|U| - 1)$ for every odd set $U \subseteq V(G)$ with $|U| \geq 3$ (because of (1.6) and $|V(G)| \geq 3$). We proceed by induction on $|U|$. If each vertex of $G[U]$ has simple degree at least two, this follows from the hypothesis. So assume that $G[U]$ has a vertex u of simple degree at most 1. Let v be its unique neighbor in $G[U]$ if such a neighbor exists; otherwise let v be an arbitrary vertex of $G[U] - u$. Then $W = U \setminus \{u, v\}$ is an odd set and $2|E(G[U])| \leq 2\Delta(G) + 2|E(G[W])| \leq 2k + k(|W| - 1) = k(|U| - 1)$ by the induction hypothesis if $|W| \geq 3$ and trivially if $|W| = 1$. This proves the lemma. ∎

Lemma 6.29 *Let G be a critical graph with $\chi'(G) = k+1$ for an integer $k \geq \Delta(G)$. Then the following statements hold:*

(a) *Every vertex of G has simple degree at least 2.*

(b) *G does not contain two distinct vertices of simple degree 2 with the same neighbors.*

(c) *G does not contain a triangle with two vertices of simple degree 2, provided that $k \geq w(G)$.*

Proof: Statement (a) is a simple consequence of Theorem 2.3. To prove (b), suppose it is false. Then there are four distinct vertices x, y, z, u such that $N_G(x) = N_G(u) = \{y, z\}$. By symmetry, we may assume that $\mu_G(y, u) \geq \mu_G(x, z)$. Clearly, there is an edge $e \in E_G(x, y)$ and a coloring $\varphi \in C^k(G - e)$. Since $k \geq \Delta(G)$, there are colors α, β such that $\alpha \in \overline{\varphi}(x)$ and $\beta \in \overline{\varphi}(y)$. By Theorem 2.1, $\alpha \neq \beta$ and $P = P_y(\alpha, \beta)$ is a path having endvertices x and y. Since x and u have simple degree 2 in G, we conclude that P contains an edge of the set $E_G(x, z)$ but no edge of the set $E_G(y, u)$. Since $\mu_G(y, u) \geq \mu_G(x, z)$, we deduce that there is an edge $e' \in E_G(y, u)$ such that the color $\gamma = \varphi(e')$ is not used for any edge in $E_G(x, z)$. Since x has simple degree 2, this implies that $\gamma \in \overline{\varphi}(x)$. But then $P_y(\gamma, \beta)$ is a path joining y with some vertex distinct from x, contradicting Theorem 2.1.

To prove (c), suppose on the contrary that there is a triangle with vertices x, y, z such that both x and y have simple degree 2 in G. Then there is an edge $e \in E_G(x, y)$ and a coloring $\varphi \in C^k(G - e)$. Since $k \geq w(G)$ (by hypothesis), $E' \doteq E(G[\{x, y, z\}])$ has at most k edges and, therefore, there is a color $\alpha \in \{1, \ldots, k\}$ that does not occur in E', because $e \in E'$. Since x and y have both simple degree 2, this implies $\alpha \in \overline{\varphi}(x) \cap \overline{\varphi}(y)$. Hence we can color e with α, contradicting $\chi'(G) = k + 1$.   ∎

Proposition 6.30 *Every graph of the second class contains a critical graph of the second class with the same chromatic index as a subgraph.*

Proof: Let G be a graph of the second class. There is a subgraph $H \subseteq G$ with $\chi'(H) = \chi'(G)$ (Proposition 1.1(a)). Then $\kappa(H)+1 \leq \kappa(G)+1 \leq \chi'(G) = \chi'(H)$, which completes the proof.   ∎

Theorem 6.31 (Seymour [283] 1990) *Every series-parallel graph G is of the first class, i.e., $\chi'(G) = \max\{\Delta(G), w(G)\}$.*

Proof: Suppose on the contrary that there exists a series-parallel graph of the second class. Among all such graphs, let G be one with $|E(G)| + |V(G)|$ minimum. Then $\chi'(G) = k + 1$ for an integer $k \geq \kappa(G)$ and, by Proposition 6.30, G is critical. Note that $\Delta(G) \geq 3$ and $|V(G)| \geq 3$. To reach a contradiction, we apply Lemma 6.27 to the underlying simple graph S of G.

If S satisfies (a) or (b) of Lemma 6.27, the required contradiction follows from Lemma 6.29. So let us assume that S satisfies one of the remaining three cases (c), (d) or (e).

According to case (e), let u_1, u_2, v_1, v_2, w be distinct vertices of G such that $N_G(w) = \{u_1, u_2, v_1, v_2\}$ and $N_G(v_i) = \{w, u_i\}$ for $i = 1, 2$. We us use the following abbreviations for the multiplicities: $a = \mu_G(u_1, v_1), b = \mu_G(w, u_1), c = \mu_G(w, v_1), d = \mu_G(w, v_2), e = \mu_G(w, u_2), f = \mu_G(u_2, v_2)$. In order to include the cases (c) and (d) in our discussion, we allow that $a = c = e = 0$ (case (c)) or $a = c = 0$ (case (d)). So in what follows we only assume that b, d and f are positive.

We deduce from inequality (2.1) that $b + c + d + e + f = d_G(w) + d_G(v_2) - \mu_G(w, v_2) \geq k + 1$. Furthermore, we have $a + b + c \leq k$ and $d + e + f \leq k$, since $w(G) \leq \kappa(G) \leq k$; and we have $b + c + d + e \leq k$, since $\Delta(G) \leq \kappa(G) \leq k$.

Let $z_1 = \max\{0, a + b + c + e - k\}, z_2 = \max\{0, b + d + e + f - k\}$ and $s = k - (b + c + d + e)$. Then $z_1 \leq e, z_2 \leq b, s \geq 0$ and

$$
a + f - z_1 - z_2 - s = \begin{cases} k - (b + e) & \text{if } z_1 > 0 \text{ and } z_2 > 0, \\ a + c & \text{if } z_1 = 0 \text{ and } z_2 > 0, \\ d + f & \text{if } z_1 > 0 \text{ and } z_2 = 0, \\ a + f - s & \text{if } z_1 = z_2 = 0. \end{cases} \tag{6.4}
$$

We have $a - z_1 \geq \min\{a, k - (b + c + e)\} \geq \min\{a, d\} \geq 0$ and $f - z_2 \geq \min\{f, k - (b + d + e)\} \geq \min\{f, c\} \geq 0$. Since $a + f - s = a + b + c + d + e + f - k \geq 0$, it follows from (6.4) that $a - z_1 + f - z_2 \geq s$. Hence there are nonnegative integers s_1 and s_2 such that $s = s_1 + s_2, s_1 \leq a - z_1$ and $s_2 \leq f - z_2$.

We now construct a graph G' from $G - v_1 - v_2 - w$ by adding two new vertices x and y, and by adding new edges such that $G - v_1 - v_2 - w = G' - x - y$, $\mu_{G'}(x, u_1) = a - z_1 - s_1, \mu_{G'}(x, u_2) = f - z_2 - s_2, \mu_{G'}(y, u_1) = b - z_2$, $\mu_{G'}(y, u_2) = e - z_1$, and $\mu_{G'}(u_1, u_2) = \mu_G(u_1, u_2) + z_1 + z_2$. Obviously, G' is a series–parallel graph and $|V(G')| + |E(G')| < |V(G)| + |E(G)|$. First, we prove the following.

Claim 6.31.1 $\kappa(G') \leq k$.

Proof: To show that $\Delta(G') \leq k$, it suffices to consider the vertices u_1, u_2, x and y, since all other vertices have the same degree in G' as in G. That each of these four vertices has degree at most k in G' can be easily checked by using (6.4).

To show that $w(G') \leq k$, we apply Lemma 6.28. So let $U' \subseteq V(G')$ be an odd set such that each vertex in $G'[U']$ has simple degree at least 2. We consider several cases. In each case we construct an odd set $U \subseteq V(G)$ and compute the value $n_U = |U'| - |U|$. Then it is not hard to see that in each case the value $m_U = 2|E(G'[U'])| - 2|E(G[U])|$ satisfies $m_U \leq kn_U$. Since $w(G) \leq k$, this gives $2|E(G'[U'])| = 2|E(G[U])| + m_U \leq k(|U| - 1) + kn_U = k(|U'| - 1)$, as required.

Case 1: $|U' \cap \{u_1, u_2\}| \leq 1$. Since each vertex in $G'[U']$ has simple degree at least 2, this implies $U' \cap \{x, y\} = \emptyset$. We put $U' = U$, which gives $m_U = n_U = 0$, since $G'[U'] = G[U]$.

Case 2: $u_1, u_2 \in U'$. If $x, y \in U'$, then let $U = U' \setminus \{x, y\}$. Then we obtain $n_U = 2$ and $m_U = 2(a - z_1 - s_1 + f - z_2 - s_2 + z_1 + z_2 + b - z_2 + e - z_1) \leq 2k$, since $s_1 + s_2 = k - (b + c + d + e), z_1 \geq a + b + c + e - k$, and $z_2 \geq b + d + e + f - k$.

If $x \in U'$ and $y \notin U'$ we put $U = U' \setminus \{x\} \cup \{w, v_1, v_2\}$. This gives $n_U = -2$ and $m_U = 2(a - z_1 - s_1 + f - z_2 - s_2 - (a+b+c+d+e+f)) = 2(-k - z_1 - z_2) \leq -2k$. If $x \notin U'$ and $y \in U'$ we let $U = U' \setminus \{y\} \cup \{w\}$. Then $n_U = 0$ and $m_U = 2(b - z_2 + e - z_1 - (b+e)) \leq 0$.

So the remaining case is $x, y \notin U'$. If $z_1 = z_2 = 0$ we put $U = U'$, which gives $m_U = n_U = 0$, since $G'[U'] = G[U]$. If $z_1 > 0$ and $z_2 > 0$, then let $U = U' \setminus \{u_1, u_2\}$. Then $n_U = 2$ and $m_U \leq 2(k - (a+b) + k - (e+f) + z_1 + z_2) = 2(b + c + d + e) \leq 2k$. If $z_1 = 0$ and $z_2 > 0$ we put $U = U' \cup \{w, v_2\}$. Then $n_U = -2$ and $m_U = 2(z_1 + z_2 - (b+d+e+f)) = -2k$. If $z_1 > 0$ and $z_2 = 0$ let $U = U' \cup \{w, v_1\}$. Then $n_U = -2$ and $m_U = 2(z_1 + z_2 - (a+b+c+e)) = -2k$. Hence the proof of the claim is complete. $\triangle$

By the choice of G, it follows that $\chi'(G') \leq \kappa(G') \leq k$, i.e., there is a coloring $\varphi' \in \mathcal{C}^k(G')$. Clearly, this coloring induces a coloring $\varphi_1 \in \mathcal{C}^k(G - w - v_1 - v_2)$. We now prove the following result, which gives the required contradiction.

Claim 6.31.2 *The coloring φ_1 can be extended to a k-edge-coloring of G.*

Proof: Let $E' = E_G(w, v_1) \cup E_G(w, v_2)$ and let (E_1, E_2) be a partition of the $z_1 + z_2$ edges of $E_{G'}(u_1, u_2) - E(G)$ such that $|E_i| = z_i$ for $i = 1, 2$. Then we can extend φ_1 to a coloring $\varphi \in \mathcal{C}^k(G - E')$ such that the following conditions hold: $\varphi(E_G(u_1, w)) = \varphi'(E_2 \cup E_{G'}(u_1, y))$, $\varphi'(E_1 \cup E_{G'}(u_1, x)) \subseteq \varphi(E_G(u_1, v_1))$, $\varphi(E_G(u_2, w)) = \varphi'(E_1 \cup E_{G'}(u_2, y))$, and $\varphi'(E_2 \cup E_{G'}(u_2, x)) \subseteq \varphi(E_G(u_2, v_2))$. Now, $|\varphi(w) \cup \varphi(v_1)| = |\varphi(w)| + |\varphi(v_1)| - |\varphi(w) \cap \varphi(v_1)| \leq a + b + e - z_1 \leq k - c$, $|\varphi(w) \cup \varphi(v_2)| \leq b + e + f - z_2 \leq k - d$ and $|\varphi(w) \cup (\varphi(v_1) \cap \varphi(v_2))| \leq b - z_2 + e - z_1 + z_1 + z_2 + s \leq k - (c+d)$. Now we can color the c edges of $E_G(v_1, w)$ with colors from the set $\{1, \ldots, k\}$, using as many colors from $\varphi(v_2)$ as possible. If the c edges can be colored using colors in $\varphi(v_2)$ only, then there are at least $k - |\varphi(w) \cup \varphi(v_2)| \geq d$ colors left to color the edges in $E_G(w, u_2)$. Otherwise, the edges in $E_G(w, v_1)$ will be colored using $\ell = |\varphi(w) \setminus (\varphi(v_1) \cup \varphi(v_2))|$ colors from $\varphi(v_2)$ and $c - \ell$ other colors. Thus the number of colors available to color the edges in $E_G(w, v_2)$ is at least $k - |\varphi(w) \cup \varphi(v_2)| - (c - |\varphi(w) \setminus (\varphi(v_1) \cup \varphi(v_2))|) = k - c - |\varphi(w) \cup (\varphi(v_1) \cap \varphi(v_2))| \geq d$. Hence in both cases we can extend φ to k-edge-coloring of G. $\triangle$

The proof of the theorem is now complete. ∎

Fernandes and Thomas [94] observed that their proof can be transformed into a linear-time algorithm to decide whether $\chi'(G) \leq k$ for a given series-parallel graph G and a given integer $k \geq 0$. Note, however, that the first part of the proof, where we apply Lemma 6.29, differs slightly from the original proof. A linear algorithm for determining the chromatic index of series-parallel graphs was first given by Zhou, Suzuki, and Nishizeki [325]. Their algorithm, however, is not based on the min–max formula in Theorem 6.31.

6.7 NOTES

It should come as no surprise that several graph theorists came up, about the same time and independently of each other, with conjectures equivalent to Goldberg's conjecture. However, to the best of our knowledge, the conjecture appeared first in print in Goldberg's paper [110], published in 1973 in Russian. In this short paper, Goldberg proves that each graph with $\chi' > (5\Delta + 2)/4$ contains a dense subgraph of order 3, and that it is thus elementary. This extended an earlier result of Vizing [298], who proved the same conclusion under the stronger assumption that the chromatic index equals Shannon's bound, i.e., $\chi' = \lfloor 3\Delta/2 \rfloor$. Goldberg's paper finishes with a conjecture saying that each graph with $\chi' > \Delta + 1$ contains a dense subgraph such that $w \geq \chi'$. The conjecture also gives a precise bound for the order of such a dense subgraph in terms of χ' and Δ (the bound follows from Proposition 1.7(a)). Finally, he added a remark saying, in our terminology, that $w \leq \chi'$. So the conjecture in Goldberg's paper [110] says that every graph with $\chi' > \Delta + 1$ is elementary. In a private communication Goldberg has informed us that he started thinking about the conjecture around 1970.

Goldberg did not define the density w in his early papers [109, 110]. As far as we know, the density w as an interesting self-contained graph parameter, appears first in Paul Seymour's paper [281]. This extensive paper, submitted in 1974 to the Proceedings of the London Mathematical Society and finally published in 1977, deals with the edge coloring problem of regular graphs and various other related problems. The paper includes a new proof of Edmonds' matching polytope theorem and of Edmonds' and Johnson's Chinese postman theorem. Studying edge colorings of regular graphs led Seymour to introduce the class of r-graphs and to propose various conjectures, including the conjecture that $\chi' \leq \max\{\Delta + 1, w\}$. A graph G is an r-**graph** if G is r-regular and for each odd set $X \subseteq V(G)$ we have $|\partial_G(X)| \geq r$ (or equivalently $\mathrm{def}_r(G[X]) \geq r$). Note that any r-graph with $r \geq 1$ must have even order, since $\partial_G(V(G)) = \emptyset$. If an r-regular graph admits an r-edge-coloring, then it must be an r-graph, since each color class forms a matching. However, the converse statement is not true, as the Petersen graph shows. Every r-graph satisfies $w \leq r$ (by (6.3)) and so $\kappa = r$, where $\kappa = \max\{\Delta, w\}$. Hence Seymour's above conjecture implies the conjecture that every r-graph admits an $(r + 1)$-edge-coloring. Seymour proved that the latter conjecture is equivalent to the conjecture that $\chi' \leq \kappa + 1$. On the other hand, it follows from (6.1) that r-graphs satisfies $\chi'^* = r$. In fact, r-graphs are the only r-regular graphs with this property. Let G be an r-graph with $r \geq 1$. It follows from the theory of linear programming that there is an optimal fractional edge coloring w of G such that $w(M)$ is rational for every matching M, where $\chi'^*(G) = r = \sum_M w(M)$. Let t denote the least common multiple of the denominators of all the values $w(M)$, let $\mathbf{x}_M$ denote the characteristic function of the matching M (so that $\mathbf{x}_M(e)$ is 1 or 0 depending on whether e belongs to M or not), and let $\mathbf{1}$ denote the all-1 function ($\mathbf{1}(e) = 1$ for all $e \in E(G)$). Then t is a positive integer and $k_M = tw(M)$ is a nonnegative integer for all matching M of G.

Since w is an optimal fractional edge coloring, we conclude that

$$\sum_M k_M \mathbf{x}_M = t\mathbf{1} \text{ and } \sum_M k_M = tr.$$

This means that there is a family of tr matchings in G using each edge t times. Clearly, this is equivalent to $\chi'(tG) = tr$. A conjecture of Seymour posed in [281] says that every r-graph G satisfies $\chi'(2G) = 2r$. Seymour called this conjecture the *Generalized Fulkerson Conjecture*, as Fulkerson [105] asked if every 3-graph G satisfies $\chi'(2G) = 6$. If true, this implies a conjecture of Berge [27], that the edges of every 3-graph can be covered by 5 matchings. That the two conjectures are in fact equivalent was proved by Mazzuoccolo [212].

An equivalent form of the Four-Color Theorem says that each bridgeless cubic graph has chromatic index 3, as observed first by Tait [290]. So Seymour's exact conjecture that every planar graph G is of the first class (i.e., $\chi'(G) = \max\{\Delta(G), w(G)\}$) may be considered as a far reaching generalization of the Four-Color Theorem. The name *Exact Conjecture* was introduced by Marcotte [210]. The conjecture appears first in Seymour's paper [281] as a remark and somewhat later in explicit form in [282] (see also [280, 283]). A consequence of Seymour's exact conjecture is that each planar r-graph has chromatic index r. For $r = 3$, this is equivalent to the Four-Color Theorem. That the statement for $r = 4$ implies the Four-Color Theorem follows from a theorem of Kotzig [179], that if G is a connected cubic graph with an even number of edges, then $\chi'(G) = 3$ if and only if $\chi'(L(G)) = 4$. A proof of this theorem can be found in Schrijver's book [277, Theorem 28.10] and, of course, in Kotzig's paper [179].

That r-graphs form an interesting class of graphs is also due to the fact that the family of graphs satisfying $\kappa \leq r$ do not only contain all r-graphs, but any graph in this family is a subgraph of an r-graph. This was discovered by Seymour [281]. Let G be a graph with $\kappa(G) \leq r$. The following construction of an r-graph H with $G \subseteq H$ is taken from [281]. If $|V(G)| \leq 2$, then $\kappa(G) = \Delta(G) \leq r$ and $H = rK_2$ is an r-graph with $G \subseteq H$. So assume that $|V(G)| \geq 3$. If we add an isolated vertex, we get $\kappa(G + K_1) = \kappa(G) \leq r$. Hence, we may assume that G has even order. If $r = 0$ take $H = G$, and if $r = 1$ take as H any 1-regular graph obtained from G by adding edges. If $r = 2$, then G is the disjoint union of paths and even cycles, since an odd cycle has density 3. Hence by adding edges to G we obtain a 2-regular graph H without odd cycles. Thus, in what follows, assume that $r \geq 3$ and G has even order at least 4. Let $d = \mathrm{def}_r(V(G)) = r|V(G)| - 2|E(G)|$. Clearly, d is even and $d \geq 0$. If $d = 0$, then G is r-regular and hence an r-graph. If $d > 0$, let G_d be an $(r-1)$-edge-connected $(r-1)$-regular graph with d vertices constructed as follows. If r is odd, put $G_d = \frac{1}{2}(r-1)C_d$. If r is even, let G_d be the graph obtained from $\frac{1}{2}(r-1)C_d$ by joining each vertex to the vertex opposite it in the cycle. Now, we obtain an r-regular graph H from the disjoint union of G and G_d by adding d edges. To complete the proof, we show that H is an r-graph. Suppose this is not true. Then there is an odd set $X \subseteq V(H)$ such that $|\partial_H(X)| \leq r - 1$. Since H is r-regular, we have $|\partial_H(X)| = \mathrm{def}_r(H[X]) = r|X| - 2|E(H[X])|$. Using a parity argument, we deduce $|\partial_H(X)| \neq r - 1$ and hence $|\partial_H(X)| \leq r - 2$. Since

H has even order, $X' = V(H) \setminus X$ is an odd set, where $\partial_H(X) = \partial_H(X')$. So we may assume that $V(G_d) \not\subseteq X$. Thus $X \cap V(G_d) = \emptyset$, because G_d is $(r-1)$-edge-connected. Hence $H[X] = G[X]$ and, since $w(G) \le r$, we then conclude from (6.3) that $|\partial_H(X)| = \operatorname{def}_r(H[X]) = \operatorname{def}_r(G[X]) \ge r$, a contradiction. This shows that H is an r-graph, and we are done.

In 1975 Lars D. Andersen submitted an M.Sc. thesis at the University of Aarhus on edge coloring of graphs. Andersen's thesis [4] was supervised by Gabriel A. Dirac and inspired by Ivan T. Jakobsen's work about critical graphs, and two of the most interesting results in the thesis settle the cases $m = 5$, respectively $m = 7$, of Jakobsen's conjecture (Conjecture 6.4). The proof for both cases is based on the critical chain method and inspired Andersen to propose the conjecture that if G is a critical graph with $\chi'(G) = k + 1 \ge \Delta(G) + 2$, then, for every edge $e \in E(G)$ and every coloring $\varphi \in C^k(G - e)$, the vertex set $V(G)$ is elementary with respect to φ. That Andersen's conjecture is equivalent to the conjecture that $\chi' \le \max\{\Delta + 1, w\}$ follows from Theorem 1.4 and the fact that the latter conjecture can be reduced to the class of critical graphs. However, Andersen did not seem to be aware of this equivalence, and there is no reference in the thesis neither to Goldberg nor to Seymour. Two years later, when part of the thesis was published in [5], Andersen had became familiar with Goldberg's work. In the paper [5] Andersen's does not only cite Goldberg's papers [109, 110, 111], but he also added a remark saying that the above conjecture is equivalent to a conjecture by Goldberg in [110], however, without mentioning the form of Goldberg's conjecture.

In 1967 Ram P. Gupta submitted a thesis [120] to the Indian Statistical Institute, Calcutta. The thesis consists of four chapters which are, more or less, independent of each other. The only chapter dealing with edge coloring of graphs is the second chapter entitled "The chromatic index of a graph". The first chapter is devoted to the covering index of a graph, while the third and fourth chapters both deals with digraphs. It is well known that the main result in the second chapter of Gupta's thesis is the theorem that $\chi' \le \Delta + \mu$. However, it seems less known that the theorem contains an addendum saying that the bound is achieved if $\mu \ge 0$ and $\Delta = 2p\mu - r$, where $r \ge 0$ and $p > \lfloor (r+1)/2 \rfloor$. That for all other possible values of Δ and μ we have $\chi' \le \Delta + \mu - 1$ is posed as a conjecture at the end of the coloring chapter. This is the only conjecture in Gupta's thesis about edge coloring of graphs. A more detailed discussion of Gupta's conjecture is given here in Chap. 7. Gupta's result concerning the sharpness of Vizing's bound was rediscovered by Scheide [267, 272] who also proved that Gupta's conjecture is implied by Goldberg's conjecture. In 1978 a paper by Gupta [121] was published in the proceedings of the graph theory conference, held at the Western Michigan University (Kalamazoo) in 1976. This paper, written in the typical style of a conference paper, deals with the cover index and the chromatic index of graphs, the two favorite topics of Gupta's thesis. While the paper gives an overview of some new results, most proofs are postponed to a later paper. Gupta [121] defines the r-**density** of a graph G by

$$w_r(G) = \max_{X \subseteq V(G), |X| = 2r+1} \left\lceil \frac{2|E(G[X])|}{r} \right\rceil,$$

where r is an nonnegative integer. Then the following two results are stated in the paper: $\chi' \leq \max\{(5\Delta + 2)/4, w_1\}$ and $\chi' \leq \max\{(7\Delta + 4)/6, w_1, w_2\}$. These results lead quite naturally to the following conjecture, also stated in the paper: For any graph G,

$$\chi'(G) \leq \max\{\frac{(2r + 3)\Delta + 2r}{2r + 2}, w_1(G), \ldots, w_r(G)\}, r = 1, 2, \ldots. \quad (6.5)$$

Obviously, Gupta's conjecture is a parameterized version of the conjecture that $\chi' \leq \max\{\Delta + 1, w\}$ (see Proposition 1.7). However, Gupta was not familiar with the results of Goldberg [110], nor with the results of Seymour [281] and Andersen [5], obtained around the same time. While the results of Goldberg [110, 114], Andersen [5], and Nishizeki and Kashiwagi [237] are obtained by means of the critical chain method, Gupta's [121] results are derived from more general statements about improper edge colorings of graphs. It is a pity that Gupta never seems to have published more on edge colorings of graphs.

Theorem 6.5 combined with Propositions 1.1 and 1.7 shows that Gupta's conjecture (6.5) holds for $r = 6$. The proof of Theorem 6.5, based on various lemmas, is taken from Scheide's thesis [269], respectively from Scheide's paper [271]. In particular, the proof of Proposition 6.10 and the results of Sect. 6.3 are from Scheide [271] with permission from Elsevier. However, balanced and normal Tashkinov trees were already introduced in Favrholdt et al. [91]. The key lemma in Scheide's proof (Lemma 6.13) is a far reaching extension of an earlier result by Favrholdt et al. [91, Theorem (O6)]. The proof is based on Tashkinov's method which may be considered as an extension of the critical chain method. As a consequence of Lemma 6.14, which is used in the proof of Theorem 6.5, we obtain the following moderate strengthening of an earlier result by McDonald [218]:

$$\text{For every graph } G, \chi'(G) \leq \max\{w(G), \Delta(G) + 1 + \frac{\Delta(G) - 3}{g_o(G) + 5}\}. \quad (6.6)$$

Clearly, it suffices to prove (6.6) for critical graphs G with $\chi'(G) = k+1 \geq \Delta(G)+2$. If G is elementary, we are done. Otherwise, $t(G) \geq h(G) + 4 \geq g_o(G) + 4$ (by Lemma 6.14) and $t(G) \leq \frac{\Delta(G)-3}{k-\Delta(G)} - 1$ (by (5.4)), which yields the required bound for $\chi'(G)$.

The presented proof of Theorem 6.31, that each series-parallel graph is of the first class, is due to Fernandes and Thomas [94]. Apart from Seymour's original proof from [283], there is a third proof of the theorem due to Marcotte [209]. For graphs with fewer than 10 vertices, the theorem follows from a result by Plantholt and Tipnis [248]. For series-parallel graphs with more that 7 vertices, Lemma 6.29 shows that there exists an essential 2-cut (or a 1-cut). Then Marcotte's decomposition result from [209] can be applied to obtain the required coloring by induction. Marcotte [207, 208] established Goldberg's conjecture for every graph not containing K_5^- as a minor. In 2001 Marcotte [210] verified that the exact conjecture is true for every graph not containing K_5^- or $K_{3,3}$ as a minor (Theorem 6.24).

CHAPTER 7

EXTREME GRAPHS

In this chapter we investigate Shannon's bound and Vizing's bound for the chromatic index and discuss the following problems: For which values of Δ and μ does there exist a graph G satisfying $\Delta(G) = \Delta$, $\mu(G) = \mu$ and $\chi'(G) = \Delta + \mu$? What is the best upper bound for the chromatic index χ' in terms of the maximum degree Δ and the maximum multiplicity μ?

7.1 SHANNON'S BOUND AND RING GRAPHS

For integers $r, s, t \geq 0$, let $T = T(r, s, t)$ denote the graph consisting of three vertices x, y, z such that $\mu_T(x, y) = r$, $\mu_T(y, z) = s$, and $\mu_T(z, x) = t$. The graph $T = T(r, s, t)$ satisfies $\chi'(T) = r + s + t$ and $\Delta(T) = \max\{r + s, s + t, t + r\}$. Hence, $\chi'(T) \geq \Delta(T) + 1$ if and only if $r, s, t \geq 1$.

The graph $S_d = T(\lfloor \frac{d}{2} \rfloor, \lfloor \frac{d}{2} \rfloor, \lfloor \frac{d+1}{2} \rfloor)$ is called a **Shannon graph** of degree d. Note that the Shannon graph S_d achieves Shannon's bound, since $\Delta(S_d) = d$ and $\chi'(S_d) = \lfloor 3d/2 \rfloor = \lfloor 3\Delta(S_d)/2 \rfloor$. Clearly, S_d is a critical graph if $d \geq 2$.

Theorem 7.1 (Vizing [298] 1965) *Let G be a graph such that $\chi'(G) = \lfloor \frac{3}{2}\Delta \rfloor$, where $\Delta = \Delta(G) \geq 4$. Then G contains a Shannon graph S_Δ of degree Δ as a subgraph.*

Proof: The graph G contains a critical subgraph H with $\chi'(H) = \chi'(G)$. From Theorem 2.4 it follows that $\lfloor \frac{3}{2}\Delta \rfloor = \chi'(G) = \chi'(H) \leq \lfloor \frac{3}{2}\Delta(H) \rfloor \leq \lfloor \frac{3}{2}\Delta \rfloor$ and, therefore, $\chi'(H) = \lfloor \frac{3}{2}\Delta \rfloor$ and $\Delta = \Delta(H)$. Furthermore, since $\Delta \geq 4$, we conclude that $\chi'(H) > \frac{5}{4}\Delta + \frac{1}{2}$, i.e., $H \in \mathcal{J}_5$. By Corollary 6.8, this implies that $|V(H)| \leq 3$. Then $|V(H)| = 3$, since otherwise $|V(H)| \leq 2$ yields $\chi'(H) = \Delta < \lfloor \frac{3}{2}\Delta \rfloor$, a contradiction. Hence $H = T(r, s, t)$ is a graph with three vertices, where $0 \leq r \leq s \leq t$. Then $\chi'(H) = r + s + t = \lfloor \frac{3}{2}\Delta \rfloor$ and $\Delta = \max\{r + s, r + t, s + t\} = s + t$. This implies, by an easy calculation, that $r = s = \lfloor \frac{\Delta}{2} \rfloor$ and $t = \lfloor \frac{\Delta+1}{2} \rfloor$ and, therefore, $H = S_\Delta$ is a Shannon graph of degree Δ.   ∎

Theorem 7.1 provides a good characterization of graphs with $\Delta \geq 4$ for which Shannon's bound is attained. By the result of Holyer [150], a good characterization of graphs with $\Delta = 3$ for which Shannon's bound is attained seems unlikely. Clearly, a connected graph G with maximum degree $\Delta \in \{1, 2\}$ achieves Shannon's bound if and only if $\Delta = 2$ and G is an odd cycle or $\Delta = 1$ and $G = K_2$.

Theorem 7.1 implies that the Shannon graph S_Δ is the only critical graph with $\chi' = \lfloor 3\Delta/2 \rfloor$, provided that $\Delta \geq 4$. This leads to the question of finding a better bound for the chromatic index of a graph with maximum degree Δ not containing S_d as a subgraph. For integers $\Delta \geq 0$ and $d \geq 2$, let

$$sh(\Delta, d) = \max\{\chi'(G) \mid \Delta(G) = \Delta \text{ and } S_d \not\subseteq G\}.$$

Berge [25] proposed the conjecture that $sh(\Delta, d) \leq \lfloor 3\Delta(G)/2 \rfloor - \lfloor \Delta/d \rfloor$, provided that $4 \leq d \leq \Delta$. Berge himself verified the conjecture for Δ and d even and Bosák [31] verified the conjecture for the case $\Delta \leq 2d$ and d even. However, as proved by Andersen [5], the actual value of $sh(\Delta, d)$ is much less than the upper bound conjectured by Berge.

Theorem 7.2 (Andersen [5] 1977) *If $\Delta \geq 0$ and $d \geq 2$ are integers, then*

$$sh(\Delta, d) = \begin{cases} \lfloor 3\Delta/2 \rfloor & \text{if } \Delta \leq d - 1, \\ \Delta - 1 + \lfloor d/2 \rfloor & \text{if } d \leq \Delta \leq 2d - 6, \\ \lfloor 5\Delta/4 + 1/2 \rfloor & \text{if } \Delta \geq 2d - 5. \end{cases}$$

Proof: First, let $\Delta \leq d - 1$. By Shannon's bound, we have $sh(\Delta, d) \leq \lfloor 3\Delta/2 \rfloor$ and this bound is obtained for graphs with maximum degree Δ and not containing S_d, namely if $\Delta \in \{0, 1\}$ by the complete graph K_Δ, and if $\Delta \geq 2$ by the Shannon graph S_Δ.

Next, let $d \leq \Delta \leq 2d - 6$. This implies that $d \geq 6$. Suppose on the contrary that there exists a graph G satisfying $\Delta(G) = \Delta$, $S_d \not\subseteq G$ and $\chi'(G) > \Delta - 1 + \lfloor d/2 \rfloor$. Since $\Delta \geq 2d - 6$, this gives

$$\chi'(G) \geq \Delta + \lfloor d/2 \rfloor \geq \Delta + \lfloor (\Delta + 6)/4 \rfloor > \lfloor 5\Delta/4 + 1/2 \rfloor,$$

i.e., $G \in \mathcal{J}_5$. Since each graph in $\mathcal{J}_5$ contains at least three vertices, it easily follows from Proposition 1.1 and Corollary 6.8 that G contains a critical subgraph $T \in \mathcal{J}_5$

satisfying $|V(T)| = 3$ and $\chi'(T) = \chi'(G)$. Hence $T = T(r, s, t)$ with $r \leq s \leq t$. Then we obviously have $\chi'(G) = \chi'(T) = r + s + t \geq \Delta + \lfloor d/2 \rfloor \geq d + \lfloor d/2 \rfloor$ and $\Delta(T) = s + t \leq \Delta \leq d$. This implies that $t \geq s \geq r \geq \lfloor d/2 \rfloor$. Furthermore, if d is odd we infer that $s \geq r \geq \lfloor d/2 \rfloor$ and $t \geq \lfloor (d+1)/2 \rfloor$. Hence we obtain that $S_d \subseteq T \subseteq G$, a contradiction.

The bound is obtained for the ring graph $T = T(\lfloor \Delta/2 \rfloor, \lfloor (\Delta+1)/2 \rfloor, \lfloor d/2 \rfloor - 1)$. Obviously, $\Delta(T) = \Delta$ and $S_d \not\subseteq T$.

Eventually, let $\Delta \geq 2d - 5$. Suppose on the contrary that there exists a graph G satisfying $\Delta(G) = \Delta$, $S_d \not\subseteq G$ and $\chi'(G) > \lfloor 5\Delta/4 + 1/2 \rfloor$. Then $G \in \mathcal{J}_5$ and, as in the previous case, G contains a graph $T = T(r, s, t)$ with $r \leq s \leq t$ as a subgraph such that $\chi'(T) = \chi'(G)$. Since $\Delta \geq 2d - 5$, we obtain that

$$\chi'(G) \geq \lfloor \tfrac{5}{4}\Delta + \tfrac{1}{2} \rfloor + 1 = \Delta + \lfloor \tfrac{1}{4}\Delta + \tfrac{3}{2} \rfloor \geq \Delta + \lfloor \tfrac{d}{2} - \tfrac{5}{4} + \tfrac{3}{2} \rfloor \geq \Delta + \lfloor \tfrac{d}{2} \rfloor.$$

This gives $\chi'(G) = \chi'(T) = r + s + t \geq \Delta + \lfloor d/2 \rfloor$. Since $\Delta(T) = s + t \leq \Delta$, this implies $t \geq s \geq r \geq \lfloor d/2 \rfloor$. Since $S_d \not\subseteq T$, we then conclude that d is odd and $\Delta \leq d - 1$. Since $\Delta \geq 2d - 5$ and $d \geq 2$, this gives $d = 3$. But then $T = C_3$ and, therefore, $\chi'(T) = \chi'(G) = 3 \leq \lfloor 5\Delta/4 + 1/2 \rfloor$, a contradiction.

The bound is obtained for all Δ and d. If $\Delta \in \{0, 1\}$ take the complete graph K_Δ. If $\Delta \geq 2$ take the ring graph H consisting of 5 vertices x, y, z, u, v such that $\mu_H(x, y) = \mu_H(y, z) = \mu_H(u, v) = \lfloor \Delta/2 \rfloor$ and $\mu_H(z, u) = \mu_H(v, x) = \lfloor (\Delta + 1)/2 \rfloor$. Then $\Delta(H) = \Delta$ and $S_d \not\subseteq H$. Furthermore, by Theorem 6.3, we conclude that $\chi'(H) = \lfloor 5\Delta/4 + 1/2 \rfloor$. ∎

As a refinement of Shannon's bound, Goldberg [114] established an upper bound for the chromatic index in terms of the maximum degree and the odd girth. He proved (Theorem 5.8) that every graph G with odd girth $g_o \geq 3$ satisfies

$$\chi'(G) \leq \Delta(G) + 1 + \left\lfloor \frac{\Delta(G) - 2}{g_o - 1} \right\rfloor. \tag{7.1}$$

Let $\mathcal{GO}(\Delta, g_o)$ denote the set of all graphs G with maximum degree Δ and odd girth g_o attaining the bound in (7.1), where $\Delta \geq 2$ is an integer and g_o is an odd integer with $g_o \geq 3$. The following characterization of ring graphs belonging to $\mathcal{GO}(\Delta, g_o)$ is due to McDonald (personal communication).

Proposition 7.3 (McDonald [217] 2010) *Let $\Delta \geq 2$ be an integer, let g_o be an odd integer with $g_o \geq 3$, and let $i \in \{0, 1, \ldots, g_o - 2\}$ be such that $\Delta - 2 \equiv i \bmod (g_o - 1)$. If R is a ring graph with g_o vertices and maximum degree Δ, then $R \in \mathcal{GO}(\Delta, g_o)$ if and only if*

$$|E(R)| \geq \frac{\Delta g_o - (i + 1)}{2}.$$

Proof: By Theorem 6.3, we have

$$\chi'(R) = \max \left\{ \Delta, \left\lceil \frac{2|E(R)|}{g_o - 1} \right\rceil \right\}$$

so $R \in \mathcal{GO}(\Delta, g_o)$ if and only if

$$\left\lceil \frac{2|E(R)|}{g_o - 1} \right\rceil = \Delta + 1 + \left\lfloor \frac{\Delta - 2}{g_o - 1} \right\rfloor.$$

Since $2|E(R)| \leq \Delta g_o$, there exists a nonnegative integer k such that $2|E(R)| = \Delta g_o - k$. Using k and i, we can rewrite $\left\lceil \frac{2|E(R)|}{g_o-1} \right\rceil$ as follows:

$$\left\lceil \frac{2|E(R)|}{g_o - 1} \right\rceil = \left\lceil \frac{\Delta g_o - k}{g_o - 1} \right\rceil = \Delta + \left\lceil \frac{\Delta - 2 - i}{g_o - 1} + \frac{2 + i - k}{g_o - 1} \right\rceil.$$

Since

$$\left\lceil \frac{\Delta - 2 - i}{g_o - 1} \right\rceil = \frac{\Delta - 2 - i}{g_o - 1} = \left\lfloor \frac{\Delta - 2}{g_o - 1} \right\rfloor,$$

we get that

$$\left\lceil \frac{2|E(R)|}{g_o - 1} \right\rceil = \Delta + \left\lfloor \frac{\Delta - 2}{g_o - 1} \right\rfloor + \left\lceil \frac{2 + i - k}{g_o - 1} \right\rceil.$$

Note that Theorem 5.8 (respectively (7.1)) implies $\left\lceil \frac{2+i-k}{g_o-1} \right\rceil \leq 1$. Hence, we have that

$$\left\lceil \frac{2|E(R)|}{g_o - 1} \right\rceil = \Delta + 1 + \left\lfloor \frac{\Delta - 2}{g_o - 1} \right\rfloor$$

if and only if $1 \leq 2 + i - k$, that is, if and only if $k \leq i + 1$. Thus the proof is complete. ∎

For integers $m_1, \ldots, m_p \geq 0$, let $R = R_p(m_1, \ldots, m_p)$ denote the ring graph consisting of $p \geq 3$ vertices $x_1, \ldots, x_p$ satisfying $\mu_R(x_i, x_{i+1}) = m_i$ for $i = 1, \ldots, p$, where the index $i + 1$ is taken modulo p, and $\mu_R(x_i, x_j) = 0$ otherwise. If m and g_o are integers such that $m \geq 1$, g_o is odd and $g_o \geq 3$, then Proposition 7.3 implies that $R_{g_o}(m, \ldots, m) \in \mathcal{GO}(2m, g_o)$ and $R_{g_o}(m, m+1, m, \ldots, m+1, m) \in \mathcal{GO}(2m + 1, g_o)$. This shows in particular, that for all possible values of Δ and g_o there is a ring graph in $\mathcal{GO}(\Delta, g_o)$.

Theorem 7.4 (McDonald [215] 2009) *Let G be a connected graph with odd girth g_o for an odd integer $g_o \geq 3$. Then, $\chi'(G) = \Delta(G) + 1 + \frac{\Delta(G)-2}{g_o-1}$ if and only if $G = mC_{g_o}$ and $(g_o - 1) \mid 2(m - 1)$.*

Proof: If $G = mC_{g_o}$ for an odd integer $g_o \geq 3$ such that $(g_o - 1) \mid 2(m - 1)$, then $\Delta(G) - 2 = 2(m-1)$ and Proposition 7.3 implies that $\chi'(G) = \Delta(G) + 1 + \frac{\Delta(G)-2}{g_o-1}$.

Now, assume that G is a connected graph with odd girth $g_o \geq 3$ satisfying $\chi'(G) = \Delta(G) + 1 + \frac{\Delta(G)-2}{g_o-1}$. Clearly, this implies that $(g_o - 1) \mid \Delta(G) - 2$. We may assume that $\Delta(G) \geq 3$, otherwise $G = C_{g_o}$ and we are done. By Proposition 1.1, there is a critical graph $H \subseteq G$ such that $\chi'(H) = \chi'(G)$. By Theorem 5.8, we obtain

$$\Delta(G) + 1 + \frac{\Delta(G) - 2}{g_o - 1} = \chi'(G) = \chi'(H) \leq \Delta(H) + 1 + \frac{\Delta(H) - 2}{g_o(H) - 1}.$$

Since $g_o(H) \geq g_o(G) = g_o$ and $\Delta(H) \leq \Delta(G)$, this implies that $g_o(H) = g_o$, $\Delta(H) = \Delta(G) \geq 3$, and

$$\chi'(H) = \Delta(H) + 1 + \frac{\Delta(H) - 2}{g_o - 1}. \tag{7.2}$$

Hence, $\chi'(H) = k + 1$ for an integer $k \geq \Delta(H) + 1$. Since H is critical, there is an edge $e \in E(H)$, a coloring $\varphi \in C^k(H - e)$, and a maximal Tashkinov tree T with respect to e and φ. From Theorem 5.3(a)(f) it follows that $|V(T)| \geq g_o$ and $\Delta(H) \geq \delta(H[V(T)]) \geq (|V(T)| - 1)(\chi'(H) - \Delta(H) - 1) + 2$. By (7.2), this implies that $\Delta(H) = \delta(H[V(T)])$ and $|V(T)| = g_o$. Since H is critical, we then conclude that $V(H) = V(T)$ and H is a regular ring graph. Since $|V(H)| = g_o$ is odd, this implies that $H = mC_{g_o}$ for an integer $m \geq 1$. Then $g_o - 1$ is a divisor of $\Delta(G) - 2 = \Delta(H) - 2 = 2m - 2$. Furthermore, since G is connected and $H \subseteq G$ is regular with $\Delta(H) = \Delta(G)$, we conclude that $H = G$. This completes the proof. ∎

It remains an open problem to characterize those graphs achieving Goldberg's bound in (7.1). Theorem 7.4 solves this problems only in the case when the quotient $(\Delta - 2)/(g_o - 1)$ is an integer. Theorem 7.1 solves this problem in the case when $g_o = 3$ and $\Delta \geq 4$.

7.2 VIZING'S BOUND AND EXTREME GRAPHS

Whereas Shannon's bound is sharp for all $\Delta \geq 1$, Vizing's bound is not sharp for all values of Δ and μ. Already Vizing [297, 298] proved that $\chi'(G) \leq \Delta(G) + \mu(G) - 1$ for every graph G satisfying $\Delta(G) = 2\mu(G) + 1 \geq 5$. Favrholdt, Stiebitz, and Toft [91] extended this result and proved that $\chi'(G) \leq \Delta(G) + \mu(G) - 1$ for every graph G satisfying $2\mu(G) + 1 \leq \Delta(G) \leq 3\mu(G) - 1$. The proof of this result is mainly based on Theorem 2.10.

Diego Scheide established in his diploma thesis [267, 272] a list of pairs (Δ, μ) for which there exists a graph G such that $\Delta(G) = \Delta$, $\mu(G) = \mu$, and $\chi'(G) = \Delta + \mu$. He also proved in [267, 272] that this list is complete, provided that Goldberg's conjecture holds. In the remarkable Ph.D thesis [120] of Ram P. Gupta, dating back to February 1967, and still of much interest, the same list of pairs is exhibited (Theorem 7.6). In the thesis Gupta conjectured that the list is complete (Conjecture 7.8). By the result of Scheide, this may be regarded as a weaker version of Goldberg's conjecture and as a first step in this direction. Gupta did not formulate Goldberg's conjecture in the thesis, but he did so, independently of Goldberg, Seymour and Andersen, in his later paper [121].

We shall now explain the results of Gupta and Scheide in more detail. A graph G such that $\chi'(G) = \Delta(G) + \mu(G)$ is said to be an **extreme graph**. We start with a simple observation.

Proposition 7.5 *Every extreme graph contains a critical extreme graph with the same maximum degree and the same maximum multiplicity.*

Proof: For the proof consider an extreme graph G. By Proposition 1.1, G contains a critical subgraph H with $\chi'(H) = \chi'(G)$. Since $H \subseteq G$, we have $\Delta(H) \leq \Delta(G)$ and $\mu(H) \leq \mu(G)$. By Vizing's bound, we then obtain $\Delta(G) + \mu(G) = \chi'(G) = \chi'(H) \leq \Delta(H) + \mu(H) \leq \Delta(G) + \mu(G)$. Clearly, this implies that $\Delta(H) = \Delta(G)$ and $\mu(H) = \mu(G)$. ∎

Theorem 7.6 (Gupta [120] 1967) *Let* Δ, μ *be positive integers satisfying*

$$2p(\mu - 1) + 2 \leq \Delta \leq 2p\mu$$

for some integer $p \geq 1$. *Then there is an extreme graph* G *with* $\Delta(G) = \Delta$ *and* $\mu(G) = \mu$.

Proof: Let d be an integer such that $2 \leq d \leq 2p$. Then there is a simple graph $G = G_{p,d}$ on $2p + 1$ vertices satisfying the following degree condition. If d is even, then G is d-regular, that is, every vertex of G has degree d in G. If d is odd, then every vertex of G has degree d in G besides one vertex v^* the degree of which is $d - 1$. In the first case we can take the $\frac{d}{2}$th power of a cycle of order $2p + 1$. In the second case we can take the graph obtained from the $\frac{d-1}{2}$th power of a cycle C of order $2p$ by adding a matching M of size $p - \frac{d-1}{2}$ and joining an additional vertex v^* to the $d - 1$ vertices of C that are not incident with the edges in M. Hence, G is a simple graph on $2p + 1$ vertices and

$$|E(G)| = \left\lfloor \frac{(2p + 1)d}{2} \right\rfloor = pd + \lfloor \tfrac{d}{2} \rfloor.$$

Next, we construct a graph $H = H_{p,d}^m$, where m is a positive integer. For the vertex set, we have $V(H) = V(G)$. For two distinct vertices u, v of H, we prescribe

$$\mu_H(u, v) = \begin{cases} m & \text{if } uv \in E(G) \\ m - 1 & \text{if } uv \notin E(G). \end{cases}$$

Obviously, $|V(H)| = 2p + 1$ and $\mu(H) = m$. If d is even, then every vertex of H has degree $2p(m - 1) + d$ in H. If d is odd, then every vertex of H has degree $2p(m - 1) + d$ in H besides the vertex v^* the degree of which is $2p(m - 1) + d - 1$. Hence, the maximum degree of H is $\Delta(H) = 2p(m - 1) + d$. For the number of edges, we then obtain

$$|E(H)| = \left\lfloor \frac{(2p + 1)(2p(m - 1) + d)}{2} \right\rfloor = p(2p + 1)(m - 1) + pd + \lfloor \tfrac{d}{2} \rfloor.$$

Consequently, we have

$$w(H) \geq \left\lceil \frac{|E(H)|}{\lfloor \frac{|V(H)|}{2} \rfloor} \right\rceil$$

$$= \left\lceil \frac{p(2p + 1)(m - 1) + pd + \lfloor \frac{d}{2} \rfloor}{p} \right\rceil$$

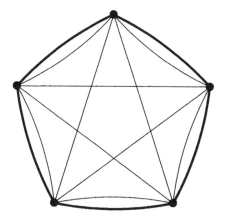

Figure 7.1 An extreme graph with $(\Delta, \mu) = (6, 2)$ and $(p, d) = (2, 2)$.

$$= 2p(m - 1) + d + m$$
$$= \Delta(H) + \mu(H).$$

From (1.5) and Theorem 2.2 it then follows that $\chi'(H) = \Delta(H) + \mu(H)$, i.e., H is an extreme graph. This proves Theorem 7.6. ∎

Corollary 7.7 [272] *Let Δ, μ be positive integers such that $\Delta \geq 2\mu(\mu - 1) + 2$. Then there is an extreme graph G with $\Delta(G) = \Delta$ and $\mu(G) = \mu$.*

Proof: Let p be the smallest integer such that $\Delta \leq 2p\mu$. Since $\Delta \geq 2\mu(\mu - 1) + 2$, this implies that $p \geq \mu$. If $p = \mu$, this yields $2p(\mu - 1) + 2 = 2\mu(\mu - 1) + 2 \leq \Delta$. If $p \geq \mu + 1$, then $2p(\mu - 1) + 2 = 2(p - 1)\mu - 2(p - \mu) + 2 \leq 2(p - 1)\mu \leq \Delta$. Hence, in both cases we have $2p(\mu - 1) + 2 \leq \Delta \leq 2p\mu$. Then, by Theorem 7.6, there is an extreme graph G with $\Delta(G) = \Delta$ and $\mu(G) = \mu$. ∎

7.3 EXTREME GRAPHS AND ELEMENTARY GRAPHS

For an integer $p \geq 1$, let

$$\mathcal{P}_1^p = \{(\Delta, \mu) \mid 2p\mu + 1 \leq \Delta \leq 2(p + 1)\mu - (2p + 1)\}.$$

Furthermore, let

$$\mathcal{P}_1 = \bigcup_{p=1}^{\mu-1} \mathcal{P}_1^p$$

and $\mathcal{P}_2 = \{(\Delta, \mu) \mid \mu \leq \Delta < 2\mu\}$. Observe that $\mathcal{P}_1^p = \emptyset$ if $p \geq \mu$. Theorem 7.6 tells us that, for every feasible tuple (Δ, μ) not contained in $\mathcal{P}_1 \cup \mathcal{P}_2$, there is an extreme graph G with $\Delta(G) = \Delta$ and $\mu(G) = \mu$. If G is a graph with $\Delta(G) < 2\mu(G)$, then

Theorem 2.4 implies that $\chi'(G) \le \Delta(G) + \frac{\Delta(G)}{2} < \Delta(G) + \mu(G)$. Hence, we only need to consider the tuples in $\mathcal{P}_1$. Gupta [120] posed the conjecture that there exists no extreme graph G with $(\Delta(G), \mu(G)) \in \mathcal{P}_1$.

Conjecture 7.8 (Gupta [110] 1967) *Let* Δ, μ *be positive integers, and let* G *be a graph with* $\Delta(G) = \Delta$ *and* $\mu(G) = \mu$ *such that*

$$2p\mu + 1 \le \Delta \le 2(p+1)\mu - (2p+1)$$

holds for some integer $p \ge 1$. *Then* $\chi'(G) \le \Delta + \mu - 1$.

The following simple lemma implies that in order to prove Gupta's conjecture it would be sufficient to show that any extreme graph with $\mu \ge 2$ is elementary. So Gupta's conjecture is a consequence of Goldberg's conjecture.

Lemma 7.9 [272] *Let* Δ, μ *be positive integers, and let* G *be a graph with* $\Delta(G) = \Delta$ *and* $\mu(G) = \mu$ *such that*

$$2p\mu + 1 \le \Delta \le 2(p+1)\mu - (2p+1)$$

holds for some integer $p \ge 1$. *If* G *is critical and elementary, then* $\chi'(G) \le \Delta + \mu - 1$.

Proof: Suppose on the contrary that $\chi'(G) \ge \Delta + \mu \ge \Delta + 1$. Since G is critical and elementary, it then follows from Proposition 1.3 that $|V(G)|$ is odd and

$$\chi'(G) = \left\lceil \frac{|E(G)|}{\left\lfloor \frac{|V(G)|}{2} \right\rfloor} \right\rceil.$$

Let v be a vertex of degree Δ in G. Since $\Delta \ge 2p\mu + 1$, we conclude that v has at least $2p + 1$ neighbors in G and, therefore, $|V(G)| \ge 2p + 2$. Since $|V(G)|$ is odd, this implies that $|V(G)| \ge 2p + 3$. Then we conclude that

$$
\begin{aligned}
\chi'(G) &= \left\lceil \frac{|E(G)|}{\left\lfloor \frac{|V(G)|}{2} \right\rfloor} \right\rceil \le \left\lceil \frac{\lfloor \frac{1}{2}|V(G)|\Delta \rfloor}{\frac{1}{2}(|V(G)| - 1)} \right\rceil \\
&= \left\lceil \frac{\frac{1}{2}(|V(G)| - 1)\Delta + \lfloor \frac{1}{2}\Delta \rfloor}{\frac{1}{2}(|V(G)| - 1)} \right\rceil \\
&= \Delta + \left\lceil \frac{\lfloor \frac{1}{2}\Delta \rfloor}{\frac{1}{2}(|V(G)| - 1)} \right\rceil \le \Delta + \left\lceil \frac{\lfloor \frac{1}{2}\Delta \rfloor}{p + 1} \right\rceil.
\end{aligned}
$$

Since $\Delta \le 2(p+1)\mu - (2p+1)$, this implies that

$$
\begin{aligned}
\chi'(G) &\le \Delta + \left\lceil \frac{\lfloor \frac{1}{2}\{2(p+1)\mu - (2p+1)\}\rfloor}{p+1} \right\rceil \\
&= \Delta + \left\lceil \frac{(p+1)\mu - p - 1}{p+1} \right\rceil = \Delta + \mu - 1,
\end{aligned}
$$

a contradiction. ∎

Corollary 7.10 [272] *Let Δ, μ be positive integers, and let G be a graph with $\Delta(G) = \Delta$ and $\mu(G) = \mu$ such that*

$$2p\mu + 1 \leq \Delta \leq 2(p+1)\mu - (2p+1)$$

holds for some integer $p \geq 1$. If Goldberg's Conjecture holds, then $\chi'(G) \leq \Delta + \mu - 1$.

Proof: Suppose on the contrary that $\chi'(G) \geq \Delta + \mu$. By Vizing's bound, this implies that $\chi'(G) = \Delta + \mu$, i.e., G is an extreme graph. Then Proposition 7.5 implies that G contains a critical subgraph H with $\Delta(H) = \Delta$, $\mu(H) = \mu$ and $\chi'(H) = \Delta + \mu$. Since $2p\mu + 1 \leq \Delta \leq 2(p+1)\mu - (2p+1)$ holds for some integer $p \geq 1$, we conclude that $\mu \geq p + 1 \geq 2$. Hence $\chi'(H) \geq \Delta(H) + 2$. If Goldberg's conjecture holds, this implies that H is elementary. From Lemma 7.9 it then follows that $\chi'(H) \leq \Delta + \mu - 1$, a contradiction. ∎

Using Theorem 6.5 (special cases of Goldberg's conjecture), we easily deduce the following result from Lemma 7.9.

Theorem 7.11 *Let Δ, μ be positive integers, and let G be a graph with $\Delta(G) = \Delta$ and $\mu(G) = \mu$ such that*

$$2p\mu + 1 \leq \Delta \leq 2(p+1)\mu - (2p+1)$$

holds for some integer $p \geq 1$. If $p \leq 6$, then $\chi'(G) \leq \Delta + \mu - 1$.

Proof: Suppose on the contrary that $\chi'(G) \geq \Delta + \mu$. By Vizing's bound, this implies that $\chi'(G) = \Delta + \mu$, i.e., G is an extreme graph. Then Proposition 7.5 implies that G contains a critical subgraph H with $\Delta(H) = \Delta$, $\mu(H) = \mu$, and $\chi'(H) = \Delta + \mu$. Since $2p\mu + 1 \leq \Delta \leq 2(p+1)\mu - (2p+1)$ and $1 \leq p \leq 6$, we conclude that

$$\mu \geq \frac{\Delta + (2p+1)}{2p+2} = \frac{\Delta - 1}{2p+2} + 1 \geq \frac{\Delta - 1}{14} + 1 > \frac{1}{14}\Delta + \frac{12}{14}.$$

Consequently, we have

$$\chi'(H) = \Delta + \mu > \frac{15}{14}\Delta + \frac{12}{14}.$$

From Theorem 6.5 it then follows that H is elementary. Since H is critical, Lemma 7.9 then implies that $\chi'(H) \leq \Delta + \mu - 1$, a contradiction. ∎

7.4 UPPER BOUNDS FOR χ' DEPENDING ON Δ AND μ

In his diploma thesis [267] (see also [274, 275]), Diego Scheide investigated the function f defined by

$$f(\Delta, \mu) = \max\{\chi'(G) \mid \Delta(G) = \Delta, \mu(G) = \mu\},$$

i.e., $f(\Delta, \mu)$ is the maximum chromatic index possible for a graph with maximum degree Δ and maximum multiplicity μ. Observe that $f(0,0) = 0$ and that besides this value the function f is defined for all pairs (Δ, μ) with $1 \leq \mu \leq \Delta$. Combining the results from Sect. 7.2 and Sect. 7.3 we obtain the following theorem.

Theorem 7.12 *For a pair* (Δ, μ) *of positive integers with* $\Delta \geq \mu$ *the following statements hold:*

(a) *If* $(\Delta, \mu) \notin \mathcal{P}_1 \cup \mathcal{P}_2$, *then* $f(\Delta, \mu) = \Delta + \mu$.

(b) *If* $(\Delta, \mu) \in \mathcal{P}_2$, *then* $f(\Delta, \mu) \leq \Delta + \mu - 1$.

(c) *If* $(\Delta, \mu) \in \mathcal{P}_1$, *then* $f(\Delta, \mu) \leq \Delta + \mu - 1$ *provided that Goldberg's conjecture is true.*

(d) *If* $1 \leq p \leq 6$ *and* $(\Delta, \mu) \in \mathcal{P}_1^p$, *then* $f(\Delta, \mu) \leq \Delta + \mu - 1$.

Shannon's bound implies that the gap between $f(\Delta, \mu)$ and Vizing's bound can be arbitrarily large if $\Delta < 2\mu$. That this is also the case if $\Delta \geq 2\mu$ was proved by Scheide in his diploma thesis [267, 274, 275]. The proof of this result shows the usefulness of the fan number, since the improvement of Vizing's bound in this case is in fact an improvement for the fan number. This shows, as a byproduct, that the improvement for the chromatic index in this case is efficiently realizable; it can be just realized by the algorithm COL presented in Sect. 2.4. To formulate the result, we introduce the following function D defined by

$$D(p,r) = \left\lfloor \frac{rp^2 + 6(r-1)p + 5r + 1}{4} \right\rfloor,$$

where p, r are positive integers.

Theorem 7.13 [274] *Let* p, r *be positive integers. Every graph* G *with*

$$p\mu(G) + r \leq \Delta(G) \leq (p+1)\mu(G) - D(p,r)$$

satisfies $\chi'(G) \leq \mathrm{Fan}(G) \leq \Delta(G) + \mu(G) - r$.

The proof of Theorem 7.13 is mainly based on the adjacency result Theorem 2.22 for fan-critical graphs and the following lemma.

Lemma 7.14 [274] *Let* p, r *be positive integers and let* H *be a fan-critical graph with* $\Delta(H) = \Delta$ *and* $\mu(H) = \mu$. *If*

$$p\mu + r \leq \Delta \leq (p+1)\mu - D(p,r),$$

then $\mathrm{fan}(H) \leq \Delta + \mu - r$.

Proof: Suppose this is false, i.e., $\mathrm{fan}(H) \geq \Delta + \mu - r + 1$. Since $D(p,r) \geq r \geq 1$ the assumption of the theorem implies that $r \leq \mu - 1$ and $\Delta \geq p\mu + r \geq 3$. Hence $\mathrm{fan}(H) \geq \Delta + 1 \geq 4$ and, therefore, we can apply Theorem 2.22.

Since H is fan-critical and $\text{fan}(H) \geq 4$, we conclude that H is connected and $|V(H)| \geq 3$. For $x \in V(H)$ define

$$\tilde{\mu}(x) = \min\{\mu_H(x, y) \mid y \in V(H), \mu_H(x, y) \geq 1\}.$$

Now, let x be a vertex of H with $d_H(x) = \Delta$ and let $y \in V(H)$ be a neighbor of x such that $\mu_H(x, y) = \tilde{\mu}(x)$. Furthermore, let N be the set of all neighbors z of x such that $z \neq y$ and $d_H(z) + \mu_H(x, z) \geq \text{fan}(H) \geq \Delta + \mu - r + 1$. Then every vertex $z \in N$ satisfies

$$\mu_H(x, z) \geq \mu - r + 1. \tag{7.3}$$

From Theorem 2.21 it follows that $r \geq \Delta + \mu - \text{fan}(H) + 1 \geq 1$. Hence, by Theorem 2.22, we obtain that

$$\begin{aligned}
|N| &\geq \left\lceil \frac{\text{fan}(H) - d_H(y) - \mu_H(x, y) + 1}{\Delta + \mu - \text{fan}(H) + 1} \right\rceil \\
&\geq \left\lceil \frac{\text{fan}(H) - d_H(y) - \mu_H(x, y) + 1}{r} \right\rceil.
\end{aligned}$$

Since $\text{fan}(H) \geq \Delta + \mu - r + 1$ and $d_H(y) \leq \Delta$, this implies that

$$|N| \geq \left\lceil \frac{(\Delta - d_H(y)) + (\mu - \mu_H(x, y)) - r + 2}{r} \right\rceil$$

and, therefore,

$$|N| \geq \left\lceil \frac{\mu - \tilde{\mu}(x) - r + 2}{r} \right\rceil. \tag{7.4}$$

For the vertex x, we have $d_H(x) = \Delta \geq p\mu + 1$. This implies that x has at least $p + 1$ neighbors. If $\mu - r + 1 \leq \tilde{\mu}(x)$, then we obtain $d_H(x) \geq (p + 1)\tilde{\mu}(x) \geq (p+1)(\mu-r+1) = (p+1)\mu - p(r-1) - r + 1 \geq (p+1)\mu - D(p,r) + 1 \geq \Delta + 1$, a contradiction. Otherwise, $\mu - r + 1 > \tilde{\mu}(x)$ and we argue as follows. From (7.3) and (7.4) we first conclude that

$$\begin{aligned}
d_H(x) &= \sum_{y \in V} \mu_H(x, y) \geq |N|(\mu - r + 1) + (p + 1 - |N|)\tilde{\mu}(x) \\
&= (p+1)(\mu - r + 1) - (p + 1 - |N|)(\mu - \tilde{\mu}(x) - r + 1) \\
&\geq (p+1)(\mu - r + 1) \\
&\quad - \left(p + 1 - \frac{\mu - \tilde{\mu}(x) - r + 2}{r}\right)(\mu - \tilde{\mu}(x) - r + 1) \\
&= (p+1)(\mu - r + 1) \\
&\quad - \frac{1}{r}[r(p+1) - 1 - (\mu - \tilde{\mu}(x) - r + 1)][\mu - \tilde{\mu}(x) - r + 1] \\
&= (p+1)(\mu - r + 1) - \frac{1}{r}\left(\frac{r(p+1) - 1}{2}\right)^2 \\
&\quad + \frac{1}{r}\left(\frac{r(p+1) - 1}{2} - (\mu - \tilde{\mu}(x) - r + 1)\right)^2.
\end{aligned}$$

Next, we conclude that

$$
\begin{aligned}
d_H(x) \;\geq\;& (p+1)(\mu - r + 1) - \frac{1}{r}\left\lfloor \left(\frac{r(p+1)-1}{2}\right)^2 \right\rfloor \\
& + \frac{1}{r}\left\lfloor \left(\frac{r(p+1)-1}{2} - (\mu - \tilde{\mu}(x) - r + 1)\right)^2 \right\rfloor \\
\geq\;& (p+1)(\mu - r + 1) - \left\lceil \frac{1}{r}\left\lfloor \left(\frac{r(p+1)-1}{2}\right)^2 \right\rfloor \right\rceil \\
=\;& (p+1)\mu - p(r-1) - r + 1 - \left\lceil \frac{1}{r}\left\lfloor \left(\frac{r(p+1)-1}{2}\right)^2 \right\rfloor \right\rceil.
\end{aligned}
$$

Hence $d_H(x) \geq (p+1)\mu - D + 1$ with

$$
D = p(r-1) + r + S \text{ and } S = \left\lceil \frac{1}{r}\left\lfloor \left(\frac{r(p+1)-1}{2}\right)^2 \right\rfloor \right\rceil.
$$

Based on a simple case analysis, we conclude that

$$
\begin{aligned}
S \;=\;&
\begin{cases}
\left\lceil \frac{1}{r}\frac{(r(p+1)-1)^2}{4} \right\rceil & \text{if } p \equiv 0 \bmod 2,\ r \equiv 1 \bmod 2, \\[6pt]
\left\lceil \frac{1}{r}\frac{(r(p+1)-1)^2 - 1}{4} \right\rceil & \text{otherwise}
\end{cases} \\[12pt]
=\;&
\begin{cases}
\left\lceil \frac{r(p+1)^2 - 2(p+1)}{4} + \frac{1}{4r} \right\rceil & \text{if } p \equiv 0 \bmod 2, \\
& \quad r \equiv 1 \bmod 2, \\[4pt]
\left\lceil \frac{r(p+1)^2 - 2(p+1)}{4} \right\rceil & \text{otherwise}
\end{cases} \\[12pt]
=\;&
\begin{cases}
\frac{r(p+1)^2 - 2(p+1)}{4} + \frac{1}{2} & \text{if } p \equiv 0 \bmod 2,\ r \equiv 0 \bmod 4, \\[3pt]
\frac{r(p+1)^2 - 2(p+1)}{4} + \frac{1}{4} & \text{if } p \equiv 0 \bmod 2,\ r \equiv 1 \bmod 4, \\[3pt]
\frac{r(p+1)^2 - 2(p+1)}{4} + \frac{3}{4} & \text{if } p \equiv 0 \bmod 2,\ r \equiv 3 \bmod 4, \\[3pt]
\frac{r(p+1)^2 - 2(p+1)}{4} & \text{otherwise}
\end{cases} \\[12pt]
=\;& \left\lfloor \frac{r(p+1)^2 - 2(p+1)}{4} + \frac{3}{4} \right\rfloor.
\end{aligned}
$$

Since $D = p(r-1) + r + S$, this implies that

$$
\begin{aligned}
D \;=\;& p(r-1) + r + \left\lfloor \frac{r(p+1)^2 - 2(p+1)}{4} + \frac{3}{4} \right\rfloor \\
=\;& \left\lfloor \frac{rp^2 + 6(r-1)p + 5r + 1}{4} \right\rfloor = D(p,r).
\end{aligned}
$$

Consequently, $d_H(x) \geq (p+1)\mu - D(p,r) + 1 \geq \Delta + 1$, a contradiction, too. This completes the proof. ∎

Proof of Theorem 7.13: Let p, r be positive integers and let G be a graph such that

$$p\mu(G) + r \le \Delta(G) \le (p+1)\mu(G) - D(p,r).$$

Since $D(p,r) \ge r \ge 1$, this implies that $r \le \mu(G) - 1$. Hence $\Delta(G) + \mu(G) - r \ge \Delta(G)$ and for the proof it suffices to show that $\mathrm{fan}(G) \le \Delta(G) + \mu(G) - r$. Suppose to the contrary that $\mathrm{fan}(G) \ge \Delta(G) + \mu(G) - r + 1$.

By Proposition 1.1, there is a fan-critical subgraph H with $\mathrm{fan}(H) = \mathrm{fan}(G)$. Let $\Delta = \Delta(H)$, $\mu = \mu(H)$, $s = \Delta(G) - \Delta$, $t = \mu(G) - \mu$, and $r' = r - s - t$. Since H is a subgraph of G, we conclude that $s, t \ge 0$. Furthermore, it follows from Theorem 2.21 that

$$r' \ge \Delta(G) + \mu(G) + 1 - \mathrm{fan}(G) - s - t = \Delta + \mu + 1 - \mathrm{fan}(H) \ge 1.$$

Then, for $D' = D(p, r') + (p+1)t$, we obtain that

$$
\begin{aligned}
D' &= \left\lfloor \frac{(r-s-t)p^2 + 6((r-s-t)-1)p + 5(r-s-t)+1}{4} \right\rfloor + (p+1)t \\
&= \left\lfloor \frac{rp^2 + 6(r-1)p + 5r + 1 - (s+t)p^2 - 6sp - 2tp - 5s - t}{4} \right\rfloor \\
&\le D(p,r).
\end{aligned}
$$

On one hand, since $p\mu(G) + r \le \Delta(G)$ and $p, t \ge 0$, we obtain that

$$
\begin{aligned}
p\mu + r' &= p(\mu(G) - t) + r - s - t \\
&= p\mu(G) + r - pt - s - t \\
&\le \Delta(G) - pt - s - t \\
&= \Delta - pt - t \le \Delta.
\end{aligned}
$$

On the other hand, since $D' = D(p, r') + (p+1)t \le D(p, r)$, we obtain that

$$
\begin{aligned}
\Delta &\le \Delta(G) \le (p+1)\mu(G) - D(p,r) \\
&= (p+1)\mu + (p+1)t - D(p,r) \\
&\le (p+1)\mu - D(p,r').
\end{aligned}
$$

Since H is fan-critical and $\mathrm{fan}(H) \ge 1$, we then conclude from Lemma 7.14 that $\mathrm{fan}(G) = \mathrm{fan}(H) \le \Delta + \mu - r' \le \Delta(G) + \mu(G) - r$, a contradiction. ∎

7.5 NOTES

The above proof of Theorem 7.1 is from the M.Sc. thesis of Andersen [4] from July 1975 at the University of Aarhus. The proof of Theorem 7.2 follows Andersen's paper [5] which was an outcome of the thesis. The proof of Theorem 7.4 is due to McDonald [215]. It is still an unsolved problem to characterize the graphs achieving

Goldberg's bound (in terms of Δ and g_o). As pointed out by McDonald [214] every multiple tC_p of and odd cycle C_p with $t \geq 1$ achieves Goldberg's bound, since $\Delta(tC_p) = 2t$, $g_o(tC_p) = p$ and $\chi'(tC_p) = 2t + \frac{2t-2}{p-1}$ (this follows easily from Theorem 6.3). Some new results related to Goldberg's bound can be found in the paper by McDonald [218].

It is remarkable that the conjectured list of pairs (Δ, μ) for which there exist graphs with $\chi' = \Delta + \mu$ was described by Ram P. Gupta already in 1967. Since the conjecture is implied by Goldberg's conjecture this may be seen as the first step in this direction. Later Gupta [121] also formulated Goldberg's conjecture (independently of Andersen, Goldberg, and Seymour), but it is not clear if he realized the connection to his earlier conjecture. The connection was to the best of our knowledge first pointed out by Scheide [267, 272]. The above proof of Theorem 7.6 is from [267, 272] and resembles the original proof by Gupta [120]. The results in Sect. 7.4 were obtained by D. Scheide in his diploma thesis [267] and first published in a series of papers by Scheide and Stiebitz [272, 274, 275]. The proofs of Theorem 7.6, Corollary 7.7, Lemma 7.9, and Corollary 7.10 are from Scheide and Stiebitz [272] with permission from Elsevier. The proofs of Theorem 7.13 and Lemma 7.14 are from Scheide and Stiebitz [274] with permission of John Wiley and Sons.

It would be of much interest to have a good characterization of the class of extreme graphs with $\mu \geq 2$. Note that Holyer's result from [150] implies that it is co-**NP**-complete to decide whether a simple graph is an extreme graph. On the other hand, Goldberg's conjecture implies that any graph with $\mu \geq 2$ is an extreme graph if and only if $w = \Delta + \mu$, and it follows from the work of Edmonds [76] that it can be decided in polynomial time if this holds or not (see Schrijver [277]). Hence one step towards a characterization of the class of extreme graphs with $\mu \geq 2$ would be to show that each such graph is elementary. But even to establish Goldberg's conjecture for the class of extreme graphs with $\mu \geq 2$ seems to be difficult. However, if μ is bounded from below by a logarithmic function of Δ this can be done, as proved by Penny Haxell and Jessica McDonald. Combining (5.4) and (5.5) we obtain the following result due to Haxell and McDonald [132]: If G is a critical graph with $\chi'(G) \geq \Delta(G) + 2$ and $\chi'(G) \neq w(G)$, then

$$\left(1 + \frac{\chi'(G) - 1 - \Delta(G)}{2\mu(G)}\right)^{\chi'(G)-1-\Delta(G)} < \frac{\Delta(G)}{\chi'(G) - 1 - \Delta(G)}. \tag{7.5}$$

As proved by Haxell and McDonald [132], this implies the following two results related to Goldberg's conjecture. While the first result provides a characterization of extreme graphs, provided μ is not too small with respect to Δ, the second result yields a partial characterization of graphs achieving Steffen's bound (Theorem 5.7).

(a) *Let G be a graph with $\mu(G) \geq \log_{4/5}(\Delta(G)) + 1$. Then $\chi'(G) = \Delta(G) + \mu(G)$ if and only if $w(G) = \Delta(G) + \mu(G)$.*

(b) *Let G be a graph with girth $g \geq 3$ and $\mu(G) \geq (g/2)(\log_{(1+1/2g)}(\Delta(G)) + 1)$. Then $\chi'(G) = \Delta(G) + \left\lceil \frac{\mu(G)}{\lfloor g/2 \rfloor} \right\rceil$ if and only if $w(G) = \Delta(G) + \left\lceil \frac{\mu(G)}{\lfloor g/2 \rfloor} \right\rceil$.*

Proof of (a): The "if" part follows from Vizing's bound and (1.5). For the proof of the "only if" part assume that $\chi'(G) = \Delta(G) + \mu(G)$, but the conclusion does not hold. Then $\chi'(G) \neq w(G)$. If $\mu(G) = 1$, then $\Delta(G) = 1$ and there is no graph satisfying $\chi'(G) = \Delta(G) + \mu(G)$. So we may assume $\mu(G) \geq 2$. Hence we can apply (7.5). This yields

$$\left(1 + \frac{\mu(G) - 1}{2\mu(G)}\right)^{\mu(G)-1} < \frac{\Delta(G)}{\mu(G) - 1}.$$

Since $\mu(G) \geq 2$, we get $(5/4)^{\mu(G)-1} < \Delta(G)$, which leads to a contradiction to the assumption on $\mu(G)$. ∎

Proof of (b): The "if" part follows from Steffen's bound (Theorem 5.7) and (1.5). The proof of the "only if" part is again by contradiction. As in the previous proof, the assumptions imply that $A = \lceil \mu(G)/\lfloor g/2 \rfloor \rceil$ satisfies $A \geq 2$. By (7.5), we obtain

$$\left(1 + \frac{A - 1}{2\mu(G)}\right)^{A-1} < \frac{\Delta(G)}{A - 1} \leq \Delta(G).$$

The assumption on $\mu(G)$ implies $\mu(G) \geq g$, and so the quantity insides the bracket is at least $1 + \frac{1}{2g}$. Thus $(1 + 1/2g)^A < \Delta$, which yields a contradiction to the asssumption on $\mu(G)$. ∎

Various families of extreme graphs are described in the proof of Theorem 7.6. For instance, if $(\Delta, \mu) = (12, 2)$ we obtain the extreme graph $G = K_{11} \cup C_{11}$ on 11 vertices. Another such extreme graph with $\Delta = 12$ and $\mu = 2$ was found by McDonald [214], namely $G = 2K_7 - e$. For $\Delta = 2p\mu$ Theorem 7.6 yields the extreme graph $G = \mu K_{2p+1}$. McDonald [214, Chap. 5] found the following two families of extreme graphs that are multiples of simple graphs. If G is the complement of a 2-regular graph on $d + 3$ vertices for an even number $d \geq 0$, then $H = tG$ with $0 < t \leq d/2$ is an extreme graph, since $\Delta(H) = td$, $\mu(G) = t$ and $w(H) \geq td + \lceil td/(d + 2) \rceil$. If G is the complement of $(d - 1)/2$ single edges and one path of length two for an odd integer $d \geq 1$, then $H = tG$ with $0 < t \leq (d - 1)/2$ is an extreme graph with $d + 2$ vertices, since $\Delta(H) = td$, $\mu(H) = t$ and $w(H) \geq td + \lceil (td - t)/(d + 1) \rceil$. The existence of these two families is in sharp contrast to the following result due to McDonald [216].

(c) *Let G be a simple connected graph with maximum degree d, and let t be an integer with $t > d/2$. Then tG is an extreme graph if and only if $G = K_{d+1}$ and d is even.*

Proof of (c): The "if" part is easy (see the proof of Theorem 7.6). For the proof of the "only if" part, suppose that tG is an extreme graph, where G is a simple connected graph with maximum degree d and t is an integer with $t > d/2$. Our goal is to show that $G = K_{d+1}$ and d is even. If $d = 0$ this is evident. If $d = 1$, then tG cannot be an extreme graph. So suppose $d \geq 2$ and hence $t \geq 2$. Then $\Delta(tG) = td$, $\mu(tG) = t$, and $\chi'(G) = td + t$. There is a critical subgraph G' of tG with $\Delta(G') = td$, $\mu(G') = t$, and $\chi'(G') = td + t \geq \Delta(G') + 2$ (Proposition 7.5). Obviously, $t > d/2$ implies $t > \sqrt{(td-1)/2}$. Hence we obtain $\chi'(G') > \Delta(G') + \sqrt{(\Delta(G') - 1)/2}$. Then Corollary 5.20 implies that G' is elementary, i.e., $w(G') = \chi'(G') = td + t$. Let $X = V(G')$ and $p = |X|$. From Proposition 1.3 it follows that p is an odd integer and $\lceil 2|E(G')|/(p-1)\rceil = td + t$. Since $td = \Delta(G') \leq \mu(G')(p-1) = t(p-1)$, we have $p \geq d + 1 \geq 3$. Combining Theorem 1.4 and Proposition 1.7, we obtain $p \leq (\chi'(G') - 3)/(\chi'(G') - 1 - \Delta(G')) = (td + t - 3)/(t - 1)$, which yields $p < d + 3$ because $d < 2t$. Hence $p = d + 1$ or $p = d + 2$.

The graph $H = (tG)[X]$ satisfies $G' \subseteq H \subseteq G$, which implies that $\Delta(H) = \Delta(G) = td$, $\mu(H) = t$, $\chi'(H) = \chi'(G) = td + t$, $w(H) = w(G) = td + t$, and $\lceil 2|E(H)|/(p-1)\rceil = w(H) = td + t$. Observe that $H = t \cdot G[X]$.

First, suppose that $p = d+1$. Then we claim that $G[X] = K_{d+1}$. Otherwise, since $d < 2t$, we would obtain $\lceil 2|E(H)|/(p-1)\rceil \leq \lceil (t(d+1)d - 2t)/d\rceil \leq t(d+1) - 1$, a contradiction. Hence, as claimed, $G[X] = K_{d+1}$. Since G is connected and $\Delta(G) = d$, it follows that $G = K_{d+1}$. Since p is odd, $d = p - 1$ is even.

Now, suppose that $p = d + 2$, which means that d is odd. For the simple graph $G_1 = G[X]$, we obtain $2|E(G_1)| \leq \Delta(G)p = d(d+2)$. Since $H = t \cdot G[X]$ satisfies $\lceil 2|E(H)|/(p-1)\rceil = t(d+1)$, we obtain $2t|E(G_1)|/(d+1) > t(d+1) - 1$, which gives $2|E(G_1)| > d(d+2) - 1$ because $d + 1 \leq 2t$. Hence $2|E(G_1)| = \Delta(G)p = d(d+2)$. Since $d(d+2)$ is odd, this is impossible. ∎

Let G be an arbitrary graph and let $t \geq 1$ be an integer. Evidently, the multiple tG of G satisfies $\Delta(tG) = t\Delta(G)$, $\mu(tG) = t\mu(G)$ and $|E(tG)| = t|E(G)|$. By (6.1), this implies that

$$\chi'^*(G) = \max\left\{\Delta(G), \max_{H \subseteq G, |V(H)| \geq 2} \frac{|E(H)|}{\left\lfloor \frac{1}{2}|V(H)|\right\rfloor}\right\} \leq \frac{\chi'(tG)}{t}$$

and it is well known (see Scheinerman and Ullman [276], respectively Sect. B.2) that

$$\chi'^*(G) = \lim_{t\to\infty} \frac{\chi'(tG)}{t}.$$

By Vizing's bound, G satisfies $\Delta(G) \leq \chi'^*(G) \leq \chi'(G) \leq \Delta(G) + \mu(G)$ and it was observed by Stahl [286] that if G is a simple connected graph, then $\chi'^*(G) = \Delta(G) + 1$ if and only if G is a complete graph having odd order. So the above mentioned result (c) of McDonald may be considered as a strengthening of the result of Stahl.

A special case of Theorem 5.6 says that every extreme graph with $\mu \geq 2$ contains a vertex triple (x, y, z) such that $\mu_T(x, y) = \mu$ and $\mu_T(y, z) + \mu_T(z, x) \geq 2\mu - 1$. Similar results showing that, with the exception of μK_3, every extreme graph with $\mu \geq 2$ must contain a specific dense graph of order five are proved by McDonald [215]. Additionally, it was proved by McDonald [215] that if $\mu \geq 2$ and $\Delta < \mu^2$, then every extreme graph except μK_3 must contain a K_5.

CHAPTER 8

GENERALIZED EDGE COLORINGS OF GRAPHS

Over the years, many generalizations of the edge color problem have been introduced and investigated in the graph theory literature. Such generalizations are typically obtained by relaxation of the coloring condition that each color class form a matching, or by imposing restriction on the colors that may be used. An example of the first type is the f-coloring problem, in which it is allowed that at each vertex v the same color may appear up to $f(v)$ times. A popular example of the second type is the list edge color problem, in which it is required to choose a color for each edge e from an individual list $L(e)$ of available colors. Other variations of the chromatic index are the total chromatic index and the cover index.

8.1 EQUITABLE AND BALANCED EDGE COLORINGS

Let G be an arbitrary graph. An **improper edge coloring** of G with color set C is a map $\varphi : E(G) \to C$ from the edge set of G into the set C. For an improper edge coloring φ, we introduce the following notation. For a subset $D \subseteq C$, let $G_{\varphi,D}$ denote the subgraph H of G satisfying $V(H) = V(G)$ and $E(H) = \{e \in E(G) \mid \varphi(e) \in D\}$. If $D = \{\alpha_1, \ldots, \alpha_p\}$, we write $G_{\varphi,\alpha_1,\ldots,\alpha_p}$ in addition to $G_{\varphi,D}$. If φ is only a **partial improper edge coloring** of G, i.e., a map $\varphi : E' \to C$

from a subset $E' \subseteq E(G)$ to the color set C, we also write $G_{\varphi,D}$ to denote the subgraph with vertex set $V(G)$ and edge set $\{e \in E' \mid \varphi(e) \in D\}$. Furthermore, for the degree function and the multiplicity function of $G_{\varphi,D}$ we write $d_{\varphi,D}$ and $\mu_{\varphi,D}$ rather than $d_{G_{\varphi,D}}$ and $\mu_{G_{D,\varphi}}$. If it is clear that we refer to the coloring φ, then we sometimes omit the index φ. The set of all improper edge colorings of G with color set $C = \{1, \ldots, k\}$ is denoted by $\mathcal{IC}^k(G)$.

Improper edge colorings become a subject of interest only when some restrictions are imposed. For an improper edge coloring φ of a graph G with color set C, we define the **vertex deviation** $vd(\varphi)$ of φ, respectively the **edge deviation** $ed(\varphi)$ of φ as follows:

$$vd(\varphi) = \max_{v \in V(G)} \max_{\alpha,\beta \in C} |d_{\varphi,\alpha}(v) - d_{\varphi,\beta}(v)|$$

$$ed(\varphi) = \max_{u,v \in V(G)} \max_{\alpha,\beta \in C} |\mu_{\varphi,\alpha}(u,v) - \mu_{\varphi,\beta}(u,v)|.$$

Then φ is said to be an **equitable edge coloring**, respectively a **nearly equitable edge coloring** if $vd(\varphi) \le 1$, respectively $vd(\varphi) \le 2$. Similarly, φ is said to be a **μ-equitable edge coloring**, respectively a **nearly μ-equitable edge coloring** if $ed(\varphi) \le 1$, respectively if $ed(\varphi) \le 2$. An improper edge coloring φ of G with color set C is called **balanced** if $||E(G_{\varphi,\alpha}| - |E(G_{\varphi,\beta})|| \le 1$ for every pair $\alpha, \beta \in C$ of colors.

The following result is folklore and it may be considered to be an equivalent version of König's theorem (Theorem 5.9). Proofs were given by several authors, including Hakimi and Kariv [127], Hilton [134], McDirmid [213] and de Werra [306].

Theorem 8.1 *Let G be a bipartite graph and let $k \ge 2$ be an integer. Then G admits an improper edge coloring with k colors that is balanced and equitable.*

Proof: With the bipartite graph G and the integer $k \ge 2$ we associate a graph $G^{(k)}$ as follows. For each vertex $v \in V(G)$ there are unique integers m_v and r_v satisfying $d_G(v) = km_v + r_v$ and $0 \le r_v \le k - 1$. To obtain the graph $G^{(k)}$ from G, split each vertex v of G into m_v vertices of degree k and one additional vertex of degree r_v. Clearly, $G^{(k)}$ is bipartite and $\Delta(G^{(k)}) \le k$, where equality holds if $k \le \Delta(G)$. By König's theorem (Theorem 5.9), $\chi'(G^{(k)}) \le k$. As proved in Sect. 2.6 (statement (b)) there is a balanced k-edge coloring $\varphi \in \mathcal{C}^k(G^{(k)})$. Clearly, φ induces a balanced equitable edge coloring of G with k colors. ∎

Theorem 8.2 (de Werra [308] 1975) *Let G be a bipartite graph and let $k \ge 2$ be an integer. Then G admits an improper edge coloring with k colors that is balanced, equitable and μ-equitable.*

Proof: Let C be a set of k colors. For an improper edge coloring φ with color set C and two distinct vertices u, v of G, we define

$$S_\varphi(u,v) = \sum_{\alpha,\beta \in C} |\mu_{\varphi,\alpha}(u,v) - \mu_{\varphi,\beta}(u,v)|,$$

where the sum ranges over all 2-element subsets of C. By Theorem 8.1, there is an improper edge coloring of G with color set C that is balanced and equitable. Among all such colorings, let φ be one for which

$$S = \sum_{u,v \in V(G)} S_\varphi(u,v)$$

is minimum. If $ed(\varphi) \leq 1$ there is nothing to prove. So assume that $ed(\varphi) \geq 2$. Then there is a vertex pair (x,y) and two colors $\alpha, \beta \in C$ such that

$$|\mu_{\varphi,\alpha}(x,y) - \mu_{\varphi,\beta}(x,y)| \geq 2.$$

We now construct a new improper edge coloring φ' of G by recoloring the edges of $G_{\varphi,\alpha,\beta}$ using the colors α and β. For each vertex pair (u,v), we first delete from $E_G(u,v)$ exactly $2\lfloor \mu_G(u,v)/2 \rfloor$ edges and color half of them with α and the other half with β. The remaining subgraph H of $G_{\varphi,\alpha,\beta}$ is a bipartite simple graph and, by Theorem 8.1, we can color the edges with α and β in such a way that we get an improper edge coloring that is balanced and equitable. The resulting improper edge coloring of G is denoted by φ'. It is easy to see that the coloring φ' remains balanced and equitable. Furthermore, for each vertex pair (u,v) we have

$$|\mu_{\varphi',\alpha}(u,v) - \mu_{\varphi',\beta}(u,v)| \leq \min\{1, |\mu_{\varphi,\alpha}(u,v) - \mu_{\varphi,\beta}(u,v)|\}.$$

Taking also all other pairs of colors into consideration, we deduce that $S_{\varphi'}(u,v) \leq S_\varphi(u,v)$ for each vertex pair (u,v) with strict inequality for the pair (x,y). Then we obtain

$$\sum_{u,v \in V(G)} S_{\varphi'}(u,v) < S,$$

which gives a contradiction.   ∎

Unfortunately, however, Theorem 8.1 does not extend to arbitrary graphs: An odd cycle has no equitable edge coloring with two colors. But, on the other hand, Hilton and de Werra [139] proved that any graph G has a nearly equitable edge coloring with k colors for each $k \geq 2$. We will discuss some strengthenings of this result. First, we prove two technical results about improper edge colorings with two colors. The first result is a simple observation; it appears in several papers (see, e.g., Hakimi and Kariv [127], Hilton [134] or de Werra [307]). An **eulerian graph**, sometimes also called an **even graph**, is a graph G such that $d_G(v)$ is even for all $v \in V(G)$.

Proposition 8.3 *Let G be a connected graph with m edges. When m is odd, let u be an arbitrary vertex of G. Then there is a balanced improper edge coloring φ of G with a set $\{\alpha, \beta\}$ of two colors such that the following conditions hold:*

(a) *If G is eulerian and m is odd, then $d_{\varphi,\alpha}(u) - d_{\varphi,\beta}(u) = 2$ and $d_{\varphi,\alpha}(v) = d_{\varphi,\beta}(v)$ for all $v \in V(G) \setminus \{u\}$.*

(b) *If G is eulerian and m is even, then $d_{\varphi,\alpha}(v) = d_{\varphi,\beta}(v)$ for all $v \in V(G)$.*

(c) *If G is not eulerian, then $|d_{\varphi,\alpha}(v) - d_{\varphi,\beta}(v)| \leq 1$ for all $v \in V(G)$.*

Proof: First assume that G is eulerian. Let u be any vertex of G. Since G is connected, there exists a closed trail C with initial vertex u such that $E(C) = E(G)$. Coloring the edges of C alternately with α and β, where we start with α, gives an edge coloring φ that is balanced and satisfies (a), respectively (b).

Now assume that G is not eulerian. Then G has an even number of vertices having odd degree in G. Join all these vertices to a new vertex u_0 by an edge and denote the resulting graph by G_0. Clearly, G_0 has a closed trail C containing all edges of G_0 and with initial vertex u_0. As before, coloring the edges of C alternately with α and β yields an edge coloring φ that is balanced and satisfies (c). ∎

Proposition 8.4 (de Werra [308] 1975) *Let G be a graph, and let φ be an improper edge coloring of G with a set $C = \{\alpha, \beta\}$ of two colors. Then there is an improper edge coloring φ' of G with color set C such that the following conditions hold:*

(a) *φ' is balanced, nearly equitable, and nearly μ-equitable.*

(b) *$|d_{\varphi',\alpha}(v) - d_{\varphi',\beta}(v)| \leq |d_{\varphi,\alpha}(v) - d_{\varphi,\beta}(v)|$ for every vertex v of G.*

(c) *$\min\{d_{\varphi,\alpha}(v), d_{\varphi,\beta}(v)\} \leq d_{\varphi',\alpha}(v), d_{\varphi',\beta}(v) \leq \max\{d_{\varphi,\alpha}(v), d_{\varphi,\beta}(v)\}$ for every vertex v of G.*

(d) *$|\mu_{\varphi',\alpha}(u,v) - \mu_{\varphi',\beta}(u,v)| \leq |\mu_{\varphi,\alpha}(u,v) - \mu_{\varphi,\beta}(u,v)|$ for every vertex pair (u,v) of G.*

(e) *$\min\{\mu_{\varphi,\alpha}(u,v), \mu_{\varphi,\beta}(u,v)\} \leq \mu_{\varphi',\alpha}(u,v), \mu_{\varphi',\beta}(u,v) \leq \max\{\mu_{\varphi,\alpha}(u,v), \mu_{\varphi,\beta}(u,v)\}$ for every vertex pair (u,v) of G.*

Proof: Let G_1 be the graph obtained from G by successively deleting as many pairs as possible of parallel edges (f, g) with $\varphi(f) = \alpha$ and $\varphi(g) = \beta$. Furthermore, let φ_1 be φ restricted to G_1. For every vertex pair (u, v) with $\mu_{G_1}(u, v) > 2$, we now delete an even number of edges between u and v such that either one or two edges remain. Let G_2 be the resulting graph. Note that there is a one-to-one correspondence between the components of G_1 and the components of G_2.

We now color the edges of every component H of G_2 with colors α and β. We distinguish two cases. If H is not eulerian or H is eulerian and has an even number of edges, then Proposition 8.3 implies that there is an edge coloring φ_H of H with color set C such that φ_H is balanced and equitable. Hence we have $|d_{\varphi_H,\alpha}(v) - d_{\varphi_H,\beta}(v)| \leq 1$ for every vertex v of H. Since $d_G(v) - d_{G_2}(v)$ is even, this implies that

$$|d_{\varphi_H,\alpha}(v) - d_{\varphi_H,\beta}(v)| \leq |d_{\varphi,\alpha}(v) - d_{\varphi,\beta}(v)|.$$

It remains to consider the case when H is eulerian and has an odd number of edges. Then the corresponding component of G_1 is also eulerian and has an odd number of edges. Consequently, $|d_{\varphi_1,\alpha}(v) - d_{\varphi_1,\beta}(v)|$ is even for every vertex v of H and there is a vertex u in H such that $|d_{\varphi_1,\alpha}(u) - d_{\varphi_1,\beta}(u)| \geq 2$. By Proposition 8.3, there is an edge coloring φ_H of H with color set C such that φ_H is balanced, $|d_{\varphi_H,\alpha}(u) - d_{\varphi_H,\beta}(u)| = 2$ and $d_{\varphi_H,\alpha}(v) = d_{\varphi_H,\beta}(v)$ for all $v \in V(H) \setminus \{u\}$. Then

$$|d_{\varphi_H,\alpha}(v) - d_{\varphi_H,\beta}(v)| \leq |d_{\varphi_1,\alpha}(v) - d_{\varphi_1,\beta}(v)| = |d_{\varphi,\alpha}(v) - d_{\varphi,\beta}(v)|$$

for all $v \in V(H)$.

Coloring all components H of G_2 in this way, we get an edge coloring φ_2 of G_2 with color set C. Clearly, since the colors α and β are interchangeable on the components of G_2, we can color the components in such a way that φ_2 is balanced. Moreover, the edge coloring φ_2 is nearly equitable and, since $\mu(G_2) \leq 2$, also nearly μ-equitable.

Eventually, we extend φ_2 to an edge coloring φ' of G by simply giving both colors α, β to every formerly deleted edge pair. Then the edge coloring φ' is balanced, nearly equitable, nearly μ-equitable, and

$$|d_{\varphi',\alpha}(v) - d_{\varphi',\beta}(v)| \leq |d_{\varphi,\alpha}(v) - d_{\varphi,\beta}(v)|$$

for every vertex $v \in V(G)$. Hence, φ' satisfies (a) and (b). Since

$$d_{\varphi',\alpha}(v) + d_{\varphi',\beta}(v) = d_{\varphi,\alpha}(v) + d_{\varphi,\beta}(v)$$

for every vertex v, (c) is a consequence of (b). Moreover, for every vertex pair (u, v) of G, we have

$$\begin{aligned} |\mu_{\varphi',\alpha}(u,v) - \mu_{\varphi',\beta}(u,v)| &\leq \mu_{G_2}(u,v) \\ &\leq \mu_{G_1}(u,v) = |\mu_{\varphi,\alpha}(u,v) - \mu_{\varphi,\beta}(u,v)|. \end{aligned}$$

Hence, the edge coloring φ' also satisfies (d). Since

$$\mu_{\varphi',\alpha}(u,v) + \mu_{\varphi',\beta}(u,v) = \mu_{\varphi,\alpha}(u,v) + \mu_{\varphi,\beta}(u,v)$$

for every vertex pair (u, v) of G, (e) follows from (d).   ∎

Lemma 8.5 (de Werra [308] 1975) *Let G be a graph, let C be a set of $k \geq 2$ colors, and let φ be an improper edge coloring of G with color set C. Then there is a balanced improper edge coloring φ' of G with color set C such that the following conditions hold:*

(a) $\min_\delta d_{\varphi,\delta}(v) \leq d_{\varphi',\gamma}(v) \leq \max_\delta d_{\varphi,\delta}(v)$ *for every vertex $v \in V(G)$ and every color $\gamma \in C$.*

(b) $\min_\delta \mu_{\varphi,\delta}(u,v) \leq \mu_{\varphi',\gamma}(u,v) \leq \max_\delta \mu_{\varphi,\delta}(u,v)$ *for every vertex pair (u,v) of G and every color $\gamma \in C$.*

Proof: Let Φ be the set of all improper edge colorings φ' of G with color set C satisfying (a) and (b). Clearly $\varphi \in \Phi$, i.e., $\Phi \neq \emptyset$. For an edge coloring $\varphi' \in \Phi$, let

$$p_1(\varphi') = \max_\gamma |E(G_{\varphi',\gamma})|, \quad p_2(\varphi') = \min_\gamma |E(G_{\varphi',\gamma})|,$$

and

$$q(\varphi') = |\{\gamma \,|\, |E(G_{\varphi',\gamma})| = p_1(\varphi')\}|.$$

Now choose an edge coloring φ' from Φ such that

(1) $p_1(\varphi') - p_2(\varphi')$ is minimum, and

(2) $q(\varphi')$ is minimum subject to (1).

Let $p_1 = p_1(\varphi'), p_2 = p_2(\varphi')$ and $q = q(\varphi')$. Our aim is to show that $p_1 - p_2 \leq 1$, i.e., φ' is balanced. Suppose, on the contrary, that $p_1 - p_2 \geq 2$. Then there are two colors α, β such that $|E(G_{\varphi',\alpha})| = p_1$ and $|E(G_{\varphi',\beta})| = p_2$. Now let $G_0 = G_{\varphi',\alpha,\beta}$. By Proposition 8.4, there is an edge coloring φ_0 of G_0 with color set $\{\alpha, \beta\}$ such that φ_0 is balanced,

$$\min\{d_{\varphi',\alpha}(v), d_{\varphi',\beta}(v)\} \leq d_{\varphi_0,\alpha}(v), d_{\varphi_0,\beta}(v) \leq \max\{d_{\varphi',\alpha}(v), d_{\varphi',\beta}(v)\}$$

for every vertex v of G, and

$$\min\{\mu_{\varphi',\alpha}(u,v), \mu_{\varphi',\beta}(u,v)\} \leq \mu_{\varphi_0,\alpha}(u,v), \mu_{\varphi_0,\beta}(u,v) \leq$$
$$\leq \max\{\mu_{\varphi',\alpha}(u,v), \mu_{\varphi',\beta}(u,v)\}$$

for every vertex pair (u,v) of G.

Now let φ'' be the edge coloring of G that is equal to φ_0 on $E(G_0)$ and equal to φ' on $E(G) \setminus E(G_0)$. Then we clearly have

$$\min_\delta d_{\varphi',\delta}(v) \leq d_{\varphi'',\gamma}(v) \leq \max_\delta d_{\varphi',\delta}(v)$$

for every vertex v and every color γ as well as

$$\min_\delta \mu_{\varphi',\delta}(u,v) \leq \mu_{\varphi'',\gamma}(u,v) \leq \max_\delta \mu_{\varphi',\delta}(u,v)$$

for every vertex pair (u,v) and every color γ. Since φ' satisfies (a) and (b), this implies that also φ'' satisfies (a) and (b). Hence $\varphi'' \in \Phi$.

For every color $\gamma \in C \setminus \{\alpha, \beta\}$, we certainly have $|E(G_{\varphi'',\gamma})| = |E(G_{\varphi',\gamma})|$. Since φ_0 is balanced and $p_1 - p_2 \geq 2$, we have $p_2 < |E(G_{\varphi'',\alpha})|, |E(G_{\varphi'',\beta})| < p_1$. Consequently, $p_1(\varphi'') \leq p_1$ and $p_2(\varphi'') \geq p_2$, where equality must hold because of (1). However, we then have $q(\varphi'') < q$, contradicting the choice of φ'. Therefore, $p_1 - p_2 \leq 1$, i.e., φ' is balanced. This completes the proof. ∎

Theorem 8.6 (de Werra [308] 1975) *Let G be a graph, let C be a set of $k \geq 2$ colors, and let φ be an improper edge coloring of G with color set C. Then there is a balanced improper edge coloring φ' of G with color set C such that the following conditions hold:*

(a) *φ' is nearly μ-equitable.*

(b) *$\min_\delta d_{\varphi,\delta}(v) \leq d_{\varphi',\gamma}(v) \leq \max_\delta d_{\varphi,\delta}(v)$ for every vertex $v \in V(G)$ and every color $\gamma \in C$.*

(c) $\min_\delta \mu_{\varphi,\delta}(u,v) \leq \mu_{\varphi',\gamma}(u,v) \leq \max_\delta \mu_{\varphi,\delta}(u,v)$ *for every vertex pair* (u,v)
of G *and every color* $\gamma \in C$.

Proof: Let Φ be the set of all balanced improper edge colorings φ' of G with color
set C satisfying (b) and (c). By Lemma 8.5, $\Phi \neq \emptyset$. To prove the theorem, we have
to show that there is an edge coloring $\varphi' \in \Phi$ with $ed(\varphi') \leq 2$. For an edge coloring
$\varphi' \in \Phi$, let

$$q(\varphi') = |\{(\gamma, \delta, u, v) \mid |\mu_{\varphi',\gamma}(u,v) - \mu_{\varphi',\delta}(u,v)| = ed(\varphi')\}|.$$

Now choose an edge coloring φ' from Φ such that

(1) $ed(\varphi')$ is minimum, and

(2) $q(\varphi')$ is minimum subject to (1).

Let $p = ed(\varphi')$ and $q = q(\varphi')$. We claim that $p \leq 2$, which proves the theorem.
Suppose, on the contrary, that $p \geq 3$. Then there are two colors α, β and two vertices
x, y such that $|\mu_{\varphi',\alpha}(x,y) - \mu_{\varphi',\beta}(x,y)| = p \geq 3$. Now let $G_0 = G_{\varphi',\alpha,\beta}$. By
Proposition 8.4, there is an edge coloring φ_0 of G_0 with color set $\{\alpha, \beta\}$ such that
φ_0 is balanced, nearly μ-equitable,

$$\min\{d_{\varphi',\alpha}(v), d_{\varphi',\beta}(v)\} \leq d_{\varphi_0,\alpha}(v), d_{\varphi_0,\beta}(v) \leq \max\{d_{\varphi',\alpha}(v), d_{\varphi',\beta}(v)\}$$

for every vertex v of G, and

$$\min\{\mu_{\varphi',\alpha}(u,v), \mu_{\varphi',\beta}(u,v)\} \leq \mu_{\varphi_0,\alpha}(u,v), \mu_{\varphi_0,\beta}(u,v)$$

and

$$\mu_{\varphi_0,\alpha}(u,v), \mu_{\varphi_0,\beta}(u,v) \leq \max\{\mu_{\varphi',\alpha}(u,v), \mu_{\varphi',\beta}(u,v)\}$$

for every vertex pair (u,v) of G.

Now let φ'' be the edge coloring of G that is equal to φ_0 on $E(G_0)$ and equal to
φ' on $E(G) \setminus E(G_0)$. Since φ' and φ_0 are balanced, φ'' is balanced, too. We clearly
have $\min_\delta d_{\varphi',\delta}(v) \leq d_{\varphi'',\gamma}(v) \leq \max_\delta d_{\varphi',\delta}(v)$ for every vertex v and every color
γ and, therefore, φ'' satisfies (b). Furthermore, $\min_\delta \mu_{\varphi',\delta}(u,v) \leq \mu_{\varphi'',\gamma}(u,v) \leq$
$\max_\delta \mu_{\varphi',\delta}(u,v)$ for every vertex pair (u,v) and every color γ and, therefore, φ''
satisfies (c). Hence $\varphi'' \in \Phi$. Moreover, this also implies that $ed(\varphi'') \leq ed(\varphi') = p$.
By the choice of φ', we then have $ed(\varphi'') = p$.

Let (u,v) be a vertex pair of G. Since φ_0 is nearly μ-equitable, we have
$|\mu_{\varphi'',\alpha}(u,v) - \mu_{\varphi'',\beta}(u,v)| \leq 2 < p$. Now let $\gamma \notin \{\alpha, \beta\}$ and, for $\varphi^* \in \{\varphi', \varphi''\}$,
let

$$q_1(\varphi^*) = |\{\delta \notin \{\alpha, \beta\} \mid |\mu_{\varphi^*,\gamma}(u,v) - \mu_{\varphi^*,\delta}(u,v)| = p\}|$$

and

$$q_2(\varphi^*) = |\{\delta \in \{\alpha, \beta\} \mid |\mu_{\varphi^*,\gamma}(u,v) - \mu_{\varphi^*,\delta}(u,v)| = p\}|.$$

Clearly, we have $q_1(\varphi'') = q_1(\varphi')$. Moreover, since

$$|\mu_{\varphi'',\alpha}(u,v) - \mu_{\varphi'',\beta}(u,v)| \leq |\mu_{\varphi',\alpha}(u,v) - \mu_{\varphi',\beta}(u,v)| \leq p$$

and

$$\mu_{\varphi'',\alpha}(u,v) + \mu_{\varphi'',\beta}(u,v) = \mu_{\varphi',\alpha}(u,v) + \mu_{\varphi',\beta}(u,v),$$

it is easy to see that $q_2(\varphi'') \leq q_2(\varphi')$. Since the inequality holds for every vertex pair (u, v), it follows that $q(\varphi'') \leq q(\varphi')$. Since the vertex pair (x, y) satisfies

$$|\mu_{\varphi'',\alpha}(x,y) - \mu_{\varphi'',\beta}(x,y)| < |\mu_{\varphi',\alpha}(x,y) - \mu_{\varphi',\beta}(x,y)| = p,$$

we deduce that $q(\varphi'') < q(\varphi')$, which gives a contradiction to the choice of φ'. This completes the proof. ∎

Theorem 8.7 (de Werra [308] 1975) *Let G be a graph, and let $k \geq 2$ be an integer. Then there is an improper edge coloring φ of G with k colors that is balanced, μ-equitable and nearly equitable.*

Proof: Let Φ be the set of all improper edge colorings φ of G with color set $C = \{1, \ldots, k\}$ that are balanced and μ-equitable. We first show that Φ is not empty. To this end, let $\ell = (e_1, \ldots, e_m)$ be a linear ordering of the edges of G such that, for every vertex pair (u, v), the edges between u and v are consecutive in ℓ. Now we get a coloring φ of G by distributing the colors $1, \ldots, k$ such that $\varphi(e_i) \equiv i (\text{mod } k)$. Clearly, the resulting edge coloring φ is balanced and μ-equitable. Hence $\Phi \neq \emptyset$. To finish the proof, we have to show that there is a coloring $\varphi \in \Phi$ such that $vd(\varphi) \leq 2$. For a coloring $\varphi \in \Phi$, let

$$q(\varphi') = |\{(\gamma, \delta, v) \,|\, |d_{\varphi,\gamma}(v) - d_{\varphi,\delta}(v)| = vd(\varphi)\}|.$$

Now choose a coloring φ from Φ such that

(1) $vd(\varphi)$ is minimum, and

(2) $q(\varphi)$ is minimum subject to (1).

Let $p = vd(\varphi)$ and $q = q(\varphi)$. We claim that $p \leq 2$, which proves the theorem. Suppose, on the contrary, that $p \geq 3$. Then there are two colors α, β and a vertex x such that $|d_{\varphi,\alpha}(x) - d_{\varphi,\beta}(x)| = p \geq 3$. Now let $G' = G_{\varphi,\alpha,\beta}$. By Proposition 8.4, there is an edge coloring φ' of G' with color set $\{\alpha, \beta\}$ that is balanced, nearly equitable, and satisfies

$$\min\{d_{\varphi,\alpha}(v), d_{\varphi,\beta}(v)\} \leq d_{\varphi',\alpha}(v), d_{\varphi',\beta}(v) \leq \max\{d_{\varphi,\alpha}(v), d_{\varphi,\beta}(v)\}$$

for every vertex v as well as

$$\min\{\mu_{\varphi,\alpha}(u,v), \mu_{\varphi,\beta}(u,v)\} \leq \mu_{\varphi',\alpha}(u,v), \mu_{\varphi',\beta}(u,v)$$

and

$$\mu_{\varphi',\alpha}(u,v), \mu_{\varphi',\beta}(u,v) \leq \max\{\mu_{\varphi,\alpha}(u,v), \mu_{\varphi,\beta}(u,v)\}$$

for every vertex pair (u, v). This implies, in particular, that φ' is μ-equitable.

Now let φ'' be the coloring of G that is equal to φ' on $E(G')$ and equal to φ on $E(G) \setminus E(G')$. Since φ and φ' are balanced and μ-equitable, φ'' is balanced and μ-equitable and, therefore, $\varphi'' \in \Phi$. Moreover, we clearly have

$$\min_\delta d_{\varphi,\delta}(v) \leq d_{\varphi'',\gamma}(v) \leq \max_\delta d_{\varphi,\delta}(v)$$

for every vertex v and every color γ. This implies that $vd(\varphi'') \leq vd(\varphi') = p$. By the choice of φ, we then have $vd(\varphi'') = p$. Let v be a vertex of G. Since φ'' is nearly equitable, we have

$$|d_{\varphi'',\alpha}(v) - d_{\varphi'',\beta}(v)| \leq 2 < p.$$

Now let $\gamma \notin \{\alpha, \beta\}$ and, for $\varphi^* \in \{\varphi, \varphi''\}$, let

$$q_1(\varphi^*) = |\{\delta \notin \{\alpha, \beta\} \,|\, |d_{\varphi^*,\gamma}(v) - d_{\varphi^*,\delta}(v)| = p\}|$$

and

$$q_2(\varphi^*) = |\{\delta \in \{\alpha, \beta\} \,|\, |d_{\varphi^*,\gamma}(v) - d_{\varphi^*,\delta}(v)| = p\}|.$$

Clearly, we have $q_1(\varphi'') = q_1(\varphi)$. Furthermore, since $|d_{\varphi'',\alpha}(v) - d_{\varphi'',\beta}(v)| \leq |d_{\varphi,\alpha}(v) - d_{\varphi,\beta}(v)| \leq p$ and $d_{\varphi'',\alpha}(v) + d_{\varphi'',\beta}(v) = d_{\varphi,\alpha}(v) + d_{\varphi,\beta}(v)$, it is easy to see that $q_2(\varphi'') \leq q_2(\varphi)$. Since this holds for every vertex v, it follows that $q(\varphi'') \leq q(\varphi)$. Since the vertex x satisfies

$$|d_{\varphi'',\alpha}(x) - d_{\varphi'',\beta}(x)| < |d_{\varphi,\alpha}(x) - d_{\varphi,\beta}(x)| = p,$$

we deduce that $q(\varphi'') < q(\varphi')$, a contradiction to the choice of φ'. This completes the proof. ∎

For a graph G and an integer $k \geq 2$, let $D_k(G)$ denote the set of all vertices $v \in V(G)$ such that $d_G(v)$ is divisible by k. If $k = \Delta(G) + 1 \geq 2$, then $D_k(G) = \emptyset$. Observe that any equitable edge coloring of a graph G with $k \geq \Delta(G)$ colors must be a proper edge coloring of G. Hence the following result implies Vizing's theorem that $\chi'(G) \leq \Delta(G) + 1$ for every simple graph G.

Theorem 8.8 (Hilton and de Werra [140] 1994) *Let G be a simple graph, and let $k \geq 2$ be an integer. If $G[D_k(G)]$ is edgeless, then G has an equitable edge coloring with k colors.*

The proof of the theorem given in Hilton and de Werra [140] is based on a sophisticated recoloring procedure. The procedure is applied to a nearly equitable edge coloring of G to start with, and an equitable edge coloring of G is obtained at the end. For improving the coloring, Kempe changes are used as a main tool. However, the Kempe changes are applied to (α, β)-trails that form a natural generalization of (α, β)-chains. We shall discuss this method in Sect. 8.4. Theorem 8.8 has been generalized as follows:

Theorem 8.9 (Zhang and Liu [319] 2011) *Let G be a simple graph, and let $k \geq 2$ be an integer. If $G[D_k(G)]$ is a forest, then G has an equitable edge coloring with k colors.*

8.2 FULL EDGE COLORINGS AND THE COVER INDEX

In this section we shall briefly discuss a type of edge coloring which was first defined and studied by R. P. Gupta in his thesis [120].

Let G be a graph, and let φ be an improper edge coloring of G with color set C. Note that φ is a proper edge coloring of G if and only if $\Delta(G_{\varphi,\alpha}) \leq 1$ for each color $\alpha \in C$. We call φ a **full edge coloring** of G if $\delta(G_{\varphi,\alpha}) \geq 1$ for each color $\alpha \in C$. Obviously, G has a full edge coloring with at least one color if and only if $\delta(G) \geq 1$. The largest number k such that G has a full edge coloring with a set C of k colors is said to be the **cover index** $\psi(G)$ of G, where $\psi(G) = 0$ if G is empty or contains an isolated vertex.

An **edge cover** of G is an edge set $M \subseteq E(G)$ such that each vertex of G is incident with at least one edge in M. Observe that the cover index of a graph G with $\delta(G) \geq 1$ is the largest number k such that the edge set of G admits a partition into k disjoint edge covers. Clearly, every graph G satisfies $\psi(G) \leq \delta(G)$. Another useful observation is that an equitable edge coloring of G with k colors is a full edge coloring if $k \leq \delta(G)$. Therefore, a simple consequence of Theorem 8.1 is the following result.

Theorem 8.10 (Gupta [120] 1967) *Each bipartite graph G satisfies $\psi(G) = \delta(G)$.*

For $k = \delta(G) - 1 \geq 2$ we certainly have $D_k(G) = \emptyset$. Hence it follows from Theorem 8.8 that every simple graph G satisfies $\psi(G) \geq \delta(G) - 1$. Gupta [120] established the following lower bound for the cover index of an arbitrary graph.

Theorem 8.11 (Gupta [120] 1967) *Each graph G satisfies $\psi(G) \geq \delta(G) - \mu(G)$.*

For a graph G and an improper edge coloring φ of G with a set C of $k \geq 1$ colors, we introduce the following notation. The **color valency** of a vertex $v \in V(G)$ with respect to φ is

$$\text{val}(v : \varphi) = |\{\alpha \in C \,|\, d_{\varphi,\alpha}(v) \geq 1\}|.$$

Clearly, each vertex v of G satisfies $\text{val}(v : \varphi) \leq \min\{k, d_G(v)\}$ and, moreover, φ is a full edge coloring of G with k colors if and only if $\text{val}(v : \varphi) = k$ for each vertex v of G.

Now let G be a graph and let $k = \delta(G) - \mu(G)$. To find a full edge coloring of G with k colors, Gupta argues as follows. Since such a coloring certainly exists if $k \leq 1$, we may assume that $k \geq 2$. Among all improper edge colorings of G with color set $C = \{1, \ldots, k\}$, let φ be one with

$$S(\varphi) = \sum_{v \in V(G)} (k - \text{val}(v : \varphi))$$

minimum. If $S(\varphi) = 0$ there is nothing to prove. So assume that $S(\varphi) \geq 1$. Then there is a vertex v such that $\text{val}(v : \varphi) \leq k - 1$. Since $d_G(v) \leq k$, this implies that there are two colors $\alpha, \beta \in C$ such that $d_{\varphi,\alpha}(v) = 0$ and $d_{\varphi,\beta}(v) \geq 2$. The main part in Gupta's proof consists in constructing an appropriate (α, β)-trail F starting at v so that interchanging the colors on F results in a coloring φ' of G with $S(\varphi') < S(\varphi)$, which gives the required contradiction.

Schrijver [277] found a shorter proof of Gupta's theorem; he gives a reduction to Vizing's theorem (Theorem 2.2). The trick is to construct an auxiliary graph H by splitting each vertex v into $d_G(v) - \delta(G) + 1$ vertices such that one vertex has degree $\delta(G)$ and the remaining vertices have degree 1 in H. Then $\Delta(H) = \delta(G)$ and Vizing's theorem implies that H admits a proper edge coloring with $\delta(G) + \mu(G)$ colors. Since there is a bijection between $E(G)$ and $E(H)$, this proper edge coloring of H yields an improper edge coloring φ of G with a set C of $\delta(G) + \mu(G)$ colors such that $\mathrm{val}(v : \varphi) \geq \delta(G)$ for each vertex $v \in V(G)$. We now partition the color set C into two sets C' and D such that $|C'| = \delta(G) - \mu(G)$ and $|D| = 2\mu(G)$. Then we orient the edges of $F = G_{\varphi,D}$ such that the in-degree of each vertex v is at least $\lfloor d_F(v)/2 \rfloor$. For a vertex v of G, let $a(v)$ be the number of colors $\alpha \in C'$ that are missing at v, i.e., $d_{\varphi,\alpha}(v) = 0$, and let $b(v)$ be the number of colors $\beta \in D$ that are present at v, i.e., $d_{\varphi,\beta}(v) \geq 1$. Then

$$b(v) + (\delta(G) - \mu(G) - a(v)) \geq \mathrm{val}(v : \varphi) \geq \delta(G),$$

which gives $b(v) \geq a(v) + \mu(G)$ and, therefore, $d_F(v) \geq b(v) = 2b(v) - b(v) \geq 2(a(v) + \mu(G)) - 2\mu(G) = 2a(v)$. Hence the in-degree of v in the orientation of F is at least $a(v)$. This implies that we can recolor the edges in F oriented towards v by using the $a(v)$ colors from C' that are missing at v. This certainly results in a full edge coloring of G with color set C'.

Gupta [121] announced several results related to the cover index without proofs. In particular, he pointed out that, apart from the minimum degree, there is a further upper bound for the cover index. For a graph G and a vertex set $X \subseteq V(G)$, let $E_G(X) = E_G(X, X) \cup \partial_G(X)$, that is, $E_G(X)$ is the set of all edges of G with at least one endvertex in X. The **co-density** $w^c(G)$ of G is defined by

$$w^c(G) = \min_{X \subseteq V(G), |X| \text{ odd}} \left\lfloor \frac{2|E_G(X)|}{|X| + 1} \right\rfloor.$$

Note that any edge cover of G must contain at least $(|X| + 1)/2$ edges from $E_G(X)$ if $|X|$ is odd. Consequently, every graph G satisfies $\psi(G) \leq w^c(G)$. Gupta [121] posed the following conjecture.

Conjecture 8.12 (Gupta [121] 1978) *Every graph G satisfies*

$$\psi(G) \geq \max\{\delta(G) - 1, w^c(G)\}.$$

The conjecture may be viewed as a counterpart to Goldberg's conjecture. Gupta [121] also announced a result supporting his conjecture; the results says that $\psi(G) \geq \max\{\lfloor (7\delta(G) + 1)/8 \rfloor, w^c(G)\}$ for every graph G, this being a counterpart to a result mentioned in Sect. 6.1 (see Corollary 6.7).

8.3   EDGE COLORINGS OF WEIGHTED GRAPHS

Let G be a graph. A function $f : V(G) \to \mathbb{N}$ from the vertex set of G to the set $\mathbb{N}$ of positive integers is called a **vertex function** of G. An **edge function** of G is a function $g : V(G) \times V(G) \to \mathbb{N}$ such that $g(u,v) = g(v,u)$ for all $(u,v) \in V(G) \times V(G)$.

Let $s, t \geq 0$ be given integers. In this and the following section we consider **weighted graphs of type** (s,t), that is, tuples $\mathcal{G} = (G, \mathbf{f}, \mathbf{g})$ consisting of a graph G, a sequence $\mathbf{f} = (f_1, \ldots, f_s)$ of s vertex functions of G, and a sequence $\mathbf{g} = (g_1, \ldots, g_t)$ of t edge functions of G. A **weighted graph parameter** is a function ρ that assigns to each weighted graph $\mathcal{G}$ a real number $\rho(\mathcal{G})$ such that $\rho(\mathcal{G}) = \rho(\mathcal{G}')$ whenever $\mathcal{G}$ and $\mathcal{G}'$ are isomorphic. Instead of $\rho(G, \mathbf{f}, \mathbf{g})$ we also write $\rho_{\mathbf{f}, \mathbf{g}}(G)$.

Let $\mathcal{G} = (G, \mathbf{f}, \mathbf{g})$ be a weighted graph of type (s, t) with $\mathbf{f} = (f_1, \ldots, f_s)$ and $\mathbf{g} = (g_1, \ldots, g_t)$. Then each vertex function f_i of G and each edge function g_j of G may be also considered as a vertex function respectively edge function of each subgraph H of G. So we denote the restriction of f_i respectively g_j to any subgraph H also by f_i respectively g_j and say that $(H, \mathbf{f}, \mathbf{g})$ is a weighted graph of type (s, t).

A weighted graph parameter ρ is called **monotone** if $\rho(H, \mathbf{f}, \mathbf{g}) \leq \rho(G, \mathbf{f}, \mathbf{g})$ for every weighted graph $(G, \mathbf{f}, \mathbf{g})$ and every subgraph H of G. Let ρ and ρ' be two weighted graph parameters. If $\rho(\mathcal{G}) \leq \rho'(\mathcal{G})$ for every weighted graph $\mathcal{G}$, then we say that ρ' is an **upper bound** for ρ, or conversely, ρ is a **lower bound** for ρ'.

Let ρ be a monotone weighted graph parameter defined for weighted graphs of type (s, t). A weighted graph $(G, \mathbf{f}, \mathbf{g})$ of type (s, t) is called **critical with respect to** ρ if $\rho(H, \mathbf{f}, \mathbf{g}) < \rho(G, \mathbf{f}, \mathbf{g})$ for every proper subgraph H of G. If $(G, \mathbf{f}, \mathbf{g})$ is critical with respect to ρ, we also say that G is $\rho_{\mathbf{f}, \mathbf{g}}$-**critical**. Furthermore, we say that $e \in E(G)$ is a $\rho_{\mathbf{f}, \mathbf{g}}$-**critical edge** of G if $\rho_{\mathbf{f}, \mathbf{g}}(G - e) < \rho_{\mathbf{f}, \mathbf{g}}(G)$. The proof of the following result is straightforward and left to the reader.

Proposition 8.13 *Let ρ and ρ' be two monotone weighted graph parameters. Then the following statements hold:*

(a) *For every weighted graph $(G, \mathbf{f}, \mathbf{g})$ there exists a subgraph H of G such that $(H, \mathbf{f}, \mathbf{g})$ is ρ-critical and $\rho(H, \mathbf{f}, \mathbf{g}) = \rho(G, \mathbf{f}, \mathbf{g})$.*

(b) *If $\rho(\mathcal{H}) \leq \rho'(\mathcal{H})$ for all weighted graphs $\mathcal{H}$ that are critical with respect to ρ, then ρ' is an upper bound for ρ.*

In what follows, we will only consider weighted graphs of type $(1, 1)$ or type $(1, 0)$. A weighted graph of type type $(1, 1)$ is briefly called a **weighted graph**, and a weighted graph of type type $(1, 0)$ is called a **vertex-weighted graph**.

Let G be a graph, let f be a vertex function of G, and let g be an edge function of G. Furthermore, let φ be an improper edge coloring of G with color set C. We say that φ is an f-**coloring** of G if $d_{\varphi, \alpha}(v) \leq f(v)$ for all $\alpha \in C$ and all $v \in V(G)$. With $C_f^k(G)$ we denote the set of all f-colorings of G with color set $C = \{1, \ldots, k\}$. The f-**chromatic index** $\chi_f'(G)$ of G is the least integer $k \geq 0$ such that $C_f^k(G) \neq \emptyset$. We call φ an fg-**coloring** of G if φ is an f-coloring of G such that $\mu_{\varphi, \alpha}(u, v) \leq g(u, v)$

for all $\alpha \in C$ and all pairs $(u,v) \in V(G)^2$. With $\mathcal{C}^k_{f,g}(G)$ we denote the set of all fg-colorings of G with color set $C = \{1, \ldots, k\}$. The fg-**chromatic index** $\chi'_{f,g}(G)$ of G is the least integer $k \geq 0$ such that $\mathcal{C}^k_{f,g}(G) \neq \emptyset$.

Note that χ'_f and $\chi'_{f,g}$ are weighted graph parameters; the first parameter is defined for vertex-weighted graphs (i.e., weighted graphs of type $(1,0)$) and the second parameter is defined for weighted graphs (i.e. weighted graphs of type $(1,1)$). Both parameters χ'_f and $\chi'_{f,g}$ are monotone. Obviously,

$$\mathcal{C}^k(G) \subseteq \mathcal{C}^k_{f,g}(G) \subseteq \mathcal{C}^k_f(G)$$

for every $k \geq 0$ and, therefore, we have

$$\chi'_f(G) \leq \chi'_{f,g}(G) \leq \chi'(G). \tag{8.1}$$

If $g(u,v) \geq \min\{f(u), f(v)\}$ for all $(u,v) \in V(G)^2$, then $\mathcal{C}^k_{f,g}(G) = \mathcal{C}^k_f(G)$ for all $k \geq 0$ and, therefore, $\chi'_f(G) = \chi'_{f,g}(G)$. Furthermore, $\chi'(G) = \chi'_f(G) = \chi'_{f,g}(G)$ provided that $f \equiv 1$, i.e., $f(v) = 1$ for all $v \in V(G)$.

The f-color problem was first defined and studied by Hakimi and Kariv [127] in 1986. The fg-color problem was introduced by Nakano, Nishizeki, and Saito [232] in 1990. Applications of the fg-color problem to multiple resource problems are discussed in references [59, 66, 184, 230, 231, 323]. How wavelet assignment problems in so-called multi-fiber WDM networks can be modeled by means of the f-color problem is explained in Koster [176].

Let G be a graph, let f be a vertex function of G, and let g be a edge function of G. Suppose that $\chi'_f(G) = k$. Then there is a coloring $\varphi \in \mathcal{C}^k_f(G)$. Since each vertex $v \in V(G)$ satisfies $d_{G_{\varphi,\alpha}}(v) \leq f(v)$ for every color $\alpha \in \{1, \ldots, k\}$, we have $d_G(v) \leq f(v)k$. Let Δ_f be the **maximum f-degree** of G defined by

$$\Delta_f(G) = \max_{v \in V(G)} \left\lceil \frac{d_G(v)}{f(v)} \right\rceil$$

if $G \neq \emptyset$ and $\Delta_f(G) = 0$ otherwise. Then, clearly, G satisfies

$$\chi'_f(G) \geq \Delta_f(G). \tag{8.2}$$

Similarly, we define the **maximum fg-degree** $\Delta_{f,g}(G)$ by

$$\Delta_{f,g}(G) = \max\left\{ \Delta_f(G), \max_{(u,v) \in V(G)^2} \left\lceil \frac{\mu_G(u,v)}{g(u,v)} \right\rceil \right\}$$

if $G \neq \emptyset$ and $\Delta_{f,g}(G) = 0$ otherwise. Then G satisfies

$$\chi'_{f,g}(G) \geq \Delta_{f,g}(G). \tag{8.3}$$

Theorem 8.14 (Nakano, Nishizeki, and Saito [232] 1990) *Every weighted graph (G, f, g) such that G is bipartite satisfies $\chi'_{f,g}(G) = \Delta_{f,g}(G)$.*

Proof: Because of (8.3) it suffices to show that $C^k_{f,g}(G) \neq \emptyset$ for $k = \Delta_{f,g}(G)$. To see this, we deduce from Theorem 8.2 that the bipartite graph G has an improper edge coloring φ with color set $C = \{1, \ldots, k\}$ such that φ is equitable and μ-equitable. We claim that $\varphi \in C^k_{f,g}(G)$. Suppose, on the contrary, that $d_{\varphi,\alpha}(v) \geq f(v) + 1$ for some vertex v and some color $\alpha \in C$. Since φ is equitable, this implies that $d_{\varphi,\gamma}(v) \geq f(v)$ for every color $\gamma \in C \setminus \{\alpha\}$, which gives

$$d_G(v) \geq f(v) + 1 + (k-1)f(v) = kf(v) + 1,$$

and hence $k < \Delta_f(G) \leq \Delta_{f,g}(G)$, a contradiction. Similarly, one can show that $\mu_{\varphi,\alpha}(u,v) \leq g(u,v)$ for every vertex pair (u,v) and every color $\alpha \in C$. ∎

If all vertex pairs (u,v) satisfy $g(u,v) = \max\{f(u), f(v)\}$, then we have $\chi'_{f,g}(G) = \chi'_f(G)$ and $\Delta_{f,g}(G) = \Delta_f(G)$. Hence Theorem 8.14 implies the following result.

Corollary 8.15 (Hakimi and Kariv [127] 1986) *Every vertex-weighted graph (G, f) such that G is bipartite satisfies $\chi'_f(G) = \Delta_f(G)$.*

Theorem 8.16 (Nakano, Nishizeki, and Saito [232] 1990) *Every weighted graph (G, f, g) such that $g(u,v) \leq \max\{f(u), f(v)\}$ for every vertex pair (u,v) of G satisfies*

$$\chi'_{f,g}(G) \leq \max_{u,v \in V(G)} \left\{ \left\lceil \frac{d_G(u)}{f(u)} + \frac{\mu_G(u,v)}{g(u,v)} \right\rceil \right\}.$$

The theorem provides a general upper bound for the fg-chromatic index. In case of $f \equiv 1$ and $g \equiv 1$, we get Ore's version (Theorem 2.5) of Vizing's bound for the ordinary chromatic index. The proof of the theorem given in [232] is constructive and resembles Vizing's fan argument. However, the recoloring arguments used in the proof are slightly more sophisticated. As a simple consequence of the theorem we obtain the following result for the f-chromatic index. We shall give an improvement of the result in the next section (Corollary 8.26)

Corollary 8.17 (Hakimi and Kariv [127] 1986) *Every vertex-weighted graph (G, f) satisfies*

$$\chi'_f(G) \leq \max_{u,v \in V(G)} \left\{ \left\lceil \frac{d_G(u) + \mu_G(u,v)}{f(u)} \right\rceil \right\}.$$

Theorem 8.18 (Nakano, Nishizeki, and Saito [232] 1990) *Let (G, f, g) be a weighted graph. Then the following statement holds:*

(a) *If $f(v) \geq 2$ for every vertex $v \in V(G)$, then*

$$\chi'_{f,g}(G) \leq \max \left\{ \max_{v \in V(G)} \left\lceil \frac{d_G(v) - 1}{f(v) - 1} \right\rceil, \max_{(u,v) \in V(G)^2} \left\lceil \frac{\mu_G(u,v)}{g(u,v)} \right\rceil \right\}. \quad (8.4)$$

(b) *If $g(u,v) \geq 2$ for every vertex pair $(u,v) \in V(G)^2$, then*

$$\chi'_{f,g}(G) \leq \max\left\{\chi'_f(G), \max_{(u,v)\in V(G)^2}\left\lceil\frac{\mu_G(u,v)-1}{g(u,v)-1}\right\rceil\right\}. \quad (8.5)$$

Proof: For the proof of (a), let k denote the right hand side of (8.4). By Theorem 8.7, there is an improper edge coloring φ of G with color set $C = \{1,\ldots,k\}$ such that φ is μ-equitable and nearly equitable. Then it is easy to show that $\varphi \in C^k_{f,g}(G)$.

For the proof of (b), let k denote the right hand side of (8.5). Then $k \geq \chi'_f(G)$ and there is a coloring $\varphi \in C^k_f(G)$. It follows from Theorem 8.6 that there is an improper edge coloring φ' of G with color set $C = \{1,\ldots,k\}$ such that φ' is nearly μ-equitable and $d_{\varphi',\gamma}(v) \leq \max_\delta d_{\varphi,\delta}(v)$ for every vertex $v \in V(G)$ and every color $\gamma \in C$. Since φ is an f-coloring of G, this implies that also φ' is an f-coloring of G. We claim that $\varphi' \in C^k_{f,g}(G)$. Suppose this is not true. Then there is a vertex pair (x,y) of G and a color $\alpha \in C$ such that $\mu_{\varphi',\alpha}(x,y) \geq g(x,y)+1$. Since φ' is nearly μ-equitable, this implies that $\mu_{\varphi',\gamma}(x,y) \geq g(x,y)-1$ for every color $\gamma \in C \setminus \{\alpha\}$, which gives

$$\mu_G(x,y) \geq g(x,y)+1+(k-1)(g(x,y)-1) = k(g(x,y)-1)+2,$$

a contradiction to the fact that $k \leq \lceil(\mu_G(x,y)-1)/(g(x,y)-1)\rceil$. ∎

Theorem 8.19 (Nakano, Nishizeki, and Saito [232] 1990) *Every weighted graph (G,f,g) such that $f(v) \geq 2$ for every vertex $v \in V(G)$ and $g(u,v) \geq 2$ for every vertex pair $(u,v) \in V(G)^2$ satisfies $\chi'_{f,g}(G) \leq \max\{D,M\}$, where*

$$D = \max_{v\in V(G)}\left\{\left\lceil\frac{\lfloor d_G(v)/2\rfloor}{\lfloor f(v)/2\rfloor}\right\rceil, \left\lceil\frac{\lceil d_G(v)/2\rceil}{\lceil f(v)/2\rceil}\right\rceil\right\}$$

and

$$M = \max_{(u,v)\in V(G)^2}\left\{\left\lceil\frac{\lfloor \mu_G(u,v)/2\rfloor}{\lfloor g(u,v)/2\rfloor}\right\rceil, \left\lceil\frac{\lceil \mu_G(u,v)/2\rceil}{\lceil g(u,v)/2\rceil}\right\rceil\right\}.$$

Proof: First, we show that we can direct the edges of G in such a way that the resulting oriented graph D satisfies the following two conditions:

(1) $|d^+(u)-d^-(u)| \leq 1$ for every vertex $u \in V(G)$, where $d^+(u)$ is the out-degree of u in D and $d^-(u)$ is the in-degree of u in D.

(2) $|\mu^+(u,v) - \mu^-(u,v)| \leq 1$ for every vertex pair $(u,v) \in V(G)^2$, where $\mu^+(u,v)$ is the number of edges in $E_G(u,v)$ oriented from u to v in D and $\mu^-(u,v)$ is the number of edges in $E_G(u,v)$ oriented from v to u in D.

To obtain the oriented graph D we proceed as follows. For each vertex pair (u,v), we first delete from $E_G(u,v)$ exactly $2\lfloor\mu_G(u,v)/2\rfloor$ edges and orient half of them from u to v and the other half from v to u. The remaining subgraph G' of G is a simple graph. We add an additional vertex w to G' and join w to each vertex of G' having odd degree in G'. Then we obtain an eulerian graph G_0 and we now orient the edges along an eulerian trail of (each component) G_0. Then a desired oriented graph D of G is obtained.

Next, we construct a new (undirected) graph H as follows: we substitute each vertex v of D by two vertices v^+ and v^- and, for each edge of D directed from u to v, we add an edge to $E_H(u^+, v^-)$. Clearly, H is bipartite and there is a one-to-one correspondence between the edges of G and the edges of H. Note that each $v \in V(G)$ satisfies $d_G(v) = d_H(v^+) + d_H(v^-)$, $d_H(v^+) = d^+(v)$, and $d_H(v^-) = d^-(v)$. For a vertex pair $(u, v) \in V(G)^2$, we have $\mu_G(u, v) = \mu_H(u^+, v^-) + \mu_H(u^-, v^+)$, $\mu_H(u^+, v^-) = \mu^+(u, v)$ and $\mu_H(u^-, v^+) = \mu^-(u, v)$. Because of (1) and (2), we then deduce that H satisfies the following conditions:

(3) $\{d_H(v^+), d_H(v^-)\} = \{\lfloor d_G(v)/2 \rfloor, \lceil d_G(v)/2 \rceil\}$, for every vertex $v \in V(G)$.

(4) $\{\mu_H(u^+, v^-), \mu_H(u^-, v^+\} = \{\lfloor \mu_G(u, v)/2 \rfloor, \lceil \mu_G(u, v)/2 \rceil \}$ for every vertex pair $(u, v) \in V(G)^2$.

Finally, we define a vertex function f' for H as well as an edge function g' of H as follows. If $d_H(v^+) \leq d_H(v^-)$ then $f'(v^+) = \lfloor f(v)/2 \rfloor$ and $f'(v^-) = \lceil f(v)/2 \rceil$ else $f'(v^+) = \lceil f(v)/2 \rceil$ and $f'(v^-) = \lfloor f(v)/2 \rfloor$. If $\mu_H(u^+, v^-) \leq \mu_H(u^-, v^+)$ then $g'(u^+, v^-) = \lfloor g(u, v)/2 \rfloor$ and $g'(u^-, v^+) = \lceil g(u, v)/2 \rceil$ else $g'(u^+, v^-) = \lceil g(u, v)/2 \rceil$ and $g'(u^-, v^+) = \lfloor g(u, v)/2 \rfloor$.

By (3) and (4), we obtain $\Delta_{f',g'}(H) = \max\{D, M\}$. Furthermore, every $f'g'$-coloring of H corresponds to an fg-coloring of G. Since H is bipartite, Theorem 8.14 implies that $\chi'_{f',g'}(H) = \Delta_{f',g'}(H)$, which gives $\chi'_{f,g}(G) \leq \max\{D, M\}$. ∎

If (G, f, g) is a weighted graph such that $f(v)$ is even for all $v \in V(G)$ and $g(u, v)$ is even for all $(u, v) \in V(G)^2$, then $\Delta_{f,g}(G) = \max\{D, M\}$, where D, M are the values defined in Theorem 8.19. Hence the theorem, combined with the inequalities (8.3) and (8.2), implies the following two results.

Corollary 8.20 (Nakano, Nishizeki, and Saito [232] 1990) *Every weighted graph* (G, f, g) *such that* $f(v)$ *is even for every vertex* $v \in V(G)$ *and* $g(u, v)$ *is even for every vertex pair* $(u, v) \in V(G)^2$ *satisfies* $\chi'_{f,g}(G) = \Delta_{f,g}(G)$

Corollary 8.21 (Hakimi and Kariv [127] 1986) *Every vertex-weighted graph* (G, f) *such that* $f(v)$ *is even for every vertex* $v \in V(G)$ *satisfies* $\chi'_f(G) = \Delta_f(G)$.

8.4 THE FAN EQUATION FOR THE CHROMATIC INDEX χ'_f

Kempe changes form an important tool for the ordinary edge color problem. As shown by Hakimi and Kariv [127], a similar recoloring technique can be applied to the f-color problem.

In the sequel, let (G, f) be a vertex-weighted graph, and let $e \in E_G(x, y)$ be an edge of G such that $\chi'_f(G - e) = k$ for some integer $k \geq \Delta_f(G)$. Furthermore, let $\varphi \in C_f^k(G - e)$ be an f-coloring of $G - e$ with color set $C = \{1, \ldots, k\}$.

For a color $\alpha \in C$ and a vertex $v \in V(G)$, the degree function of $G_{\varphi,\alpha}$ is denoted by $d_{\varphi,\alpha}(v)$, and we put $\bar{d}_{\varphi,\alpha}(v) = f(v) - d_{\varphi,\alpha}(v)$. Then define two color sets:

$$\varphi(v) = \{\alpha \mid d_{\varphi,\alpha}(v) = f(v)\} \text{ and } \overline{\varphi}(v) = \{1, \ldots, k\} \setminus \varphi(v).$$

We call $\varphi(v)$ the set of **colors present** at v and $\overline{\varphi}(v)$ the set of **colors missing** at v. Note that $\alpha \in \overline{\varphi}(v)$ if and only if $\bar{d}_{\varphi,\alpha}(v) \geq 1$. For a subset X of $V(G)$, define

$$\overline{\varphi}(X) = \bigcup_{v \in X} \overline{\varphi}(v).$$

If $X = \{v_1, \ldots v_p\}$, then we also write $\overline{\varphi}(v_1, \ldots, v_p)$ instead of $\overline{\varphi}(X)$. If $\alpha \in \overline{\varphi}(X)$, we say that α is **missing** at X with respect to φ. If $\sum_{v \in X} \bar{d}_{\varphi,\alpha}(v) \geq 2$, we say that α is **missing twice** at X with respect to φ. The set X is called **elementary** with respect to φ if no color is missing twice at X with respect to φ. The set X is called **closed** with respect to φ if, for every colored edge $e' \in \partial_G(X)$, the color $\varphi(e')$ is present at every vertex of X, i.e., $\varphi(e') \in \varphi(v)$ for every $v \in X$. Finally, the set X is called **strongly closed** with respect to φ if X is closed with respect to φ and $\varphi(e_1) \neq \varphi(e_2)$ for every two distinct colored edges $e_1, e_2 \in \partial_G(X)$.

Let $S = (v_0, e_1, v_1, \ldots, v_{p-1}, e_p, v_p)$ be a sequence such that $v_0, \ldots, v_p$ are vertices of G and $e_1, \ldots, e_p$ are edges of G. The integer $p \geq 0$ is then called the **length** of S. For the vertex $v_i \in V(S)$, we define $Sv_i = (v_0, e_1, \ldots, e_i, v_i)$ and $v_i S = (v_i, e_{i+1}, \ldots, v_p)$. We say that S is a **trail** in G if the edges are distinct and $e_i \in E_G(v_{i-1}, v_i)$ for $1 \leq i \leq p$. If $v_0 = u$ and $v_p = w$, then we say that S **connects** u to w or S **starts** in u and **terminates** in w. The vertices u and w are called the **endvertices** of the trail S, u being the **initial vertex** and w its **terminal vertex**; the vertices $v_1, \ldots, v_{p-1}$ are its **internal vertices**. A trail is **closed** if its initial and terminal vertices are identical; otherwise it is called **open**.

Let α, β be two distinct colors. A sequence $C = (v_0, e_1, v_1, \ldots, e_p, v_p)$ is called an (α, β)-**trail** with respect to $\varphi \in C_f^k(G - e)$ if $p \geq 1$ and the following conditions hold:

(1) C is a trail in G such that $\varphi(e_i) = \beta$ if $i \in \{1, \ldots, p\}$ is odd, and $\varphi(e_i) = \alpha$ if $i \in \{1, \ldots, p\}$ is even.

(2) $\alpha \in \overline{\varphi}(v_0)$ and, if C is a closed trail of odd length, α is missing twice at v_0 with respect to φ.

(3) If C has even length, then $\beta \in \overline{\varphi}(v_p)$, otherwise $\alpha \in \overline{\varphi}(v_p)$.

Now consider an arbitrary (α, β)-trail C with respect to φ. If we interchange the colors α and β on C, then we obtain a new coloring $\varphi' \in C_f^k(G - e)$. In what follows, we briefly say that the coloring φ' is obtained from φ by **recoloring** C and we write $\varphi' = \varphi/C$. Observe that C becomes an (β, α)-trail with respect to $\varphi' = \varphi/C$ such that $d_{\varphi',\gamma}(v) = d_{\varphi,\gamma}(v)$ whenever $v \in V(G) \setminus \{v_0, v_p\}$ and $\gamma \in \{1, \ldots, k\}$ or $v \in \{v_0, v_p\}$ and $\gamma \in \{1, \ldots, k\} \setminus \{\alpha, \beta\}$. For $v \in \{v_0, v_p\}$ and $\gamma \in \{\alpha, \beta\}$, the difference $t_{v,\gamma} = d_{\varphi',\gamma}(v) - d_{\varphi,\gamma}(v)$ is depending on whether the length of C is odd or even and whether C is open or closed. For instance, if C is open and has even length, then $t_{v_0,\alpha} = 1, t_{v_0,\beta} = -1, t_{v_p,\alpha} = -1$ and $t_{v_p,\beta} = 1$.

In the case when $f \equiv 1$, every (α, β)-trail is an (α, β)-chain, and every (α, β)-chain is either an open (α, β)-trail, or an open (β, α)-trail, or an even cycle whose edges are colored alternately with α and β.

For a vertex $u \in V(G)$ and two distinct colors α, β, let $R_u(\alpha, \beta, \varphi)$ denote the set of all vertices w such that there exists an (α, β)-trail with respect to φ with initial vertex u and terminal vertex w.

Lemma 8.22 *Let (G, f) be a vertex-weighted graph, let $\varphi \in C_f^k(G-e)$ be a coloring for an integer $k \geq \Delta_f(G)$, and let α, β be two distinct colors. Then the following statements hold:*

(a) *For each vertex u with $\alpha \in \overline{\varphi}(u)$ and for each edge $g \in E_G(u)$ with $\varphi(g) = \beta$, there is an (α, β)-trail C with respect to φ with initial vertex u and $g \in E(C)$. Moreover, if C is closed, then C has even length and $\beta \in \overline{\varphi}(u)$, or C has odd length and α is missing twice at u with respect to φ.*

(b) *Let C be an open (α, β)-trail with respect to φ of even length with initial vertex u and terminal vertex x, and let $u' \in V(G) \setminus \{x\}$ be a vertex. Suppose that $\bar{d}_{\varphi,\beta}(x) = 1$, $\beta \in \varphi(u) \cap \varphi(u')$, $\alpha \in \varphi(x)$, $\alpha \in \overline{\varphi}(u')$, and α is missing twice at u' with respect to φ if $u = u'$. Then there exists an (α, β)-trail C' with respect to φ with initial vertex u' and terminal vertex $x' \neq x$ such that C and C' are edge disjoint.*

(c) *Let x, y be two distinct vertices such that $\bar{d}_{\varphi,\beta}(x) = 1$, $\alpha \in \varphi(x)$, $\beta \in \varphi(y)$, $\alpha \in \overline{\varphi}(y)$, and $R_y(\alpha, \beta, \varphi) = \{x\}$. Then $\bar{d}_{\varphi,\alpha}(y) = 1$. Moreover, for any vertex $y' \in V(G) \setminus \{x, y\}$ such that $\beta \in \varphi(y')$ and $\alpha \in \overline{\varphi}(y')$ we have $R_{y'}(\alpha, \beta, \varphi) \neq \{x\}$.*

Proof: For an (α, β)-trail C with respect to φ, let $d_{C,\gamma}(v)$ denote the number of edges of C incident to v and colored with γ, where $v \in V(G)$ and $\gamma \in \{\alpha, \beta\}$.

For the proof of (a), assume that u is a vertex and $g \in E_G(u)$ is an edge satisfying $\alpha \in \overline{\varphi}(u)$ and $\varphi(g) = \beta$. Let $C = (v_0, e_1, v_1, \ldots, e_p, v_p)$ be a trail in G of maximum length such that $v_0 = u$, $e_1 = g$, $\varphi(e_i) = \beta$ if $i \in \{1, \ldots, p\}$ is odd, and $\varphi(e_i) = \alpha$ if $i \in \{1, \ldots, p\}$ is even. Clearly, $p \geq 1$ and every vertex $v \in V(C) \setminus \{v_0, v_p\}$ satisfies $d_{C,\alpha}(v) = d_{C,\beta}(v)$. We claim that C is an (α, β)-trail with respect to φ. To prove this, we distinguish two cases.

First, assume that $p \geq 1$ is odd. Then $\varphi(e_p) = \beta$ and, by the choice of C, we have $d_{\varphi,\alpha}(v_p) = d_{C,\alpha}(v_p)$. If $v_p = v_0 = u$, then $d_{C,\beta}(u) = d_{C,\alpha}(u) + 2$. This implies $d_{\varphi,\alpha}(u) = d_{C,\beta}(u) - 2 \leq f(u) - 2$, which gives $\bar{d}_{\varphi,\alpha}(u) = f(u) - d_{\varphi,\alpha}(u) \geq 2$ and we are done. If otherwise $v_p \neq v_0$, then $d_{C,\beta}(v_p) = d_{C,\alpha}(v_p) + 1$, which implies $\bar{d}_{\varphi,\alpha}(y_p) \geq 1$ and we are done, too.

Now, assume that $p \geq 1$ is even. Then $\varphi(e_p) = \alpha$ and, by the choice of C, we have $d_{C,\beta}(v_p) = d_{\varphi,\beta}(v_p)$. If $v_p = v_0 = u$, then $d_{C,\beta}(u) = d_{C,\alpha}(u)$. Since $\bar{d}_{\varphi,\alpha}(u) \geq 1$, this implies that $\bar{d}_{\varphi,\beta}(u) \geq 1$ and we are done. If otherwise $v_p \neq v_0$, then $d_{C,\alpha}(v_p) = d_{C,\beta}(v_p) + 1$. This implies that $\bar{d}_{\varphi,\beta}(v_p) \geq 1$ and we are done, too. This completes the proof of statement (a).

For the proof of (b), let C be an open (α, β)-trail of even length with initial vertex u and terminal vertex x, and let $u' \in V(G) \setminus \{x\}$ be a vertex. Suppose that $\bar{d}_{\varphi,\beta}(x) = 1$, $\beta \in \varphi(u) \cap \varphi(u')$, $\alpha \in \varphi(x)$, $\alpha \in \bar{\varphi}(u')$, and α is missing twice at u' with respect to φ if $u = u'$. Then we have

(1) $d_{C,\alpha}(v) = d_{C,\beta}(v)$ for every vertex $v \in V(G) \setminus \{x, u\}$, $d_{C,\alpha}(x) = d_{C,\beta}(x) + 1$, and $d_{C,\beta}(u) = d_{C,\alpha}(u) + 1$.

Furthermore, we conclude that there is an edge $g \in E_G(u') \setminus E(C)$ such that $\varphi(g) = \beta$. Now, let $C' = (v_0, e_1, v_1, \ldots, e_p, v_p)$ be a trail in G of maximum length such that $v_0 = u'$, $e_1 = g$, $e_i \notin E(C)$ for $i \in \{1, \ldots, p\}$, $\varphi(e_i) = \beta$ if $i \in \{1, \ldots, p\}$ is odd, and $\varphi(e_i) = \alpha$ if $i \in \{1, \ldots, p\}$ is even. Clearly, we have $p \geq 1$, $E(C) \cap E(C') = \emptyset$, and $d_{C',\alpha}(v) = d_{C',\beta}(v)$ for every vertex $v \in V(G) \setminus \{v_0, v_p\}$.

We claim that C' is an (α, β)-trail with respect to φ such that $v_p \neq x$. For the proof, we distinguish two cases.

Case 1: $p \geq 1$ *is odd.* Then $\varphi(e_p) = \beta$ and, by the choice of C', we have $d_{C,\alpha}(v_p) + d_{C',\alpha}(v_p) = d_{\varphi,\alpha}(v_p) = f(v_p) - \bar{d}_{\varphi,\alpha}(v_p)$.

First, assume that $\alpha \notin \bar{\varphi}(v_p)$, i.e., $\bar{d}_{\varphi,\alpha}(v_p) = 0$. Since $\alpha \in \bar{\varphi}(u) \cap \bar{\varphi}(u')$, this implies that $v_p \notin \{u, u'\}$. Then $v_p \neq v_0$ and, therefore, $d_{C',\beta}(v_p) = d_{C',\alpha}(v_p) + 1$. Moreover, we have $f(v_p) = d_{C,\alpha}(v_p) + d_{C',\alpha}(v_p)$. If $v_p = x$, then it follows from (1) that $f(v_p) = d_{C,\alpha}(v_p) + d_{C',\alpha}(v_p) = (d_{C,\beta}(v_p) + 1) + (d_{C',\beta}(v_p) - 1) = d_{C,\beta}(v_p) + d_{C',\beta}(v_p)$, contradicting $\bar{d}_{\varphi,\beta}(x) = 1$. If $v_p \neq x$, then $v_p \in V(G) \setminus \{x, u, u'\}$ and, by (1), we obtain that $f(v_p) = d_{C,\alpha}(v_p) + d_{C',\alpha}(v_p) = d_{C,\beta}(v_p) + (d_{C',\beta}(v_p) - 1) \leq d_{\varphi,\beta}(v_p) - 1 \leq f(v_p) - 1$, a contradiction, too.

Now, assume that $\alpha \in \bar{\varphi}(v_p)$, that is $\bar{d}_{\varphi,\alpha}(v_p) \geq 1$. Since $\alpha \in \varphi(x)$, this implies that $v_p \neq x$. If $v_p \neq v_0$, then C' is an (α, β)-trail and we are done. If $v_p = v_0 = u'$ and $u' = u$, then, by assumption, $\bar{d}_{\varphi,\alpha}(v_p) \geq 2$. Hence C' is an (α, β)-trail and we are done, too. Otherwise, we have $v_p = v_0 = u'$ and $u \neq u'$. Then $d_{C',\beta}(v_p) = d_{C',\alpha}(v_p) + 2$ and, since $v_p \notin \{u, x\}$, (1) implies that $d_{C,\beta}(v_p) = d_{C,\alpha}(v_p)$. Hence, we obtain $d_{\varphi,\alpha}(v_p) = d_{C,\alpha}(v_p) + d_{C',\alpha}(v_p) = d_{C,\beta}(v_p) + (d_{C',\beta}(v_p) - 2) \leq d_{\varphi,\beta}(v_p) - 2 \leq f(v_p) - 2$. Thus α is missing twice at v_p and, therefore, C' is an (α, β)-trail and we are done. Clearly, this settles Case 1.

Case 2: $p \geq 1$ *is even.* Then $\varphi(e_p) = \alpha$ and, by the choice of C', we have $d_{C,\beta}(v_p) + d_{C',\beta}(v_p) = d_{\varphi,\beta}(v_p) = f(v_p) - \bar{d}_{\varphi,\beta}(v_p)$.

First, we claim that $v_p \neq x$. If not, we would obtain $v_p = x \neq u' = v_0$ and, therefore, $d_{C',\alpha}(v_p) = d_{C',\beta}(v_p) + 1$. By (1), this would imply that $d_{\varphi,\beta}(x) = d_{\varphi,\beta}(v_p) = d_{C,\beta}(v_p) + d_{C',\beta}(v_p) = (d_{C,\alpha}(v_p) - 1) + (d_{C',\alpha}(v_p) - 1) \leq f(v_p) - 2$, which contradicts the hypothesis $\bar{d}_{\varphi,\beta}(x) = 1$. Hence, as claimed, $v_p \neq x$.

If $\beta \in \bar{\varphi}(v_p)$, then C' is an (α, β)-trail and we are done. So suppose that $\beta \notin \bar{\varphi}(v_p)$. Then $f(v_p) = d_{\varphi,\beta}(v_p) = d_{C,\beta}(v_p) + d_{C',\beta}(v_p)$. If $v_p \notin \{u, u'\}$ then $d_{C',\alpha}(v_p) = d_{C',\beta}(v_p) + 1$. By (1), this implies that

$$f(v_p) = d_{C,\beta}(v_p) + d_{C',\beta}(v_p) = d_{C,\alpha}(v_p) + d_{C',\alpha}(v_p) - 1 \leq f(v_p) - 1,$$

which is impossible. If $v_p = u$ and $u \neq u'$, then $d_{C',\alpha}(v_p) = d_{C',\beta}(v_p) + 1$. By (1), this implies that $f(v_p) = d_{C,\beta}(v_p) + d_{C',\beta}(v_p) = (d_{C,\alpha}(v_p) + 1) + (d_{C',\alpha}(v_p) - 1) = d_{C,\alpha}(v_p) + d_{C',\alpha}(v_p) \leq d_{\varphi,\alpha}(v_p) = d_{\varphi,\alpha}(u)$, contradicting $\alpha \in \overline{\varphi}(u)$. If $v_p = u = u'$, then $d_{C',\alpha}(v_p) = d_{C',\beta}(v_p)$. By (1), this implies that $f(v_p) = d_{C,\beta}(v_p) + d_{C',\beta}(v_p) = (d_{C,\alpha}(v_p) + 1) + d_{C',\alpha}(v_p) \leq d_{\varphi,\alpha}(v_p) + 1 = d_{\varphi,\alpha}(u') + 1$, which contradicts the hypothesis $d_{\varphi,\alpha}(u') \geq 2$. It remains to consider the case that $v_p = u'$ and $u \neq u'$. Then $d_{C',\beta}(v_p) = d_{C',\alpha}(v_p)$. Since $v_p \notin \{x, u\}$, (1) implies that $f(v_p) = d_{\varphi,\beta}(v_p) = d_{C,\beta}(v_p) + d_{C',\beta}(v_p) = d_{C,\alpha}(v_p) + d_{C',\alpha}(v_p) \leq d_{\varphi,\alpha}(v_p)$, which gives $f(u') = d_{\varphi,\alpha}(u')$, contradicting the hypothesis $\alpha \in \overline{\varphi}(u')$. This settles Case 2.

For the proof of (c), let x, y be two distinct vertices such that $\bar{d}_{\varphi,\beta}(x) = 1$, $\alpha \in \varphi(x)$, $\beta \in \varphi(y)$, $\alpha \in \overline{\varphi}(y)$, and $R_y(\alpha, \beta, \varphi) = \{x\}$. This implies, in particular, that there is an (α, β)-trail C with respect to φ with initial vertex y and terminal vertex x. Moreover, C is open and has even length. Then $\bar{d}_{\varphi,\alpha}(y) = 1$, since otherwise, by (b), there would be an (α, β)-trail C' with initial vertex y and terminal vertex $x' \neq x$, contradicting $R_y(\alpha, \beta, \varphi) = \{x\}$. Now, let $y' \in V(G) \setminus \{x, y\}$ be a vertex such that $\beta \in \varphi(y')$ and $\alpha \in \overline{\varphi}(y')$. Then (b) implies that there is an (α, β)-trail C' with initial vertex y and terminal vertex $x' \neq x$. Hence $R_{y'}(\alpha, \beta, \varphi) \neq \{x\}$. ∎

We continue this section with some basic facts about elementary sets that are useful in the further investigation. The proofs of these statements are straightforward and are left to the reader.

Proposition 8.23 *Let (G, f) be a vertex-weighted graph with maximum f-degree Δ_f, let $e \in E_G(x, y)$ be an edge of G, and let $\varphi \in C_f^k(G - e)$ be a coloring for an integer $k \geq \Delta_f$. Then the following statements hold:*

(a) *Let $v \in V(G)$ be a vertex such that no color is missing twice at v with respect to φ. Then*

$$|\overline{\varphi}(v)| = \begin{cases} kf(v) - d_G(v) + 1 & \text{if } e \in E_G(v), \\ kf(v) - d_G(v) & \text{if } e \notin E_G(v). \end{cases}$$

(b) *Let $v \in V(G)$ be a vertex such that no color is missing twice at v with respect to φ. If $k \geq \Delta_f + 1$, then*

$$|\overline{\varphi}(v)| \geq \begin{cases} f(v) + 1 & \text{if } e \in E_G(v), \\ f(v) & \text{if } e \notin E_G(v). \end{cases}$$

Furthermore, if there is an edge $e' \in E_G(v)$ such that $\varphi(e') \in \overline{\varphi}(v)$, then $f(v) \geq 2$ and there is a color $\gamma \in \overline{\varphi}(v)$ such that $\gamma \neq \varphi(e')$.

(c) *If $X \subseteq V(G)$ is an elementary set with respect to φ, then*

$$|X \cap \{x, y\}| + \sum_{v \in X} (kf(v) - d_G(v)) \leq |\overline{\varphi}(X)|.$$

Let (G, f) be a vertex-weighted graph, let $e \in E_G(x, y)$ be an edge, and let $\varphi \in C^k_f(G - e)$ be an f-coloring for some integer $k \geq 0$. A **multi-fan** at x with respect to e and φ is a sequence $F = (e_1, y_1, \ldots, e_p, y_p)$ with $p \geq 1$ consisting of edges $e_1, \ldots, e_p$ and vertices $y_1, \ldots, y_p$ satisfying the following two conditions:

(F1) The edges $e_1, \ldots, e_p$ are distinct, $e_1 = e$, and $e_i \in E_G(x, y_i)$ for $i = 1, \ldots, p$.

(F2) For every edge e_i with $2 \leq i \leq p$, there is a vertex y_j with $1 \leq j < i$ such that $\varphi(e_i) \in \overline{\varphi}(y_j)$.

Let $F = (e_1, y_1, \ldots, e_p, y_p)$ be a multi-fan at x. Since the vertices of F need not be distinct, the set $V(F) = \{y_1, \ldots, y_p\}$ may have cardinality smaller than p. For $z \in V(F)$, let $\mu_F(x, z) = |E_G(x, z) \cap \{e_1, \ldots, e_p\}|$. The next result from Scheide and Stiebitz [273] is a generalization of Theorem 2.1.

Theorem 8.24 *Let (G, f) be a weighted graph with $\chi'_f(G) = k + 1$ for an integer $k \geq \Delta_f(G)$, and let $e \in E_G(x, y)$ be a χ'_f-critical edge. Furthermore, let $F = (e_1, y_1, \ldots, e_p, y_p)$ be a multi-fan at x with respect to e and $\varphi \in C^k_f(G - e)$. Then the following statements hold:*

(a) $\{x, y\}$ *is elementary with respect to* φ.

(b) $\overline{\varphi}(x) \cap \overline{\varphi}(y_i) = \emptyset$ *for* $i = 1, \ldots, p$.

(c) *If* $\alpha \in \overline{\varphi}(y_i)$ *for* $1 \leq i \leq p$ *and* $\beta \in \overline{\varphi}(x)$, *then* $R_{y_i}(\alpha, \beta, \varphi) = \{x\}$.

(d) $V(F) \cup \{x\}$ *is elementary with respect to* φ.

(e) *If F is a maximal multi-fan at x with respect to e and φ, then*

$$\sum_{z \in V(F)} \big[f(x)\, d_G(z) + \mu_F(x, z) - k\, f(x)\, f(z) \big] = f(x) + 1.$$

Furthermore, $|V(F)| \geq 2$ provided that $f(x) \geq f(y)$ or $\mu_G(x, y) = 1$.

Proof: In order to prove (a), suppose that $\{x, y\}$ is not elementary with respect to φ. Then there is a color α such that $\alpha \in \overline{\varphi}(x) \cap \overline{\varphi}(y)$ or $\overline{\varphi}(x) \cap \overline{\varphi}(y) = \emptyset$ and α is missing twice at x or at y. In the former case, we can color e with α, contradicting $\chi'_f(G) = k + 1$. In the latter case, since $e \in E_G(x, y)$ is the uncolored edge, we may assume that α is missing twice at x. Since y is incident with e and $k \geq \Delta_f(G)$, Proposition 8.23(a) implies that there is a color $\beta \in \overline{\varphi}(y)$ By assumption, $\beta \neq \alpha$ and $\alpha \in \varphi(y)$. Then Lemma 8.22(a) implies that there is a (β, α)-trail C with respect to φ with initial vertex y. Then $\varphi' = \varphi/C \in C^k_f(G - e)$ and, since α is missing twice at x with respect to φ and β is missing at y with respect to φ, we have $\alpha \in \overline{\varphi}'(x) \cap \overline{\varphi}'(y)$. Then we can color e with α, contradicting $\chi'_f(G) = k + 1$. This completes the proof of (a).

In order to prove (b), assume that it is false and choose φ and F such that there is a color $\alpha \in \overline{\varphi}(x) \cap \overline{\varphi}(y_i)$ with i as small as possible. By (a), we have $i \geq 2$.

By (F2), for the color $\beta = \varphi(e_i)$, there is a vertex y_j with $1 \le j < i$ such that $\beta \in \overline{\varphi}(y_j)$. Recolor e_i with color α. This results in a new coloring $\varphi' \in C_f^k(G - e)$. Then $Fy_j = (e_1, y_1, \ldots, e_j, y_j)$ is a multi-fan at x with respect to e and φ' where $\beta \in \overline{\varphi}'(x) \cap \overline{\varphi}'(y_j)$, contradicting the minimality of i. This proves statement (b).

In order to prove (c), assume that there is a counterexample. Choose i smallest. By (a) and (b), β is not missing twice at x and $\beta \in \varphi(y_j)$ for $j = 1, \ldots, p$. Since $\alpha \in \overline{\varphi}(y_i)$ and $R_{y_i}(\alpha, \beta, \varphi) \neq \{x\}$, it then follows from Lemma 8.22(a) that there is an (α, β)-trail C with respect to φ with initial vertex y_i and terminal vertex $z \neq x$. Then $z \notin \{y_1, \ldots, y_{i-1}\}$, since otherwise we would have $\alpha \in \overline{\varphi}(z)$ (by (b)) and, therefore, there would be an (α, β)-trail with respect to φ with initial vertex $z \in \{y_1, \ldots, y_{i-1}\}$ and terminal vertex $y_i \neq x$, contradicting the minimality of i.

First assume that $E(C) \cap \{e_1, \ldots, e_i\} = \emptyset$. Then the coloring $\varphi' = \varphi/C \in C_f^k(G - e)$ satisfies $\varphi'(e_j) = \varphi(e_j)$ for $j = 2, \ldots, i$, $\overline{\varphi}'(y_j) = \overline{\varphi}(y_j)$ for $j = 1, \ldots, i-1$, $\overline{\varphi}'(x) = \overline{\varphi}(x)$ and $\beta \in \overline{\varphi}'(y_i)$. Consequently, $F' = (e_1, y_1, \ldots, e_i, y_i)$ is a multi-fan at x with respect to e and φ', where $\beta \in \overline{\varphi}'(x) \cap \overline{\varphi}'(y_i)$, contradicting statement (b).

Now assume that $E(C) \cap \{e_1, \ldots, e_i\} \neq \emptyset$. Then there is a smallest $j \le i$ such that $e_j \in E(C)$. Since $\beta \notin \overline{\varphi}(y_1, \ldots, y_i)$, it follows from (F2) that $\varphi(e_j) = \alpha$ and, therefore, $j \ge 2$. Then, by (F2), there is an $\ell \in \{1, \ldots, j-1\}$ such that $\alpha \in \overline{\varphi}(y_\ell)$. The trail C has the form $C = (v_0, f_1, v_1, \ldots, f_q, v_q)$ with $v_0 = y_i$ and $v_q = z$. Note that $\varphi(f_1) = \beta$ and hence $f_1 \notin \{e_1, \ldots, e_i\}$. Then $e_j = f_h$ and $x \in \{v_{h-1}, v_h\}$ with $2 \le h \le q$. If $x = v_{h-1}$, then $C_1 = v_{h-1}C$ is a (β, α)-trail with respect to φ with initial vertex x and terminal vertex z. Then $\varphi' = \varphi/C_1 \in C_f^k(G - e)$ and $F' = (e_1, y_1, \ldots, e_\ell, y_\ell)$ is a multi-fan at x with respect to e and φ' such that $\alpha \in \overline{\varphi}'(x) \cap \overline{\varphi}'(y_\ell)$, a contradiction to (b). If $x = v_h$, then $C_2 = Cv_h$ is a (α, β)-trail with respect to φ with initial vertex y_i and terminal vertex x. Then $\varphi'' = \varphi/C_2 \in C_f^k(G - e)$ and $F' = (e_1, y_1, \ldots, e_\ell, y_\ell)$ is a multi-fan at x with respect to e and φ'' such that $\alpha \in \overline{\varphi}''(x) \cap \overline{\varphi}''(y_\ell)$, a contradiction to (b). This completes the proof of (c).

In order to prove (d), assume that it is false. By (a) and (b), this implies that there is a color α that is missing twice at $V(F)$ with respect to φ. Since x is incident with the uncolored edge and $k \ge \Delta_f(G)$, Proposition 8.23(a) implies that there is a color $\beta \in \overline{\varphi}(x)$. If α is missing at a vertex $y \in V(F)$ then, by (c), we have $R_y(\alpha, \beta, \varphi) = \{x\}$. Then it follows from Lemma 8.22(c) that α is not missing twice at y. Consequently, we have $\alpha \in \overline{\varphi}(y_i) \cap \overline{\varphi}(y_j)$ with $y_i \neq y_j$. Then (c) implies that $R_{y_i}(\alpha, \beta, \varphi) = R_{y_i}(\alpha, \beta, \varphi) = \{x\}$, a contradiction to Lemma 8.22(c). This completes the proof of (d).

For the proof of (e), assume that F is maximal. Let $\Gamma = \overline{\varphi}(V(F))$ be the set of colors that are missing at $V(F)$, and let $E' = \{e' \in E_G(x) \setminus \{e\} \mid \varphi(e') \in \Gamma\}$. First, we claim that $E' = E(F) \setminus \{e\}$. By (F2), we have $E(F) \setminus \{e\} \subseteq E'$. Conversely, if $e' \in E'$ then, since we allow a multi-fan to have multiple edges, the maximality of F implies that $e' \in E(F)$. This proves the claim. By (d), we have $\Gamma \subseteq \varphi(x)$ and, therefore, $p - 1 = |E(F) \setminus \{e\}| = |E'| = f(x)|\Gamma|$. By (d), $V(F)$ is elementary with respect to φ. Since $V(F) \cap \{x, y\} = \{y\}$, Proposition 8.23(c) implies that

$$|\Gamma| = |\overline{\varphi}(V(F))| \geq \sum_{z \in V(F)} (kf(z) - d_G(z)) + 1.$$

Consequently, we obtain

$$p - 1 = f(x)|\Gamma| = f(x)\big(1 + \sum_{z \in V(F)} (kf(z) - d_G(z))\big).$$

Since $p = \sum_{z \in V(F)} \mu_F(x, z)$, this implies

$$\sum_{z \in V(F)} \big[f(x)\, d_G(z) + \mu_F(x, z) - k\, f(x)\, f(z)\big] = f(x) + 1.$$

Eventually, assume that $f(x) \geq f(y)$ or $\mu_G(x, y) = 1$. Since $k \geq \Delta_f(G)$ and the uncolored edge e belongs to $E_G(x, y)$, there is a color $\alpha \in \overline{\varphi}(y)$. By (a), α is not missing twice at $\{x, y\}$. Since $f(x) \geq f(y)$ or $\mu_G(x, y) = 1$, this implies that there is an edge $e' \in E_G(x, y')$ with $\varphi(e') = \alpha$ where $y' \neq y = y_1$. Since F is maximal, $y' \in V(F)$ and, therefore, $\{y', y_1\} \subseteq V(F)$ and $|V(F)| \geq 2$. This completes the proof of (e) and hence of the theorem. ∎

We shall refer to the equation in statement (e) of Theorem 8.24 as the **weighted fan equation**. Obviously, if $f \equiv 1$ the weighted fan equation reduces to the ordinary fan equation.

Theorem 8.25 Let (G, f) be a vertex-weighted graph with $\chi'_f(G) = k + 1$ for an integer $k \geq \Delta_f(G)$, and let $e \in E_G(x, y)$ be a χ'_f-critical edge such that $f(x) \geq f(y)$ or $\mu_G(x, y) = 1$. Then there is a vertex set Z such that $|Z| \geq 2$, $y \in Z \subseteq N_G(x)$, and

$$\sum_{z \in Z} \big[f(x)\, d_G(z) + \mu_G(x, z) - k\, f(x)\, f(z)\big] \geq f(x) + 1. \tag{8.6}$$

Furthermore, there is a vertex $z \in Z$ such that

$$\left\lceil \frac{d_G(z)}{f(z)} + \frac{\mu_G(x, z)}{f(x)\, f(z)} \right\rceil \geq k + 1 = \chi'_f(G) \tag{8.7}$$

Proof: Since $\mu_F(x, z) \leq \mu_G(x, z)$, (8.6) is an immediate consequence of the weighted fan equation. If $a_1, \ldots, a_m$ is a nonincreasing sequence of $m \geq 2$ integers with $\sum_{i=1}^m a_i \geq 2$, then, clearly, $a_1 \geq 1$ and $a_1 + a_2 \geq 2$. Since $f(x) \geq 1$, it then follows from (8.6) that there is a vertex $z \in Z$ such that $f(x)d_G(z) + \mu_G(x, z) - kf(x)f(z) \geq 1$. Clearly, this implies that the vertex z satisfies (8.7). ∎

The following result, due to Scheide and Stiebitz [273], which improves the bound obtained by Hakimi and Kariv (Corollary 8.17), is an immediate consequence of (8.7) and Proposition 8.13.

Corollary 8.26 *Every vertex-weighted graph* (G, f) *satisfies*

$$\chi'_f(G) \le \max_{u,v \in V(G)} \left\{ \left\lceil \frac{d_G(u)}{f(u)} + \frac{\mu_G(u, v)}{f(u)f(v)} \right\rceil \right\}.$$

For a simple vertex-weighted graph (G, f), it follows from the inequality (8.2) and Corollary 8.17 that

$$\Delta_f(G) \le \chi'_f(G) \le \Delta_f(G) + 1.$$

This gives a classification problem for simple weighted graphs. To investigate this classification problem, the following counterpart to Vizing's Adjacency Lemma might be useful.

Theorem 8.27 (Zhang and Liu [318]) *Let* (G, f) *be a vertex-weighted graph, where G is a simple graph and $\chi'_f(G) = k+1$ for an integer $k \ge \Delta_f(G)$. Let $e \in E_G(x, y)$ be a χ'_f-critical edge. Then x is adjacent to at least $f(x)(\Delta_f(G)f(y) - d_G(y) + 1)$ vertices $z \ne y$ such that $d_G(z) = f(z)\Delta_f(G)$.*

Proof: Since $\mu_G(x, y) = 1$, it follows from Theorem 8.25 that there is vertex set Z such that $|Z| \ge 2$, $y \in Z \subseteq N_G(x)$, and

$$\sum_{z \in Z} \left[f(x)\, d_G(z) + 1 - k\, f(x)\, f(z) \right] \ge f(x) + 1.$$

We refer to this inequality as the weighted fan inequality. For $z \in Z$, let $b(z) = f(x)d_G(z) + 1 - \Delta_f(G)f(x)f(z)$. Then every vertex $z \in Z$ satisfies $b(z) \le 1$ where equality holds if and only if $d_G(z) = f(z)\Delta_f(G)$. Let $Z' = Z \setminus \{y\}$ and $U = \{z \in Z' \mid b(z) = 1\}$. Using the weighted fan inequality, we now deduce that

$$
\begin{aligned}
|U| \ge \sum_{z \in Z'} b(z) &\ge f(x) + 1 - b(y) \\
&= f(x) + 1 - (f(x)(d_G(y) - \Delta_f(G)f(y)) + 1) \\
&= f(x)(\Delta_f(G)f(y) - d_G(y) + 1),
\end{aligned}
$$

which proves the theorem. ∎

Let (G, f) be an arbitrary vertex-weighted graph. For a subgraph H of G, define

$$f(H) = \sum_{v \in V(H)} f(v).$$

Apart from the maximum f-degree $\Delta_f(G)$ there is another trivial lower bound for the f-chromatic index. Let us consider an f-coloring $\varphi \in C_f^k(G)$ with $k = \chi'_f(G)$ and let $C = \{1, \dots, k\}$. For each color $\alpha \in C$, we have $d_{G_{\varphi,\alpha}}(v) \le f(v)$ for all $v \in V(G)$ and, therefore, $|E(G_{\varphi,\alpha})| \le \lfloor \frac{1}{2}f(G) \rfloor$. Since $|C| = k$, this gives

$|E(G)| \leq k \lfloor \frac{1}{2} f(G) \rfloor$ and, therefore, $\chi'_f(G) \geq \lceil |E(G)| / \lfloor \frac{1}{2} f(G) \rfloor \rceil$. Furthermore, we can replace G in the right-hand side of this inequality by any subgraph H with at least two vertices. This leads to the parameter w_f defined by

$$w_f(G) = \max_{H \subseteq G, |V(H)| \geq 2} \left\lceil \frac{|E(H)|}{\lfloor \frac{1}{2} f(H) \rfloor} \right\rceil$$

if $|V(G)| \geq 2$ and $w_f(G) = 0$ otherwise. We call $w_f(G)$ the f-**density** of G. Then, clearly, G satisfies

$$\chi'_f(G) \geq w_f(G). \tag{8.8}$$

Using the f-density one can see that the bound in Corollary 8.26 is best possible. Let (G, f) be the vertex-weighted graph with $G = t^3 K_t$ and $f(v) = t$ for each vertex $v \in V(G)$, where $t \geq 3$ is an odd integer. From Corollary 8.17 or Theorem 8.19 we obtain $\chi'_f(G) \leq t^3$, whereas Corollary 8.26 implies that $\chi'_f(G) \leq t^3 - t^2 + t$. For the f-density of G we obtain

$$w_f(G) \geq \left\lceil \frac{t^3 t (t-1)/2}{\lfloor t^2/2 \rfloor} \right\rceil = \left\lceil \frac{t^3 t (t-1)}{t^2 - 1} \right\rceil = \left\lceil \frac{t^4}{t+1} \right\rceil = t^3 - t^2 + t.$$

Consequently, $\chi'_f(G) = w_f(G) = t^3 - t^2 + t$ and the bound in Corollary 8.26 is sharp for (G, f).

Nakano, Nishizeki, and Saito [231] used critical chains as test objects in order to prove that every vertex-weighted graph (G, f) satisfies

$$\chi'_f(G) \leq \max\{\frac{9}{8} \Delta_f(G) + \frac{6}{8}, w_f(G)\}. \tag{8.9}$$

They also proposed the following conjecture in [231] concerning the f-chromatic index of a graph G. This conjecture may be viewed as a natural generalization of Goldberg's conjecture.

Conjecture 8.28 (Nakano, Nishizeki, and Saito [231] 1988) *Every vertex-weighted graph (G, f) satisfies*

$$\chi'_f(G) \leq \max\{\Delta_f(G) + 1, w_f(G)\}. \tag{8.10}$$

Since both Δ_f and w_f are lower bounds of χ'_f, Conjecture 8.28 implies that $\chi'_f(G) \in \{\Delta_f(G), \Delta_f(G) + 1, w_f(G)\}$ and the difficulty in determining $\chi'_f(G)$ is only to distinguish between the two cases $\chi'_f(G) = \Delta_f(G)$ and $\chi'_f(G) = \Delta_f(G) + 1$.

As shown in Scheide and Stiebitz [273], one can also use Kierstead paths and Tashkinov trees as test objects for the f-coloring problem. Based on these kinds of test objects, the following bound was established in [273].

Theorem 8.29 *Every vertex-weighted graph (G, f) with $\Delta_f(G) \geq 1$ satisfies*

$$\chi'_f(G) \leq \max\{\Delta_f(G) + \sqrt{(\Delta_f(G) - 1)/2}, w_f(G)\}. \tag{8.11}$$

We conclude this section with some basic facts about χ'_f-critical graphs satisfying $\chi'_f \geq \Delta_f + 1$.

Proposition 8.30 *Let* (G, f) *be a vertex-weighted graph. Suppose that* G *is* χ'_f-*critical and* $\chi'_f(G) = k + 1$ *for an integer and* $k \geq \Delta_f(G)$. *Then the following statements hold:*

(a) $|V(G)| \geq 3$.

(b) $f(v) \leq d_G(v) - 1$ *for every vertex* $v \in V(G)$.

(c) *If* $\chi'_f(G) = w_f(G)$, *then*

$$w_f(G) = \left\lceil \frac{|E(G)|}{\lfloor \frac{1}{2} f(G) \rfloor} \right\rceil.$$

Furthermore, $f(G) \geq 3$ *is odd and* $|E(G)| = k \left(\frac{1}{2}(f(G) - 1) \right) + 1$.

(d) G *is connected and bridgeless.*

Proof: Clearly, $|V(G)| \leq 2$ implies that $\chi'_f(G) = \Delta_f(G)$. This proves (a). For the proof of (b), suppose that there is a vertex v such that $f(v) \geq d_G(v)$. If $d_G(v) = 0$, then $\chi'_f(G - v) = \chi'_f(G)$, a contradiction. Hence, there is an edge $e \in E_G(v, w)$ for some vertex $w \in V(G)$. This implies, in particular, that $k \geq \Delta_f(G) \geq 1$. Furthermore, there is a coloring $\varphi \in C_f^k(G - e)$. Since $k \geq \Delta_f(G) \geq \lceil d_G(w)/f(w) \rceil$, we conclude that there is a color $\alpha \in \{1, \ldots, k\}$ such that $d_{\varphi, \alpha}(w) < f(w)$. Since $f(v) \geq d_G(v)$, we have $d_{\varphi, \alpha}(v) \leq d_G(v) - 1 < f(v)$. Hence, if we color e with α, the resulting coloring belongs to $C_f^k(G)$, contradicting the hypothesis $\chi'_f(G) = k + 1$. This proves (b).

For the proof of (c), assume that $\chi'_f(G) = w_f(G)$. Since G is χ'_f-critical, we get that $w_f(H) \leq \chi'_f(H) < \chi'_f(G) = w_f(G)$ for every proper subgraph H of G. Clearly, this implies that $w_f(G) = \lceil |E(G)| / \lfloor \frac{1}{2} f(G) \rfloor \rceil$. By (a), we have $f(G) \geq |V(G)| \geq 3$. If $f(G)$ is even, we obtain

$$\chi'_f(G) = w_f(G) = \left\lceil \frac{|E(G)|}{\lfloor \frac{1}{2} f(G) \rfloor} \right\rceil = \left\lceil \frac{\sum_{v \in V(G)} d_G(v)}{f(G)} \right\rceil$$

$$= \left\lceil \sum_{v \in V(G)} \left(\frac{d_G(v)}{f(v)} \cdot \frac{f(v)}{f(G)} \right) \right\rceil \leq \Delta_f(G),$$

a contradiction. This proves that $f(G) \geq 3$ is odd. To count the number of edges in G, choose an edge $e \in E(G)$. Since G is χ'_f-critical, there is a coloring $\varphi \in C_f^k(G - e)$. For any color $\alpha \in \{1, \ldots, k\}$, we then have $|E(G_\alpha)| \leq \lfloor \frac{1}{2} f(G) \rfloor$. Furthermore, equality holds for every color α, since otherwise $|E(G)| - 1 \leq k\lfloor\frac{1}{2}f(G)\rfloor - 1$ implying

$$\left\lceil \frac{|E(G)|}{\lfloor\frac{1}{2}f(G)\rfloor} \right\rceil \leq k < \chi'_f(G) = w_f(G),$$

a contradiction. Hence $|E(G_{\varphi,\alpha})| = \lfloor\frac{1}{2}f(G)\rfloor$ for every color α and, therefore, $|E(G)| = k\lfloor\frac{1}{2}f(G)\rfloor + 1 = k\left(\frac{1}{2}(f(G) - 1)\right) + 1$. This completes the proof of (c).

Finally, we prove (d). Since G is χ'_f-critical, G is certainly connected. So suppose, on the contrary, that G is not bridgeless. Then there is an edge $e \in E_G(x,y)$ such that $G - e$ is the vertex disjoint union of two subgraphs, say G_1 and G_2. Let $X = V(G_1)$ and $Y = V(G_2)$. Since G is connected, x and y belong to different graphs, say $x \in X$ and $y \in Y$. By (a), we have $d_G(z) \geq f(z) + 1 \geq 2$ for $z \in \{x,y\}$, which implies that $|X| \geq 2$ and $|Y| \geq 2$. Since G is χ'_f-critical, there is a coloring $\varphi_1 \in C^k_f(G[X \cup \{y\}])$ as well as a coloring $\varphi_2 \in C^k_f(G[Y \cup \{x\}])$. By permuting colors if necessary, we may choose the two colorings such that $\varphi_1(e) = \varphi_2(e)$. Then $\varphi = \varphi_1 \cup \varphi_2$ is obviously an f-coloring of G, i.e., $\varphi \in C^k_f(G)$, a contradiction to the hypothesis $\chi'_f(G) = k + 1$. This proves (d). ∎

8.5 DECOMPOSING GRAPHS INTO SIMPLE GRAPHS

Recall from Sect.4.3 that the one-factorization conjecture (Conjecture 4.16) asserts that every Δ-regular simple graph on $2n$ vertices and $\Delta \geq n$ is 1-factorable, i.e., $\chi'(G) = \Delta$. In 2001 Plantholt and Tipnis [249] proposed the following natural generalization to graphs having multiple edges.

Conjecture 8.31 (Multigraph One-factorization Conjecture) *For every positive integer s, every Δ-regular graph of order $2n$ satisfying $\mu(G) \leq s$ and $\Delta \geq sn$ is 1-factorable.*

In what follows, we shall discuss several results supporting the multigraph one-factorization conjecture.

Lemma 8.32 (Plantholt and Tipnis [247]) *Every connected eulerian graph G contains a spanning tree T such that $d_T(v) \leq \frac{1}{2}d_G(v) + 1$.*

Proof: Since G is connected and eulerian, there is an Euler tour in G, that is, a closed trail $C = (v_0, e_1, v_1, \ldots, v_{p-1}, e_p, v_p)$ containing each edge of G exactly once. We now define an edge set F recursively as follows. We begin with $F = \emptyset$. For $i = 1, \ldots, p$, we add e_i to F if and only if the resulting edge set contains no cycle of G. Since G is connected and $E(G) = E(C)$, the final set F is the edge set of an spanning tree T of G. Observer that for every $i \in \{0, \ldots, p - 1\}$, the graph with vertex set $\{v_0, \ldots, v_i\}$ and edge set $F \cap \{e_1, \ldots, e_i\}$ is a tree. Now, let v be an arbitrary vertex of G and let $I = \{i | v_i = v\}$. If $v \neq v_0$, then $i = \min(I)$ satisfies $i \geq 1$, $e_i \in F$, and $e_j \notin F$ for all $j \in I \setminus \{i\}$. This gives $d_T(v) \leq d_G(v)/2 + 1$. If $v = v_0$, then $e_1 \in F$ and $e_j \notin F$ for all $j \in I \setminus \{1\}$. This gives $d_T(v) \leq d_G(v)/2$. So we are done. ∎

A **decomposition** of a graph G is a collection $G_1, \ldots, G_p$ of edge disjoint spanning subgraphs of G (i.e., $V(G_i) = V(G)$ for $i = 1, \ldots, p$) such that each edge of G belongs to exactly one of the subgraphs. There is a one-to-one correspondence between the decomposition of G into p subgraphs and the improper edge colorings of G with color set $C = \{1, \ldots, p\}$.

Lemma 8.33 (Plantholt and Tipnis [247]) *Let G be a Δ-regular graph of order $2n$ such that $\mu(G) \leq 2$ and $\Delta = 2r$ for an integer $r \geq n + 1$. Then G has a decomposition into a Hamilton cycle of G and two simple $(r - 1)$-regular subgraphs of G.*

Proof: First, construct a graph G' by deleting all pairs of parallel edges from G. Then G' is an eulerian simple graph not necessarily connected. It follows from Lemma 8.5 that G contains a spanning forest T such that $d_T(v) \leq d_{G'}(v)/2 + 1$. Let H be the underlying simple of G, that is, the graph obtained from G by deleting only one edge from each pair of parallel edges. Clearly, T is a spanning subgraph of H. Eventually, let $H' = H - E(T)$. Then H' is a simple graph on $2n$ vertices and we claim that $\delta(H) \geq n$. To this end, let u be an arbitrary vertex of G. Let a denote the number of neighbors v of u such that $\mu_G(u, v) = 2$ and let b denote the number of neighbors v of u such that $\mu_G(u, v) = 1$. Then $d_G(u) = 2a + b = 2r$, $d_H(u) = a + b$ and $d_{G'}(u) = b$, which gives $d_T(u) \leq d_{G'}(u)/2 + 1 = b/2 + 1$, and hence $d_{H'}(u) = d_H(u) - d_T(u) \geq a + b - (b/2 + 1) = a + b/2 - 1 = r - 1 \geq n$. This shows that $\delta(H') \geq n$ as claimed. Dirac's criterion now tells us that H' contains a Hamilton cycle C. Clearly, C is a Hamilton cycle of G and, therefore, $G'' = G - E(C)$ is a $(2r - 2)$-regular graph. To complete the proof, we will now construct an improper edge coloring φ of G'' with two colors α, β such that $G_{\varphi, \gamma}$ is a simple $(s - 1)$ regular graph for $\gamma \in \{\alpha, \beta\}$. Again, first delete from G'' all pairs of parallel edges. Then color one edge of each such pair with α and the other edge with β, which results in a partial coloring φ'. Let G_1 denote the remaining spanning subgraph of G''. First observe that G_1 is a simple graph, and T is a spanning subgraph of G_1. Clearly, the Hamilton cycle C is edge disjoint from T and joins the components of T. In particular, G is connected. For every edge $e \in E(C)$ that joins two distinct components of T, there is an edge $e' \notin E(C)$ that is parallel to e' in G, which implies that e belongs to G_1. Then we deduce that G_1 is connected. Furthermore, we conclude that G_1 is an eulerian graph with an even number of edges. It follows now from Proposition 8.3(b) that there exists an improper edge coloring φ_1 of G_1 with two colors α, β such that $d_{\varphi_1, \alpha}(v) = d_{\varphi_1, \beta}(v)$ for all $v \in V(G_1)$. Then $\varphi = \varphi_1 \cup \varphi'$ is an improper edge coloring of G'' with two colors α, β such that $G_{\varphi, \gamma}$ is a simple $(r - 1)$ regular graph for $\gamma \in \{\alpha, \beta\}$. This proves the lemma.   ∎

Theorem 8.34 (Plantholt and Tipnis [247]) *Let G be a Δ-regular graph of order $2n$ such that $\mu(G) \leq s$ for an even integer $s \geq 2$ and $\Delta = rs$ for an integer $r \geq n + 1$. Then G has a decomposition into $s/2$ Hamilton cycles of G and s simple $(r - 1)$-regular subgraphs of G.*

Proof: Get a weighted graph (G, f, g) by letting $f(v) = 2r$ for all $v \in V(G)$ and $g(u, v) = 2$ for all $g(u, v) \in V(G)^2$. Then Corollary 8.20 tells us that $\chi'_{f,g}(G) = \Delta_{f,g}(G) = s/2$. So there is a fg-coloring $\varphi \in C^k_{f,g}(G)$ with $k = s/2$. This implies that, for each color $\alpha \in \{1, \ldots, k\}$, $G_{\varphi,\alpha}$ is $2r$-regular graph with $\mu(G_{\varphi,\alpha}) \leq 2$. It follows from Lemma 8.33 that $G_{\varphi,\alpha}$ has a decomposition into a Hamilton cycle of G and two simple $(r-1)$-regular graphs. This gives the desired decomposition of G. ∎

Theorem 8.35 (Plantholt and Tipnis [247]) *Suppose that there exist a constant $c \geq 1$ such that every simple Δ'-regular graph of order $2n$ and with $\Delta' \geq cn$ is 1-factorable. Furthermore, let G be a Δ-regular graph of order $2n$ such that $\mu(G) \leq s$ for an even integer $s \geq 2$ and $\Delta \geq s\lceil cn + 1 \rceil$. Then G is 1-factorable.*

Proof: There are integers r, p such that $\Delta = rs + p$ and $0 \leq p \leq s - 1$. By hypothesis, $\Delta = rs + p \geq s\lceil cn + 1 \rceil$ which implies $r \geq \lceil cn + 1 \rceil \geq cn + 1$. Since $c \geq 1$ and $\mu(G) \leq s$, we deduce that the underlying simple graph G_1 of G satisfies $\Delta(G') \geq r \geq n + 1$. Hence G_1 contains a Hamilton cycle and therefore a perfect matching. As long as $p \geq 1$, we can therefore delete perfect matchings from G until the resulting graph G' is rs-regular. It follows from Theorem 8.34 that G' has a decomposition into into $s/2$ Hamilton cycles of G and s simple $(r-1)$-regular subgraphs of G. Since $r - 1 \geq cn$, the hypothesis of the theorem implies that each of these subgraphs is 1-factorable. This implies that G is 1-factorable. ∎

Corollary 8.36 (Plantholt and Tipnis [247])

(a) *Let G be a Δ-regular graph of order $2n$ such that $\mu(G) \leq s$ for an even integer $s \geq 2$ and $\Delta \geq s\lceil (\sqrt{7} - 1)n + 1 \rceil$. Then G is 1-factorable.*

(b) *For every $\varepsilon > 0$, there is an integer $N(\varepsilon)$ such that every Δ-regular graph G of order $2n$ with $n > N(\varepsilon)$ is 1-factorable, provided that there is an even integer $s \geq 2$ such that $\mu(G) \leq s$ and $\Delta \geq s(1 + \varepsilon)n$.*

Proof: Statement (a) follows from Theorem 8.35 and Theorem 4.17. For the proof of (b), let $\varepsilon > 0$ be given. By Theorem 4.21, there is an integer $M(\varepsilon)$ such that every Δ-regular simple graph G satisfying $|V(G)| = 2n$, $n > M(\varepsilon)$, and $\Delta \geq (1 + \varepsilon/3)n$ is 1-factorable. Let $N(\varepsilon) = \max\{M(\varepsilon), \lceil 3/\varepsilon \rceil\}$. Now, let G be a Δ-regular graph of order $2n$ such that $n \geq N(\varepsilon)$, $\mu(G) \leq s$ for an even integer $s \geq 2$ and $\Delta \geq s(1+\varepsilon)n$. This gives $\Delta > s(1+\varepsilon/3)n + 2s$. So the underlying simple graph of G has a Hamilton cycle and we can delete perfect matchings from G until the resulting graph G' is Δ'-regular such that $\Delta' = rs$ and $r \geq (1+\varepsilon/3)n + 1$. It follows from Theorem 8.34 that G' has a decomposition into into a $s/2$ Hamilton cycles of G and s simple $(r-1)$-regular subgraphs of G, where each of these subgraphs has order $2n$ with $n > M(\varepsilon)$ and $r - 1 \geq (1 + \varepsilon/3)n$. It follows from Theorem 4.21 that each simple subgraph in the decomposition of G is 1-factorable, which implies that G is 1-factorable. ∎

Plantholt and Tipnis [249] extended their results to include the case when the multiplicity is bounded from above by an odd number.

Theorem 8.37 (Plantholt and Tipnis [249])

(a) *Let G be a Δ-regular graph of order $2n$ such that $\mu(G) \leq s$ for an odd integer $s \geq 1$ and $\Delta \geq s\lceil(\sqrt{7} - 1)n + 2\rceil + 1$. Then G is 1-factorable.*

(b) *For every $\varepsilon > 0$, there is an integer $N(\varepsilon)$ such that every Δ-regular graph G of order $2n$ with $n > N(\varepsilon)$ is 1-factorable, provided that there is an odd integer $s \geq 1$ such that $\mu(G) \leq s$ and $\Delta \geq s(1 + \varepsilon)n + n$.*

We conclude this section with a related result du to Vaughan [296]. The advantage of Vaughan's result is that its proof does not rely on Theorem 4.21. So it provides a self-contained proof for an asymptotic version of the multigraph one-factorization conjecture.

Theorem 8.38 (Vaughan [296] 2010) *For every integer $s \geq 1$ and every real $\varepsilon > 0$, there is an integer $N = N(s, \varepsilon)$ such that for all $n \geq N$ every Δ-regular graph G of order $2n$ with $\mu(G) \leq s$ and $\Delta \geq s(1 + \varepsilon)n$ is 1-factorable.*

The proof of the theorem given in Vaughan [296] relies on two lemmas. To formulate the first lemma, we need a further notation. For a graph G, a vertex $u \in V(G)$ and a vertex set $X \subseteq V(G)$, let $d_G(u : X)$ denote the degree of u in the subgraph of G induced by $X \cup \{u\}$.

Lemma 8.39 (Vaughan [296] 2010) *For every integer $s \geq 1$, there is an integer $M = M(s)$ such that for all $n \geq M$ every graph G of order $2n$ and with $\mu(G) \leq s$ has a partition (S, T) of its vertex set satisfying $|S| = |T| = n$ and*

$$|d_G(u : S) - d_G(u : T)| < n^{2/3} \tag{8.12}$$

for all $v \in V(G)$.

The proof of the lemma given in [296] uses a probabilistic argument. First a labeling of the $2n$ vertices is fixed, say $x_1, \ldots, x_n, y_1, \ldots, y_n$. Next, a random partition S, T of $V(G)$ is chosen as follows. For each $i \in \{1, \ldots, n\}$, we chosen one vertex of $V_i = \{x_i, y_i\}$ to be in S and the other one to be in T with equal probability. For a fixed vertex v, let X_i be the random variable defined by

$$X_i = \frac{d_G(v : S \cap V_i)) - d_G(v : T \cap V_i))}{s}.$$

Then the variables are mutually independent and we have $E(X_i) = 0$ and $|X_i| \leq 1$ $(i = 1, \ldots, n)$. Furthermore, $X = X_1 + \cdots + X_n$ satisfies

$$sX = d_G(v : S) - d_G(v : T).$$

Chernoff's bound now tells us that $Pr(|X| > a) < 2e^{-a^2/2n}$. We then obtain that

$$Pr(|d_G(v : S) - d_G(v : T)| > n^{2/3}) = Pr(|X| > s^{-1}n^{2/3}) < 2e^{-b},$$

where $b = s^{-2}n^{1/3}/2$. Let p denote the probability that there is a vertex v that does not satisfy (8.12). Since we have $2n$ vertices, $p < 4ne^{-b}$, which certainly tends to to 0 as n tends to infinity. So if n is large enough there exists a partition (S, T) of $V(G)$ such that $|S| = |T|$ and every vertex $v \in V(G)$ satisfies (8.12).

The next lemma provides a general condition for a regular graph of even order to be 1-factorable. The proof is based on a algorithm and uses Kempe changes as a main tool.

Lemma 8.40 (Vaughan [296] 2010) *Let G be a regular graph of order $2n$ and with $\mu(G) \leq s$, where $n^{5/6} > 3s > 0$. Suppose that there exists a partition (S, T) of $V(G)$ satisfying $|S| = |T|$, $\min\{d_G(v : S), d_G(v : T)\} > sn/2 + 14sn^{5/6}$, and*

$$\max\{\Delta(G[S]), \Delta(G[T])\} - \min\{\delta(G[S]), \delta(G[T])\} < n^{2/3}.$$

Then G is 1-factorable.

Proof of Theorem 8.38:   Choose n large enough so that $n^{5/6} > 3s$, $n \geq M(s)$ and $n^{1/6} > 29/\varepsilon$. Now, let G be a Δ-regular graph of order $2n$ with $\mu(G) \leq s$ and $\Delta \geq s(1 + \varepsilon)n$. It follows from Lemma 8.39 that there is a partition (S, T) of $V(G)$ such that $|S| = |T| = n$ and $|d_G(v : S) - d_G(v : T)| < n^{2/3}$ for all $v \in V(G)$. Since $d_G(v : S) + d_G(v : T) = \Delta$ for all $v \in V(G)$, this implies that

$$\frac{\Delta - n^{2/3}}{2} < d_G(v : U) < \frac{\Delta + n^{2/3}}{2}$$

for $U = S, T$ and for all $v \in V(G)$, which yields

$$\frac{\Delta - n^{2/3}}{2} < \delta(G[U]) \leq \Delta(G[U]) < \frac{\Delta + n^{2/3}}{2}$$

for $U = S, T$, and hence

$$\max\{\Delta(G[S]), \Delta(G[T])\} - \min\{\delta(G[S]), d(G[T])\} < n^{2/3}.$$

Since $n^{1/6} > 29/\varepsilon$, we obtain $\Delta \geq s(1 + \varepsilon)n > sn + 28sn^{5/6} + n^{2/3}$. This implies that, for $U = S, T$ and for all $v \in V(G)$, we have $d_G(v : U) > sn/2 + 14sn^{5/6}$. It now follows from Lemma 8.40 that G is 1-factorable.   ∎

8.6 NOTES

The results in Sect. 8.3 about the chromatic index of weighted and vertex-weighted graphs are mainly obtained by applying results about equitable edge colorings from Sect. 8.1 and by using a splitting operation. However, the proof of Theorem 8.16 given in Nakano, Nishizeki, and Saito [232], which is left out here, uses the Kempe chain method and resembles Vizing's fan argument. The trick is to choose the (α, β)-trails not in the graph $G' = G_{\varphi, \alpha, \beta}$ but in a reduced graph obtained from G' by deleting as many as possible pairs (e, f) of parallel edges with $\varphi(e) = \alpha$ and $\varphi(f) = \beta$.

Lemma 8.22 seems not to be new; similar results can be found in the paper by Hakimi and Kariv [127], in the papers by Nakano, Nishizeki, and Saito [231, 232], and perhaps in several other papers dealing with edge color problems for weighted or vertex-weighted graphs.

As observed by Zhou and Nishizeki [323] upper bounds for the chromatic index of a vertex-weighted graph (G, f) can be obtained by using the following splitting operation. For each vertex v of G, replace v with $f(v)$ copies and attach the edges that are incident to v in G equally to the copies of v. The resulting graph G_f, which might not be unique, is called an f-splitting of G. Since any ordinary edge-coloring of G_f induces an f-coloring of G, we have $\chi'_f(G) \leq \chi'(G_f)$. Graphs G satisfying $\chi'_f(G) < \chi'(G_f)$ can be found in [323]. Clearly, $\Delta_f(G) = \Delta(G_f)$ and Vizing's theorem (Theorem 2.2) implies that $\chi'_f(G) \leq \Delta_f(G) + \mu(G_f)$. For a simple graph G, this gives $\chi'_f(G) \leq \Delta_f(G) + 1$. We obviously have $\mu_G(G_f) \geq M(G, f)$ with $M(G, f) = \max_{(u,v)} \mu_G(u, v)/(f(u)f(v))$. So the best upper bound for $\chi'_f(G)$ that can be obtained by this splitting argument is $\Delta_f(G) + M(G, f)$. This bound is at most one more than the bound in Theorem 8.16 .

Zhou and Nishizeki [324] showed that the f-color problem can be reduced to the ordinary edge color problem for simple graphs. That is, given a vertex-weighted graph (G, f) and a positive integer k, one can construct, in polynomial time, a simple graph $G_{f,k}$ such that $\chi'_f(G) \leq k$ if and only if $\chi'(G_{f,k}) \leq k$. While any edge coloring algorithm for the simple graph $G_{f,k}$ with $\Delta(G_{f,k})$ colors yields an f-coloring of G with k colors, however, an edge coloring algorithm for $G_{f,k}$ with $\Delta(G_{f,k}) + 1$ colors does not imply any approximate f-coloring of G.

Conjecture 8.28 by Nakano, Nishizeki, and Saito [231] seems to be an appropriate extension of Goldberg's conjecture to vertex-weighted graphs. The upper bound in the conjecture is closely related to the fractional f-chromatic index (Appendix B), which also supports the conjecture. We do not know whether there exists a similar bound for the fg-chromatic index, or whether every weighted graph (G, f, g) satisfies $\chi'_{f,g}(G) \leq \max\{\Delta_{f,g}(G) + 1, w_f(G)\}$.

The f-color problem seems to be very similar to the ordinary edge color problem. A result concerning the chromatic index whose proof mainly relies on Kempe changes can quite often be transformed into a similar result about the f-chromatic index. Theorem 8.24 and Theorem 8.29 are two examples.

The proof of Lemma 8.32 about spanning trees in eulerian connected graphs follows an approach by El-Zanati, Plantholt, and Tipnis [79]. The proofs of Lemma 8.33, Theorems 8.34 and 8.35, and of Corollary 8.36 are due to Plantholt and Tipnis [247]. Improvements of these results are discussed in the paper by El-Zanati, Plantholt, and Tipnis [79] and in the paper by Plantholt and Tipnis [249].

CHAPTER 9

TWENTY PRETTY EDGE COLORING CONJECTURES

In this chapter we shall present a selection of unsolved edge coloring problems. To make the chapter self-contained, we shall recall the definitions of the necessary graph parameters. The reader is assumed to have a basic knowledge about graphs. A graph G may have multiple edges and $\mu(G)$ denotes the maximum multiplicity, that is, the maximum number of edges with the same two endvertices. The edge coloring problem is to color the edges of a graph G in such a way that each edge receives a color and adjacent edges (i.e., edges having a common endvertex) receive different colors. The minimum number of colors needed in such an edge coloring of G is called the chromatic index $\chi'(G)$. There are two trivial lower bounds for $\chi'(G)$ at hand: The maximum degree $\Delta(G)$, that is, the maximum number of edges sharing a common endvertex, and the density $w(G)$ defined by

$$w(G) = \max_{H \subseteq G, |V(H)| \geq 2} \left\lceil \frac{|E(H)|}{\lfloor \frac{1}{2}|V(H)| \rfloor} \right\rceil,$$

where $w(G) = 0$ if $|V(G)| \leq 1$. The density is closely related to the k-deficiency $\mathrm{def}_k(G) = \sum_{v \in V(G)}(k - d_G(v))$. As observed by Seymour [280] any graph G with $|V(G)| \geq 3$ satisfies $w(G) \leq k$ if and only if $\mathrm{def}_k(G[X]) \geq k$ for every set

Graph Edge Coloring: Vizing's Theorem and Goldberg's Conjecture,
First Edition. By M. Stiebitz, D. Scheide, B. Toft, and L. M. Favrholdt
Copyright © 2012 John Wiley & Sons, Inc.

$X \subseteq V(G)$ with $|X|$ odd and $|X| \geq 3$ (see (6.3)). By a result of Holyer [150] the determination of the chromatic index is an NP-hard optimization problem, even for 3-regular simple graphs, where the chromatic index is either 3 or 4.

♣   1.   GOLDBERG'S CONJECTURE

Every graph G satisfies $\chi'(G) \leq \max\{\Delta(G) + 1, w(G)\}$.   ∎

Comment: This conjecture was proposed independently by Goldberg [110] and by Seymour [281] in the 1970s. A detailed account of the history of the conjecture is given in Sect. 6.7. Goldberg's conjecture, discussed at length in this book, can be stated in different ways:

1. Every graph G satisfies $\chi'(G) \in \{\Delta(G), \Delta(G) + 1, w(G)\}$.

2. Every graph G satisfies $\chi'(G) \leq \max\{\frac{(2r+3)\Delta(G)+2r}{2r+2}, w_1(G), \ldots, w_r(G)\}$ for every $r \in \mathbb{N}$, where

$$w_r(G) = \max_{H \subseteq G, |V(H)|=2r+1} \left\lceil \frac{|E(H)|}{r} \right\rceil$$

   (Gupta [121]).

3. Let G be a critical graph (that is, $\chi'(H) < \chi'(G)$ for every proper subgraph H of G), let $\chi'(G) = k + 1$ for an integer $k \geq \Delta(G) + 1$, let $e \in E(G)$ be an edge, and let φ be an edge coloring of $G - e$ with k colors. Then no color is missing at two distinct vertices of G with respect to φ (Andersen [5]). A color is present at a vertex x (with respect to φ) if one of the edges incident with x is colored with this color, otherwise the color is missing at x.

4. Every critical graph G with $\chi'(G) \geq \Delta(G) + 2$ is of odd order and satisfies $2|E(G)| = (\chi'(G) - 1)(|V(G)| - 1) + 2$ (Conjecture 1.6).

Goldberg's conjecture is one of the most important conjectures in edge coloring theory. It generalizes Vizing's theorem in a strong way by showing that there are only three possible values for the chromatic index of an arbitrary graph. While it is not clear whether the density $w(G)$ can be computed in polynomial time, the parameter $\kappa(G) = \max\{\Delta(G), w(G)\}$ can be determined in polynomial time. This follows from Edmonds' matching polytope theory and the theory of antiblocking polyhedra, see the book by Schrijver [277]. So Goldberg's conjecture supports the following conjecture due to Hochbaum, Nishizeki, and Shmoys [146]:

- There is a polynomial time algorithm that colors the edges of any graph G using at most $\chi'(G) + 1$ colors.   ∎

Results related to Goldberg's conjecture are discussed in Chap. 5 and Chap. 6. In particular, Goldberg's conjecture is confirmed for graphs having order at most 15 and also for graphs having maximum degree at most 15 (Theorem 6.22). Gupta's version of Goldberg's conjecture is known to be true for $1 \leq r \leq 6$ (Corollary 6.7). Furthermore, it is known that every graph G satisfies

$$\chi'(G) \leq \max\{\Delta(G) + \sqrt{(\Delta(G) - 1)/2}, w(G)\}$$

(Corollary 5.20). Whether Goldberg's conjecture holds for all planar graphs is unknown.

♣   **2. SEYMOUR'S r-GRAPH CONJECTURE**

If G is an r-regular graph such that $|\partial_G(X)| \geq r$ for every set $X \subseteq V(G)$ with $|X|$ odd, then G satisfies $\chi'(G) \leq r + 1$.   ∎

Comment: This conjecture is due to Seymour [280]. Recall that $\partial_G(X)$ denotes the set of edges in G joining a vertex of X with a vertex of $\overline{X} = V(G) \setminus X$. An r-regular graph such that $|\partial_G(X)| \geq r$ for every set $X \subseteq V(G)$ with $|X|$ odd is said to be an r-graph. If an r-regular graph has an edge coloring with r colors, then it must be an r-graph. However, the converse statement is not true. The Petersen graph P is a 3-graph with $\chi'(P) = 4$. For an r-regular graph we have $|\partial_G(X)| = \mathrm{def}_r(G[X])$ for all $X \subseteq V(G)$. This implies that any r-graph G satisfies $w(G) \leq r$. Hence Seymour's r-graph conjecture is implied by Goldberg's conjecture. Seymour [280] proved that every graph G with $\kappa(G) = \max\{\Delta(G), w(G)\} \leq r$ is contained in an r-graph as a subgraph. Hence Seymour's r-graph conjecture is equivalent to:

- Every graph G satisfies $\chi'(G) \leq \max\{\Delta(G), w(G)\} + 1$.   ∎

So Seymour's r-graph conjecture may be considered as a direct relaxation of Goldberg's conjecture. From Theorem 6.22 it follows that the above conjecture is true for $r \leq 15$. As proved by Seymour $\kappa(G)$ is equal to the upper integer part of the fractional chromatic index $\chi'^*(G)$. Consequently, any r-graph G satisfies $\chi'^*(G) = r$. Since the fractional chromatic index can be expressed as the optimal value of a linear program, it follows that if G is a graph with $\chi'^*(G) = q$, then q is a rational number and there is an integer t such that there exists a family of tq matchings in G using each edge exactly t times. The last statement is equivalent to $\chi'(tG) = tq$, where tG denotes the graph obtained from G by replacing each edge in G by t edges joining the same two endvertices. Hence, for any graph G there is an integer t such that $\chi'(tG) = t\chi'^*(G) = \chi'^*(tG) = \kappa(tG)$ (see also Problem ♣ 6). Seymour's r-graph conjecture suggests that if G is an r-graph then, for all $t \geq 1$, either $\chi'(tG) = \kappa(tG)$ or $\chi'(tG) = \kappa(tG) + 1$.

♣   **3. JAKOBSEN'S CRITICAL GRAPH CONJECTURE**

Let G be a critical graph, and let

$$\chi'(G) > \frac{m}{m-1} \Delta(G) + \frac{m-3}{m-1}$$

for an odd integer $m \geq 3$. Then $|V(G)| \leq m - 2$. ∎

Comment: This conjecture was posed by Jakobsen [156] at the graph theory conference in Keszthely (Hungary) in 1973 and published in 1975 in the proceedings of the conference. Shannon proved that every graph G satisfies $\chi'(G) \leq 3\Delta(G)/2$. Hence the first interesting case of Jakobsen's conjecture is $m = 5$. For the history of the conjecture and the series of partial results obtained, the reader is referred to Sect. 1.4 and Sect. 6.1. The conjecture is known to be true for $m \leq 15$ (Corollary 6.8). It was proved by Andersen [5] that Jakobsen's conjecture is implied by Goldberg's conjecture (Sect. 6.1).

♣   **4. GUPTA'S CONJECTURE**

Let G be any graph with $\mu(G) = \mu$ such that $\Delta(G)$ cannot be expressed in the form $2p\mu - r$, where $r \geq 0$ and $p > \lfloor (r+1)/2 \rfloor$. Then $\chi'(G) \leq \Delta(G) + \mu(G) - 1$. ∎

Comment: This conjecture was formulated by Gupta in his Ph.D. thesis [120] from 1967. Clearly, the conjecture is related to Vizing's theorem that $\chi'(G) \leq \Delta(G) + \mu(G)$ for every graph G. It is well known that Gupta [120] gave an independent proof of Vizing's bound. On the other hand, it is not well known that Gupta's theorem has an addendum that the bound is best possible if $\mu(G) \geq 0$ and $\Delta(G) = 2p\mu(G) - r$, where $r \geq 0$ and $p > \lfloor (r+1)/2 \rfloor$. So Gupta's conjecture claims that these are the only cases where Vizing's bound can be achieved. Scheide [267] investigated the function

$$f(\Delta, \mu) = \max\{\chi'(G) \mid \Delta(G) = \Delta, \ \mu(G) = \mu\}.$$

He rediscovered Gupta's list of pairs (Δ, μ) for which $f(\Delta, \mu) = \Delta + \mu$ and proved that $f(\Delta, \mu) \leq \Delta + \mu - 1$ for all other feasible pairs (Δ, μ), provided that Goldberg's conjecture is true (Sect. 7.3). By Scheide's result, Gupta's conjecture about Vizing's bound is implied by Goldberg's conjecture. Because of this it would be interesting to establish Goldberg's conjecture for the class of graphs achieving Vizing's bound:

• Prove that every graph G with $\chi'(G) = \Delta(G) + \mu(G)$ and $\mu(G) \geq 2$ satisfies $w(G) = \Delta(G) + \mu(G)$. ∎

The conjecture that this is possible was posed by Haxell and McDonald [132]. In the same paper they also gave some partial results (Sect. 7.5). A graph G satisfying $\chi'(G) = \Delta(G) + \mu(G)$ is said to be an extreme graph. Holyer's result [150] implies that it is co-NP-complete to decide whether a simple graph is an extreme graph. On the other hand, Goldberg's conjecture implies that a graph G with multiple edges is an extreme graph if and only if $w(G) = \Delta(G) + \mu(G)$, and whether this equation holds can be verified in polynomial time using the result of Edmonds [76] (see the book by Schrijver [277]). There is another function related to f, namely the function g defined by

$$g(\Delta, \mu) = \max\{w(G) \mid \Delta(G) = \Delta, \ \mu(G) = \mu\}.$$

Clearly, we have $g(\Delta, \mu) \leq f(\Delta, \mu) \leq \Delta + \mu$ for all feasible pairs (Δ, μ), i.e., $\Delta \geq \mu \geq 1$ or $\Delta = \mu = 0$. Goldberg's conjecture implies that $g(\Delta, \mu) = f(\Delta, \mu)$, provided that $f(\Delta, \mu) \geq \Delta + 2$. In [272] it is shown that $g(\Delta, 1) = f(\Delta, 1) = \Delta + 1$ unless $\Delta \leq 1$. This supports the following conjecture:

- $g(\Delta, \mu) = f(\Delta, \mu)$ for every feasible pair (Δ, μ) with $\Delta \geq 2$. ∎

Shannon's bound that $\chi'(G) \leq \lfloor 3\Delta(G)/2 \rfloor$ for every graph G was already investigated by Vizing [298], who proved that equality holds if and only if G contains a graph S consisting of three vertices x, y, z such that $\mu_S(x, y) = \lfloor \frac{d}{2} \rfloor$, $\mu_S(y, z) = \lfloor \frac{d}{2} \rfloor$ and $\mu_S(x, z) = \lfloor \frac{d+1}{2} \rfloor$, where $d = \Delta(G)$ (note that $w(S) = w_1(S) = \lfloor 3d/2 \rfloor$). Goldberg [109] extended Shannon's bound and established a bound for the chromatic index χ' in terms of the maximum degree Δ and the odd girth g_o, namely

$$\chi'(G) \leq \Delta(G) + 1 + \left\lfloor \frac{\Delta(G) - 2}{g_o(G) - 1} \right\rfloor. \tag{9.1}$$

For an integer $\Delta \geq 2$ and an odd integer $g_o \geq 3$, let $\mathcal{GO}(\Delta, g_o)$ denote the set of all graphs G with maximum degree Δ and odd girth g_o attaining the bound in (9.1). A graph G is defined to be a ring graph if its underlying simple graph is a cycle. It is known that $\mathcal{GO}(\Delta, g_o)$ contains a ring graph of order g_o (Proposition 7.3). If $G \in \mathcal{GO}(\Delta, g_o)$ and $\Delta \geq g_o + 1$, then $\chi'(G) \geq \Delta(G) + 2$ and Goldberg's conjecture implies that $w(G) = \chi'(G)$. This supports the following conjecture:

- Let Δ, g_o be integers such that g_o is odd and $\Delta \geq g_o + 1 \geq 4$. Then every graph $G \in \mathcal{GO}(\Delta, g_o)$ contains as a subgraph a ring graph $R \in \mathcal{GO}(\Delta, g_o)$ having order g_o. ∎

A first step towards a proof of the conjecture would be to show that every critical graph in $\mathcal{GO}(\Delta, g_o)$ is a ring graph of order g_o, provided that $\Delta \geq g_o + 1 \geq 4$ and g_o is odd. McDonald [215] verified the conjecture when $\Delta - 2$ is divisible by $g_o - 1$ (Theorem 7.4). That the conjecture is true for $g_o = 3$ follows from Vizing's characterization of graphs achieving Shannon's bound (Theorem 7.1). A proof for the case $g_o = 5$ and $\Delta \geq 10$ is given at the end of this problem (Theorem 9.1). Let $g(G)$ denote the girth of G, that is, the smallest integer $\ell \geq 3$ such that $C_\ell \subseteq G$. Define

$$st(\Delta, \mu, g) = \max\{\chi'(G) \mid \Delta(G) = \Delta, \mu(G) = \mu, g(G) = g\}.$$

Steffen [287] proved (Theorem 5.7) that

$$st(\Delta, \mu, g) \leq \Delta + \left\lceil \frac{\mu}{\lfloor g/2 \rfloor} \right\rceil. \tag{9.2}$$

For an odd integer $g \geq 3$, the graph $G = \mu C_g$ has girth g, maximum degree $\Delta = 2\mu$, maximum multiplicity μ, and chromatic index $st(\Delta, \mu, g)$ (by Theorem 6.3). On the other hand, it is unknown whether equality in (9.2) can hold for an even number $g \geq 4$.

- For which triples (Δ, μ, g) is $st(\Delta, \mu, g) = \Delta + \left\lceil \frac{\mu}{\lfloor g/2 \rfloor} \right\rceil$? ∎

- Let g be an odd integer and let μ be an integer with $\mu \geq g \geq 3$. Is there a graph G with $g(G) = g$ and $\mu(G) = \mu$ achieving Steffen's bound but not containing a ring graph of order g with the same chromatic index as G? ∎

Theorem 9.1 *Let $\Delta \geq 10$ be an integer, and let G be a graph with $\Delta(G) = \Delta$ and $g_o(G) = 5$. Then $G \in \mathcal{GO}(\Delta, 5)$ if and only if G contains a ring graph $H \in \mathcal{GO}(\Delta, 5)$ as a subgraph.*

Proof: The backwards implication is evident. So assume that $G \in \mathcal{GO}(\Delta, 5)$. By definition of the class $\mathcal{GO}(\Delta, 5)$, we have $\Delta(G) = \Delta$, $g_o(G) = 5$, and

$$\chi'(G) = \Delta(G) + 1 + \left\lfloor \frac{\Delta - 2}{4} \right\rfloor. \tag{9.3}$$

Then there is a critical graph $H \subseteq G$ such that $\chi'(H) = \chi'(G)$ (Proposition 1.1). Since $H \subseteq G$, we have $\Delta(H) \leq \Delta(G) = \Delta$ and $g_o(H) \geq g_o(G) = 5$. From (9.1), respectively Theorem 5.8, and (9.3) it follows that

$$\begin{aligned}
\Delta + 1 + \left\lfloor \frac{\Delta - 2}{4} \right\rfloor &= \chi'(G) \\
&= \chi'(H) \quad \leq \quad \Delta(H) + 1 + \left\lfloor \frac{\Delta(H) - 2}{g_o(H) - 1} \right\rfloor \\
&\leq \quad \Delta + 1 + \left\lfloor \frac{\Delta - 2}{4} \right\rfloor,
\end{aligned}$$

which implies that $\Delta(H) = \Delta$ and

$$\chi'(H) = \Delta + 1 + \left\lfloor \frac{\Delta - 2}{4} \right\rfloor. \tag{9.4}$$

Next, we claim that H is elementary, i.e., $\chi'(H) = w(H)$. Suppose this is false, i.e., $\chi'(H) > w(H)$. Since $g_o(H) \geq 5$, we deduce from the upper bound of the chromatic index in (6.6) that

$$\chi'(H) \leq \Delta + 1 + \frac{\Delta - 3}{g_o(H) + 5} \leq \Delta + 1 + \left\lfloor \frac{\Delta - 3}{10} \right\rfloor.$$

Since $\Delta \geq 10$, this is a contradiction to (9.4). This proves the claim that H is elementary. Since H is critical and $\chi'(H) \geq \Delta(H) + 2$, this implies that $X = V(H)$

is elementary with respect to a coloring $\varphi \in C^k(H - e)$, where $e \in E(H)$ is an arbitrary edge and $k = \chi'(H) - 1$ (Theorem 1.4). By Proposition 1.7(a), this implies that

$$|X| \le \frac{k-2}{k-\Delta} = 1 + \frac{\Delta - 2}{k - \Delta}.$$

Furthermore, we obtain that $|X|$ is odd (Proposition 1.3). Eventually, we claim that $|X| \le g_o = 5$. Suppose this is false. Then $|X| \ge 7$ and we deduce from the above inequality that

$$\chi'(H) \le \Delta + 1 + \left\lfloor \frac{\Delta - 2}{6} \right\rfloor.$$

Since $\Delta \ge 10$, this give a contradiction to (9.4). Hence H is a critical graph with at most 5 vertices. Since $g_o(H) \ge 5$ and $\chi'(H) \ge \Delta(H) + 2$, this implies that $g_o(H) = 5$ and H is a ring graph belonging to $\mathcal{GO}(\Delta, 5)$. This completes the proof of the theorem. ∎

♣ 5. SEYMOUR'S EXACT CONJECTURE

Every planar graph G satisfies $\chi'(G) = \max\{\Delta(G), w(G)\}$. ∎

Comment: This conjecture is due to Seymour; it appears in [280, 281] as a question and in [282] as an explicit conjecture. An equivalent form of the conjecture is that every planar graph G satisfies $\chi'(G) = \lceil \chi'^*(G) \rceil$. The special case $\Delta(G) = 3$ of the conjecture is due to Herbert Grötzsch (see Seymour [282]). Grötzsch's conjecture is equivalent to the conjecture that a planar graph has a 3-edge-coloring if and only if it has a fractional 3-edge-coloring.

Seymour [283] proved the exact conjecture for series-parallel graphs (Theorem 6.31), and Marcotte [210] verified the conjecture for the class of graphs not containing $K_{3,3}$ or K_5^- as a minor (Theorem 6.24).

An interesting implication of Seymour's exact conjecture is that every planar r-graph admits an r-edge coloring and is therefore 1-factorable, since $w \le r$ (see Sect. 6.7). While the cases $r = 0, 1, 2$ are trivial, the case $r = 3$ states that every bridgeless cubic planar graph has chromatic index 3. By a result of Tait [290] this is equivalent to the Four-Color Theorem. The cases $r = 4$ and $r = 5$ were proved by Guenin [119]. As observed by Seymour [281], the case $r = 4$ implies the Four-Color Theorem by a result of Kotzig [179] (see Sect. 6.6). The next case $r = 6$ was solved by Dvořák, Kawarabayashi, and Král' [75] in 2011. The proofs for the values $r \ge 4$ are based on reductions to the previous case, i.e., the Four-Color Theorem is assumed.

Kilakos and Shepherd [170] proved that if Seymour's exact conjecture is true, then every r-graph without a T_2 minor has an r-edge-coloring, where T_2 is the simple graph obtained from the Petersen graph by deleting one vertex and then contracting an edge incident to a vertex of degree 2 (or from $K_{3,3}$ by subdividing two independent edges). This results supports the following conjecture:

- An r-graph without a Petersen graph minor satisfies $\chi'(G) = r$ ∎

The special case $r = 3$ was proposed by Tutte [295]. The generalization is due to László Lovász (see Schrijver [277] and Goddyn [108]; Schrijver refers to Lovász's paper [198] about the matching lattice and Goddyn [108] refers to a private communication by Lovász, Sebö, and Seress). Kilakos and Shepherd [170] have suggested the strengthening of Seymour's exact conjecture that $\chi'(G) = \max\{\Delta(G), w(G)\}$ for all graphs that do not contain the Petersen graph with one vertex deleted as a minor. Observe that the automorphism group of the Petersen graph P is edge transitive, so the Petersen graph minus a vertex, denoted P^*, as well as the Petersen graph minus an edge, denoted P^-, are well defined. The graph P^* is minor minimal with respect to property that $\chi' > \lceil\chi'^*\rceil$. Kilakos and Shepherd [171] proved a relaxation of Tutte's conjecture that $\chi'(G) = 3$ for every 3-graph not containing P^- as a minor. A solution of Tutte's conjecture was announced by Robertson, Sanders, Seymour, and Thomas (private communication, see also http://people.math.gatech.edu/ thomas/FC/generalize.html).

♣ 6. THE BERGE–FULKERSON CONJECTURE

Every bridgeless cubic graph G contains a family of six perfect matchings such that each edge of G is contained in precisely two of the matchings.   ∎

Comment: The conjecture is attributed to Berge [280], but it first appeared in print in Fulkerson's paper [105]. Recall that 3-regular graphs are also referred to as cubic graphs, and an edge of a graph whose deletion increases the number of its components is called a bridge. An equivalent form of the conjecture is that $\chi(2G) = 6$ for every bridgeless cubic graph. The conjecture was generalized by Seymour [281] to the generalized Fulkerson conjecture:

• Every r-graph G satisfies $\chi(2G) = 2r$.   ∎

Since every r-graph G has fractional chromatic index $\chi'^*(G) = r$, there are integers t such that $\chi'(tG) = tr$ (Theorem B.6). Seymour's conjecture suggests that $t = 2$ suffices.

The following seemingly weaker version of the Berge–Fulkerson conjecture was proposed by Berge (personal communication from F. Jaeger in 1994). It is easy to see that this Berge's conjecture is implied by the Berge–Fulkerson conjecture. That the two conjectures are in fact equivalent was proved by Mazzuoccolo [212] in 2011.

• Every bridgeless cubic graph G contains a family of five perfect matchings such that each edge of G is contained in at least one of the matchings.   ∎

On one hand, Frink's proof [104] of Petersen's theorem [241] tells us that every edge of a cubic bridgeless graph is contained in a perfect matching (see also König's book [175]). On the other hand, it is unknown whether Berge's conjecture holds, even with five replaced by any fixed number. For an integer $k \geq 1$, define

$$m_k(G) = \max \left\{ \frac{|\bigcup_{i=1}^k M_i|}{|E(G)|} \mid M_1, \ldots, M_k \text{ are perfect matchings of } G \right\}.$$

Evidently, $m_1(G) = \frac{1}{3}$ and Berge's conjecture suggests that $m_5(G) = 1$. For the Petersen graph P, we have

$$m_2(P) = \frac{3}{5}, m_3(P) = \frac{4}{5}, m_4(P) = \frac{14}{15}, m_5(P) = 1,$$

and Patel [239] made the following conjecture:

- Every bridgeless cubic graph G satisfies $m_k(G) \geq m_k(P)$ for $1 \leq k \leq 5$. ∎

Patel [239] proved that this conjecture is implied by the Berge–Fulkerson conjecture. Hence, by the result of Mazzuoccolo [212], the two conjectures are equivalent. Lower bounds for $m_k(G)$ for $2 \leq k \leq 5$ were established by Kaiser, Král', and Norine [162]. In particular, they verified the case $k = 2$ of the conjecture (Theorem B.9). The paper by Fouquet and Vanherpe [101] provides some partial results related to the perfect matching index of bridgeless cubic graphs, that is, the minimum number of perfect matchings that covers all edges.

For the Petersen graph P, only 12 edges can be covered with three perfect matchings, however 13 edges can be covered with three matchings. Rizzi [257] proved that if G is a simple graph with $\Delta(G) \leq 3$, then there are three matchings M_1, M_2 and M_3 such that $|M_1 \cup M_2 \cup M_3| \geq \frac{6}{7}|E(G)|$. The max edge t-coloring problem is to color as many edges of a graph G in a proper way using t colors; the complexity of this problem was studied by Feige, Ofek, and Wieder [92].

There is another attractive formulation of the Berge–Fulkerson conjecture in terms of even graphs, that is, graphs in which all vertices have even degree. That the following conjecture is equivalent to the Berge–Fulkerson conjecture was proved by Jaeger [152]:

- Every bridgeless graph G contains a family of six even subgraphs such that each edge of G is contained in exactly four of them. ∎

Hao et al. [131] proved that a cubic graph G satisfies $\chi'(2G) = 6$ if and only if there are two edge disjoint matchings M_1 and M_2 such that $\chi'(G - M_i) \leq 3$ for $i = 1, 2$ and $M_1 \cup M_2$ is the edge set of a cycle. They used this to verify the Berge–Fulkerson conjecture for certain families of snarks.

The relation between the Berge–Fulkerson conjecture and shortest cycle cover problems has been investigated by Fan and Raspaud [86]. They also drew attention to the following implication of the Berge–Fulkerson conjecture which has become known as the Fan–Raspaud conjecture:

- Every bridgeless cubic graph contains three perfect matchings with empty intersection. ∎

If a cubic graph has a family of six perfect matchings covering each edge twice, then the intersection of any three of them must be empty. On the other hand, even the existence of an integer $k \geq 3$ such that every bridgeless cubic graph contains k perfect matchings with empty intersection is not clear.

♣   **7. THE PETERSEN COLORING CONJECTURE**

Let G be a bridgeless cubic graph. Then it is possible to color the edges of G, using the edges of the Petersen graph as colors, in such a way that any three mutually adjacent edges of G are colored with three edges that are mutually adjacent in the Petersen graph. ■

Comment: The conjecture was posed by Jaeger [152] in a survey paper about nowhere-zero flow problems as a coloring conjecture in the above form and as a unifying flow conjecture which implies all "reasonable" symmetric flow conjectures, where the set B of flow values is a subset of an additive group satisfying $0 \in B$ and $B = -B$. As noted by Jaeger, the conjecture would imply the Berge–Fulkerson conjecture and the cycle double cover conjecture (CDCC), that every bridgeless graph has a family of cycles such that every edge of the graph is contained in precisely two of the cycles, proposed independently by Szekeres [288] and Seymour [279]. It is known that the CDCC is equivalent to its restriction to cubic graphs.

A cubic graph G which satisfies the condition of the above conjecture is said to have a Petersen coloring. A general framework for such colorings was proposed by D. Archdeacon. Consider a Steiner triple system (respectively a partial Steiner triple system) $\mathcal{S}$, that is, a finite set of 3-element sets such that each 2-element subset of $S = \bigcup \mathcal{S}$ is contained in exactly (respectively in at most) one set of $\mathcal{S}$. The elements of S are called points, the sets of $\mathcal{S}$ are called blocks, triples or lines, and the number of points is called the order of $\mathcal{S}$. By an $\mathcal{S}$-coloring of a cubic graph G we mean map $\varphi : E(G) \to \bigcup \mathcal{S}$ such that for every vertex $x \in V(G)$ the set $\{\varphi(e) \mid e \in E_G(x)\}$ is a block of $\mathcal{S}$. The smallest Steiner triple system $\mathcal{S}_3$ consists of 3 points and one block, and a cubic graph G admits an $\mathcal{S}_3$-coloring if and only if G admits a 3-edge-coloring. With the Petersen graph P we can associate the partial Steiner triple system $\mathcal{S}_{15}^{10} = \{E_P(v) \mid v \in V(P)\}$ with 15 points and 10 blocks. Thus a cubic graph G has a Petersen coloring if and only if G has an $\mathcal{S}_{15}$-coloring.

Holroyd and Škoviera [149] proved that every bridgeless cubic graph admits an $\mathcal{S}$-coloring for every Steiner triple system of order greater than 3. On the other hand, several interesting conjectures can be reformulated as $\mathcal{S}$-coloring conjectures choosing an appropriate partial Steiner triple system $\mathcal{S}$. Many such systems come from geometrical configurations. The smallest Steiner triple system of order greater than 3 is the famous Fano plane $\mathcal{S}_7$ having order 7 and consisting of seven lines. The partial Steiner triple system $\mathcal{S}_7^4 = \{\{a, b, c\}, \{a, d, e\}, \{a, f, g\}, \{c, e, g\}\}$ can be extended to the Fano plane. Investigating Fano colorings of cubic graphs led Máčajová and Škoviera [206] to propose the following four-line conjecture:

- Every bridgeless cubic graph admits an $\mathcal{S}_7^4$-coloring   ■

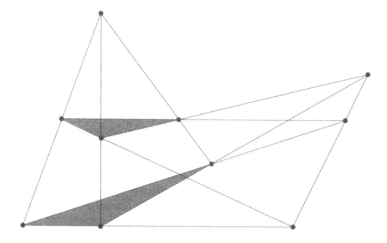

Figure 9.1   Desargues configuration.

Máčajová and Škoviera [206] proved that the four-line conjecture is equivalent to the Fan–Raspaud conjecture (see Problem ♣ 6). Another interesting partial Steiner triple system S_{10} with 10 points and 10 lines can be obtained from the Desargues configuration which arises in the famous Theorem of Desargues from projective geometry (Figure 9.1). As discovered by Král' et al. [182] the same configuration arises in graph theory in connection with the 5-cycle double cover conjecture (5-CDCC), a well known strengthening of the CDCC. The 5-CDCC asserts that each bridgeless graph admits a 5-cycle double cover, that is, a family of five even subgraphs (i.e., all degrees are even) such that each edge belongs to exactly two of them. The 5-CDCC implies the CDCC and it is known to be equivalent to its restriction to cubic graphs. Král' et al. [182] proved that a cubic graph G has a 5-cycle double cover if and only if G admits an S_{10}-coloring.

The partial Steiner triple system S_{15}^{10} associated with the Petersen graph can be extended to a partial Steiner triple system S_{15} with 15 points and 15 lines forming the so-called Cremona-Richmond configuration (see [182]). Král' et al. [182] proved that a cubic graph G satisfies the condition of the Berge–Fulkerson conjecture if and only if G admits an S_{15}-coloring. The paper [182] provides an excellent systematic approach to edge colorings of cubic graphs based on configurations with 3-element blocks.

Circular edge coloring form another relaxation of ordinary edge colorings. A (p, q)-edge-coloring of a graph G is a coloring of edges with colors from the set $\{1, \ldots, p\}$ such that any two adjacent edges receive colors a and b with $q \leq |a - b| \leq p - q$. The infimum of the ratio p/q such that G has a (p, q)-edge-coloring is the circular chromatic index of G denoted $\chi'^c(G)$. It is known that the infimum is in fact a minimum and that every graph G satisfies $\chi'(G) = \lceil \chi'^c(G) \rceil$, see the survey by Zhu [326]. While the fractional chromatic index of any bridgeless cubic graph is 3, the circular chromatic index of a bridgeless cubic graph G is 3 if and only if $\chi'(G) = 3$.

Zhu [326] asked whether there exists snarks with circular chromatic index close to 4. Afshani et al. [1] proved that every bridgeless cubic graph G satisfies $\chi'^c(G) \leq 11/3$ with equality holding for the Petersen graph. Kaiser et al. [163] showed that for every $\varepsilon > 0$, there is a g such that every bridgeless cubic graph G with girth at least g satisfies $\chi'^c(G) \leq 3 + \varepsilon$. As noted by Král' et al. [181] the Petersen graph is the only bridgeless cubic graph that is known to have circular chromatic index greater than 7/2 and they offered the following conjecture:

- Every bridgeless cubic graph G different from the Petersen graph satisfies $\chi'^c(G) < 11/3$, and even $\chi'^c(G) \leq 7/2$. ∎

The following result supporting the above conjecture was obtained by Král' et al. [181]. If G is a cubic graph with a 2-factor composed of cycles of lengths different from 3 and 5, then $\chi'^c(G) \leq 7/2$. It is known that a bridgeless cubic graph G has a 2-factor containing a cycle of length different from 5 if and only if G is not the Petersen graph (Theorem B.10).

♣ 8. VIZING'S PLANAR GRAPH CONJECTURE

Every simple planar graph G with maximum degree $\Delta(G) = 6$ satisfies $\chi'(G) = 6$. ∎

Comment: The conjecture suggests an affirmative answer to a question raised by Vizing [299]. It follows from Euler's formula that the conjecture is implied by Seymour's exact conjecture. For a simple graph G, Vizing's general bound $\Delta(G) + \mu(G)$ implies that $\Delta(G) \leq \chi'(G) \leq \Delta(G) + 1$. This leads to a natural way of classifying simple graphs into two classes. A graph G is of class one if G is simple and $\chi'(G) = \Delta(G)$, and of class two if G is simple and $\chi'(G) = \Delta(G) + 1$. In 1964–1965 Vizing published three papers about edge colorings. While the first paper [297] is dealing with (multi)graphs, the two following papers [298, 299] are mainly devoted to the classification problem for simple graphs. Vizing [298] showed that a simple planar graph is of class one provided $\Delta(G) \geq 10$. In [299] the bound was improved to $\Delta(G) \geq 8$. Then Vizing raised the question whether there is a planar graph of class two with $\Delta = 7$ or $\Delta = 6$. He also noticed that the answer is positive for $2 \leq \Delta \leq 5$. That all planar simple graphs with $\Delta = 7$ are of class one was proved, independently, by Grünewald [116], by Sanders and Zhao [264], and by Zhang [317] (see Theorem 4.59). So the remaining unsolved case is $\Delta = 6$.

The coloring number $\mathrm{col}(G)$ of a simple nonempty graph G is the least positive integer k such that G has a vertex order in which each vertex is preceded by fewer than k of its neighbors. Equivalently, $\mathrm{col}(G)$ is the maximum minimum degree of the subgraphs of G plus 1. Vizing [299] proved that a simple graph G is of class one if $\Delta(G) \geq 2\,\mathrm{col}(G) - 2$ (Theorem 4.3). Let $s(k)$ denote the least positive integer such that any simple graph G satisfying $\mathrm{col}(G) \leq k + 1$ and $\Delta(G) \geq s(k)$ is of class one. Vizing's result says that $s(k) \leq 2k$. Vizing noticed that $s(0) = 0, s(1) = 1, s(2) = 4$ (take the Petersen graph with one vertex deleted), and $s(3) = 6$. Furthermore, he posed the conjecture that for k large enough his estimate $2k$ may be improved.

- Determine the behavior of the function $s = s(k)$. ▪

Since all planar simple graphs have coloring number at most 6, it follows from Vizing's result that there is no planar graph of class two with $\Delta(G) \geq 10$. Furthermore, it follows that for every closed surface **S** (i.e., a compact, connected 2-dimensional manifold without boundary), the function

$$\Delta(\mathbf{S}) = \max\{\Delta(G) \mid G \text{ is a graph of class two that is embeddable in } \mathbf{S}\}$$

exists and is well-defined. Vizing's planar graph conjecture suggests that for the sphere $\mathbf{S}_0$, we have $\Delta(\mathbf{S}_0) = 5$. A possible generalization of this conjecture is the following:

- $\Delta(\mathbf{S}) \leq \left\lfloor 3 + \sqrt{13 - 6\varepsilon} \right\rfloor$ for every surface **S** with Euler characteristic $\varepsilon \leq 0$. ▪

Sanders and Zhao [266] proved that $\Delta(\mathbf{S}) = 6$ if $\varepsilon(\mathbf{S}) = 0$. Luo and Zhao [202] proved that $\Delta(\mathbf{S}) = 7$ if $\varepsilon(\mathbf{S}) = -1$ and $\Delta(\mathbf{S}) = 8$ if $\varepsilon(\mathbf{S}) = -2, -3$. If **S** is a surface distinct from the Klein bottle of Euler characteristic $\varepsilon \leq 1$, then $\Delta(\mathbf{S}) \geq H(\varepsilon) - 1$, where $H(\varepsilon) = \lfloor (7 + \sqrt{49 - 24\varepsilon})/2 \rfloor$ is the Heawood number (Proposition 4.60). For the suggested upper bound $J(\varepsilon) = \left\lfloor 3 + \sqrt{13 - 6\varepsilon} \right\rfloor$ we either have $J(\varepsilon) = H(\varepsilon)$ or $J(\varepsilon) = H(\varepsilon) - 1$.

♣ 9. VIZING'S 2-FACTOR CONJECTURE

Let G be a simple graph. If G is critical and $\chi'(G) = \Delta(G) + 1$, then G contains a 2-factor, that is, a 2-regular subgraph H with $V(H) = V(G)$. ▪

Comment: The conjecture is due to Vizing [298]. Recall that a graph G is critical if $\chi'(H) < \chi'(G)$ for every proper subgraph H of G. So the conjecture suggests that any critical graph of class two has a 2-factor.

The main motivation for Vizing to study critical graphs of class two is the classification problem for simple graphs. Vizing [299] proved that if G is a graph of class two with maximum degree Δ, then G contains a critical graph of class two with maximum degree Δ' as a subgraph for every Δ' satisfying $2 \leq \Delta' \leq \Delta$ (Proposition 4.2). In particular, every graph of class two contains a critical graph of class two with the same maximum degree as a subgraph.

Grünewald and Steffen [118] established Vizing's 2-factor conjecture for critical graphs of class two with many edges. In particular, they proved that a critical graph G of class two has a 2-factor if G is overfull, i.e., $|E(G)| > \Delta(G)\lfloor |V(G)|/2 \rfloor$.

♣ 10. VIZING'S INDEPENDENCE NUMBER CONJECTURE

Let G be a simple graph on n vertices. If G is critical and $\chi'(G) = \Delta(G) + 1$, then G cannot have more that $n/2$ pairwise nonadjacent vertices ▪

Comment: The conjecture is due to Vizing [298]. The conjecture suggests that every critical class two graph of order n has independence number at most $\frac{1}{2}n$. That this holds with $\frac{1}{2}$ replaced by $\frac{3}{5}$ was proved by Woodall [311] in 2011 (Theorem 4.44). Observe that Vizing's independence number conjecture would be a simple consequence of Vizing's 2-factor conjecture.

♣ 11. VIZING'S AVERAGE DEGREE CONJECTURE

Let $G = (V, E)$ be a simple graph with maximum degree Δ. If G is critical and $\chi'(G) = \Delta + 1$, then $|E| \geq \frac{1}{2}((\Delta - 1)|V| + 3)$. ∎

Comment: The conjecture is due to Vizing [300]. Among Vizing's conjectures dealing with structural problems of critical class two graphs, his average degree conjecture is without doubt the most popular one. For a detailed discussion of the many partial results related to this conjecture, the reader is referred to Sect. 4.5 and Sect. 4.10. The conjecture is known to be true for $2 \leq \Delta \leq 6$. Up to now, the best lower bound for the number of edges in a critical graph $G = (V, E)$ of class two with maximum degree Δ is due to Woodall [309], namely $|E| \geq \frac{1}{3}(\Delta + 1)|V|$ (Theorem 4.40). Woodall [310] proposed the following strengthening of Vizing's average degree conjecture:

- If $G = (V, E)$ is a critical graph of class two with maximum degree Δ, then

$$|E| \geq \begin{cases} \frac{1}{2}(\Delta - 1 + \frac{3}{\Delta+1})|V| & \text{if } \Delta \text{ is even,} \\ \\ \frac{1}{2}(\Delta - 1 + \frac{4}{\Delta+1})|V| & \text{if } \Delta \text{ is odd.} \end{cases}$$

∎

Woodall also noticed that it seems unlikely that the sharp lower bound for the number of edges in a critical graph of class two is a multiple of its order.

Since a graph G with n vertices, m edges, and maximum degree Δ has independence number at most $n - m/\Delta$, Vizing's average degree conjecture supports Vizing's independence number conjecture, at least asymptotically. The suggested upper bound for $\Delta(S)$ (see Problem ♣ 8) would also follow from Vizing's average degree conjecture (Proposition 4.60).

♣ 12. HILTON'S OVERFULL CONJECTURE

If G is a simple graph satisfying $\Delta(G) > \frac{1}{3}|V(G)|$ and $\chi'(G) = \Delta(G) + 1$, then $\chi'(G) = w(G)$. ∎

Comment: Not much progress has been made, since the conjecture was raised by A. J. W. Hilton at the graph theory conference at Sandbjerg, Denmark, June 1985. The conjecture appeared in print in Chetwynd and Hilton [51, 52].

A graph H is called overfull if $|E(H)| \geq \Delta(H)\lfloor|V(H)|/2\rfloor + 1$. Any simple overfull graph has odd order and is of class two. Furthermore, a simple graph G is a class two graph with $\chi'(G) = w(G)$ if and only if G contains an overfull subgraph H with $\Delta(H) = \Delta(G)$ (Proposition 4.13). Hence an attractive reformulation of the conjecture is the following:

- Let G be a simple graph with n vertices and $\Delta(G) > n/3$. Then G is a class two graph if and only if G contains an overfull subgraph H with $\Delta(H) = \Delta(G)$. ∎

The graph P^*, the Petersen graph with one vertex deleted, has order $n = 9$ and satisfies $w(P^*) = \Delta(P^*) = n/3 = 3$ and $\chi'(P^*) = \Delta(P^*) + 1$. So the threshold $n/3$ for the maximum degree cannot be lowered. It is known that overfull subgraphs are necessary for class two graphs when $\Delta \geq n - 3$. This seems, up to know, the best known threshold. Plantholt [246] verified the overfull conjecture for simple graphs of even order and high minimum degree. Many interesting consequences of Hilton's overfull conjecture are described by Hilton and Johnson [138] (Sect. 4.3).

A possible extension of the overfull conjecture to arbitrary graphs having multiple edges would be the following conjecture (the multigraph overfull conjecture):

- If G is a graph such that $\Delta(G) > \frac{1}{3}\mu(G)|V(G)|$ and $\chi'(G) \geq \Delta(G) + 1$, then $\chi'(G) = w(G)$. ∎

♣ **13. ONE-FACTORIZATION CONJECTURE**

If G is a simple Δ-regular graph on $2n$ vertices satisfying $\Delta \geq n$, then $\chi'(G) = \Delta$, i.e., G has a one-factorization. ∎

Comment: The conjecture appeared first in a paper by Chetwynd and Hilton [49] in 1985, but it "was going around" in the early 1950s, Hilton [136] was told by G. A. Dirac. Note that for an r-regular graph G, we have $\chi'(G) = r$ if and only if G has a one-factorization, that is, G is the edge-disjoint union of linear factors (i.e., 1-factors). The conjecture is best possible, at least when n is odd. To see this, take the disjoint union of two complete graphs on n vertices. Hilton [135] showed that the one-factorization conjecture is implied by the overfull conjecture (Lemma 4.15).

That the one-factorization conjecture holds if $\Delta \geq (\sqrt{7} - 1)n$ was proved by Chetwynd and Hilton [53] and, independently, by Niessen and Volkmann [236] (Theorem 4.17). Perković and Reed [240] verified the conjecture for $\Delta \geq (1 + \varepsilon)n$ and n large (Theorem 4.21).

Fifteen years before the one-factorization conjecture was published by Chetwynd and Hilton, Nash-Williams [233] proposed a much stronger conjecture, which is also known as the hamiltonian factorization conjecture:

- Let G be a Δ-regular graph on $2n$ vertices, where $\Delta \geq n$. Then G has a hamiltonian factorization, i.e., G is the edge-disjoint union of $\Delta/2$ Hamilton cycles if Δ is even or $(\Delta-1)/2$ Hamilton cycles and a linear factor if Δ is odd.   ∎

A natural generalization (the multigraph one-factorization conjecture) was posed by Plantholt and Tipnis [249] in 2001:

- If G is a Δ-regular graph on $2n$ vertices such that $\mu(G) \leq s$ and $\Delta \geq sn$, then G has a one-factorization.   ∎

The conjecture is implied by the multigraph overfull conjecture (Lemma 4.15). Plantholt and Tipnis [247] proved the following result. Suppose that $c \geq 1$ is a constant such that $\chi'(H) = \Delta(H)$ for every regular simple graph H satisfying that $|V(H)|$ is even and $\Delta(H) \geq c|V(H)|/2$. Then every Δ-regular graph on $2n$ vertices has a one-factorization, provided that $\mu(G) \leq s$ for an even integer s and $\Delta \geq s\lceil cn+1\rceil$. A similar result for odd s was obtained subsequently by Plantholt and Tipnis [249]. They also proved [249] an asymptotic version of the multigraph one-factorization conjecture using a corresponding theorem of Perkovi ć and Reed [240] for simple graphs. In 2010, Vaughan [296] found a self-contained elegant proof that for all positive integers s and all $\varepsilon > 0$, there is an integer N such that for all $n \geq N$ any Δ-regular graph on $2n$ vertices has a one-factorization if $\mu(G) \leq s$ and $\Delta \geq s(1+\varepsilon)n$.

♣   14.   THE LIST-COLORING CONJECTURE

Let G be a graph with $\chi'(G) = k$ and let each edge of G be assigned a list of k colors. Then it is possible to color the edges of G in such a way that each edge receives a color from its list and adjacent edges receive different colors.   ∎

Comment: The conjecture, suggested independently by various researchers, including V. G. Vizing, M. O. Albertson, K. Collins, A. Tucker, R. P. Gupta, B. Bollob ás, and A. J. Harris, is the most popular problem in edge coloring theory. The conjecture appeared first in print in a paper by Bollob ás and Harris [29].

The study of list coloring problems for graphs was initiated in the 1970s by Vizing [301] and, independently, by Erd ős, Rubin, and Taylor [84]. The list-chromatic index $\chi'^{\ell}(G)$ of a graph G is the least number k such that whenever we assign a list of k colors to each edge of G, there is a coloring of the edges of G such that each edge receives a color from its list and adjacent edges receive different colors. By definition, every graph G satisfies $\chi'(G) \leq \chi'^{\ell}(G)$ and the list-coloring conjecture suggests that equality holds. In contrast to that, the gap between the chromatic number and the list chromatic number, where we color the vertices of a graph such that adjacent vertices receive different colors, can be arbitrarily large, even for bipartite graphs, as observed by Vizing [301] and by Erd ős, Rubin, and Taylor [84].

One of the most spectacular results about edge colorings from the 1990s is Galvin's [107] verification of the list-coloring conjecture for the class of bipartite graphs, that is, the proof that $\chi'(G) = \chi'^{\ell}(G)$ for every bipartite graph G. This particular case of the list-coloring conjecture is due to J. Dinitz (see Chetwynd and Häggkvist [46]). In 1997, Borodin, Kostochka, and Woodall [30] obtained an attractive generalization of Galvin's landmark theorem, allowing lists of different lengths (an edge $e \in E_G(x, y)$ has a list of length $\max\{d_G(x), d_G(y)\}$). Based on this they proved that

$$\chi'^{\ell}(G) \leq \left\lfloor \frac{3}{2}\Delta(G) \right\rfloor$$

for all graphs G. Kahn [160] verified the list-coloring conjecture for the class of simple graphs, at least asymptotically. He proved that every simple graph G satisfies $\chi'^{\ell}(G) = \Delta(G) + o(\Delta(G))$. The following conjecture posed by Bojan Mohar (see his internet page) deals with list-colorings of critical class two graphs, that is, simple graphs G such that $\chi'(G) = \Delta(G) + 1$ and $\chi'(H) < \chi'(G)$ whenever H is a proper subgraphs of G.

- Suppose that G is a critical graph of class two and for each edge of G there is a list of $\Delta(G)$ colors. Then the edges of G can be colored such that each edge receives a color from its list and adjacent edges receive different colors, unless all lists are equal to each other. ∎

♣ 15. BEHZAD'S TOTAL COLORING CONJECTURE

Let G be a simple graph with maximum degree Δ. Then it is possible to color the edges and vertices of G with $\Delta + 2$ colors in such a way that no two adjacent or incidents elements in $V(G) \cup E(G)$ receive the same color. ∎

Comment: There has been some controversy about the priority to this conjecture, as described by Soifer [285]. The conjecture appeared in Behzad's Ph.D. thesis [18] at Michigan State University in 1965. Behzad also conjectured that the chromatic index of a simple graph is at most $\Delta(G) + 1$, being unaware that Vizing [297, 298] had obtained and published this result, even in a multigraph version, in 1964. In connection with his research in 1964 Vizing also, independently of Behzad, proposed the total graph coloring conjecture, and he shared it with the participants in the graph theory seminars, led by A. A. Zykov, at the Academy of Sciences in Akademgorodok, outside Novosibirsk. Vizing used the term simultaneous coloring of vertices and edges for this new type of coloring. This is well documented by Soifer [285]. Jensen and Toft [157] had a similar documentation (obtained from participants in the seminar). Therefore Jensen and Toft correctly gave joint credit for the conjecture to Behzad and Vizing, but without giving the details of their documentation. They referred to Vizing's two papers to indicate his role; however, the two papers of Vizing do not contain the conjecture, nor any mention of simultaneous colorings.

At the international symposium of graph theory in Rome in June 1966 both M. Behzad and A. A. Zykov were among the more than 150 participants. In the proceedings from the meeting Behzad and Chartrand [19] stated the total graph coloring conjecture and they mentioned Vizing's theorem for simple graphs [297]. In the same proceedings also Zykov [328] mentioned Vizing's theorem, but in the multigraph version. Zykov did not mention simultaneous colorings in the proceedings of the Rome symposium, but he did so for simple graphs at the meeting in Manebach in Germany in May 1967 [329]. For the proceedings of that meeting, Zykov also prepared a 160-page-long list of all graph theory publications from the period 1963–1967, including Behzad's thesis [18] as entry number B48 and [19] as B49 (probably Zykov had these entries from the proceedings of the meeting in Rome).

Finally, in 1968 Vizing published a paper [300], containing a wealth of problems, including the total graph coloring conjecture for graphs, that $\Delta(G) + \mu(G) + 1$ colors suffice in a simultaneous coloring of the vertices and edges of a graph G of maximum degree $\Delta(G)$ and maximum multiplicity $\mu(G)$. A natural stronger conjecture is that $\chi'(G) + 1$ colors suffice for all graphs. Vizing [300] uses the term chromatic index to denote the least number of colors needed in a simultaneous coloring of vertices and edges! This number is now commonly called the total chromatic number, denoted χ'', and a simultaneous coloring of the vertices and edges of a graph is called a total coloring, following Behzad.

A further strengthening when $\chi' \geq \Delta + 3$ was proposed by Goldberg [114] in 1984:

- A graph G with $\chi'(G) \geq \Delta(G) + 3$ satisfies $\chi''(G) = \chi'(G)$. ■

Observe that Goldberg's conjecture (Problem ♣ 1) says that any graph G with $\chi'(G) \geq \Delta(G) + 2$ satisfies $\chi'(G) = w(G)$. Goldberg [114] noticed that any critical graph G with $\chi'(G) = w(G)$ satisfies $\chi''(G) = \chi'(G)$. First, we can take a $(k-1)$-edge-coloring of $G - e$, where e is an edge of G and $k = \chi'(G)$. Then it is known (Theorem 1.4) that none of the $k-1$ colors is missing at two distinct vertices of G. So we can color each vertex v of G, using a color that is missing at v, and we can color e with color k, which obviously results in a total coloring of G with k colors.

One may also ask, if every graph satisfies $\chi''(G) \leq \max\{\Delta(G) + 2, w(G)\}$, as an analogue of Goldberg's conjecture (Problem ♣ 1).

It is folklore that any graph G satisfies $\chi''(G) \leq \chi'^{\ell}(G) + 2$. In 1997, Borodin, Woodall, and Kostochka [30] used list coloring arguments to show that $\chi''(G) \leq 3\Delta/2 + 1$ for graphs G with maximum degree $\Delta \geq 2$. One year later, Molloy and Reed [227] used probabilistic arguments to show that $\chi''(G) \leq \Delta + 10^{26}$ for simple graphs G with maximum degree Δ.

Clearly, there is a list variant for the total chromatic number, and Borodin, Kostochka, and Woodall [30] proposed the conjecture that the total list chromatic number equals the total chromatic number.

♣ 16. FIAMČIK'S ACYCLIC EDGE COLORING CONJECTURE

Let G be a simple graph with maximum degree Δ. Then the edges of G can be colored with $\Delta + 2$ colors such that any two adjacent edges receive different colors and any cycle contains edges of at least three distinct colors.   ∎

Comment: The conjecture is due to Fiamčik [95]. An edge coloring of a graph may be viewed as an partition of the edge set into matchings called color classes. An acyclic edge coloring of a graph G is an edge coloring such that the union of any two color classes contains no cycle with at least three vertices. The acyclic chromatic index of G, denoted $a\chi'(G)$, is the least number of colors in an acyclic edge coloring of G. Fiamčik's conjecture is that $a\chi'(G) \leq \Delta(G) + 2$ for all simple graphs G. We suggest that this can be generalized to graphs having multiple edges.

- Prove that $a\chi'(G) \leq \chi'(G) + 1$ for all graphs G.   ∎

Alon et al. [2] proved that $a\chi'(G) \leq 64\Delta(G)$ for all simple graphs. This was improved to $a\chi'(G) \leq 16\Delta(G)$ by Molloy and Reed [228] using a similar probabilistic argument. In a subsequent paper Alon et al. [3] proved that almost all Δ-regular simple graphs satisfy $a\chi'(G) \leq \Delta + 2$. They also noticed that all known simple graphs G with $a\chi'(G) > \Delta(G) + 1$ are subgraphs of K_{2n} that have at least $2n^2 - 2n + 2$ edges. Furthermore, they pointed out that the problem of determining $a\chi'(K_{2n+1})$ is closely related to the perfect one-factorization conjecture due to Kotzig [180]:

- For any $n \geq 2$, the complete graph K_{2n} can be decomposed into $2n - 1$ linear factors such that the union of any two of these linear factors forms a Hamilton cycle of K_{2n}.   ∎

A perfect one-factorization of K_{2n+2} can be used to show that $a\chi'(K_{2n+1}) = \Delta(K_{2n+1}) + 1$. The above conjecture remains unsolved in general, although Kotzig [180] himself established it when n is a prime. Furthermore, Anderson [7, 8] verified the perfect one-factorization conjecture when $2n - 1$ is an odd prime, see the book by Wallis [305] about one-factorizations.

Cohen et al. [67] proved that $a\chi'(G) \leq \Delta(G) + 25$ for all planar simple graphs G, and they proposed the following conjecture:

- There exists Δ_0 such that $a\chi'(G) = \Delta(G)$ for all planar simple graph with $\Delta(G) \geq \Delta_0$.   ∎

The conjecture is an analogue to Vizing's planar graph conjecture. The bound $\Delta + 25$ has been improved to $\Delta + 12$ by Basavaraju and Chandran [15]. Obviously, there is a list coloring variant for the acyclic edge coloring and the main question is whether the acyclic list chromatic index can be larger than the acyclic chromatic index.

♣   **17. THE STRONG EDGE COLORING CONJECTURE**

Let G be a graph with maximum degree Δ. Then the edges of G can be colored with $\lfloor 5\Delta^2/4 \rfloor$ colors in such a way that two adjacent vertices in G are not incident to two distinct edges of the same color.   ∎

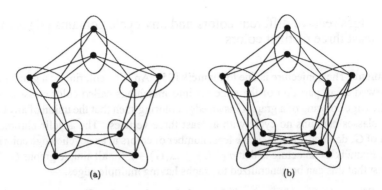

Figure 9.2   Graphs with (a) $\Delta = 4$, $s\chi' = 20$ and (b) $\Delta = 5$, $s\chi' = 25$.

Comment: The conjecture was proposed by P. Erdős and J. Nešetřil during a discussion at a seminar in Prague in 1985 (see the papers by Faudree et al. [88] and by Chung et al. [64]). Erdős [81] formulated the conjecture only for the class of simple graphs, but Faudree et al. [89] extended the conjecture to graphs having multiple edges.

A strong edge coloring of a graph G is an edge coloring in which every color class is an induced matching, that is, any two vertices belonging to distinct edges with the same color are not adjacent. The strong chromatic index $s\chi'(G)$ is the minimum number of colors in a strong edge coloring of G. As in Vizing's theorem, the strong edge coloring conjecture suggests an upper bound for the strong chromatic index in terms of the maximum degree. Let $si = si(\Delta)$ be the function defined by

$$si(\Delta) = \begin{cases} 5\Delta^2/4 & \text{if } \Delta \text{ is even,} \\ 5\Delta^2/4 - \Delta/2 + 1/4 & \text{if } \Delta \text{ is odd.} \end{cases}$$

The following construction of a graph G_Δ ($\Delta \geq 2$) satisfying $\Delta(G_\Delta) = \Delta$ and $s\chi'(G_\Delta) = si(\Delta)$ appears in the paper by Faudree et al. [89], but it was without doubt known to P. Erdős and J. Nešetřil. For Δ even, G_Δ is obtained by replacing each vertex of a cycle C_5 on 5 vertices by an independent set of cardinality $\Delta/2$, and joining two vertices by an edge if the corresponding vertices of the cycle are adjacent. For Δ odd, G_Δ is obtained in a similar way, where two adjacent vertices of C_5 are replaced by independent sets of cardinality $(\Delta + 1)/2$ and the three other vertices of C_5 are replaced by independent sets of cardinality $(\Delta - 1)/2$ (Fig. 9.2). Evidently, G_Δ has maximum degree Δ and contains no induced matching of size 2 and, therefore, $s\chi'(G_\Delta) = |E(G_\Delta)| = si(\Delta)$. This motivated Faudree et al. [89] to propose the following strengthening of the conjecture by P. Erdős and J. Nešetřil:

- Every graph G satisfies $s\chi'(G) \leq si(\Delta(G))$. ∎

Chung et al. [64] proved the following result supporting the strong edge coloring conjecture. Let $\Delta \geq 2$ and let G be a connected graph with maximum degree at most Δ such that G is $2K_2$-free, that is, G contains no induced matching of size 2. Then $|E(G)| \leq si(\Delta)$, where equality holds if and only if $G = G_\Delta$.

It is not hard to show that $s\chi'(G) \leq 2\Delta^2(G) - 2\Delta(G) + 1$ for all graphs G. This follows by using a greedy coloring strategy or by using the facts that $s\chi'(G)$ equals the chromatic number of $L(G)^2$ and $\Delta(L(G)^2) \leq 2\Delta(G)(\Delta(G) - 1)$, where $L(G)$ denotes the line graph of G and H^2 denotes the square[3] of H. Consequently, the strong edge coloring conjecture is true for $\Delta \leq 2$. That every graph with maximum degree at most 3 satisfies $s\chi'(G) \leq si(3) = 10$ was proved by Andersen [6] and, independently, by Horák, Qing, and Trotter [151]. Molloy and Reed [226] used probabilistic arguments to show that any simple graph G satisfies $s\chi'(G) \leq 1.998\Delta^2(G)$, provided that $\Delta(G)$ is sufficiently large.

For a graph G, let $am(G)$ denote the maximum number of edges in G such that each pair of them has a common endvertex or is joined by an edge in G. Equivalently, $am(G)$ is the largest order of a complete graph contained in $L(G)^2$ as a subgraph. Evidently, $s\chi'(G) \geq am(G)$ for all graphs G. Hence the strong edge coloring conjecture by Faudree et al. implies the conjecture that $am(G) \leq si(\Delta(G))$ for all graphs G. But even this seemingly weaker conjecture due to Faudree et al. [89] is still open. Faudree et al. [89] proved that every bipartite graph G satisfies $am(G) \leq \Delta(G)^2$, and they posed the following conjecture:

- Every bipartite graph G satisfies $s\chi'(G) \leq \Delta(G)^2$. ∎

The complete bipartite graph $K_{d,d}$ shows that the suggested bound in the conjecture is best possible because $s\chi'(K_{d,d}) = \Delta(K_{d,d})^2 = d^2$.

For strong edge coloring there also exists a list coloring variant, and the main question is whether the strong list chromatic index can be larger than the strong chromatic index.

♣ 18. THE INTERVAL EDGE COLORING CONJECTURE

A coloring of the edges of a graph with colors 1, 2, 3, ... is called an interval coloring if the colors received by the edges incident to a vertex are distinct and form an interval of integers for all vertices of the graph. For G bipartite with bipartition (X, Y), let all vertices in X have the same degree d_X and all vertices in Y have the same degree d_Y. Does G allow an interval coloring? ∎

Comment: Interval colorings were first looked at by Asratyan and Kamalian [12, 13] in 1987. Independently of Asratyan and Kamalian interval colorings also arose as a mathematical problem from a practical scheduling problem at Saint Canute High School in Odense, Denmark, around 1990 (private communication from Jesper Bang-Jensen).

Without the condition on the degrees the answer to the posed question is no. The first example showing this was due to Sevast'yanov [278], and it has maximum degree

---

[3]The square H^2 is the simple graph with vertex set $V(H)$, where two vertices are joined by an edge in H^2 if and only if they are joined by an edge or a path of length 2 in H.

$\Delta(G) = 21$. An example with $\Delta(G) = 12$ is inspired by a similar example by P. Erdős (private communication). Let the bipartite graph G with bipartition (X, Y) represent the affine plane with 9 points and 12 lines, i.e. the set X is the set of points and Y the set of lines and $(x, y) \in E(G)$ with $x \in X$ and $y \in Y$ if and only if the point x is on the line y. Add an extra vertex x' to X and join it to all 12 vertices of Y. The obtained graph has $\Delta(G) = 12$ and does not have an interval coloring. For $d_X = 2$ and $d_Y \geq 2$ the answer to the posed problem is yes, as proved by Hanson, Loten, and Toft [129]. But even for the case $d_X = 3$ and $d_Y = 4$ the problem is open. For general bipartite graphs with $\Delta(G) = 12$ the answer is no, as indicated above, and for $\Delta(G) = 3$ it is yes, as observed by Hansen [130]. The problem is open for general bipartite graphs G for all values of $\Delta(G)$ where $4 \leq \Delta(G) \leq 11$.

♣   19.   VIZING'S INTERCHANGE CONJECTURE

An edge coloring of any graph G with $\chi'(G)$ colors can always be obtained from an arbitrary edge coloring of G by a sequence of Kempe changes and suppressing empty color classes.   ∎

Comment: This conjecture suggests an affirmative answer to a question raised by Vizing in his second paper [298] from 1965 about edge coloring of graphs and in his survey paper [300] about unsolved problems in graph theory from 1968.

Let G be an arbitrary graph. Let φ be any k-edge-coloring of G, i.e., an edge coloring of G with color set $C = \{1, \ldots, k\}$ such that each edge receives a color from C and adjacent edges receive different colors. To obtain a new edge coloring choose two distinct colors $\alpha, \beta \in C$ and let H denote the subgraph of G with $V(H) = V(G)$ and $E(H) = \{e \in E(G) \mid \varphi(e) \in \{\alpha, \beta\}\}$. Obviously, every component of H is a path or an even cycle whose edges are colored alternately by α or β. If we interchange the colors α and β on a component of H, we obtain a new k-edge-coloring φ' of G. This operation is usually called a Kempe change. Two k-edge-colorings φ and φ' of G are said to be Kempe equivalent, written $\varphi \sim_k \varphi'$, if $\varphi' = \varphi$ or φ' can be obtained from φ by a sequence of Kempe changes, possibly involving more than one pair of colors in successive Kempe changes. Evidently, "$\sim_k$" is an equivalence relation on the set of all k-edge-colorings of G and we denote by $Kc(G : k)$ the number of equivalence classes.

Note that an optimal edge coloring of G, that is, an $\chi'(G)$-edge-coloring is a k-edge-coloring for all integers $k \geq \chi'(G)$. Vizing [298, 300] asked whether for every k-edge-coloring φ of G there is an optimal edge coloring φ' such that $\varphi \sim_k \varphi'$. On one hand, Vizing [297, 298] proved that for every k-edge-coloring of G with $k \geq \Delta(G) + \mu(G)$, there is an $(\Delta(G) + \mu(G))$-edge-coloring φ' of G such that $\varphi \sim_k \varphi'$. On the other hand, he found a simple graph G with $\chi'(G) = \Delta(G) = 4$ such that $Kc(G : k) \geq 2$.

Mohar [225] proved that every simple graph G satisfies $Kc(G : k) = 1$ if $k \geq \chi'(G) + 2$. For simple planar graphs G, he improved this result to $Kc(G : k) = 1$ if $k \geq \chi'(G) + 1$. McDonald, Mohar, and Scheide [219] proved that if G is a graph with maximum degree $\Delta \leq 4$, then $Kc(G : \Delta+1) = 1$ if $\Delta \leq 3$ and $Kc(G : 6) = 1$ if $\Delta = 4$. They also pointed out that $Kc(K_5, 5) = 6$ and $Kc(K_{2p-1} : 2p-1) > 1$ if p is a prime number. This follows from the fact that K_{2p} has a perfect one-factorization (see Problem ♣ 16).

♣   20.   GUPTA'S DENSITY CONJECTURE

Let G be a graph and let

$$k = \min\left\{\delta(G) - 1, \min_{X \subseteq V(G), |X| \text{ odd}} \left\lfloor \frac{2|E_G(X)|}{|X| + 1} \right\rfloor\right\},$$

where $\delta(G)$ is the minimum degree of G and $E_G(X)$ is the set of edges of G with at least one endvertex in X. Then it is possible to color the edges of G with k colors (where incident edges may be colored by the same color) such that each vertex v of G is incident with edges of all k colors.   ■

Comment: This conjecture is due to Gupta [121] and may be viewed as a counterpart to Goldberg's conjecture. In his thesis, Gupta [120] investigated in addition to the chromatic index also the cover index for graphs. An edge set $F \subseteq E(G)$ is called an edge cover of G if every vertex of G is incident with at least one edge in F. The cover index of a graph G, denoted $\psi(G)$, is the largest number k such that $E(G)$ contains k disjoint edge covers. There are two immediate upper bounds for the cover index of a graph G: The minimum degree $\delta(G)$ and the co-density $w^c(G)$ defined by

$$w^c(G) = \min_{X \subseteq V(G), |X| \text{ odd}} \left\lfloor \frac{2|E_G(X)|}{|X| + 1} \right\rfloor.$$

Note that any edge cover of G must contain at least $(|X| + 1)/2$ edges from the edge set $E_G(X)$ if $|X|$ is odd. So $\psi(G) \leq \min\{\delta(G), w^c(G)\}$ for all graphs G, and Gupta's conjecture is that $\psi(G) \geq \min\{\delta(G) - 1, w^c(G)\}$. This is obviously equivalent to the statement that $\psi(G) \in \{\delta(G), \delta(G) - 1, w^c(G)\}$ for all graphs G. Gupta [120] proved that $\psi(G) \geq \delta(G) - \mu(G)$ for all graphs G, and that the bound is best possible if $\mu(G) \geq 1$ and $\delta(G) = 2p\mu(G) - r$, where $r \geq 0$ and $p > \lfloor (r - 1)/2 \rfloor + \mu(G)$. This led Gupta [120] to propose the following conjecture in his thesis:

- Let G be any graph with $\mu(G) = \mu$ such that $\delta(G)$ cannot be expressed in the form $2p\mu - r$, where $r \geq 0$ and $p > \mu(G) + \lfloor (r - 1)/2 \rfloor$. Then $\psi(G) \geq \delta(G) - \mu(G) + 1$.   ■

APPENDIX A

VIZING'S TWO FUNDAMENTAL PAPERS

A.1 ON AN ESTIMATE OF THE CHROMATIC CLASS OF A p-GRAPH

VADIM. G. VIZING

Diskretnyi Analiz 3 (1964) Novosibirsk, 25–30.

At the moment there is no practical effective algorithm for a minimal edge-coloring of a multigraph, thus it is interesting to estimate the chromatic class using more visible graph parameters. This paper deals with this problem.

A multigraph is a finite nonoriented multigraph without loops [1]. It is called a p-graph if it has at most p parallel edges. A 1-graph is just a graph. A multigraph with colored edges is said to be properly colored if the edges from the same vertex are always colored differently. The smallest number of colors needed to color the multigraph G properly is called the chromatic class of G and denoted $q(G)$.

The maximum degree in G we denote $\sigma(G)$. This is the maximum number of edges from a vertex. Of course, for every multigraph G we have $q(G) \geq \sigma(G)$. There is also a trivial upper bound $q(G) \leq 2\sigma(G) - 1$.

C. E. Shannon [2] proved that any multigraph G satisfies $q(G) \leq \lfloor \frac{3}{2}\sigma(G) \rfloor$, where the parenthesis denotes the lower integer part. It is possible for each m to construct a multigraph G with $\sigma(G) = m$ and $q(G) = \lfloor \frac{3}{2}m \rfloor$, but even so it is possible to get a better upper bound for $q(G)$ by introducing a second parameter for multigraphs.

First we describe some lemmas from Shannon [2].

Let us suppose that the multigraph G has been properly colored, and let s and t denote two different colors. An (s, t)-path is a set of edges of G forming a connected subgraph, each edge colored s or t.[1] We call an (s, t)-path maximal if it is not a proper part of another (s, t)-path. We say that an (s, t)-path is recolored if the colors s and t are interchanged on the edges of the (s, t)-path.

Lemma A.1 *A properly colored multigraph is still properly colored after recoloring a maximal (s, t)-path.*

A color s is missing at vertex x in a properly colored multigraph if no edge at x has color s.

Lemma A.2 *Let x, y and z be three different vertices in a properly colored multigraph G. Suppose that in each of x, y and z either the color s or the color t is missing. Then at least one of x, y and z is not contained in the same (s, t)-path as any of the two other vertices.*

Theorem A.3 *If m is the maximum degree in the p-graph G, then $q(G) \leq m + p$.*

Proof: We denote the colors by numbers from 1 to $m + p$. We shall show how one can properly color the edges of the p-graph G using these colors.

Let (a, b) be an uncolored edge between vertices a and b. Since the degree of any vertex is at most m there are at least p colors missing at each vertex.[2]

Let A and B respectively denote the set of colors missing at a and at b.

If $A \cap B \neq \emptyset$, then the edge (a, b) may be colored by a color missing at both a and b.

Suppose therefore that $A \cap B = \emptyset$. We associate to each colored edge between a and a neighbor x of a a color missing at x in such a way that different colored edges between a and x are associated with different missing colors at x. This is possible because there are at least p missing colors at each vertex.

Let $s_0, \beta_1, ..., \beta_{p-1}$ be colors not present at b. We may assume that the color s_0 is not associated with any edge between a and b. Since $s_0 \in B$ and $A \cap B = \emptyset$ it follows that s_0 does not belong to A, hence there is an edge (a, x_1) colored s_0. Clearly $x_1 \neq b$. Let s_1 be the color related to (a, x_1). If $s_1 \in A$, then the edge (a, x_1) may be recolored by color s_1, and then the edge (a, b) can get the color s_0. If $s_1 \notin A$, then there is an edge (a, x_2) from a colored s_1.

---

[1] Such a 2-colored (s, t)-path might be a cycle, or it might contain only one edge, or no edges.
[2] From the set of colors $1, 2, ..., m + p$.

Let $(a, x_1), (a, x_2), ..., (a, x_k)$ $(k \geq 2)$ be a sequence of different edges in the multigraph G, all from the vertex a. Let their colors be $s_0, s_1, ..., s_{k-1}$. The colors s_j $(j = 0, 1, 2, ..., k - 1)$ are all different and not in A. For each $i = 1, 2, ..., k - 1$ we assume that s_i is the color related to the edge (a, x_i).[3]

Let s_k be the color related to the edge (a, x_k). There are two possible cases:

1. $s_k \in A$. Then $s_k \neq s_j$ for all $j = 0, 1, ..., k - 1$. We recolor the edge (a, x_k) by s_k, the edge (a, x_{k-1}) by s_{k-1}, ... , the edge (a, x_1) by s_1. This is still a proper coloring, and then the edge (a, b) may be colored by the color s_0.

2. $s_k \notin A$. Then there are the following possibilities (a), (b), and (c):

(a) $s_k = s_0$. Then $x_k \neq x_1$ since the color s_0 is present at x_1 as the edge (a, x_1) is colored s_0. Also $x_k \neq b$ since the color s_0 is not related to edges parallel to (a, b). We chose a color $t \in A$ and get the following:

- In vertex a the color t is missing.

- In vertex b the color s_0 is missing.

- In vertex x_k the color s_0 is missing.

By Lemma A.2, at least one of the three vertices a, b, x_k is not joined to any of the other two vertices by an (s_0, t)-path. If a and b are not in the same (s_0, t)-path, then recolor the maximal (s_0, t)-path starting with the edge (a, x_1) at a. After the recoloring the color s_0 is missing at both a and b. Then (a, b) may be colored s_0. If x_k is not in the same (s_0, t)-path as a or b, then recolor the maximal (s_0, t)-path starting at x_k. Then the color $t \in A$ is missing at x_k. If we relate this color to the edge (a, x_k), we are back in Case 1 and can recolor and color the edge (a, b).

(b) $s_k = s_i$ for $1 \leq i \leq k - 2$. (Note that $s_k = s_{k-1}$ is impossible because (a, x_k) is colored s_{k-1}.) Then $x_i \neq x_k$ since otherwise two parallel edges (a, x_i) and (a, x_k) are related to the same color s_i, which is impossible.

Let $t \in A$. We get the following:

- In vertex a the color t is missing.

- In vertex x_k the color s_i is missing.

- In vertex x_i the color s_i is missing.

We use again Lemma A.2. If x_k is not in the same (s_i, t)-path as a or x_i, then recolor the maximal (s_i, t)-path starting at x_k. Then $t \in A$ is missing at x_k. We relate t with the edge (a, x_k) and get back to Case 1. In the same way we get back to Case 1 if x_i is not in the same (s_i, t)-path as a or x_k.

---

[3] Note that we may have that $x_\ell = x_r$ for $\ell \neq r$; $1 \leq \ell, r \leq k$. But even then the edges (a, x_ℓ) and (a, x_r) are different parallel edges, because the edges (a, x_i) for $i = 1, 2, ..., k$ are all different.

So assume that a is not connected by an (s_i, t)-path to x_k or x_i. The first edge in a maximal (s_i, t)-path starting in a is (a, x_{i+1}) colored s_i. After recoloring the path the color s_i is missing at a, and s_i is related to the edge (a, x_i). (The color s_i is still missing at the vertex x_i, since a is not connected to x_i by an (s_i, t)-path.) We are again back in Case 1.

(c) $s_k \neq s_j$ for all $j = 0, 1, ..., k-1$. In this case an edge (a, x_{k+1}) is colored by the color s_k and it is different from the edges $(a, x_1), (a, x_2), ..., (a, x_k)$. We now repeat the argument for the longer sequence of edges $(a, x_1), (a, x_2), ..., (a, x_k), (a, x_{k+1})$. Since the graph is finite we get either Case 1 or Case 2 (a or b) after a finite number of steps. ∎

The theorem gives an upper bound for the chromatic class of a ρ-graph G, and in case $p < \lfloor \frac{\sigma(G)}{2} \rfloor$ it is better than Shannon's bound $q(G) \leq \lfloor \frac{3}{2}\sigma(G) \rfloor$. A natural question may be asked: Is it possible for each m and p with $p \leq \lfloor \frac{m}{2} \rfloor$ to obtain a p-graph G with $\sigma(G) = m$ and $q(G) = m + p$? In case $p = 1$ the answer is yes:

Corollary A.4 *If m is the maximum degree in the graph G, then $m \leq q(G) \leq m+1$. Moreover, for each $m \geq 2$ there is a graph G with $\sigma(G) = m$ and $q(G) = m + 1$*

Proof: We shall only prove the second part.

Let $m \geq 2$. Let H be a graph with $2m$ vertices $x_1, x_2, ..., x_m, y_1, y_2, ..., y_m$ and edges (x_i, y_j) for $i = 1, 2, ..., m$ and $j = 1, 2, ..., m$. Add a new vertex z outside H, remove the edge (x_1, y_1) from H and join z to x_1 and y_1 by edges. The new graph G has $2m + 1$ vertices, $m^2 + 1$ edges, and $\sigma(G) = m$. Because of the odd number $(2m + 1)$ of edges, at most m edges may be given the same color in a proper edge coloring of G. This means that $q(G) \geq \frac{m^2+1}{m} = m + \frac{1}{m} > m$, implying that $q(G) = m + 1$. ∎

When $p \geq 2$ our bound is not always the best possible. One can show that for each $p \geq 2$ the chromatic class of a p-graph G with $\sigma(G) = m = 2p + 1$ is at most $m + p - 1$. However, if $m = 2kp$ ($k \geq 1$) and G has $2k + 1$ vertices for which all pairs of vertices are joined by p edges, then $q(G) = m + p$.

Hence, the best possible upper bound for the chromatic class of a p-graph G with $p \leq \lfloor \frac{\sigma(G)}{2} \rfloor$ depends on the relationship between $\sigma(G)$ and p. Perhaps I will investigate this question later.

Finally, the author would like to express heartfelt thanks to A. A. Zykov for assistance and valuable advise.

REFERENCES

1. BERGE, C. (1962). *Graph Theory and its Applications* (in Russian). IL, Moscow.

2. SHANNON, C. E. (1960). A theorem on coloring the lines of a graph (in Russian). IL, Moscow, pages 249–253.

A.2   CRITICAL GRAPHS WITH A GIVEN CHROMATIC CLASS

V.G. VIZING

Diskretnyi Analiz 5 (1965) Novosibirsk, 9–17.

In this paper we use the notation from Vizing [1] and the book by C. Berge [2]. In Vizing [1] it was proved that if $\sigma(G)$ is the maximum degree of a vertex of G then the chromatic class $q(G)$ is either $\sigma(G)$ or $\sigma(G) + 1$. Each of the two possibilities depends on the structure of the graph. We do not have criteria that can help us to determine the chromatic class using visible properties of the graph. But we can investigate two directions:

1. What can one say about the properties of a graph if it is known that its chromatic class is one bigger than the maximum degree?

2. If some structural or numerical characteristics of a graph is known, what is its chromatic class?

For the research in the first direction it is natural to introduce the definition of a critical graph.

DEFINITION OF A CRITICAL GRAPH OF DEGREE m AND ITS PROPERTIES

A graph G is called critical of degree m, where $m \geq 2$ is an integer, when:

(a) G is connected;

(b) $\sigma(G) = m$;

(c) $q(G) = m + 1$;

(d) The chromatic class of a graph obtained by removing any edge from G is equal to m.

The following Lemma is of interest because of its use in induction proofs for critical graphs.

Lemma A.5 *Let G be a graph with $\sigma(G) = m$ and $q(G) = m + 1$. Then for any k satisfying $m \geq k \geq 2$ there exists a critical graph of degree k as a subgraph of G.*

Proof: This is obvious for $k = m$. Let $m > k \geq 2$ and G be critical of degree m. Color all edges of G with m colors except an edge (a, b). We have $\delta(a) \neq \emptyset$, $\delta(b) \neq \emptyset$, and since $q(G) = m + 1$, $\delta(a) \cap \delta(b) = \emptyset$. (Here and in what follows, $\delta(x)$ denotes the set of colors missing at the vertex x.) Let the colors s_1 and s_2, respectively, belong to $\delta(a)$ and $\delta(b)$ ($s_1 \neq s_2$). Remove now from G all edges colored with $m - k$ of the colors different from s_1 and s_2. The chromatic class of the remaining graph H of G is $k + 1$, therefore $\sigma(H) \geq k$. On the other hand, H does not contain any vertex of degree $k + 1$ in H so $\sigma(H) = k$. This means that there is a critical subgraph of H of degree k and therefore also of G.

The Lemma has been proved.   ■

In Vizing [3] the following properties of critical graphs were given:

PROPERTY I. A critical graph of degree m cannot have a separating vertex.

PROPERTY II. The sum of the degrees of two adjacent vertices in a critical graph of degree m is $\geq m + 2$.

PROPERTY III. In a critical graph of degree m each vertex is adjacent to at least two vertices of degree m.

We shall now obtain a common generalization of Properties II and III. For this we define a fan-sequence of edges.

Let us have a graph with colored edges (it does not matter if all edges are colored or not). A sequence of different edges $(a, x_1), (a, x_2), \ldots, (a, x_n)$ $(n \geq 1)$, at the vertex a and colored $s_1, s_2, \ldots, s_n$, is called a fan-sequence at a, starting with (a, x_1), if the color s_2 is missing at x_1, the color s_3 is missing at $x_2, \ldots$, the color s_n is missing at x_{n-1}.

REMARK. In a fan-sequence, all edges are incident with the same vertex; hence the colors $s_1, s_2, \ldots, s_n$ are all different.

Theorem A.6 *In a critical graph of degree m each vertex incident with a vertex of degree k is in addition also incident with $m - k + 1$ vertices of degree m.*

Proof: Let G be a critical graph of degree m, $\sigma(b) = k \leq m$, where $\sigma(b)$ is the degree of the vertex b and (a, b) is an edge of G. We shall prove that a is adjacent to $m - k + 1$ vertices of degree m, different from b.

We color the edges of G with m colors, except the edge (a, b). We have $|\delta(b)| = m - k + 1$ and $|\delta(a)| \geq 1$. Since $q(G) = m + 1$, we have $\delta(a) \cap \delta(b) = \emptyset$.

Using the same method as in [1], we can show that no fan-sequence at a, starting with an edge having a color from $\delta(b)$, can contain an edge (a, x) such that there is a missing color at x from $\delta(a)$ or there is a missing color at x being the color of an earlier edge of the fan-sequence.

We shall now prove that if $|\delta(b)| \geq 2$, then two fan-sequences at vertex a, starting with two different edges colored with colors from $\delta(b)$ are edge-disjoint. For a contradiction, assume that $(a, x_1), \ldots, (a, x_r)$ and $(a, y_1), \ldots, (a, y_\ell)$ are two fan-sequences at a, colored $s_1, \ldots, s_r$ and $s'_1, \ldots, s'_\ell$, where $(s_1 \in \delta(b)$, $s'_1 \in \delta(b)$, $s_1 \neq s'_1)$, and only (a, x_r) and (a, y_ℓ) are equal. Obviously, either $r \geq 2$ or $\ell \geq 2$ (both cases are possible). Let us assume that $r \geq 2$. Then $s_r = s'_\ell$ belongs to $\delta(x_{r-1})$. Let $t \in \delta(a)$ and change the colors in the maximum (s_r, t)-chain starting the vertex x_{r-1} (see Vizing [1]). If the edge (a, x_r) does not get the color t, then the fan-sequence $(a, x_1), \ldots, (a, x_{r-1})$ contains the edge (a, x_{r-1}) with $\delta(x_{r-1}) \cap \delta(a) \neq \emptyset$. If the edge (a, x_r) gets the color t, then when $\ell = 1$ we have $s_r = s'_\ell \in \delta(b) \cap \delta(a)$; and when $\ell \geq 2$ the fan-sequence $(a, y_1), \ldots, (a, y_{\ell-1})$ contains the edge $(a, y_{\ell-1})$ with $\delta(y_{\ell-1}) \cap \delta(a) \neq \emptyset$. This contradicts $q(G) = m + 1$.

We finish the proof as follows.

For each edge with a as end-vertex and having one of the finitely many colors from $\delta(b)$, construct a maximal fan-sequence at the vertex a starting with the edge. Then if (a, x) is the last vertex of the fan, then $\delta(x) = \emptyset$ and $\sigma(x) = m$. Since fan-sequences at a starting with different edges cannot have common edges, it follows that a is adjacent to at least $|\delta(b)| = m - k + 1$ different vertices of degree m, these vertices being also different from b.

Theorem A.6 has been proved.   ∎

Theorem A.6 obviously generalize properties II and III of critical graphs. Consequently the theorem proves Hypothesis 2 from Vizing [3].

Theorem A.7 *A critical graph of degree m contains an elementary cycle of length $\geq m + 1$.*

Proof: We proceed by contradiction. Let G be a critical graph of degree m for which the length of any elementary cycle is no more than m.

By $\mu = [a_1, a_2, \ldots, a_n]$ we denote a longest elementary path in G. Let a_ℓ be the vertex of μ with the maximum index which is adjacent to a_1 (this means that a_1 and a_ℓ are adjacent, $a_\ell \in \mu$, but for all $j > \ell$ the vertex a_j of the path μ is not adjacent to a_1).

Similarly, let a_r be the vertex of the path μ with the maximum index which is adjacent to a_2.

Since μ is an elementary path of maximum length, all vertices of G adjacent to a_1 belong to μ. Thus if $\sigma(a_1) = m$, then $\ell \geq m + 1$ and the length of the elementary cycle $[a_1, a_2, \ldots, a_\ell, a_1]$ is larger than m. This means that $\sigma(a_1) = k < m$ ($k \geq 2$). By Theorem A.6, the vertex a_2 is adjacent to at least $m - k + 1$ vertices of degree m. If a vertex a' is adjacent to a_2 and $\sigma(a') = m$, then $a' \in \mu$, because otherwise the elementary sequence $[a', a_2, \ldots, a_n]$ would be of maximum length which is not possible since $\sigma(a') = m$. Next, if a_1 is adjacent to a_i ($2 \leq i \leq n$), then $\sigma(a_{i-1}) < m$, because the elementary path $[a_{i-1}, a_{i-2}, \ldots, a_1, a_i, a_{i+1}, \ldots, a_n]$ has maximum length.

Now let us compare ℓ and r. If $\ell \geq r$, then as mentioned above, among the vertices $a_1, a_2, \ldots, a_n$ we can find k vertices of degree $< m$ and $m - k + 1$ vertices of degree m. Consequently the length of the elementary cycle $[a_1, a_2, \ldots, a_\ell, a_1]$ is not smaller than $k + m - k + 1 = m + 1$.

Assume now that $r > \ell$. If a_1 is not adjacent to a_3, then among the vertices $a_3, a_4, \ldots, a_r$ we can find at least $m - k + 1$ vertices of degree m and $k - 1$ vertices of degree $< m$, and consequently the length of the elementary cycle $[a_2, a_3, \ldots, a_r, a_2]$ is at least $m + 1$. If on the other hand a_1 is adjacent to a_3, then among the vertices $a_1, a_2, \ldots, a_r$ we have $m - k + 1$ vertices of degree m and k vertices of degree $< m$. Then the elementary cycle $[a_2, a_1, a_3, a_4, \ldots, a_r, a_2]$ contains at least $m + 1$ vertices.

Theorem A.7 has been proved. ∎

It would be of interest to obtain the best possible lower bound for the maximum length of an elementary cycle in a critical graph of degree m, taking into account also the number of vertices of the graph.

Using Theorem A.6, we shall now obtain a lower bound on the number of edges in a critical graph of degree m.

Theorem A.8 *In a critical graph of degree m the number of edges is $\geq (3m^2 + 6m - 1)/8$.*

Proof: Let G be a critical graph of degree m, and let k be the minimum degree of the vertices of G. Then G obviously contains at least $m - k + 2$ vertices of degree m. As the number of vertices of the graph is at least $m + 1$ the number of edges is at least $((m - k + 2)m + (k - 1)k)/2$.

The minimum of this expression is obtained for $k = (m + 1)/2$ (this may be proved by differentiation). If we insert $(m + 1)/2$ instead of k we get that the number of edges is at least $(3m^2 + 6m - 1)/8$.

Theorem A.8 has been proved. ∎

In Vizing [3] it is conjectured that the number of edges in a critical graph of degree m is $> m^2/2$ (Hypothesis I). The author has been unable to prove or disprove this conjecture. In the particular case when m is even and the number of vertices is $m + 1$, the problem may be formulated as follows:

Suppose we have an odd number n of elements. Prove or disprove that every set of $(n - 1)^2/2$ unordered pairs, each consisting of two of the elements, can be divided into $n - 1$ groups such that the pairs in each group have no common elements.

A METHOD FOR CLASSIFICATION OF GRAPHS

We define that a graph G belongs to the class L_k, where k is an integer ≥ 0, if every subgraph of G has a vertex of degree at most k. It follows that if $G \in L_k$, then $G \in L'_k$ for $k' \geq k$. On the other hand, a graph of maximum degree m belongs to L_m.

Theorem A.9 *If $G \in L_k$ and $\sigma(G) \geq 2k$, then $q(G) = \sigma(G)$.*

Proof: Suppose the theorem is not true and that $G \in L_k$ ($k \geq 1$), $\sigma(G) = m \geq 2k$, but $q(G) = m + 1$.

We may assume that G is critical of degree m. Let X denote the set of all vertices of G and S the subset of all vertices of degree $\leq k$. Then since $\sigma(G) \geq 2k$ the set $X \setminus S$ is nonempty. By Theorem A.6, every vertex from $X \setminus S$ with a neighbor from S, also has at least $m - k + 1$ neighbors of degree m. Every vertex of degree m is in $X \setminus S$ and $m - k + 1 \geq k + 1 > k$. Thus the subgraph spanned by $X \setminus S$ has no vertex of degree $\leq k$. This contradicts $G \in L_k$.

Theorem A.9 has been proved. ∎

Let $S(k)$ ($k \geq 0$ is an integer) denote the least natural number such that any graph $G \in L_k$ with $\sigma(G) \geq S(k)$ satisfies $q(G) = \sigma(G)$. Theorem A.9 says that $S(k) \leq 2k$. It is easy to show that $S(0) = 0, S(1) = 1, S(2) = 4$, and $S(3) = 6$. But the author conjectures that for k large enough the estimate of Theorem A.9 may be improved. It would be interesting to investigate more thoroughly the class L_k and the function $S(k)$.

Theorem A.10 *If G is planar and $\sigma(G) \geq 8$, then $q(G) = \sigma(G)$.*

Proof: Because Lemma A.5 it is enough to consider the case $\sigma(G) = 8$ to prove Theorem A.10. So let G be a planar graph with $\sigma(G) = 8$ and $q(G) = 9$. We may assume that G is critical of degree 8.

Let n_i ($0 \leq i \leq 8$) denote the number of vertices of G of degree i. Since G is critical, $n_0 = 0$ and $n_1 = 0$. Since G is planar, $\sum_{i=2}^{8} i n_i \leq 6(\sum_{i=2}^{8} n_i - 2).$[4] It follows that

$$2n_8 + n_7 \leq n_5 + 2n_4 + 3n_3 + 4n_2 - 12. \tag{A.1}$$

We denote by $n_8(i_1, i_2, i_3, i_4)$ the number of vertices of degree 8, joined to i_1 vertices of degree 2, i_2 vertices of degree 3, i_3 vertices of degree 4, and i_4 vertices of degree 5. It follows from Theorem A.6 that if $i_1 + i_2 + i_3 + i_4 > 0$, then $i_1 + i_2 + i_3 + i_4 \leq \ell$, where ℓ is the smallest natural number for which $i_\ell > 0$.

We denote by $n_7(j_1, j_2, j_3)$ the number of vertices of degree 7, joined to j_1 vertices of degree 3, j_2 vertices of degree 4, and j_3 vertices of degree 5. It follows again from Theorem A.6 that if $j_1 + j_2 + j_3 > 0$, then $j_1 + j_2 + j_3 \leq r$, where r is the smallest natural number for which $j_r > 0$. We have

$$2n_2 = n_8(1, 0, 0, 0) \tag{A.2}$$

$$3n_3 = n_8(0, 1, 0, 0) + n_8(0, 1, 1, 0) + n_8(0, 1, 0, 1) \\ + 2n_8(0, 2, 0, 0) + n_7(1, 0, 0). \tag{A.3}$$

Since every vertex of degree 5 is joined to at least two vertices of degree 8, we have:

$$2n_5 \leq n_8(0, 1, 0, 1) + n_8(0, 0, 1, 1) + 2n_8(0, 0, 1, 2) + n_8(0, 0, 2, 1) \\ + n_8(0, 0, 0, 1) + 2n_8(0, 0, 0, 2) + 3n_8(0, 0, 0, 3) + 4n_8(0, 0, 0, 4). \tag{A.4}$$

By Property II, a vertex of degree 4 can only be joined to vertices of degree ≥ 6; and by Theorem A.6, any vertex of degree 4 with a neighbor of degree 6 is also joined to three vertices of degree 8. Therefore, if we denote by n_4' the number of vertices of degree 4 with a neighbor of degree 6, then

$$3n_4' + 4(n_4 - n_4') = n_8(0, 1, 1, 0) + n_8(0, 0, 1, 0) \\ + n_8(0, 0, 1, 1) + n_8(0, 0, 1, 2) + 2n_8(0, 0, 2, 0) + 2n_8(0, 0, 2, 1) \\ + 3n_8(0, 0, 3, 0) + n_7(0, 1, 0) + n_7(0, 1, 1) + 2n_7(0, 2, 0).$$

[4]Here we use the fact from Euler's polyhedron formula that the sum of the degrees of a planar graph is at most $6(n - 2)$.

Since every vertex of degree 4 is joined to at most two vertices of degree 7, we have $2(n_4 - n_4') \geq 2n_7(0, 2, 0)$, hence $n_4 - n_4' \geq n_7(0, 2, 0)$. Therefore

$$3n_4 \leq n_8(0, 1, 1, 0) + n_8(0, 0, 1, 0) + n_8(0, 0, 1, 1)$$
$$+n_8(0, 0, 1, 2) + 2n_8(0, 0, 2, 0) + 2n_8(0, 0, 2, 1) + 3n_8(0, 0, 3, 0) \qquad \text{(A.5)}$$
$$+n_7(0, 1, 0) + n_7(0, 1, 1) + n_7(0, 2, 0).$$

From (A.2), (A.3), (A.4), and (A.3) we get

$$n_5 + 2n_4 + 3n_3 + 4n_2 \leq 2n_8 + n_7,$$

which contradicts (A.1).

Theorem A.10 has been proved. ∎

The author has not solved the problem if there exists a planar graph G with $\sigma(G) = 7$ or $\sigma(G) = 6$ which has $q(G) = \sigma(G) + 1$.

For each integer m, where $2 \leq m \leq 5$, it is easy to obtain a planar graph with $\sigma(G) = m$ and $q(G) = m + 1$.

REFERENCES

1. VIZING, V. G. (1964). On an estimate of the chromatic class of a p-graph. (in Russian), *Diskret. Analiz*, 3:25–50.

2. BERGE, C. (1962). *Graph Theory and its Applications* (in Russian). IL, Moscow.

3. VIZING, V. G. (1965). The chromatic class of a multigraph. (in Russian), *Kibernetika* (Kiev), 3:29–39.

Notes

Vadim G. Vizing was born in Kiev, Ukraine, on March 25, 1937. After the war, when he was 10, his family was forced to move to Siberia because his mother was half-German. He began his studies at the University of Tomsk in 1954 and graduated in 1959. From Tomsk he then moved to Moscow to the famous Steklow Institute to study for a Ph.D. He did not finish this program, however, and moved back to Siberia, to work at the Mathematical Institute of the Academy of Sciences in Akademgorodok outside Novosibirsk, where he stayed until 1968. This was an extremely fruitful period, where he was influenced and helped by the leading Russian graph theorist A. A. Zykov, and he obtained a Ph.D. in 1966.

In 1964 Vizing obtained his fundamental results on edge colorings of graphs, published in Vizing [297, 298, 299]. The main theorem soon became known also in the West, as Zykov [327] mentioned it (added in proof), together with other results by Novosibirsk mathematicians, including Vizing, in the proceedings from the meeting in Smolenice in Czechoslovakia in 1963, published jointly by the Czechoslovak Academy of Sciences in Prague and Academic Press in New York and London in

1964/65. Zykov [328] mentioned again the results of Vizing at the meeting in Rome in 1966. The theory was still in its beginnings, and during these years Vizing [300] considered extensions and related unsolved problems.

In 1968 Vizing moved back to Ukraine, where since 1974 he has lived in Odessa. He is now retired from a position as teacher of mathematics at the Academy for Food Technology, and he is still active in research in graph theory, see the interview by Gutin and Toft [122] from 2000.

In his papers Vizing uses the term chromatic class for the edge-chromatic number of a graph. It is unusual to denote a graph parameter by the word class. The terminology seems due to Sainte-Laguë [259, 260, 261], who used the term rank for the vertex-chromatic number and class for the edge-chromatic number. Sainte-Laguë's terminology for the edge-chromatic number was adopted by Claude Berge in his influential first book on graphs [24]. However, in the English translation of Berge's book the translator Alison Doig (now Alison Harcourt), then at the London School of Economics, changed the terminology from class to index. The term index thereafter became commonly used and is now the preferred term by most authors.

During the fruitful period 1964–1965 in Akademgorodok Vizing produced the three fundamental papers on edge coloring of which the first and last appear above in English translations (we thank the Russian Academy of Sciences in Novosibirsk for permission to publish these translations). In the first paper [297] Vizing dealt with graphs with multiple edges, and he introduced fans to obtain the inequality $\chi' \leq \Delta + \mu$. Moreover he briefly discussed the question for which values of Δ and μ equality is possible. In this connection he mentioned without proof that:

Theorem A.11 $\chi'(G) \leq \Delta(G) + \mu(G) - 1$ *for all graphs with* $\mu(G) \geq 2$ *and* $\Delta(G) = 2\mu(G) - 1$.

A proof of Theorem A.11 appeared in the second paper [298], which already in 1965 was translated and published in English. In that paper the main result of Vizing [297] and its proof was repeated, critical simple graphs were introduced and the first version of the adjacency lemma was obtained (that each vertex of a critical simple graph has at least two neighbors of degree Δ). Moreover, he treated König's Theorem on bipartite graphs, and Shannon's Theorem was proved together with a characterization of the graphs achieving Shannon's bound with equality. The classification of simple graphs into Class 1 and Class 2 graphs was presented, and the result that all planar simple graphs with $\Delta(G) \geq 10$ are Class 1, i.e., $\chi'(G) = \Delta(G)$, was proved by an elegant simple proof, using just that any subgraph of a planar graph has minimum degree at most 5. Also, in the second paper Vizing considered the problem of repeated recolorings of maximal connected 2-colored subgraphs to get from one edge coloring to another given coloring. And it treated complementary graphs.

In the third paper the adjacency lemma was extended, again considering critical simple graphs, to what we today know as Vizing's Adjacency Lemma. The proof again used fans, now defined in a more general way than in Vizing [297]. In addition, fundamental results about cycle lengths and the number of edges in critical graphs were obtained. In the third paper again the classification of simple graphs into Class

1 and Class 2 graphs were considered, and the result that planar simple graphs with $\Delta(G) \geq 8$ are Class 1 was proved, using a rather complicated analysis based on Euler's Theorem.

In 1968, at the end of Vizing's period in Akademgorodok he published the paper [300] containing a wealth of unsolved graph theory problems, including several about edge colorings. That paper was already the same year translated and published in English.

Theorem A.11 above was also obtained by Gupta [120], who conjectured for exactly which values of $\Delta(G)$ and $\mu(G)$ the inequality $\chi'(G) \leq \Delta(G) + \mu(G) - 1$ holds (see Gupta's Conjecture ♣ 4 in Chap. 9). Without knowing Gupta's work in detail Favrholdt, Stiebitz, and Toft [91] took the first steps in the direction of the present book. This preliminary version appeared as a technical report at the University of Southern Denmark in 2006. That same year Diego Scheide, in his Diplomarbeit at the Technical University of Ilmenau, supervised by Stiebitz, proved that Gupta's Conjecture follows from Goldberg's Conjecture, thus relating these two fundamental questions. Scheide's proof was included in Favrholdt et al. [91].

APPENDIX B

FRACTIONAL EDGE COLORINGS

Fractional graph theory, introduced by Berge [26] in 1978, deals with real-valued analogues of traditional integral graph parameters and concepts. If a graph parameter ρ can be expressed as the optimal value of an integer program, a fractional graph parameter ρ^* associated with ρ can be defined as the linear programming relaxation of this integer program.

B.1 THE FRACTIONAL CHROMATIC INDEX

In what follows, let (G, f) be an arbitrary weighted graph, that is, G is a graph and $f : V(G) \to \mathbb{N}$ a vertex function.

An f-**matching** of G is defined to be an edge set $M \subseteq E(G)$ such that each vertex $v \in V(G)$ satisfies $|M \cap E_G(v)| \leq f(v)$. The set of all f-matchings of G is denoted by $\mathcal{M}_f(G)$. Clearly, if $\varphi \in \mathcal{C}_f^k(G)$ is an f-coloring of G, then the color class $E_{\varphi,\alpha} = \{e \in E(G) \mid \varphi(e) = \alpha\}$ is an f-matching of G for every color $\alpha \in \{1, \ldots, k\}$. Hence, an f-coloring of G can be viewed as a partition of $E(G)$ into f-matchings, and the f-**chromatic index** $\chi_f'(G)$ is the least possible number of classes in such a partition.

Graph Edge Coloring: Vizing's Theorem and Goldberg's Conjecture,
First Edition. By M. Stiebitz, D. Scheide, B. Toft, and L. M. Favrholdt
Copyright © 2012 John Wiley & Sons, Inc.

A **fractional f-coloring** of G is a function $w : \mathcal{M}_f(G) \to [0,1]$ such that every edge $e \in E(G)$ satisfies

$$\sum_{M \in \mathcal{M}_f(G):e \in M} w(M) = 1.$$

Let $\mathcal{R}_f(G)$ denote the set of all fractional f-colorings of G. For a fractional coloring $w \in \mathcal{R}_f(G)$, we call

$$\sum_{M \in \mathcal{M}_f(G)} w(M)$$

the **value** of w. The minimum value over all fractional f-colorings of G is the **fractional f-chromatic index** of G denoted $\chi_f'^*(G)$, i.e.,

$$\chi_f'^*(G) = \min \left\{ \sum_{M \in \mathcal{M}_f(G)} w(M) \mid w \in \mathcal{R}_f(G) \right\}.$$

The minimum exists, since this is an LP-problem bounded from below. A subset of an f-matching is itself an f-matching, and the f-chromatic index $\chi_f'(G)$ is the smallest number k such that $E(G)$ can be covered by k f-matchings. The linear relaxation of this formulation leads to the set $\mathcal{R}_f'(G)$ of all functions $w : \mathcal{M}_f(G) \to \mathbb{R}^{\geq 0}$ such that every edge $e \in E(G)$ satisfies

$$\sum_{M \in \mathcal{M}_f(G):e \in M} w(M) \geq 1.$$

Theorem B.1 *Every weighted graph (G, f) satisfies*

$$\chi_f'^*(G) = \min \left\{ \sum_{M \in \mathcal{M}_f(G)} w(M) \mid w \in \mathcal{R}_f'(G) \right\}.$$

Proof: Let $\mathcal{R}_f'^o(G)$ denote the set of all functions $w \in \mathcal{R}_f'(G)$ having minimum value $\sum_{M \in \mathcal{M}_f(G)} w(M)$. Since $\mathcal{R}_f(G) \subseteq \mathcal{R}_f'(G)$, it suffices to show that $\mathcal{R}_f(G) \cap \mathcal{R}_f'^o(G) \neq \emptyset$. To this end, define a function $p : \mathcal{R}_f'^o(G) \to \mathbb{N} \cup \{0\}$ by

$$p(w) = \left| \left\{ e \in E(G) \mid \sum_{M \in \mathcal{M}_f(G):e \in M} w(M) = 1 \right\} \right|.$$

Let $w' \in \mathcal{R}_f'^o(G)$ be a function such that $p(w')$ is maximum. If $p(w') = |E(G)|$, then $w' \in \mathcal{R}_f(G)$ and we are done. So suppose that $p(w') < |E(G)|$. Then there exists an edge $e_0 \in E(G)$ such that $\sum_{M \in \mathcal{M}_f(G):e_0 \in M} w'(M) > 1$. Let $\mathcal{R}^0$ be the set of all functions $w \in \mathcal{R}_f'^o(G)$ such that $p(w) = p(w')$ and $\sum_{M \in \mathcal{M}_f(G):e_0 \in M} w(M) > 1$. Clearly, w' belongs to $\mathcal{R}^0$. Now define a function $q : \mathcal{R}^0 \to \mathbb{N} \cup \{0\}$ by

$$q(w) = |\{M \in \mathcal{M}_f(G) \mid e_0 \in M, w(M) = 0\}|.$$

To reach a contradiction, we choose a function $w \in \mathcal{R}^0$ such that $q(w)$ is maximum. Since $w \in \mathcal{R}'^o_f(G)$, there is a f-matching $M_0 \in \mathcal{M}_f(G)$ such that $e_0 \in M_0$ and $w(M_0) > 0$. We now define a new function $\tilde{w} : \mathcal{M}_f(G) \to \mathbb{R}$ by

$$\tilde{w}(M) = \begin{cases} w(M) - \epsilon & \text{if } M = M_0, \\ w(M) + \epsilon & \text{if } M = M_0 \setminus \{e_0\}, \\ w(M) & \text{otherwise}, \end{cases}$$

where $\epsilon := \min \left\{ \sum_{M \in \mathcal{M}_f(G):e_0 \in M} w(M) - 1, w(M_0) \right\}$. Clearly, $\tilde{w}(M) \geq 0$ for all $M \in \mathcal{M}_f(G)$ and we have

$$\sum_{M \in \mathcal{M}_f(G)} \tilde{w}(M) = \sum_{M \in \mathcal{M}_f(G)} w(M).$$

Furthermore, every edge $e \neq e_0$ of G satisfies

$$\sum_{M \in \mathcal{M}_f(G):e \in M} w(M) = \sum_{M \in \mathcal{M}_f(G):e \in M} \tilde{w}(M) \geq 1$$

and the edge e_0 satisfies

$$\sum_{M \in \mathcal{M}_f(G):e_0 \in M} \tilde{w}(M) = \sum_{M \in \mathcal{M}_f(G):e_0 \in M} w(M) - \epsilon \geq 1.$$

Consequently, $\tilde{w}$ belongs to the set $\mathcal{R}'^o_f(G)$. If

$$\epsilon = \sum_{M \in \mathcal{M}_f(G):e_0 \in M} w(M) - 1,$$

then $p(\tilde{w}) = p(w) + 1$, a contradiction. Otherwise, $\epsilon = w(M_0)$ and, therefore, $\tilde{w} \in \mathcal{R}^0$ and $q(\tilde{w}) = q(w) + 1$, a contradiction too. This completes the proof of the theorem. ∎

Finally, let $\mathcal{R}^*_f(G)$ be the set of all functions $w: \mathcal{M}_f(G) \to [0, 1]$ such that every edge $e \in E(G)$ satisfies

$$\sum_{M \in \mathcal{M}_f(G):e \in M} w(M) \geq 1.$$

Evidently, we have

$$\mathcal{R}_f(G) \subseteq \mathcal{R}^*_f(G) \subseteq \mathcal{R}'_f(G)$$

and Theorem B.1 implies that

$$\chi'^*_f(G) = \min \left\{ \sum_{M \in \mathcal{M}_f(G)} w(M) \mid w \in \mathcal{R}^*_f(G) \right\}.$$

B.2   THE MATCHING POLYTOPE

Let (G, f) be an arbitrary vertex-weighted graph, i.e., $f : V(G) \to \mathbb{N}$. For a set $U \subseteq V(G)$, define $f(U) = \sum_{u \in U} f(u)$. The set $\mathcal{V}(G)$ of all functions $\mathbf{x} : E(G) \to \mathbb{R}$ form a real vector space with respect to the addition of functions and the multiplication of a function by a real number. If $m = |E(G)|$, then $\mathcal{V}(G)$ has dimension m and is isomorphic to the standard vector space $\mathbb{R}^m$. For a vector $\mathbf{x} \in \mathcal{V}(G)$ and a set $F \subseteq E(G)$, let $\mathbf{x}(F) = \sum_{e \in F} \mathbf{x}(e)$. A **polytope** in $\mathcal{V}(G)$ is the convex hull of finitely many vectors of $\mathcal{V}(G)$. For an edge set $F \subseteq E(G)$, let $\mathbf{x}_F : E(G) \to \mathbb{R}$ be the function defined by

$$\mathbf{x}_F(e) = \begin{cases} 1 & \text{if } e \in F, \\ 0 & \text{if } e \notin F. \end{cases}$$

This function is usually called the **incidence vector** or the **characteristic function** of F. The f-**matching polytope** $\mathcal{P}_f(G)$ of G is then defined as the convex hull of the incidence vectors of all f-matchings in G, i.e.,

$$\mathcal{P}_f(G) = \text{conv}(\{\mathbf{x}_F \mid F \in \mathcal{M}_f(G)\}).$$

Note that $\mathcal{P}_f(G)$ contains the null-vector $\mathbf{x}_\emptyset = \mathbf{0}$. If $f(v) = 1$ for all $v \in V(G)$, we write $\mathcal{P}(G)$ rather than $\mathcal{P}_f(G)$. So $\mathcal{P}(G)$ is the ordinary **matching polytope** of G. Furthermore, let $\mathcal{P}_{\text{perf}}(G)$ denote the **perfect matching polytope** of G, that is, the convex hull of the incidence vectors of all perfect matchings of G. Note that $\mathcal{P}_{\text{perf}}(G) \neq \emptyset$ if and only if G has a perfect matching.

In his pioneering work of 1965, Edmonds [76] gave a full description of the matching polytope of a given graph by a finite system of linear inequalities. Even if the number of constraints is exponential in the size of the graph, the description of the matching polytope has become a very useful tool in combinatorial optimization. Over the years, several different proofs and enhancements of the matching polytope theorem has been given; we refer the reader to the book by Schrijver [277]. The matching polytope theorem follows from a description of the perfect matching polytope, also given by Edmonds [76].

Theorem B.2 (Edmonds' perfect matching polytope theorem [76] 1965) *For any graph G, a vector $\mathbf{x} \in \mathcal{V}(G)$ belongs to $\mathcal{P}_{\text{perf}}(G)$ if and only if $\mathbf{x}$ satisfies the following systems of linear inequalities:*

(a) $\forall e \in E(G) : 0 \leq \mathbf{x}(e) \leq 1$ *(capacity constraint)*

(b) $\forall v \in V(G)$: $\mathbf{x}(E_G(v)) = 1$ *(degree equation)*

(c) $\forall U \subseteq V(G)$, $|U|$ *is odd:* $\mathbf{x}(\partial_G(U))) \geq 1$ *(odd cut constraint)*

Theorem B.3 (Edmonds' matching polytope theorem [76] 1965) *For any graph G, a vector $\mathbf{x} \in \mathcal{V}(G)$ belongs to $\mathcal{P}(G)$ if and only if $\mathbf{x}$ satisfies the following systems of linear inequalities:*

(a) $\forall e \in E(G) : 0 \leq \mathbf{x}(e) \leq 1$ *(capacity constraint)*

(b) $\forall v \in V(G)$: $\mathbf{x}(E_G(v)) \leq 1$ *(degree constraint)*

(c) $\forall U \subseteq V(G)$, *where* $|U|$ *is odd:* $\mathbf{x}(E(G[U])) \leq \frac{|U|-1}{2}$ *(blossom constraint)*

A description of the f-matching polytope by a system of linear inequalities was given by Edmonds and Johnson [77] in 1970.

Theorem B.4 (Edmonds and Johnson [77] 1970) *Let (G, f) be a weighted graph. A vector $\mathbf{x} \in \mathcal{V}(G)$ belongs to $\mathcal{P}_f(G)$ if and only if $\mathbf{x}$ satisfies the following systems of linear inequalities:*

(a) $\forall e \in E(G)$: $0 \leq \mathbf{x}(e) \leq 1$ *(capacity constraint)*

(b) $\forall v \in V(G)$: $\mathbf{x}(E_G(v)) \leq f(v)$ *(weighted degree constraint)*

(c) $\forall U \subseteq V(G), \forall F \subseteq \partial_G(U)$, *where* $f(U) + |F|$
    is odd: $\mathbf{x}(E(G[U])) + \mathbf{x}(F) \leq \frac{f(U)+|F|-1}{2}$ *(weighted blossom constraint)*

Note that this result does not immediately imply the matching polytope theorem. For $f \equiv 1$, some of the weighted blossom constraints are redundant, while a blossom constraint for a subset U of $V(G)$ with odd cardinality corresponds to a weighted blossom constraint for the set U and the empty edge set $F \subseteq \partial_G(U)$. Whether in general all weighted blossom constrains are really necessary seems not clear.

From Edmonds' matching polytope theorem, a combinatorial characterization of the fractional edge chromatic index can be easily derived. For a graph G, define

$$\kappa^*(G) = \max \left\{ \Delta(G), \max_{H \subseteq G, |V(H)| \geq 2} \frac{|E(H)|}{\left\lfloor \frac{1}{2}|V(H)| \right\rfloor} \right\}. \tag{B.1}$$

Observe that a graph with at most two vertices satisfies $\kappa^*(G) = \Delta(G)$. In searching for a subgraph $H \subseteq G$ achieving the maximum in (B.1) we can clearly restrict H to be an induced subgraph and to have odd order, since for a subgraph H of even order we have $|E(H)|/\lfloor |V(H)|/2 \rfloor \leq \Delta(H) \leq \Delta(G)$. So (B.1) reduces to

$$\kappa^*(G) = \max \left\{ \Delta(G), \max_{X \subseteq V(G), |X| \geq 3 \text{ odd}} \frac{2|E(G[X])|}{|X| - 1} \right\}. \tag{B.2}$$

Theorem B.5 (Seymour [280] 1979, Stahl [286] 1979) *Every graph G satisfies $\chi'^*(G) = \kappa^*(G)$.*

Proof: The statement is evident if G is edgeless. So suppose $E(G) \neq \emptyset$. Let $\mathcal{M}$ denote the set of all matchings of G.

To see that $\chi'^*(G) \geq \kappa^*(G)$, choose a fractional edge coloring w of G with minimum value. For a vertex v of maximum degree in G, we then obtain

$$
\begin{aligned}
\chi'^*(G) &= \sum_{M \in \mathcal{M}} w(M) \geq \sum_{M \in \mathcal{M}} w(M) |M \cap E_G(v)| \\
&= \sum_{e \in E_G(v)} \sum_{M \in \mathcal{M}: e \in M} w(M) = \sum_{e \in E_G(v)} 1 = \Delta(G).
\end{aligned}
$$

Moreover, for each $H \subseteq G$ with $|V(H)| \geq 2$, we obtain

$$
\begin{aligned}
\chi'^*(G) &= \sum_{M \in \mathcal{M}} w(M) \geq \sum_{M \in \mathcal{M}} w(M) \frac{|M \cap E(H)|}{\lfloor \frac{1}{2} |V(H)| \rfloor} \\
&= \frac{1}{\lfloor \frac{1}{2} |V(H)| \rfloor} \sum_{e \in E(H)} \sum_{M \in \mathcal{M}: e \in M} w(M) = \frac{|E(H)|}{\lfloor \frac{1}{2} |V(H)| \rfloor}.
\end{aligned}
$$

This proves $\chi'^*(G) \geq \kappa^*(G)$.

To see that $\chi'^*(G) \leq \kappa^*(G)$, let $\mathbf{x} \in \mathcal{V}(G)$ be the function with $\mathbf{x}(e) = 1/\kappa^*(G)$ for all $e \in E(G)$. Obviously, $\mathbf{x}(E_G(v)) \leq 1$ for each $v \in V(G)$ and $\mathbf{x}(E(G[U])) \leq \lfloor \frac{1}{2} |U| \rfloor$ for each $U \subseteq V(G)$ with $|U| \geq 2$. Hence $\mathbf{x}$ belongs to $\mathcal{P}(G)$, that is, $\mathbf{x}$ is a convex combination of incidence vectors of matchings. Let $\mathbf{1} \in \mathcal{V}(G)$ be the all-one vector. Then $\mathbf{1} = \kappa^*(G) \cdot \mathbf{x} = \sum_{M \in \mathcal{M}} \lambda_M \mathbf{x}_M$ for some $\lambda_M \geq 0$ with $\sum_{M \in \mathcal{M}} \lambda_M = \kappa^*(G)$, implying that the function $w : \mathcal{M} \to [0,1]$ with $w(M) = \lambda_M$ for each $M \in \mathcal{M}$ is a fractional edge coloring of G. Hence, we obtain $\chi'^*(G) \leq \sum_{M \in \mathcal{M}} w(M) = \kappa^*(G)$. ∎

Theorem B.6 *Let G be a graph and let $\chi'^*(G) = r$. Then r is rational and for every positive integer t we have $\chi'(tG) \geq tr$, where equality holds if and only if tr is an integer and there exist a family of tr matchings in G using each edge exactly t times. Furthermore, there are infinitely many positive integers t such that $\chi'(tG) = tr$.*

Proof: If G is edgeless, the theorem is obviously true. So assume that $E(G) \neq \emptyset$. By (B.1) and Theorem B.5, we obtain that r is rational and $\chi'(tG) \geq \chi'^*(tG) = \kappa^*(tG) = t\kappa^*(G) = t\chi'^*(G) = tr$. Clearly, $\chi'(tG) = tr$ if and only if tr is an integer and the edge set of tG can be partitioned into tr matchings. This is equivalent to the statement that there exists a family of tr matchings in G using each edge exactly t times, because tG is obtained from G by replacing each edge of G by t parallel edges and no matching of tG contains two parallel edges.

The fractional edge chromatic index is the optimal value of an linear program with integer coefficients, hence there is an optimal fractional edge coloring w of G such that $w(M)$ is rational for every matching M, where $\chi'^*(G) = r = \sum_M w(M)$. Let s denote the least common multiple of the denominators of all the values $w(M)$. Then s is a positive integer and $k_M = sw(M)$ is a nonnegative integer for all matching M of G. Since w is an optimal fractional edge coloring, we conclude that

$$
\sum_M k_M \mathbf{x}_M = s\mathbf{1} \text{ and } \sum_M k_M = sr.
$$

This means that there is a family of sr matchings in G using each edge s times. Clearly, this is equivalent to $\chi'(sG) = sr$. If $t = ks$ for a positive integer k, then there is a family of $tr = ksr$ matchings in G using each edge $t = ks$ times, and so $\chi'(tG) = tr$. This completes the proof.   ∎

As a corollary of Theorem B.6 we obtain the following characterization of the fractional chromatic index of an arbitrary graph G:

$$\chi'^*(G) = \min_{t \geq 1} \frac{\chi'(tG)}{t} = \lim_{t \to \infty} \frac{\chi'(tG)}{t}. \tag{B.3}$$

Because of Theorem B.6 it is sufficient to show that the limit in (B.3) exists. This follows from the fact that $\chi'((s+t)G) \leq \chi'(sG) + \chi'(tG)$ for all $s, t \in \mathbb{N}$ and Fekete's Lemma [93]. This lemma says that if the sequence $g : \mathbb{N} \to \mathbb{R}^{\geq 0}$ is **subadditive**, that is, $g(s + t) \leq g(s) + g(t)$ for all $s, t \in \mathbb{N}$, then the limit $\lim_{n \to \infty} g(n)/n$ exists and is equal to the infimum of $g(n)/n$ for $n \in \mathbb{N}$. A proof of Fekete's Lemma can be found in the book by Scheinerman and Ullman [276, Lemma A.4.1], see also the book of Jensen and Toft [157, Problem 5.3].

In what follows we will discuss three interesting applications of Edmonds' matching polytope theorem and Edmonds' perfect matching polytope theorem. The first application is a result due to O. Marcotte. This result is used in the proof of Theorem 6.24.

Theorem B.7 (Marcotte [208] 1990) *Every graph G with $w(G) \leq \Delta(G)$ and $\Delta(G) \geq 1$ contains a matching M such that $\Delta(G - M) = \Delta(G) - 1$.*

Proof: Let $\Delta = \Delta(G)$ and let $\mathbf{x} = \frac{1}{\Delta}\mathbf{1}$. Clearly, $\mathbf{x}$ satisfies the capacity constraints and the degree constrains of Theorem B.3, where for each major vertex the degree constraint holds with equality. Since $\Delta \geq w(G)$, the vector $\mathbf{x}$ also satisfy the blossoms constraints. So $\mathbf{x}$ belongs to the matchin polytope $\mathcal{P}(G)$ and is therefore a convex combination of the matching vectors. If a matching vector $\mathbf{x}_M$ occurs in such a convex combination with a positive coefficient, then $\mathbf{x}_M$ also satisfies the degree constraints for each major vertex with equality, that is, each major vertex of G is an endvertex of some edge belonging to the matching M. So $\Delta(G - M) = \Delta(G) - 1$. ∎

Let G be a graph. The standard **scalar product** in the vector space $\mathcal{V}(G)$ is denoted by $\circ$, and the set of perfect matchings of G is denoted by $\mathcal{M}_{\mathrm{perf}}(G)$. Recall that an odd set means a set with odd cardinality. The proof of the following statement uses Theorem B.2 and is based on standard arguments from convex analysis.

Proposition B.8 (Kaiser, Král', and Norine [162] 2005) *Let G be a graph with at least one edge and let $\mathbf{x} \in \mathcal{P}_{\mathrm{perf}}(G)$ be a vector. Then there are $\ell \geq 1$ perfect matchings $M_1, \ldots, M_\ell \in \mathcal{M}_{\mathrm{perf}}(G)$ and positive integers $\lambda_1, \ldots, \lambda_\ell$ such that*

$$\mathbf{x} = \lambda_1 \mathbf{x}_{M_1} + \cdots \lambda_\ell \mathbf{x}_{M_\ell} \text{ and } \lambda_1 + \cdots \lambda_\ell = 1,$$

where every such convex combination of perfect matchings satisfies the following statements:

(a) *If $U \subseteq V(G)$ is an odd set such that $\mathbf{x}(\partial_G(U)) = 1$, then $\mathbf{x}_{M_i}(\partial_G(U)) = 1$ for $i = 1, \ldots, \ell$.*

(b) *For every vector $\mathbf{c} \in \mathcal{V}(G)$ there is a perfect matching $M \in \{M_1, \ldots, M_\ell\}$ such that $\mathbf{c} \circ \mathbf{x}_M \geq \mathbf{c} \circ \mathbf{x}$.*

Let G be a cubic bridgeless graph. Observe that G is a 3-graph, that is, G is 3-regular and $|\partial_G(U)| \geq 3$ for every odd set $U \subseteq V(G)$. If $U \subseteq V(G)$ satisfies $|\partial_G(U)| = 3$, then a simple parity argument shows that U is odd, and we then call $F = \partial_G(U)$ a 3-edge-cut of G. By Petersen's theorem [241], G has a perfect matching. For an integer $k \geq 1$, define

$$m_k(G) = \max\left\{ \frac{|\bigcup_{i=1}^k M_i|}{|E(G)|} \mid M_1, \ldots, M_k \in \mathcal{M}_{\mathrm{perf}}(G) \right\}.$$

Evidently, $m_1(G) = \frac{1}{3}$ and a conjecture of C. Berge suggests that $m_5(G) = 1$. Lower bounds for $m_k(G)$ for $2 \leq k \leq 5$ were established by Kaiser, Král', and Norine [162].

Theorem B.9 (Kaiser, Král', and Norine [162] 2005) *Every cubic bridgeless graph G satisfies $m_2(G) \geq \frac{3}{5}$, where equality holds for the Petersen graph.*

Proof: That the Petersen graph P satisfies $m_2(P) = 3/5$ follows from the fact that any two perfect matchings of P have exactly one edge in common.

So let G be an arbitrary cubic bridgeless graph and let $\mathbf{x} = \frac{1}{3}\mathbf{1} \in \mathcal{V}(G)$. By Theorem B.2, $\mathbf{x} \in \mathcal{P}_{\mathrm{perf}}(G)$ and $\mathbf{x}(F) = 1$ for each 3-edge-cut F of G. Hence, by Proposition B.8, there is a perfect matching M_1 of G intersecting each 3-edge-cut in a single edge. We now define a vector $\mathbf{y} \in \mathcal{V}(G)$ by $\mathbf{y}(e) = 1/5$ if $e \in M_1$ and $\mathbf{y}(e) = 2/5$ otherwise. Since M_1 contains exactly one edge of each 3-edge-cut, it follows that $\mathbf{y} \in \mathcal{P}_{\mathrm{perf}}(G)$. Let $\mathbf{c} = 1 - \mathbf{x}_{M_1}$. Again, by Proposition B.8, there is a perfect matching M_2 of G such that

$$\mathbf{c} \circ \mathbf{x}_{M_2} \geq \mathbf{c} \circ \mathbf{y} = \frac{2}{5} \cdot \frac{2}{3}|E(G)| = \frac{4}{15}|E(G)|.$$

Since $\mathbf{c} \circ \mathbf{x}_{M_2} = |M_2 \setminus M_1|$, it follows that $|M_1 \cup M_2| = (\frac{1}{3} + \frac{4}{15})|E(G)| = \frac{3}{5}|E(G)|$. This shows that $m_2(G) \geq 3/5$ as required. ∎

Using a similar approach, Kaiser, Král', and Norine [162] proved that

$$m_k(G) \geq 1 - \prod_{i=1}^k \frac{i+1}{2i+1}$$

for every cubic bridgeless graph G and every integer $k \geq 1$. For the Petersen graph P, we have

$$m_2(P) = \frac{3}{5}, m_3(P) = \frac{4}{5}, m_4(P) = \frac{14}{15}, m_5(P) = 1.$$

The third application of the perfect matching polytope theorem deals with the Petersen graph. In 2007 V. V. Mkrtchyan posted as an open question on the Open Problem Garden webpage whether any cubic bridgeless graph different from the Petersen graph contains a 2-factor such that at least one of its cycles is not a 5-cycle, that is, a cycle of length 5. Within one day M. DeVos found an affirmative answer and put an outline of a proof on the Open Problem Garden. His proof is based on standard arguments, particularly on Tutte's 1-factor theorem, and was published by DeVoss, Mkrtchyan, and Petrosyan [68]. Shortly after the problem was posed by Mkrtchyan, D. Král' (oral communication) also found a solution, but with a shorter proof. The secret of Král's proof is to show, by means of the perfect matching polytope theorem, that a connected cubic bridgeless graph without a desired 2-factor has so many 5-cycles that it can be only the Petersen graph.

Theorem B.10 *Let G be a connected cubic bridgeless graph. Then every 2-factor of G is the disjoint union of 5-cycles if and only if G is the Petersen graph.*

Proof: It is not hard to check that each 2-factor of the Petersen graph is the disjoint union of two 5-cycles. So assume that G is a connected cubic bridgeless graph of order n such that every 2-factors of G is the disjoint union of 5-cycles. We first show that the number of 5-cycles in G is at least $6n/5$, from which we then conclude that G must be the Petersen graph.

To count the number of 5-cycles in G we apply Edmonds' perfect matching polytope theorem (Theorem B.2) to the vector $\mathbf{x} = \frac{1}{3}\mathbf{1} \in \mathcal{V}(G)$. By Theorem B.2, $\mathbf{x} \in \mathcal{P}_{\mathrm{perf}}(G)$. Then there are $\ell \geq 1$ perfect matchings $M_1, \ldots, M_\ell \in \mathcal{M}_{\mathrm{perf}}(G)$ and positive integers $\lambda_1, \ldots, \lambda_\ell$ such that

$$\mathbf{x} = \lambda_1 \mathbf{x}_{M_1} + \cdots \lambda_\ell \mathbf{x}_{M_\ell} \text{ and } \lambda_1 + \cdots \lambda_\ell = 1. \tag{B.4}$$

For an edge set $M \subseteq E(G)$ define $\overline{M} = E(G) \setminus M$. If M is a perfect matching of G, then $\overline{M}$ is (the edge set of) a 2-factor of G and the hypothesis implies therefore that $\overline{M}$ is the disjoint union of 5-cycles, where the number of these 5-cycles is $n/5$.

For each i $(i = 1, \ldots, \ell)$ the $n/5$ 5-cycles of $\overline{M}_i$ are each given the weight λ_i. The total weight w given to all 5-cycles is then

$$w = (\lambda_1 + \cdots + \lambda_\ell)\frac{n}{5} = \frac{n}{5}.$$

Consider a fixed 5-cycle C of G. The total weight given to C is

$$p = \sum_{i \in I} \lambda_i,$$

where $I = \{i | E(C) \subseteq \overline{M}_i\}$. To estimate p, consider $F = \partial_G(V(C))$, which is a set of three or five edges. If $|F| = 3$, then we deduce from (B.4)

$$\frac{3}{3} = \mathbf{x}(F) = \sum_{i=1}^{\ell} \lambda_i \mathbf{x}_{M_i}(F) \geq \sum_{i \in I} 3\lambda_i + \sum_{i \notin I} \lambda_i \geq 3p + (1 - p) = 2p + 1,$$

which gives $p \leq 0$, i.e., $p = 0$. If, on the other hand, $|F| = 5$ then we deduce from (B.4) that

$$\frac{5}{3} = \mathbf{x}(F) = \sum_{i=1}^{\ell} \lambda_i \mathbf{x}_{M_i}(F) \geq \sum_{i \in I} 5\lambda_i + \sum_{i \notin I} \lambda_i \geq 5p + (1 - p) = 4p + 1,$$

which gives $p \leq 1/6$. Let c_n denote the number of 5-cycles of G. Then $(1/6)c_n \geq w = n/5$, i.e., $c_n \geq 6n/5$.

In a connected bridgeless graph G any vertex v is contained in at most six 5-cycles, with equality only if there are exactly six vertices of distance 2 from v joined to each other by six edges, implying that there are no vertices of distance 3 from v, i.e., $|V(G)| = 10$. The total number of 5-cycles is therefore $\leq 6n/5$. Since we deduced above that $c_n \geq 6n/5$, it follows that $c_n = 6n/5$ and that all vertices v are contained in exactly six 5-cycles. Then $|V(G)| = 10$ and it follows easily that G is the Petersen graph. ∎

B.3 A FORMULA FOR $\chi_f'^*$

Let (G, f) be an arbitrary vertex-weighted graph. Recall from Chap. 8 that there are two lower bounds for the f-chromatic index $\chi_f'(G)$, namely $\Delta_f(G)$ and $w_f(G)$. The **fractional maximum f-degree** of G is defined by

$$\Delta_f^*(G) = \max_{v \in V(G)} \frac{d_G(v)}{f(v)}$$

and the **fractional f-density** of G is defined by

$$w_f^*(G) = \max_{U \subseteq V(G), f(U) \geq 3 \, \text{odd}} \frac{2|E(G[U])|}{f(U) - 1},$$

where $w_f^*(G) = 0$ if $f(U) = 1$ or $f(U)$ is even for all $U \subseteq V(G)$. Furthermore, let us define

$$\kappa_f^*(G) = \max\{\Delta_f^*(G), w_f^*(G)\}.$$

Clearly, $\lceil \kappa_f^*(G) \rceil = \max\{\Delta_f(G), w_f(G)\} \leq \chi_f'(G)$ and $\chi_f'^*(G) \leq \chi_f'(G)$. Next, let us introduce a variation of the f-density. Let $T(G)$ denote the set of all tuples (U, F) satisfying $\emptyset \neq U \subseteq V(G)$, $F \subseteq \partial_G(U)$, and $f(U) + |F|$ is odd and ≥ 3. We then define

$$\tilde{w}_f^*(G) = \max_{(U,F) \in T(G)} \frac{2(|E(G[U])| + |F|)}{f(U) + |F| - 1}$$

if $T(G) \neq \emptyset$ and $\tilde{w}_f^*(G) = 0$ otherwise. Furthermore, we define

$$\tilde{\kappa}_f^*(G) = \max\{\Delta_f^*(G), \tilde{w}_f^*(G)\}.$$

Clearly, $w_f^*(G) \leq \tilde{w}_f^*(G)$ and so $\kappa_f^*(G) \leq \tilde{\kappa}_f^*(G)$. If $\Delta_f^*(G) \leq 1$, then obviously $\chi_f'^*(G) = \chi_f'(G) = \Delta_f(G)$.

Theorem B.11 *Every graph G with $\Delta^*_f(G) \geq 1$ satisfies $\chi'^*_f(G) = \tilde{\kappa}^*_f(G)$.*

Proof: First we show that $\chi'^*_f(G) \geq \tilde{\kappa}^*_f(G)$. To this end, we consider a fractional f-coloring $w \in \mathcal{R}_f(G)$ with minimum value. For every f-matching $M \in \mathcal{M}_f(G)$ and every vertex $v \in V(G)$, we have $|M \cap E_G(v)| \leq f(v)$. Let $v \in V(G)$ be a vertex such that $\Delta^*_f(G) = d_G(v)/f(v)$. Then we obtain

$$
\begin{aligned}
\chi'^*_f(G) &= \sum_{M \in \mathcal{M}_f} w(M) \geq \sum_{M \in \mathcal{M}_f} w(M) \frac{|M \cap E_G(v)|}{f(v)} \\
&= \sum_{e \in E_G(v)} \sum_{M \in \mathcal{M}_f : e \in M} \frac{w(M)}{f(v)} \\
&= \sum_{e \in E_G(v)} \frac{1}{f(v)} = \frac{d_G(v)}{f(v)} = \Delta^*_f(G).
\end{aligned}
$$

To see that $\chi'^*_f(G) \geq \tilde{w}^*_f(G)$, we may assume that $T(G) \neq \emptyset$, since otherwise we have $\tilde{w}^*_f(G) = 0 \leq \chi'^*_f(G)$. Then we choose a tuple $(U, F) \in T(G)$ such that

$$
\tilde{w}^*_f(G) = \frac{2(|E(G[U])| + |F|)}{f(U) + |F| - 1}.
$$

For the edge set $E = E(G[U]) \cup F$ and an arbitrary f-matching $M \in \mathcal{M}_f(G)$ we have

$$
2|M \cap E| \leq \sum_{u \in U} f(u) + |M \cap F| \leq f(U) + |F|.
$$

Since $2|M \cap E|$ is even and $f(U) + |F|$ is odd (because of $(U, F) \in T(G)$), we have $2|M \cap E| \leq f(U) + |F| - 1$. Then we deduce that

$$
\begin{aligned}
\chi'^*_f(G) &= \sum_{M \in \mathcal{M}} w(M) \geq \sum_{M \in \mathcal{M}} w(M) \frac{2|M \cap E|}{f(U) + |F| - 1} \\
&= \sum_{e \in E} \sum_{M \in \mathcal{M} : e \in M} w(M) \frac{2}{f(U) + |F| - 1} \\
&= \sum_{e \in E} \frac{2}{f(U) + |F| - 1} = \frac{2|E(G[U])| + |F|}{f(U) + |F| - 1} = \tilde{w}^*_f(G).
\end{aligned}
$$

This proves that $\chi'^*_f(G) \geq \tilde{\kappa}^*_f(G)$. It remains to show that $\chi'^*_f(G) \leq \tilde{\kappa}^*_f(G)$. To see this, let $\mathbf{x} \in \mathcal{V}(G)$ be the function with $\mathbf{x}(e) = 1/\tilde{\kappa}^*_f(G)$ for all $e \in E(G)$. We claim that $\mathbf{x} \in \mathcal{P}_f(G)$. Clearly, $\mathbf{x}$ satisfies the capacity constraints. From $\tilde{\kappa}^*_f(G) \geq \Delta^*_f(G)$ we obtain

$$
\mathbf{x}(E_G(v)) = \frac{d_G(v)}{\tilde{\kappa}^*_f(G)} \leq \frac{\Delta^*_f(G) f(v)}{\tilde{\kappa}^*_f(G)} \leq f(v)
$$

for all $v \in V(G)$. So $\mathbf{x}$ satisfies the weighted degree constraints. For any tuple $(U, F) \in T(G)$ we obtain

$$
\mathbf{x}(E(G[U])) + \mathbf{x}(F) = \frac{|E(G[U])| + |F|}{\tilde{\kappa}^*_f(G)} \leq \frac{f(U) + |F| - 1}{2},
$$

since $\tilde{\kappa}_f^*(G) \geq \tilde{w}_f^*(G)$. So $\mathbf{x}$ satisfies the weighted blossom constraints. Then Theorem B.4 implies that $\mathbf{x}$ belongs to $\mathcal{P}_f(G)$, that is, $\mathbf{x}$ is a convex combination of incidence vectors of f-matchings. This gives

$$1 = \tilde{\kappa}_f^*(G) \cdot \mathbf{x} = \sum_{M \in \mathcal{M}_f} \lambda_M \mathbf{x}_M$$

for some $\lambda_M \geq 0$ with

$$\sum_{M \in \mathcal{M}} \lambda_M = \tilde{\kappa}_f^*(G).$$

This implies that the function $w : \mathcal{M}_f(G) \to [0, 1]$ with $w(M) = \lambda_M$ for each $M \in \mathcal{M}_f$ is a fractional f-coloring of G. Then we obtain $\chi_f'^*(G) \leq \sum_{M \in \mathcal{M}} w(M) = \tilde{\kappa}_f^*(G)$ and the proof is complete. ∎

Every graph G satisfies $\kappa_f^*(G) \leq \tilde{\kappa}_f^*(G) = \chi_f'^*(G)$ and we are now interested in conditions for equality. To find at least a partial answer the following inequality can be used. Let $a_1, \ldots, a_n, b_1, \ldots, b_n$ be positive real numbers with $n \geq 1$. Then

$$\frac{a_1 + \cdots + a_n}{b_1 + \cdots + b_n} \leq \max_{1 \leq i \leq n} \frac{a_i}{b_i} \tag{B.5}$$

With $B = b_1 + \cdots + b_n$ and $M = \max_{1 \leq i \leq n} \frac{a_i}{b_i}$ we obtain

$$\frac{\sum_{i=1}^n a_i}{\sum_{i=1}^n b_i} = \sum_{i=1}^n \frac{a_i}{b_i} \frac{b_i}{B} \leq \sum_{i=1}^n M \frac{b_i}{B} = M,$$

which proves (B.5).

In what follows, let (G, f) be an arbitrary vertex-weighted graph. If $T(G) = \emptyset$, then we clearly have $\tilde{\kappa}_f^*(G) = \Delta_f^*(G)$ and, therefore, $\tilde{\kappa}_f^*(G) = \kappa_f^*(G)$. So assume that $T(G) \neq \emptyset$. Then there is a tuple $(U, F) \in T(G)$ such that

$$\tilde{w}_f^*(G) = \frac{2(|E(G[U])| + |F|)}{f(U) + |F| - 1}.$$

By the definition of $T(G)$, $\emptyset \neq U \subseteq V(G)$, $F \subseteq \partial_G(U)$, and $f(U) + |F|$ is odd and ≥ 3. Then we distinguish two cases.

Case 1: $f(U)$ *is odd.* Then $|F|$ is even. If $|F| = 0$, we obtain

$$\tilde{w}_f^*(G) = \frac{2|E(G[U])|}{f(U) - 1} \leq w_f^*(G).$$

Otherwise, $|F| \geq 2$ and, using (B.5), we obtain

$$\tilde{w}_f^*(G) \leq \max\{\frac{2|E(G[U])|}{f(U) - 1}, 2\} \leq \max\{w_f^*(G), 2\}.$$

Case 2: $f(U)$ *is even.* Then $|F|$ is odd. If $|F| \geq 3$, then based on (B.5) we deduce that

$$
\begin{aligned}
\tilde{w}_f^*(G) &= \frac{2(|E(G[U]| + |F|)}{f(U) + |F| - 1} = \frac{\sum_{u \in U} d_{G[U]}(u) + 2|F|}{f(U) + |F| - 1} \\
&= \frac{\sum_{u \in U} d_G(u) + |F|}{f(U) + |F| - 1} \leq \max\left\{ \frac{\sum_{u \in U} d_G(u)}{\sum_{u \in U} f(u)}, \frac{|F|}{|F| - 1} \right\} \\
&\leq \max\left\{ \max_{u \in U} \frac{d_G(u)}{f(u)}, \frac{|F|}{|F| - 1} \right\} \leq \max\left\{ \Delta_f^*(G), \frac{3}{2} \right\}.
\end{aligned}
$$

It remains to consider the case $|F| = 1$. Then we have

$$
\tilde{w}_f^*(G) = \frac{2|E(G[U])| + 2}{f(U)}.
$$

If $|\partial_G(U)| \geq 2$, then

$$
2|E(G[U])| + 2 = \sum_{u \in U} d_{G[U]}(u) + 2 \leq \sum_{u \in U} d_G(u).
$$

By (B.5), this gives

$$
\tilde{w}_f^*(G) \leq \frac{\sum_{u \in U} d_G(u)}{\sum_{u \in U} f(u)} \leq \Delta_f^*(G).
$$

Otherwise, $|\partial_G(U)| = 1$ and the only edge in F is a cut-edge (or bridge) of G. Since $f(U)$ is even, we have $f(U) \geq 2$. Based on (B.5) we then obtain that

$$
\begin{aligned}
\tilde{w}_f^*(G) &= \frac{\sum_{u \in U} d_{G[U]}(u) + 2}{f(U)} = \frac{\sum_{u \in U} d_G(u) + 1}{\sum_{u \in U} f(u)} \\
&\leq \Delta_f^*(G) + \frac{1}{f(U)} \leq \Delta_f^*(G) + \frac{1}{2}.
\end{aligned}
$$

Hence, the following result is proved.

Proposition B.12 *If* (G, f) *is a vertex-weighted graph, then*

$$
\tilde{w}_f^*(G) \leq \max\{\Delta_f^*(G) + \frac{1}{2}, w_f^*(G), 2\}.
$$

Furthermore, $\tilde{w}_f^*(G) \leq \max\{\Delta_f^*(G), w_f^*(G), 2\}$ *provided that* G *is bridgeless.*

Since any χ_f'-critical graph with $\chi_f' \geq \Delta_f + 1$ is bridgeless (Proposition 8.30), Proposition B.12 implies the following result:

Corollary B.13 *Let* (G, f) *be a vertex-weighted graph with* $\chi_f'(G) \geq \Delta_f(G) + 1$ *and* $\Delta_f^*(G) \geq 2$. *If* G *is* χ_f'-critical, *i.e.,* $\chi_f'(H) < \chi_f'(G)$ *for every proper subgraph* H *of* G, *then* $\chi_f'^*(G) = \kappa_f^*(G)$.

Notes

Fractional graph theory has developed into an important and powerful method in combinatorics and combinatorial optimization. The first monograph on the subject was written by Berge [26] in 1978. Another comprehensive monograph, providing a rational, rather than an integral, approach to the theory of graphs, was written by Scheinerman and Ullman [276] in 1997.

The formula for the fractional chromatic index in Theorem B.5 can be easily derived from Edmonds' matching polytope theorem. This was first noticed by Stahl [286] in 1979. However, Seymour [281] found a different proof of the formula and used this result to give a purely combinatorial proof of Edmonds' matching polytope theorem.

Pulleyblank and Edmonds [250] characterized the facets of the matching polytope. In particular, they proved that a blossom constraint for a set U is a facet of $\mathcal{P}(G)$ if and only if $G[U]$ is **2-connected** (i.e., $G[U]$ is connected and has no cut vertex) and **factor-critical** (i.e., $G[U] - v$ has a perfect matching for every vertex $v \in U$). As observed by Marcotte [210], the above characterization implies the following strengthening of Lemma 6.28 by Fernandes and Thomas [94]: Let G be a graph with $\Delta(G) \leq k$ for some integer k. Then $\kappa(G) = \max\{\Delta(G), w(G)\} \leq k$ if and only if $2|E(G[U])| \leq k(|U| - 1)$ for every vertex set $U \subseteq V(G)$ such that $G[U]$ is 2-connected and factor-critical.

Corollary B.13 can easily be extended to arbitrary χ'_f-critical graphs with $\Delta^*_f \geq 2$. A χ'_f-critical graph G with $\chi'^*(G) = \Delta_f(G) \geq 2$ satisfies $\chi'^*_f(G) = \Delta^*_f(G) = \kappa^*_f(G)$ and $\chi'_f(G) = \lceil \chi'^*_f(G) \rceil$. It is also easy to show that a vertex weighted graph (G, f) with $\Delta_f(G) \leq 2$ satisfies $\chi'_f(G) = \max\{\Delta_f(G), w_f(G)\}$. Combining these results with Theorem 8.29, it follows that every weighted graph (G, f) satisfies

$$\chi'_f(G) \leq \lceil \chi'^*_f(G) \rceil + \sqrt{\lceil \chi'^*_f(G) \rceil / 2}.$$

Zhang, Yu, and Liu [320] proved a strengthening of Theorem B.4 characterizing the f-matching polytope of a graph G. They proved that if $f(v) \geq d_G(v)$ for every vertex v of G, then a vector $\mathbf{x} \in \mathcal{V}(G)$ belongs to $\mathcal{P}_f(G)$ if and only if $\mathbf{x}$ satisfies the capacity constraints, the weighted degree constraints for all $v \in V(G)$, and the weighted blossom constraints for all tuples $(U, F) \in T(G)$ with $F = \emptyset$. As a consequence, they deduced that $\chi'^*_f(G) = \kappa^*_f(G)$ for every weighted graph (G, f) with $f(v) \geq d_G(v)$ for all $v \in V(G)$.

A fractional chromatic index for the fg-color problem can be also defined by relaxation of the integer program defining $\chi'_{f,g}$. However, a combinatorial characterization of the fractional chromatic index seems not known.

REFERENCES

1. AFSHANI, P., GHANDEHARI, M., GHANDEHARI, M., HATAMI, H., TUSSERKANI, R. and ZHU, X. (2005). Circular chromatic index of graphs with maximum degree 3. *J. Graph Theory*, 49:325–335.

2. ALON, N., MCDIARMID, C. J. H. and REED, B. (1991). Acyclic coloring of graphs. *Random Structures and Algorithms*, 2:277–288.

3. ALON, N., SUDAKOV, B. and ZAKS, A. (2001). Acyclic edge colorings of graphs. *J. Graph Theory*, 37:157–167.

4. ANDERSEN, L. D. (1975). *Edge-Colourings of Simple and Non-Simple Graphs*. M.Sc. Thesis, University of Aarhus.

5. ANDERSEN, L. D. (1977). On edge-colourings of graphs. *Math. Scand.*, 40:161–175.

6. ANDERSEN, L. D. (1992). The strong chromatic index of cubic graphs is at most 10. *Discrete Math.*, 108:231–252.

7. ANDERSON, B. A. (1973). Finite topologies and Hamiltonian paths. *J. Combin. Theory Ser. B*, 14:87–93.

8. ANDERSON, B. A. (1977). Symmetry groups of some perfect 1-factorization of complete graphs. *Discrete Math.*, 18:227–234.

9. APPEL, K. AND HAKEN, W. (1977). Every planar map is 4-colorable. Part I: Discharging. *Illinois J. Math*, 21:429–490.

10. APPEL, K., HAKEN, W. AND KOCH, J. (1977). Every planar map is 4-colorable. Part II: Reducibility. *Illinois J. Math*, 21:491–567.

Graph Edge Coloring: Vizing's Theorem and Goldberg's Conjecture,
First Edition. By M. Stiebitz, D. Scheide, B. Toft, and L. M. Favrholdt
Copyright © 2012 John Wiley & Sons, Inc.

11. ASRATIAN, A. S. (2009). A note on transformations of edge colourings of bipartite graphs. *J. Combin. Theory Ser. B,* 99:814–818.

12. ASRATIAN, A. S. and KAMALIAN, R. R. (1987). Interval colorings of the edges of a multigraph (in Russian). *Applied Mathematics Erevan University,* 5:25–34 & 130–131.

13. ASRATIAN, A. S. and KAMALIAN, R. R. (1994). Investigation of interval edge-colorings of graphs. *J. Combin. Theory Ser. B,* 62:34–43.

14. BALISTER, P. N., KOSTOCHKA, A. V., LI, H. and SCHELP, R. H. (2004). Balanced edge colorings. *J. Combin. Theory Ser. B,* 90:3–20.

15. BASAVARAJU, M. and CHANDRAN, L. S. (2009). Acyclic edge coloring of planar graphs., arXiv:0908.2237v1 [cs.DM].

16. BAZGAN, C., HARKAT-BENHAMDINE, A., LI, H. and WOŹNIAK, M. (1999). On the vertex-distinguishing proper edge-colorings of graphs. *J. Combin. Theory Ser. B,* 75:288–301.

17. BAZGAN, C., HARKAT-BENHAMDINE, A., LI, H. and WOŹNIAK, M. (2001). A note on the vertex-distinguishing proper coloring of graphs with large minimum degree. *Discrete Math.,* 236:37–42.

18. BEHZAD, M. (1965). *Graphs and Their Chromatic Numbers.* Ph.D. Thesis, Michigan State University.

19. BEHZAD, M. and CHARTRAND, G. (1967). An introduction to total graphs. In: *Theory of Graphs, International Symposium, Rome 1966,* pages 31–33. Gordon and Breach, New York, and Dunod, Paris.

20. BEINEKE, L. W. (1968). On derived graphs and digraphs. In: Sachs, H., Voß, H. J. and Walther, H., editors, *Beiträge zur Graphentheorie* (Internationales Kolloquium in Manebach vom 9.- 12. Mai 1967), pages 17–23. B. G. Teubner-Verlag, Leipzig.

21. BEINEKE, L. W. (1970). Characterizations of derived graphs. *J. Combin. Theory,* 9:129–135.

22. BEINEKE, L. W. and FIORINI, S. (1976). On small graphs critical with respect to edge colourings. *Discrete Math.,* 16:109–121.

23. BEINEKE, L. W. and WILSON, R. J. (1973). On the edge-chromatic number of a graph. *Discrete Math.,* 5:15–20.

24. BERGE, C. (1958). *Théorie des graphes et ses application.* Dunod, Paris 1958. Russian translation IL, Moscow 1962 ; English translation Meuthen, London 1962, Wiley, New York 1962, Dover, New York 2001.

25. BERGE, C. (1970). *Graphes et hypergraphes.* Dunod, Paris.

26. BERGE, C. (1978). *Fractional Graph Theory.* Indian Statistical Institute Lecture Notes, No. 1, Macmillan, New Delhi.

27. BERGE, C. (1979). Multicolorings of graphs and hypergraphs. In: Ramachandra Rao, editior, *Proceedings of the Symposium on Graph Theory* (Calcutta, 1976), Indian Statistical Institute Lecture Notes, No. 4, pages 1–9. Macmillan, New Delhi.

28. BERGE, C. and FOURNIER, J. C. (1991). A short proof for a generalization of Vizing's theorem. *J. Graph Theory,* 15:333–336.

29. BOLLOBÁS, B. and HARRIS, A. J. (1985). List-colourings of graphs. *Graphs Combin.,* 1:115–127.

30. BORODIN, O. V., KOSTOCHKA, A. V. and WOODALL, D. R. (1997). List edge and list total colourings of multigraphs. *J. Combin. Theory Ser. B*, 71:184–204.

31. BOSÁK, J. (1972). Chromatic index of finite and infinite graphs. *Czechoslovak Math. J.*, 22:272–290.

32. BRINKMANN, G., CHOUDUM, S. A., GRÜNEWALD, S. and STEFFEN, E. (2000). Bounds for the independence number of critical graphs. *Bull. London Math. Soc.*, 32:137–140.

33. BROOKS, R. L. (1941). On colouring the nodes of a network. *Proc. Cambridge Philos. Soc.*, 37:194–197.

34. BU, Y. and WANG, W. (2006). Some sufficient conditions for a planar graph of maximum degree six to be class I. *Discrete Math.*, 306:1440–1445.

35. BURRIS, A. C. and SCHELP, R. H. (1997). Vertex-distinguishing proper edge-colorings. *J. Graph Theory*, 26:73–82.

36. CAPRARA, A. and RIZZI, R. (1998). Improving a family of approximation algorithms to edge color multigraphs. *Inform. Process. Lett.*, 68:11–15.

37. CARIOLARO, D. (2004). *The 1-factorization Problem and Some Related Conjectures*. Ph.D. Thesis, University of Reading.

38. CARIOLARO, D. (2006). On fans in multigraphs. *J. Graph Theory*, 51:301–318.

39. CARIOLARO, D. (2008). An adjacency lemma for critical multigraphs. *Discrete Math.*, 308:4791–4795.

40. CARIOLARO, D. (2009). A theorem in edge colouring. *Discrete Math.*, 309:4208–4209.

41. CARIOLARO, D. and CARIOLARO, G. (2003). Colouring the petals of a graph. *Electron. J. Combin.*, 10:#R6.

42. CARIOLARO, D. and HILTON, A. J. W. (2009). An application of Tutte's theorem to 1-factorization of regular graphs of high degree. *Discrete Math.*, 309:4736–4745.

43. CATLIN, P. A. (1979). Hajós' graph-coloring conjecture: variations and counterexamples. *J. Combin. Theory Ser. B*, 26:268–274.

44. CHEN, G., YU, X. and ZANG, W. (2011). Approximating the chromatic index of multigraphs. *J. Combin. Optim.*, 21:219–246.

45. CHETWYND, A. G. (1984). *Edge-Colourings of Graphs*. Ph.D. Thesis, The Open University, Milton Keynes.

46. CHETWYND, A. G. and HÄGGKVIST, R. (1989). A note on list colourings. *J. Graph Theory*, 13:87–95.

47. CHETWYND, A. G. and HILTON, A. J. W. (1984). Partial edge-colourings of complete graphs or of graphs that are nearly complete. In: Bollobás, B., editor, *Graph Theory and Combinatorics* (Proc. Cambridge Conference in honor of Paul Erdős, 1983), pages 81–97. Academic Press, London.

48. CHETWYND, A. G. and HILTON, A. J. W. (1984). The chromatic index of graphs of even order with many edges. *J. Graph Theory*, 8:463–470.

49. CHETWYND, A. G. and HILTON, A. J. W. (1985). Regular graphs of high degree are 1-factorizable. *Proc. London Math. Soc.*, 50:193–206.

50. CHETWYND, A. G. and HILTON, A. J. W. (1986). Critical star multigraphs. *Graphs Combin.*, 2:209–221.

51. CHETWYND, A. G. and HILTON, A. J. W. (1986). Star multigraphs with three vertices of maximum degree. *Math. Proc. Cambridge Philos. Soc.,* 100:303–317.

52. CHETWYND, A. G. and HILTON, A. J. W. (1989). The edge chromatic class of graphs with maximum degree at least $|V| - 3$. In: Andersen, L. D., Jakobsen, I. T., Thomassen, C., Toft, B. and Vestergard, P. D., editors, *Graph Theory in Memory of G. A. Dirac,* volume 41 of *Ann. Discrete Math.,* pages 91–110. North-Holland, Amsterdam.

53. CHETWYND, A. G. and HILTON, A. J. W. (1989). 1-factorizing regular graphs of high degree – an improved bound. *Discrete Math.,* 75:103–112.

54. CHETWYND, A. G. and HILTON, A. J. W. (1989). A Δ-subgraph condition for a graph to be class 1. *J. Combin. Theory Ser. B,* 46:37–45.

55. CHETWYND, A. G., HILTON, A. J. W. and HOFFMAN, D. G. (1989). On the Δ-subgraph of graphs which are critical with respect to the chromatic index. *J. Combin. Theory Ser. B,* 46:240–245.

56. CHETWYND, A. G. and WILSON, R. J. (1981). Snarks and supersnarks. In: *The Theory and Application of Graphs* (Kalamazoo, Michigan, 1980), pages 215–241. John Wiley & Sons, New York.

57. CHETWYND, A. G. and WILSON, R. J. (1983). The rise and the fall of the critical graph conjecture. *J. Graph Theory,* 7:153–157.

58. CHEW, K. H. (1997). On Vizing's theorem, adjacency lemma and fan argument generalized to multigraphs. *Discrete Math.,* 171:283–286.

59. CHOI, H. and HAKIMI, S. L. (1987). Scheduling file transfers for trees and odd cycles. *SIAM J. Comput.,* 16:162–168.

60. CHOUDUM, S. A. (1977). Chromatic bounds for a class of graphs. *Quart. J. Math., Oxford,* 28:257–270.

61. CHOUDUM, S. A. and KAYATHRI, K. (1999). An extension of Vizing's adjacency lemma on edge chromatic critical graphs. *Discrete Math.,* 206:97–103.

62. CHUDNOVSKY, M. and FRADKIN A. O. (2008). Hadwiger's conjecture for quasi-line graphs. *J. Graph Theory,* 59:17–33.

63. CHUDNOVSKY, M., ROBERTSON, N., SEYMOUR, P. and THOMAS, R. (2006). The strong perfect graph theorem. *Ann. Math.,* 164:51–229.

64. CHUNG, F. R. K., GYÁRFÁS, A., TUZA, ZS. and TROTTER, W. T. (1990). The maximum number of edges in $2K_2$-free graphs of bounded degree. *Discrete Math.,* 81:129–135.

65. CLARK, L. H. and HAILE, D. (1997). Remarks on the size of critical edge-chromatic graphs. *Discrete Math.,* 171:287–293.

66. COFFMAN, E. G., GAREY, J. M. R., JOHNSON, D. S. and LAPAUGH, A. S. (1985). Scheduling file transfers. *SIAM J. Comput.,* 14:744–780.

67. COHEN, N., HAVET, F. and MÜLLER, T. (2009). Acyclic edge-colouring of planar graphs. Extended abstract. In: Nešetřil, J. and Raspaud, A., editors, EuroComb 2009, *Electr. Notes in Discr. Math.,* 34:417-421.

68. DEVOS, M., MKRTCHYAN, V. V. and PETROSYAN, S. S. (2008). 5-cycles and the Petersen graph. arXiv:0801.3714v1 [cs:DM]

69. DIESTEL, R. (2010). *Graph Theory*. Fourth edition. Springer, Berlin.

70. DING, D. (2009). The edge version of Hadwiger's conjecture. *Discrete Math.*, 309:1118–1122.

71. DIRAC, G. A. (1951). *On the Colouring of Graphs: Combinatorial Topology of Linear Complexes*. Ph.D. Thesis, King's College, University of London.

72. DIRAC, G. A. (1951). Note on the colouring of graphs. *Math. Z.*, 54:347–353.

73. DIRAC, G. A. (1952). A property of 4-chromatic graphs and some remarks on critical graphs. *J. London Math. Soc.*, 27:85–92.

74. DUFFIN, R. J. (1965). Topology of series-parallel networks. *J. Math. Anal. Appl.*, 10:303–318.

75. DVOŘÁK, Z., KAWARABAYASHI, K. and KRÁL', D. (2011). Packing six T-joins in plane graphs. http://iti.mff.cuni.cz/series/2010/514.pdf, Preprint.

76. EDMONDS, J. (1965). Maximum matching and a polyhedron with 0-1 vertices. *J. Res. Nat. Bur. Standards Sect. B*, 69B:125–130.

77. EDMONDS, J. and JOHNSON, E. L.(1970). Matching: A well-solved class of integer linear programs. In: Guy, R., Hanani, H., Sauer, N., Schonheim, J., editors, *Combinatorial Structures and Their Applications* (Proceedings of the Calgary International Conference on Combinatorial Structures and Their Applications 1969), pages 69–87. Gordon and Breach, New York.

78. EHRENFEUCHT, A., FABER, V. and KIERSTEAD, H. A (1984). A new method of proving theorems on chromatic index. *Discrete Math.*, 52:159–164.

79. EL-ZANATI, S. I., PLANTHOLT, M. J. and TIPNIS, S. K. (1995). Factorization of regular multigraphs into regular graphs. *J. Graph Theory*, 19:93–105.

80. ERDŐS, P (1959). Graph theory and probability. *Canad. J. Math.*, 11:34–38.

81. ERDŐS, P (1988). Problems and results in combinatorial analysis and graph theory. *Discrete Math.*, 72:81–92.

82. ERDŐS, P. and FAJTLOWICZ, S. (1981). On the conjecture of Hajós. *Combinatorica*, 1:141–143.

83. ERDŐS, P. and HAJNAL, A. (1966). On the chromatic number of graphs and set-systems. *Acta Math. Acad. Sci. Hungar.*, 17:61–99.

84. ERDŐS, P., RUBIN, A. L. and TAYLOR, H. (1979). Choosability in graphs. Proc. West-Coast Conf. on Combinatorics, Graph Theory and Computing. *Congr. Numer.*, XXVI:125–157.

85. ERDŐS, P. and WILSON, R. J. (1977). On the chromatic index of almost all graphs. *J. Combin. Theory Ser. B*, 23:255–257.

86. FAN, G. and RASPAUD, A. (1994). Fulkerson's conjecture and circuit covers. *J. Combin. Theory Ser. B*, 61:133–138.

87. FAN, H. and FU, H. (2004). The edge-coloring of graphs with small genus. *Ars Combin.*, 73:219–224.

88. FAUDREE, R. J., GYÁRFÁS, A., SCHELP, R. H. and TUZA, ZS. (1989). Induced matchings in bipartite graphs. *Discrete Math.*, 78:83–87.

89. FAUDREE, R. J., SCHELP, R. H., GYÁRFÁS, A. and TUZA, ZS. (1990). The strong chromatic index of a graph. *Ars Combin.*, 29-B:205–211.

90. FAVRHOLDT, L. M. (1998). *Kantfarvning af Grafer*. M.Sc. Thesis, Odense University.

91. FAVRHOLDT, L. M., STIEBITZ, M. and TOFT, B. (2006). Graph edge colouring: Vizing's Theorem and Goldberg's Conjecture. Preprint: DMF-2006-10-003, IMADA-PP-2006-20, University of Southern Denmark.

92. FEIGE, U., OFEK, E. and WIEDER, U. (2002). Approximating maximum edge coloring in multigraphs. *Lecture Notes in Comput. Sci.*, 2462:108–121.

93. FEKETE, M. (1923). Über die Verteilung der Wurzeln bei gewissen algebraischen Gleichungen mit ganzzahligen Koeffizienten. *Math. Z.*, 17:228–249.

94. FERNANDES, C. G. and THOMAS, R. (2000). Edge-coloring series-parallel multigraps. arXiv:1107.5370v1 [cs.DS].

95. FIAMČIK, J. (1978). The acyclic chromatic class of a graph (in Russian). *Math. Slovaca*, 28:139–145.

96. FINCK, H.-J. and SACHS, H. (1969). Über eine von H. S. Wilf angegebene Schranke für die chromatische Zahl endlicher Graphen. *Math. Nachr.*, 39:373–386.

97. FIOL, M. A. (1980). *3-Grafos Criticos*. Doctoral Thesis, Universidad Politèchnica de Barcelona.

98. FIORINI, S. (1974). *The Chromatic Index of Simple Graphs*. Ph.D. Thesis, The Open University, England.

99. FIORINI, S. (1975). Some remarks on a paper by Vizing on critical graphs. *Math. Proc. Cambridge Philos. Soc.*, 77:475–483.

100. FIORINI, S. and WILSON, R. J. (1977). *Edge-Colourings of Graphs*. Research Notes in Mathematics, Vol. 16, Pitman, London.

101. FOUQUET, J. L. and VANHERPE, J. M. (2009). On the perfect matching index of bridgeless cubic graphs. arXiv:0904.1296v1.

102. FOURNIER, J. C. (1973). Coloration des arêtes d'un graphe. *Cahiers Centre Études Recherche Opér.*, 15:311–314.

103. FOURNIER, J. C. (1977). Méthode et théorème général de coloration des arêtes d'un multigraphe. *J. Math. Pures Appl.*, 56:437–453.

104. FRINK, O. (1926). A proof of Petersen's theorem. *Ann. Math.*, 27:491–493.

105. FULKERSON, D. R. (1971). Blocking and anti-blocking pairs of polyhedra. *Mathematical Programming*, 1:168–194.

106. GARDNER, M. (1976). Mathematical games. *Sci. Amer.*, 234:126–130.

107. GALVIN, F. (1995). The list chromatic index of a bipartite multigraph. *J. Combin. Theory Ser. B*, 63:153–158.

108. GODDYN, L. A. (1993). Cones, lattices and Hilbert bases of circuits and perfect matchings. In: Robertson, N. and Seymour, P., editors, *Graph Structure Theory* (Proceedings Joint Summer Research Conference on Graph Minors, Seattle, Washington, 1991), pages 419–439. Contemporary Mathematics, volume 147, American Mathematical Society, Providence, Rhode Island.

109. GOLDBERG, M. K. (1972). Odd cycles of a multigraph with a large chromatic class (in Russian). *Upravljaemye Sistemy*, 10:45–47.

110. GOLDBERG, M. K. (1973). On multigraphs of almost maximal chromatic class (in Russian). *Diskret. Analiz*, 23:3–7.

111. GOLDBERG, M. K. (1974). A remark on the chromatic class of a multigraph (in Russian). *Vyčisl. Mat. i. Vyčisl. Tehn. (Kharkow)*, 5:128–130.

112. GOLDBERG, M. K. (1977). Structure of multigraphs with restriction on the chromatic class (in Russian). *Diskret. Analiz*, 30:3–12.

113. GOLDBERG, M. K. (1981). Construction of class 2 graphs with maximum vertex degree 3. *J. Combin. Theory Ser. B*, 31:282–291.

114. GOLDBERG, M. K. (1984). Edge-coloring of multigraphs: recoloring technique. *J. Graph Theory*, 8:123–137.

115. GRÜNEWALD, S. (2000). Chromatic-index critical multigraphs of order 20. *J. Graph Theory*, 33:240–245.

116. GRÜNEWALD, S. (2000). *Chromatic Index Critical Graphs and Multigraphs*. Doctoral Thesis, Universität Bielefeld, Bielefeld.

117. GRÜNEWALD, S. and STEFFEN, E. (1999). Chromatic-index critical graphs of even order. *J. Graph Theory*, 30:27–36.

118. GRÜNEWALD, S. and STEFFEN, E. (2004). Independent sets and 2-factors in edge-chromatic-critical graphs. *J. Graph Theory*, 45:113–118.

119. GUENIN, B. (2011). Packing T-joins and edge coloring in planar graphs. Manuscript.

120. GUPTA, R. P. (1967). *Studies in the Theory of Graphs*. Ph.D. Thesis, Tata Institute of Fundamental Research, Bombay.

121. GUPTA, R. P. (1978). On the chromatic index and the cover index of a multigraph. In: Donald, A. and Eckmann, B., editors, *Theory and Applications of Graphs* (Proceedings, Michigan May 11–15, 1976), volume 642 of *Lecture Notes in Mathematics*, pages 91–110. Springer, Berlin.

122. GUTIN, G. and TOFT, B. (2000). Interview with Vadim G. Vizing. *European Math. Soc. Newsletter*, 38:22–23.

123. GYÁRFÁS, A. (1988). Problems from the world surrounding perfect graphs. *Zastos. Mat.*, 19:413–431.

124. HADWIGER, H. (1943). Über eine Klassifikation der Streckenkomplexe. *Vierteljschr. Naturforsch. Ges. Zürich*, 88:133–142.

125. HAILE, D. (1999). Bounds on the size of critical edge-chromatic graphs. *Ars Combin.*, 53:85–96.

126. HAJÓS, G. (1961). Über eine Konstruktion nicht n-färbbarer Graphen. *Wiss. Z. Martin Luther Univ. Halle-Wittenberg, Math.-Natur. Reihe*, 10:116–117.

127. HAKIMI, S. L. and KARIV, O. (1986). A generalization of edge-colouring in graphs. *J. Graph Theory*, 10:139–154.

128. HAKIMI, S. L. and SCHMEICHEL, E. F. (1999). Improved bounds for the chromatic index of graphs and multigraphs. *J. Graph Theory*, 32:311–326.

129. HANSON, D., LOTEN, C. O. M. and TOFT, B. (1998). On interval colourings of bi-regular bipartite graphs. *Ars Combin.*, 50:23–32.

130. HANSEN, H. M. (1992). *Scheduling with Minimum Waiting Periods* (in Danish). M.Sc. Thesis, Odense University.

131. HAO, R., NIU, J., WANG, X., ZHANG, C. Q. and ZHANG, T. (2009). A note on Berge-Fulkerson coloring. *Discrete Math.*, 309:4235–4240.

132. HAXELL, P. and MCDONALD, J. (2011). On characterizing Vizing's edge colouring bound. *J. Graph Theory,* doi: 10.1002/jgt.20571.

133. HEAWOOD, P. J. (1890). Map colour theorem. *Quart. J. Pure Appl. Math.,* 24:332–338.

134. HILTON, A. J. W. (1975). Colouring the edges of a multigraph so that each vertex has at most j, or at least j, edges of each colour on it. *J. London Math. Soc.,* 12:122–128.

135. HILTON, A. J. W. (1987). Recent progress on edge-colouring graphs. *Discrete Math.,* 64:303–307.

136. HILTON, A. J. W. (1989). Two conjectures on edge-colouring. *Discrete Math.,* 74:61–64.

137. HILTON, A. J. W. and JACKSON, B. (1987). A note concerning the chromatic index of multigraphs. *J. Graph Theory,* 11:267–272.

138. HILTON, A. J. W. and JOHNSON, P. D. (1987). Graphs which are vertex-critical with respect to the edge-chromatic number. *Math. Proc. Cambridge Philos. Soc.,* 102:211–221.

139. HILTON, A. J. W. and DE WERRA D. (1982). Sufficient conditions for balanced and for equitable edge-coloring of graphs. O. R. Working paper 82/3. Dépt. of Math., Ecole Polytechnique Fédérate de Lausanne, Switzerland.

140. HILTON, A. J. W. and DE WERRA D. (1994). A sufficient condition for equitable edge-colorings of simple graphs. *Discrete Math.,* 128:179–201.

141. HILTON, A. J. W. and WILSON, R. J. (1989). Edge-colorings of graphs: a progress report. In Cabobianco, M. F. et al., editors, *Graph Theory and its Application: East and West* (Jinan, 1986), pages 241–249. Ann. New York Acad. Sci., 576, New York.

142. HILTON, A. J. W. and ZHAO, C. (1992). The chromatic index of a graph whose core has maximum degree two. Special volume to mark the centennial of Julius Petersen's "Die Theorie der regulären graphs", Part II. *Discrete Math.,* 101:135–147.

143. HILTON, A. J. W. and ZHAO, C. (1996). On the edge-colouring of graphs whose core has maximum degree two. *J. Combin. Math. Combin. Comp.,* 21:97–108.

144. HILTON, A. J. W. and ZHAO, Y. (1997). Vertex-splitting and chromatic index critical graphs. Second International Colloquium on Graphs and Optimization (Leukerbad, 1994). *Discrete Appl. Math.,* 76:205–211.

145. HIND, H. and ZHAO, Y. (1998). Edge colorings of graphs embeddable in a surface of low genus. *Discrete Math.,* 190:107–114.

146. HOCHBAUM, D. S., NISHIZEKI, T. and SHMOYS, D. B. (1986). A better than "best possible" algorithm to edge color multigraphs. *J. Algorithms,* 7:79–104.

147. HOFFMAN, D. G., MITCHEM, J. and SCHMEICHEL, E. F. (1992). On edge-coloring graphs. *Ars Combin.,* 33:119–128.

148. HOFFMAN, D. G. and RODGER, C. A. (1988). Class one graphs. *J. Combin. Theory Ser. B,* 44:372–376.

149. HOLROYD, F. and ŠKOVIERA, M. (2004). Colouring of cubic graphs by Steiner triple systems. *J. Combin. Theory Ser. B,* 91:57–66.

150. HOLYER, I. (1981). The NP-completeness of edge-colouring. *SIAM J. Comput.,* 10:718–720.

151. HORÁK, P., QING, H. and TROTTER, W. T. (1993). Induced matchings in cubic graphs. *J. Graph Theory,* 17:151–160.

152. JAEGER, F. (1988). Nowhere-zero flow problems. In: Beineke, L. W. and Wilson, R. J., editors, *Selected Topics in Graph Theory.* Vol. 3, pages 71–95. Academic Press, New York.

153. JAVDEKAR, M. (1980). Note on Choudum's "Chromatic bound for a class of graphs". *J. Graph Theory,* 4:265–267.

154. JAKOBSEN, I. T. (1973). Some remarks on the chromatic index of a graph. *Arch. Math. (Basel),* 24:440–448.

155. JAKOBSEN, I. T. (1974). On critical graphs with chromatic index 4. *Discrete Math.,* 9:265–276.

156. JAKOBSEN, I. T. (1975). On critical graphs with respect to edge-coloring. In: Hajnal, A., Rado, R. and Sós, V., editors, *Infinite and Finite Sets* (Colloq., Keszthely, 1973; dedicated to P. Erdős on his 60th birthday), Vol. II, pages 927–934. Colloq. Math. Soc. János Bolyai, Vol. 10, North-Holland, Amsterdam.

157. JENSEN, T. R. and TOFT, B. (1995). *Graph Coloring Problems.* Wiley-Interscience Series in Discrete Mathematics and Optimization, John Wiley & Sons, New York.

158. JENSEN, T. R. and TOFT, B. (1995). Choosability versus chromaticity — the plane unit distance graph has a 2-chromatic subgraph of infinite list-chromatic number. *Geombinatorics,* 5:45–64.

159. JUVAN, M., MOHAR, B. and THOMAS, R. (1999). List edge-colorings of series parallel graphs. *Electron. J. Combin.,* 6:#R42.

160. KAHN, J. (1996). Asymptotically good list-colorings. *J. Combin. Theory Ser. A,* 73:1–59.

161. KAHN, J. (1996). Asymptotics of the chromatic index for multigraphs. *J. Combin. Theory Ser. B,* 68:233–254.

162. KAISER, T., KRÁL', D. and NORINE, S. (2006). Unions of perfect matchings in cubic graphs. In: Klazar, M., Kratochvíl, J., Matousek, J., Thomas, R. and Valtr, P., editors. *Topics in Discrete Mathematics* (Dedicated to Jarik Nešetril on the occasion of his 60th birthday), pages 225–230. Springer, Berlin.

163. KAISER, T., KRÁL', D. and ŠKREKOVSKI, R. (2004). A revival of the girth conjecture. *J. Combin. Theory Ser. B,* 92:41–53.

164. KAYATHRI, K. (1994). On the size of edge-chromatic critical graphs. *Graphs Combin.,* 10:139–144.

165. KEMPE, A. B. (1879). On the geographical problem of four colours. *Amer. J. Math.,* 2:193–200.

166. KIERSTEAD, H. A. (1984). On the chromatic index of multigraphs without large triangles. *J. Combin. Theory Ser. B,* 36:156-160.

167. KIERSTEAD, H. A. (1989). Applications of edge coloring of multigraphs to vertex coloring of graphs. *Discrete Math.,* 74:117–124.

168. KIERSTEAD, H. A. and SCHMERL, J. H. (1983). Some applications of Vizing's theorem to vertex colorings of graphs. *Discrete Math.,* 45:277–285.

169. KIERSTEAD, H. A. and SCHMERL, J. H. (1986). The chromatic number of graphs which induce neither $K_{1,3}$ nor $K_5 - e$. *Discrete Math.*, 58:253–262.

170. KILAKOS, K. and SHEPHERD, F. B. (1996). Subdivisions and the chromatic index of r-graphs. *J. Graph Theory*, 22:203–212.

171. KILAKOS, K. and SHEPHERD, F. B. (1996). Excluding minors in cubic graphs. *Combin. Probab. Comput.*, 5:57–78.

172. KING, A. D., REED, B. A. and VETTA, A. (2007). An upper bound for the chromatic number of line graphs. *European J. Combin.*, 28:2182–2187.

173. KOCHOL, M., KRIVOŇÁKOVÁ, N. and SMEJOVÁ, S. (2005). Edge-coloring of multigraphs. *Discrete Math.*, 300:229–234.

174. KÖNIG, D. (1916). Über Graphen und ihre Anwendungen auf Determinantentheorie und Mengenlehre. *Math. Ann.*, 77:453–465.

175. KÖNIG, D. (1936). *Theorie der Endlichen und Unendlichen Graphen.* Akademische Verlagsgesellschaft M.B.H., Leipzig. Reprinted by Chealsea 1950 and by B. G. Teubner 1986. English translation published by Birkhäuser 1990.

176. KOSTER, M. C. A. (2005). Wavelet Assignment in Multi-fiber WDM networks by generalized edge coloring. In *Proc. INOC 2005*, pages 60–66.

177. KOSTOCHKA, A. V. and STIEBITZ, M. (2006). Edge colouring: a new fan equation. Manuscript.

178. KOSTOCHKA, A. V. and STIEBITZ, M. (2008). Partitions and edge colouring of multigraphs. *Electron. J. Combin.*, 15:#N25.

179. KOTZIG, A. (1957). On the theory of finite regular graphs of degree three and four (in Slovak). *Časopis pro Pěstování Matematiky*, 82:76–92.

180. KOTZIG, A. (1964). Hamilton graphs and Hamilton circuits. In: Fiedler, M., editor, *Theory of Graphs and Its Application* (Proceedings of the Symposium held in Smolenice in June 1963), pages 62–82. Publishing House of the Czechoslovak Academy of Sciences (Prague) and Academic Press (New York and London).

181. KRÁL', D., MÁČAJOVÁ, E., MAZÁK, J. and SERENI, J.-S. (2010). Circular edge-colorings of cubic graphs with girth six. *J. Combin. Theory Ser. B*, 100:351–358.

182. KRÁL', D., MÁČAJOVÁ, E., PANGRÁC, O., RASPAUD, A., SERENI, J.-S. and ŠKOVIERA, M. (2009). Projective, affine, and abelian colorings of cubic graphs. *European J. Combin.*, 30:53–69.

183. KRÁL', D., SERENI, J.-S. and STIEBITZ, M. (2007). Personal communication.

184. KRAWCZYK, H. and KUBALE, M. (1985). An approximation algorithm for diagnostic test scheduling in multicomputer systems. *IEEE Trans. Comput.*, C-34:869–872.

185. KRUSENSTJERNA-HAFSTRØM, U. and TOFT, B. (1980). Special subdivisions of K_4 and 4-chromatic graphs. *Monatsh. Math.*, 89:101–110.

186. KURATOWSKI, C. (1930). Sur le problème des courbes gauches en topologie. *Fund. Math.*, 15:271–283.

187. KURT, O. (2009). *On the Edge Coloring of Graphs.* Ph.D. Thesis, The Ohio State University.

188. LAM, P., LIU, J., SHIU, W. and WU, J. (1999). Some sufficient conditions for planar graphs to be of class 1. *Congr. Numer.*, 136:201–205.

189. LASKAR, R. and HARE, W. (1972). Chromatic number of certain graphs. *J. London Math. Soc.,* 4:489–492.

190. LI, S and LI, X. (2009). Edge coloring of graphs with small maximum degrees. *Discrete Math.,* 309:4843–4852.

191. LI, X. (2005). Average degrees of critical graphs. *Ars Combin.,* 74:303–322.

192. LI, X. (2006). Size of critical graphs with small maximum degrees. *Graphs Combin.,* 22:503–513.

193. LI, X. (2010). A new lower bound on the average degree of edge critical graphs. Manuscript.

194. LI, X. (2011). A lower bound on critical graphs with maximum degree of 8 and 9. *Ars Combin.,* 98:241–257.

195. LI, X. and LUO, R. (2003). Edge coloring of embedded graphs with large girth. *Graphs Combin.,* 19:393–401.

196. LI, X., LUO, R. and NIU, J. (2006). A note on class one graphs with maximum degree six. *Discrete Math.,* 306:1450–155.

197. LI, X., LUO, R., NIU, J. and ZHANG, X. (2006). Edge coloring of graphs with small maximum degrees. Manuscript.

198. LOVÁSZ, L. (1987). Matching structure and the matching lattice. *J. Combin. Theory Ser. B,* 43:187–222.

199. LUO, R., MIAO, L. and ZHAO, Y. (2009). The size of edge chromatic critical graphs with maximum degree 6. *J. Graph Theory,* 60:149–171.

200. LUO, R. and ZHANG, C. (2004). Edge coloring of graphs with small average degree. *Discrete Math.,* 275:207–218.

201. LUO, R. and ZHAO, Y. (2006). A note on Vizing's independence number conjecture of edge chromatic critical graphs. *Discrete Math.,* 306:1788–1790.

202. LUO, R. and ZHAO, Y. (2008). Finding the exact bound of the maximum degrees of class two graphs embeddable in a surface of characteristic $\epsilon \in \{-1, -2, -3\}$. *J. Combin. Theory Ser. B,* 98:707–720.

203. LUO, R. and ZHAO, Y. (2009). An application on Vizing-like adjacency lemmas to Vizing's independence number conjecture of edge chromatic critical graphs. *Discrete Math.,* 309:2925–2929..

204. LUO, R. and ZHAO, Y. (2010). Finding $\Delta(\Sigma)$ for surface Σ of characteristic $\chi(\Sigma) = -5$. *J. Graph Theory,* DOI: 10.1002/jgt.20548.

205. LUO, R. and ZHAO, Y. (2010). A new upper bound for the independence number of edge chromatic critical graphs. *J. Graph Theory,* DOI: 10.1002/jgt.20552.

206. MÁČAJOVÁ, E. and ŠKOVIERA, M. (2005). Fano colourings of cubic graphs and the Fulkerson Conjecture. *Theoret. Comput. Sci.,* 349:112–120.

207. MARCOTTE, O. (1986). On the chromatic index of multigraphs and a conjecture of Seymour, (I). *J. Combin. Theory Ser. B,* 41:306–331.

208. MARCOTTE, O. (1990). On the chromatic index of multigraphs and a conjecture of Seymour, (II). In: Cook, W. and Seymour, P., editors, *Polyhedral Combinatorics.* DIMACS Ser. in Discrete Math. Theoret. Comput. Sci., Vol. 1, pages 245–279. American Mathematical Society, Providence, Rhode Island.

209. MARCOTTE, O. (1990). Exact edge-colorings of graphs without prescribed minors. In: Cook, W. and Seymour, P., editors, *Polyhedral Combinatorics*. DIMACS Ser. in Discrete Math. Theoret. Comput. Sci., Vol. 1, pages 235–243. American Mathematical Society, Providence, Rhode Island.

210. MARCOTTE, O. (2001). Optimal edge-colorings for a class of planar multigraphs. *Combinatorica*, 21:361–394.

211. MATULA, D. W. (1968). A min-max theorem for graphs with application to graph coloring. *SIAM Rev.*, 10:481–482.

212. MAZZUOCCOLO, G. (2011). The equivalence of two conjectures of Berge and Fulkerson. *J. Graph Theory*, DOI: 10.1002/jgt.20545.

213. MCDIARMID, C. J. H. (1972). The solution of a timetabling problem. *J. Inst. Math. Appl.*, 9:23–34.

214. MCDONALD, J. M. (2009). *Multigraphs with High Chromatic Index*. Ph.D. Thesis, University of Waterloo.

215. MCDONALD, J. M. (2009). Achieving maximum chromatic index in multigraphs. *Discrete Math.*, 309:2077–2084.

216. MCDONALD, J. M. (2010). On multiples of simple graphs and Vizing's theorem. *Discrete Math.*, 310:2212–2214.

217. MCDONALD, J. M. (2010). Personal communication.

218. MCDONALD, J. M. (2011). On a theorem of Goldberg. *J. Graph Theory*, 68:8–21.

219. MCDONALD, J. M., MOHAR, B. and SCHEIDE, D. (2011). Kempe equivalence of edge-colorings subcubic and subquartic graphs. *J. Graph Theory*, doi: 10.1002/jgt.20613.

220. MEL'NIKOV, L. S. (1970). The chromatic class and the location of a graph on a closed surface (in Russian). *Mat. Zametki*, 7:671–681. English translation in *Math. Notes*, (1970), 7:405–411.

221. MIAO, L. (2011). On the independence number of edge chromatic critical graphs. *Ars Combin.*, 98:471–481.

222. MIAO, L. and PANG, S. (2008). On the size of edge-coloring critical graphs with maximum degree 4. *Discrete Math.*, 308:5856–5859.

223. MIAO, L., PANG, S. and WU, J. (2003). An upper bound on the number of edges of edge-coloring critical graphs with high maximum degree. *Discrete Math.*, 271:321–325.

224. MIAO, L. and WU, J. (2002). Edge-coloring critical graphs with high degree. *Discrete Math.*, 257:169–172.

225. MOHAR, B. (2006). Kempe equivalence of colorings. In Bondy, J. A. et al., editors, *Graph Theory in Paris* (Proceedings of a Conference in Memory of Claude Berge), pages 287–297. Birkhäuser, Basel.

226. MOLLOY, M. and REED, B. (1997). A bound on the strong chromatic index of a graph. *J. Combin. Theory Ser. B*, 69:103–109.

227. MOLLOY, M. and REED, B. (1998). A bound on the total chromatic number. *Combinatorica*, 18:241–280.

228. MOLLOY, M. and REED, B. (1998). Further algorithmic aspects of the Local Lemma. In: Proceedings of the 30th Annual ACM Symposium on Theory of Computing (STOC, Dallas, Texas, 1998), pages 524–529.

229. MOLLOY, M. and REED, B. (2002). *Graph Coloring and the Probabilistic Method.* Algorithms and Combinatorics, Vol. 23, Springer, Berlin.

230. NAKANO, S. and NISHIZEKI, T. (1993). Scheduling file transfers under port and channel constraints. *Intern. J. Found. Comput. Sci.,* 4:101–115.

231. NAKANO, S., NISHIZEKI, T. and SAITO, N. (1988). On the f-coloring of multigraphs. *IEEE Trans. Circuits and System,* 35:345–353.

232. NAKANO, S., NISHIZEKI, T. and SAITO, N. (1990). On the fg-coloring of graphs. *Combinatorica,* 10:67–80.

233. NASH-WILLIAMS, C. ST. J. A. (1971). Hamiltonian arcs and circuits. In: Capobianco, M., Frechen, J. B. and Krolik, M., editors, *Recent Trends in Graph Theory* (Proceedings of the First New York City Graph Theory Conference held on June 11, 12, and 13, 1970), volume 186 of *Lecture Notes in Mathematics,* pages 197–211. Springer, Berlin.

234. NASERASR, R. and ŠKREKOVSKI, R. (2003). The Petersen graph is not 3-edge-colorable - a new proof. *Discrete Math.,* 268:325–326.

235. NIESSEN, T. (1994). How to find overfull subgraphs in graphs with large maximum degree. 2nd Twente Workshop on Graphs and Combinatorial Optimization (Enschede, 1991). *Discrete Appl. Math.,* 51:117–125.

236. NIESSEN, T. and VOLKMANN, L. (1990). Class 1 conditions depending on the minimum degree and the number of vertices of maximum degree. *J. Graph Theory,* 14:225–246.

237. NISHIZEKI, T. and KASHIWAGI, K. (1990). On the 1.1-edge-coloring of multigraphs. *SIAM J. Discrete Math.,* 3:391–410.

238. ORE, O. (1967). *The Four-Colour Problem.* Academic Press, New York.

239. PATEL, V. (2006). Unions of perfect matchings in cubic graphs and implications of the Berge-Fulkerson conjecture. CDAM Research Report LSE-CDAM-2006-06.

240. PERKOVIĆ, L. and REED, B. (1997). Edge coloring regular graphs of high degree. Graphs and combinatorics (Marseille, 1995). *Discrete Math.,* 165/166:567–578.

241. PETERSEN, J. (1891). Die Theorie der regulären graphs. *Acta Math.,* 15:193–220.

242. PETERSEN, J. (1898). Sur le théorème de Tait. *L'Intermédiaire des Mathématiciens,* 15:225–227.

243. PLANTHOLT, M. (1981). The chromatic index of graphs with a spanning star. *J. Graph Theory,* 5:45–153.

244. PLANTHOLT, M. (1983). The chromatic index of graphs with large maximum degree. *Discrete Math.,* 47:91–96.

245. PLANTHOLT, M. (1999). A sublinear bound on the chromatic index of multigraphs. *Discrete Math.,* 202:201–213.

246. PLANTHOLT, M. (2004). Overfull conjecture for graphs with high minimum degree. *J. Graph Theory,* 47:73–80.

247. PLANTHOLT, M. and TIPNIS, S. K. (1991). Regular multigraphs of high degree are 1-factorizable. *J. London Math. Soc.,* 44:393–400.

248. PLANTHOLT, M. and TIPNIS, S. K. (1997). The chromatic index of multigraphs of order at most 10. *Discrete Math.,* 177:185–193.

249. PLANTHOLT, M. and TIPNIS, S. K. (2001). All regular multigraphs of even order and high degree are 1-factorable. *Electron. J. Combin.*, 8(1):#R41.

250. PULLEYBLANK, W. and EDMONDS, J. (1974). Facets of 1-matching polyhedra. In: Berge, C., Ray-Chaudhuri, D, editors, *Hypergraph Seminar* (Proceedings Working Seminar on Hypergraphs, Columbus, Ohio, 1972), pages 214–242. Springer, Berlin.

251. RANDERATH, B. (1998). *The Vizing Bound for the Chromatic Number Based on Forbidden Pairs*. Doctoral Thesis, RWTH Aachen, Shaker Verlag, Aachen.

252. REED, B. (1998). ω, δ, and χ. *J. Graph Theory,* 27:177–212.

253. REED, B. and SEYMOUR, P. (2004). Hadwiger's conjecture for line graphs. *European J. Combin.,* 25:873–876.

254. RINGEL, G. (1954). Bestimmung der Maximalzahl der Nachbargebiete von nichtorientierbaren Flächen. *Math. Ann.,* 127:181–214.

255. RINGEL, G. and YOUNGS, J. W. T. (1968). Solution of the Heawood map-coloring problem. *Proc. Nat. Acad. Sci. U.S.A.,* 60:438–445.

256. RIZZI, R. (1989). König's edge coloring theorem without augmenting paths. *J. Graph Theory,* 29:87.

257. RIZZI, R. (2009). Approximating the maximum 3-edge-colorable subgraph problem. *Discrete Math.,* 309:4166–4170.

258. ROBERTSON, N., SANDERS, D. P., SEYMOUR, P. and THOMAS, R. (1997). The four colour theorem. *J. Combin. Theory Ser. B,* 70:2–44.

259. SAINTE-LAGUË, A (1923). Les réseaux. *Annales de la faculté des sciences de Toulouse,* 15:27–86.

260. SAINTE-LAGUË, A (1924). Les réseaux. Privately printed thesis, 63 pages, Toulouse.

261. SAINTE-LAGUË, A. (1926). Les résaux ou graphes. *Mémorial des Sciences Mathématiques,* Fascicule, volume XVIII, Gauthier-Villars, Paris.

262. SANDERS, P. and STEURER, D. (2008). An asymptotic approximation scheme for multigraph edge coloring. *ACM Trans. Algorithms,* 4(2):Article 21.

263. SANDERS, D. P. and ZHAO, Y. (2000). Coloring edges of embedded graphs. *J. Graph Theory,* 35:197–205.

264. SANDERS, D. P. and ZHAO, Y. (2001). Planar graphs of maximum degree seven are class I. *J. Combin. Theory Ser. B,* 83:201–212.

265. SANDERS, D. P. and ZHAO, Y. (2002). On the size of edge chromatic critical graphs. *J. Combin. Theory Ser. B,* 86:408–412.

266. SANDERS, D. P. and ZHAO, Y. (2003). Coloring edges of graphs embedded in a surface of characteristic zero. *J. Combin. Theory Ser. B,* 87:254–263.

267. SCHEIDE, D. (2007). *Kantenfärbungen von Multigraphen*. Diploma Thesis, TU Ilmenau, Ilmenau.

268. SCHEIDE, D. (2007). On a 15/14-edge-colouring of multigraphs. Preprint: DMF-2007-09-007, IMADA-PP-2007-11, University of Southern Denmark.

269. SCHEIDE, D. (2008). *Edge Colourings of Multigraphs*. Doctoral Thesis, TU Ilmenau, Ilmenau.

270. SCHEIDE, D. (2009). A polynomial-time $(\Delta + \sqrt{(\Delta - 1)/2})$-edge colouring algorithm. Preprint: DMF-2009-03-0071, IMADA-PP-2009-04, University of Southern Denmark.

271. SCHEIDE, D. (2010). Graph edge colouring: Tashkinov trees and Goldberg's conjecture. *J. Combin. Theory Ser. B,* 100:68–96.

272. SCHEIDE, D. and STIEBITZ, M. (2009). On Vizing's bound for the chromatic index of a multigraph. *Discrete Math.,* 309:4920–4925.

273. SCHEIDE, D. and STIEBITZ, M. (2009). Approximating the f-chromatic index of multigraphs. Preprint: DMF-2009-05-005, IMADA-PP-2009-10, University of Southern Denmark.

274. SCHEIDE, D. and STIEBITZ, M. (2010). Vizing's colouring algorithm and the fan number. *J. Graph Theory,* 65:115–138.

275. SCHEIDE, D. and STIEBITZ, M. (2011). The maximum chromatic index of multigraphs with given Δ and μ. *Graphs Combin.,* DOI 10.1007/s00373-011-1068-4.

276. SCHEINERMAN, E. R. and ULLMAN, D. H. (1997). *Fractional Graph Theory, A Rational Approach to the Theory of Graphs.* Wiley-Interscience Series in Discrete Mathematics and Optimization, John Wiley & Sons, New York.

277. SCHRIJVER, A. (2003). *Combinatorial Optimization: Polyhedra and Efficiency.* Algorithms and Combinatorics, Vol. 24, Springer, Berlin.

278. SEVAST'YANOV, S. V. (1990). The interval colorability of the edges of a bipartite graph (in Russian). *Metody Diskret. Analiz.,* 50:61-72, 86.

279. SEYMOUR, P. (1979). Sums of circuits. In: Bondy, J. A. and Murty, U. S. R., editors, *Graph Theory and Related Topics* (Proceedings Conference, Waterloo, Ontario, 1977), pages 341–355. Academic Press, New York.

280. SEYMOUR, P. (1979). Some unsolved problems on one-factorizations of graphs. In: Bondy, J. A. and Murty, U. S. R., editors, *Graph Theory and Related Topics* (Proceedings Conference, Waterloo, Ontario, 1977), pages 367–368. Academic Press, New York.

281. SEYMOUR, P. (1979). On multicolorings of cubic graphs, and conjectures of Fulkerson and Tutte. *Proc. London Math. Soc.,* 38:423–460.

282. SEYMOUR, P. (1981). On Tutte's extension of the four-color problem. *J. Combin. Theory Ser. B,* 31:82–94.

283. SEYMOUR, P. (1990). Colouring series-parallel graphs. *Combinatorica,* 10:379–392.

284. SHANNON, C. E. (1949). A theorem on coloring the lines of a network. *J. Math. Phys.,* 28:148–151.

285. SOIFER, A. (2009). *The Mathematical Coloring Book,* Springer, Berlin.

286. STAHL, S. (1979). Fractional edge colorings. *Cahiers Centre Études Rech. Opér.,* 21:127–131.

287. STEFFEN, E. (2000). A refinement of Vizing's theorem. *Discrete Math.,* 218:289–291.

288. SZEKERES, G. (1973). Polyhedral decompositions of cubic graphs. *Bull. Austral. Math. Soc.,* 8:367–387.

289. SZEKERES, G. and WILF, H. S. (1968). An inequality for the chromatic number of a graph. *J. Combin. Theory,* 4:1–3.

290. TAIT, P. G. (1878-1880). On the colouring of maps. *Proc. Roy. Soc. Edinburgh Sect. A*, 10:501–503,729.

291. TASHKINOV, V. A. (2000). On an algorithm for the edge coloring of multigraphs (in Russian). *Diskretn. Anal. Issled. Oper. Ser. 1*, 7:72–85.

292. THOMASSEN, C. (2005). Some remarks on Hajós' conjecture. *J. Combin. Theory Ser. B*, 93:95–105.

293. THOMASSEN, C. (2007). Hajós' conjecture for line graphs. *J. Combin. Theory Ser. B*, 97:156–157.

294. TOFT, B. (1996). A survey of Hadwiger's Conjecture. *Congr. Numer.*, 115:249–283.

295. TUTTE, W. (1966). On the algebraic theory of graph colorings. *J. Combin. Theory Ser. B*, 23:15–50.

296. VAUGHAN, E. R. (2010). An asymptotic version of the multigraph 1-factorization conjecture. arXiv:1010.5192v1 [math.CO].

297. VIZING, V. G. (1964). On an estimate of the chromatic class of a p-graph (in Russian). *Diskret. Analiz*, 3:25–30.

298. VIZING, V. G. (1965). The chromatic class of a multigraph (in Russian). *Kibernetika* (Kiev), 3:29–39. English translation in: *Cybernetics and System Analysis*, 1:32–41.

299. VIZING, V. G. (1965). Critical graphs with a given chromatic class (in Russian). *Diskret. Analiz*, 5:9–17.

300. VIZING, V. G. (1968). Some unsolved problems in graph theory (in Russian). *Uspekhi Mat. Nauk*, 23:117–134. English translation in: *Russian Mathematical Surveys*, 23:125–141.

301. VIZING, V. G. (1976). Colouring the vertices of a graph in prescribed colours (in Russian). *Diskret. Analiz*, 29:3–10.

302. WAGNER, K. (1937). Über eine Eigenschaft der ebenen Komplexe. *Math. Ann.*, 114:570–590.

303. WAGNER, K. (1937). Über eine Erweiterung des Satzes von Kuratowski. *Deutsche Math.*, 2:280–285.

304. WAGNER, K. (1960). Bemerkungen zur Hadwiger Vermutung. *Math. Ann.*, 141:433–451.

305. WALLIS, W. (1997). *One-Factorizations*. Kluwer Academic Publisher, Dordrecht.

306. DE WERRA, D. (1971). Equitable colorations of graphs. *Rev. Francaise Informat. Rech. Opér.*, 5:3–8.

307. DE WERRA, D. (1975). How to color a graph. In: Roy, B., editor *Combinatorial Programming: Methods and Applications*, pages 305–325. Reidel Publishing Company, Dordrecht, Holland.

308. DE WERRA, D. (1975). A few remarks on chromatic scheduling. In: Roy, B., editor *Combinatorial Programming: Methods and Applications*, pages 337–342. Reidel Publishing Company, Dordrecht, Holland.

309. WOODALL, D. R. (2007). The average degree of an edge-chromatic critical graph II. *J. Graph Theory*, 56:194–218.

310. WOODALL, D. R. (2008). The average degree of an edge-chromatic critical graph. *Discrete Math.*, 308:803–819.

311. WOODALL, D. R. (2011). The independence number of an edge-chromatic critical graph. *J. Graph Theory,* 66:98–103.

312. WU, J. and WU, Y. (2011). Edge colorings of planar graphs with maximum degree five. *Ars Combin.,* 98:183–191.

313. YAN, Z. and ZHAO, Y. (2000). Edge colorings of embedded graphs. *Graphs Combin.,* 16:245–256.

314. YAP, H. P. (1981). A construction of chromatic index critical graphs. *J. Graph Theory,* 5:159–163.

315. YAP, H. P. (1981). On graphs critical with respect to edge-colourings. *Discrete Math.,* 37:289–296.

316. YAP, H. P. (1986). *Some Topics in Graph Theory.* London Math. Soc. Lecture Note Ser., Vol. 108. Cambridge University Press.

317. ZHANG, L. (2000). Every planar graph with maximum degree 7 is class I. *Graphs Combin.,* 16:467–495.

318. ZHANG, X. and LIU, G. (2008). Some graphs of class 1 for f-colorings. *Appl. Math. Lett.,* 21:23–29.

319. ZHANG, X. and LIU, G. (2011). Equitable edge-colorings of simple graphs. *J. Graph Theory,* 66:175–197.

320. ZHANG, X., YU, J. and LIU, G. (2010). On the fractional chromatic index of a graph. *Intern. J. Computer Math.,* 87:3359–3369.

321. ZHAO, Y. (2004). New lower bounds for the size of edge chromatic critical graphs. *J. Graph Theory,* 46:81–92.

322. ZHOU, G. (2003). A note on graphs of class I. *Discrete Math.,* 263:339–345.

323. ZHOU, X. and NISHIZEKI, T. (1999). Edge-coloring and f-coloring for various classes of graphs. *J. Graph Algorithms Appl.,* 3:18 pp. (electronic).

324. ZHOU, X. and NISHIZEKI, T. (1999). Decompositions to degree-constrained subgraphs are simply reducible to edge-colorings. *J. Combin. Theory Ser. B,* 75:270–287.

325. ZHOU, X., SUZUKI, H. and NISHIZEKI, T. (1996). A linear algorithm for edge-coloring series-parallel multigraphs. *J. Algorithms,* 20:174–201.

326. ZHU, X. (2001). Circular chromatic number: a survey. *Discrete Math.,* 229:371–410.

327. ZYKOV, A. A. (1964). Graph theoretical results of Novosibirsk mathematicians. In: Fiedler, M., editor, *Theory of Graphs and Its Application* (Proceedings of the Symposium held in Smolenice in June 1963), pages 151–153. Publishing House of the Czechoslovak Academy of Sciences (Prague) and Academic Press (New York and London).

328. ZYKOV, A. A. (1967). On some new results of Soviet Mathematicians. In: *Theory of Graphs, International Symposium, Rome 1966,* pages 415–416. Gordon and Breach, New York, and Dunod, Paris.

329. ZYKOV, A. A. (1968). In: Sachs, H., Voß, H. J. and Walther, H., editors, *Beiträge zur Graphentheorie* (Internationales Kolloquium in Manebach vom 9.- 12. Mai 1967), page 228. B. G. Teubner-Verlag, Leipzig.

Symbol Index

Name Index

Subject Index

WILEY SERIES IN
DISCRETE MATHEMATICS AND OPTIMIZATION

AARTS AND KORST • Simulated Annealing and Boltzmann Machines: A Stochastic Approach to Combinatorial Optimization and Neural Computing

AARTS AND LENSTRA • Local Search in Combinatorial Optimization

ALON AND SPENCER • The Probabilistic Method, Third Edition

ANDERSON AND NASH • Linear Programming in Infinite-Dimensional Spaces: Theory and Application

ARLINGHAUS, ARLINGHAUS, AND HARARY • Graph Theory and Geography: An Interactive View E-Book

AZENCOTT • Simulated Annealing: Parallelization Techniques

BARTHÉLEMY AND GUÉNOCHE • Trees and Proximity Representations

BAZARRA, JARVIS, AND SHERALI • Linear Programming and Network Flows

BRUEN AND FORCINITO • Cryptography, Information Theory, and Error-Correction: A Handbook for the 21st Century

CHANDRU AND HOOKER • Optimization Methods for Logical Inference

CHONG AND ŻAK • An Introduction to Optimization, Third Edition

COFFMAN AND LUEKER • Probabilistic Analysis of Packing and Partitioning Algorithms

COOK, CUNNINGHAM, PULLEYBLANK, AND SCHRIJVER • Combinatorial Optimization

DASKIN • Network and Discrete Location: Modes, Algorithms and Applications

DINITZ AND STINSON • Contemporary Design Theory: A Collection of Surveys

DU AND KO • Theory of Computational Complexity

ERICKSON • Introduction to Combinatorics

GLOVER, KLINGHAM, AND PHILLIPS • Network Models in Optimization and Their Practical Problems

GOLSHTEIN AND TRETYAKOV • Modified Lagrangians and Monotone Maps in Optimization

GONDRAN AND MINOUX • Graphs and Algorithms (Translated by S. Vajdā)

GRAHAM, ROTHSCHILD, AND SPENCER • Ramsey Theory, Second Edition

GROSS AND TUCKER • Topological Graph Theory

HALL • Combinatorial Theory, Second Edition

HOOKER • Logic-Based Methods for Optimization: Combining Optimization and Constraint Satisfaction

IMRICH AND KLAVŽAR • Product Graphs: Structure and Recognition

JANSON, LUCZAK, AND RUCINSKI • Random Graphs

JENSEN AND TOFT • Graph Coloring Problems

KAPLAN • Maxima and Minima with Applications: Practical Optimization and Duality

LAWLER, LENSTRA, RINNOOY KAN, AND SHMOYS, Editors • The Traveling Salesman Problem: A Guided Tour of Combinatorial Optimization

LAYWINE AND MULLEN • Discrete Mathematics Using Latin Squares

LEVITIN • Perturbation Theory in Mathematical Programming Applications

MAHMOUD • Evolution of Random Search Trees

MAHMOUD • Sorting: A Distribution Theory

MARTELLI • Introduction to Discrete Dynamical Systems and Chaos

MARTELLO AND TOTH • Knapsack Problems: Algorithms and Computer Implementations

McALOON AND TRETKOFF • Optimization and Computational Logic

MERRIS • Combinatorics, Second Edition

MERRIS • Graph Theory

MINC • Nonnegative Matrices

MINOUX • Mathematical Programming: Theory and Algorithms (Translated by S. Vajdā)

MIRCHANDANI AND FRANCIS, Editors • Discrete Location Theory

NEMHAUSER AND WOLSEY • Integer and Combinatorial Optimization

NEMIROVSKY AND YUDIN • Problem Complexity and Method Efficiency in Optimization (Translated by E. R. Dawson)

Printed and bound by CPI Group (UK) Ltd, Croydon, CR0 4YY

16/04/2025

14658592-0006